FOREWORD

The predecessor to this volume "Common Defects in Buildings" was prepared in 1976 by the late Henry Eldridge and was based on his experience of investigations for BRE. Since its publication, the diagnosis and cure of building defects has become an important activity for the building industry, where previously it was just the province of a few specialists. At the same time the quantity of published advice on defects has greatly increased. Many building periodicals devote regular features to the subject and several publications have been produced dealing exclusively with building defects. We, in the PSA, are aware of no other UK publications that deal with the diagnosis and cure of defects over the whole range of buildings.

PSA commissioned Haverstock Associates to bring "Common Defects in Buildings" up-to-date and assisted by the Agency's Building Advisory Branch, the volume has been revised completely to cover more types of construction and many defects that were unknown, or under-estimated 12 years ago. The methodical approach to defects investigation which was pioneered by Henry Eldridge is however as relevant now as in 1976. The authors have developed this approach to deal with the much larger volume of information that this new book contains.

The correct diagnosis of defects reduces not only the incidence of failure in repair works, but also provides vital and potentially cost saving data to the designer, materials and component manufacturer and builder. Maintenance work is by its nature expensive and the need to rectify avoidable faults is particularly wasteful when one bears in mind the findings of a BRE survey which estimated that it costs five times as much to rectify faults during construction than at the design stage. This has been recognised in PSA for some time and is also being addressed by the introduction of Quality Assurance into the design process.

Because of the range of its activities, PSA has perhaps one of the most comprehensive views on through-life building costs covering design, construction, operation and maintenance. As its Chief Executive I therefore, recommend this book to those engaged on maintenance work and to those responsible for creating new buildings. It is a valuable aid to both practitioner and student and I am sure that it will be held in the same high regard as its predecessor.

Sir Gordon Manzie KCB
Chief Executive
Property Services Agency
Department of the Environment

CONTENTS

CONTENTS

continued

DEPARTMENT OF THE ENVIRONMENT
Property Services Agency
Directorate of Building Development

Defects in Buildings

LONDON: HER MAJESTY'S STATIONERY OFFICE

First published 1989
(Replaces 'Common Defects in Buildings' published in 1976)
ISBN 0 11 671525 1

The price of this publication has been set to make a contribution to the costs incurred by the PSA in preparing the text and figures.

This publication was prepared for the Property Services Agency Directorate of Building and Quantity Surveying by Sylvester Bone of Haverstock Associates assisted by Helen Heard and David Horsfall.

The authors gratefully acknowledge the assistance of the Building Research Station who have supplied photographs and other illustrations. In many cases further details about the illustrations will be found in the publication referred to in the text.

HMSO
BOOKS

HMSO publications are available from:

HMSO Publications Centre
(Mail and telephone orders only)
PO Box 276, London, SW8 5DT
Telephone orders 01–873 9090
General enquiries 01–873 0011
(queuing system in operation for both numbers)

HMSO Bookshops
49 High Holborn, London, WC1V 6HB 01–873 0011 (Counter service only)
258 Broad Street, Birmingham, B1 2HE 021–643 3740
Southey House, 33 Wine Street, Bristol, BS1 2BQ (0272) 264306
9–21 Princess Street, Manchester, M60 8AS 061–834 7201
80 Chichester Street, Belfast, BT1 4JY (0232) 238451
71 Lothian Road, Edinburgh, EH3 9AZ 031–228 4181

HMSO's Accredited Agents
(see Yellow Pages)

and through good booksellers

CONTENTS

continued

CONTENTS

continued

CONTENTS

continued

CONTENTS

continued

CONTENTS

continued

INTRODUCTION

This book is intended for all those who control the management, maintenance and repair of buildings. It sets out to help with the recognition and the cure of commonly occurring defects in all types of buildings. It does not give specific advice on how to avoid defects in the design and construction of new projects. Lessons from defective construction in the past are often relevant to new projects but the application of these lessons is covered elsewhere and is beyond the scope of this publication.

As the original title *Common Defects* suggested this book is concerned with defects that recur in many different buildings. The ones that building surveyors and architects are likely to meet in day to day repair and renovation work. Unique forms of construction and one-off faults have not been included. Defects that require the specialist advice of a structural engineer have also been omitted except to point out where this advice is likely to be needed.

Buildings and the materials they are made from deteriorate at a more or less predictable rate, often depending on their exposure to weather or the use to which they are subjected. Eventually the deterioration reaches a point when a repair becomes necessary because occupants or adjacent materials are affected. Most constructions can be expected to fail after a certain time, but this book is not primarily concerned with failures that are predictable as part of the normal processes of deterioration. The defects examined here are those that occur unexpectedly, those that were not foreseen when the building was designed and constructed and when the composition of the materials was formulated. In the analysis of building failures, both natural deterioration that can be predicted and effects that were not intended can produce the same symptoms. To this extent the text does involve instances of natural deterioration.

Historic buildings require special consideration. It may be important to preserve the original building fabric after it has deteriorated and even repairs made at different periods may be considered part of the historical record.

Many failures in historic buildings are related to the 'natural' deterioration discussed above and are therefore outside the present scope. Those defects that are covered in this book should be given extra thought when they are found in historic buildings. The advice here on how to cure a defect may well be inappropriate in the context of an ancient structure.

While investigating defects, faults and omissions in the original construction often come to light. These are frequently unconnected with the symptom that prompted the investigation. Examination of a roof space or cavity may reveal that fire stops or insulation have been wholly or partly omitted. Building owners must be told of such discoveries, but this book is not intended to help with the exposure of such faults if they have not caused visible symptoms or malfunctions that require investigation.

Any book on building defects is inevitably incomplete. As methods of construction, building materials and the uses to which buildings are put, evolve, new defects are discovered that are related to these changes. When the new defects are first discovered they may be wrongly diagnosed;

consequently the remedies prescribed for them fail. Further experience leads to a better understanding and more effective remedies. Eventually industry learns to avoid the materials, methods or uses that led to the defects in the first place. The problems then become obsolete and of historical interest only. New defects emerging and old defects becoming obsolete create a moving target. A book on defects can only aim to record those defects current at a particular point in time. This revision of Henry Eldridge's original work on common defects in building brings in many defects that were not well known in 1972, it leaves out others that are no longer significant. These changes are the direct result of technical evolution. However, the relevance of the approach that Eldridge developed during his years with BRE has not changed. His book was the first to adopt a methodical approach to the diagnosis of defects. This revision follows his example with each defect being described under the symptoms, investigation, diagnosis and cure.

How to use the book

Defects in Buildings is primarily intended to be used for reference when a defect is being investigated. Commonly, defects come up for investigation when a symptom is reported. It may be a damp patch, a crack, a fallen cladding unit, or a piece of ironmongery that breaks for no apparent reason. The location of the visible symptom is usually known, as are the materials involved. The starting point is therefore:

1 The symptom that triggers the investigation.

2 The location of the symptoms.

3 The materials involved.

The contents of the book is arranged to allow a reader to move from this starting point to the consideration of what other information may be needed to build up a more reliable picture of what has gone wrong. Other symptoms may be less obvious, or may even become apparent only when the investigator carries out specific tests, uses particular measuring instruments or obtains historical records. The second stage of the investigation is to collect and record all the relevant information.

The description of each defect starts with an examination of the information that should be collected about defects in a particular category. It then explains how this information can be measured and analysed to diagnose the cause of the defect. Possible causes and their symptoms are discussed in turn together with ways to check whether the right diagnosis has been made.

At the end of each defect description there is a section giving advice on how to cure defects once they have been identified. This advice is in general terms and does not discuss the pros and cons of proprietary solutions.

The reader must always be aware that particular circumstances may have produced the exceptional defect and that the same symptoms can be caused by different faults. Diagnosis should not be a mechanical process. All this book can do is to review the symptoms of common defects. The reader must constantly watch for abnormal and additional symptoms or unusual forms of construction.

CHAPTER 1 CAUSES OF DETERIORATION

Agents	Acting outside the building		Acting inside the building	
	Atmosphere	Ground	Occupancy	Design consequences
1 Mechanical agents				
1.1 Gravitation	Snow loads, rain water loads	Ground pressure, water pressure	Live loads	Dead loads
1.2 Forces and imposed or restrained deformations	Ice formation pressure, thermal and moisture expansion	Subsidence, slip	Handling forces, indentation	Shrinkage, creep, forces and imposed deformations
1.3 Kinetic energy	Wind, hail, external impacts, sand-storm	Earthquakes	Internal impacts, wear	Water hammer
1.4 Vibration and noises	Wind, thunder, aeroplanes, explosions, traffic, machinery noises	Traffic and machinery vibrations	Noise and vibration from music, dancers, domestic appliances	Services noises and vibrations
2 Electro-magnetic agents				
2.1 Radiation	Solar radiation, radioactive radiation	Radioactive radiation	Lamps, radioactive radiation	Radiating surface
2.2 Electricity	Lightning	Stray currents	–	Static electricity, electrical supply
2.3 Magnetism	–	–	Magnetic fields	Magnetic fields
3 Thermal agents	Heat frost, thermal shock	Ground heat, frost	User emitted heat, cigarette	Heating, fire
4 Chemical agents				
4.1 Water and solvents	Air humidity, condensations, precipitations	Surface water, ground water	Water sprays, condensation, detergents, alcohol	Water supply, waste water, seepage
4.2 Oxidizing agents	Oxygen, ozone, oxides of nitrogen	Positive electro-chemical potentials	Disinfectant, bleach	Positive electro-chemical potentials
4.3 Reducing agents	–	Sulphides	Agents of combustion, ammonia	Agents of combustion, negative electro-chemical potentials
4.4 Acids	Carbonic acid, bird droppings, sulphuric acid	Carbonic acid, humic acids	Vinegar, citric acid, carbonic acid	Sulphuric acid, carbonic acid
4.5 Bases	–	Lime	Sodium hydroxide potassium hydroxide, ammonium hydroxide	Sodium hydroxide, cement
4.6 Salts	Salty fog	Nitrates, phosphates, chlorides, sulphates	Sodium chloride	Calcium chloride, Sulphates, plaster
4.7 Chemically neutral	Neutral dust	Limestone, silica	Fat, oil, ink, neutral dust	Fat, oil, neutral dust
5 Biological agents				
5.1 Vegetable and microbial	Bacteria, seeds	Bacteria, moulds, fungi, roots	Bacteria, house plants	–
5.2 Animal	Insects, birds	Rodents, termites, worms	Domestic animals	–

* This table is based on table 4 of ISO 6241.

Table 1 Classification of agents and the effects they have.

CHAPTER 1 CAUSES OF DETERIORATION

Buildings are generally considered to be very durable commodities. In the United Kingdom we have many examples of buildings that have survived for several hundreds of years. Many people have the impression that were it not for property developers, buildings left without attention would stay standing for ever. However, those buildings that have survived are very much in the minority compared to those that have perished, and the majority of those that are still standing today are only with us because they have been subject to care, and more importantly to constant repair and maintenance, throughout their lives.

From the time the construction process finishes, a building starts, albeit usually very slowly, to decay. This fact is often, if not taken account of, at least recognised in design, through the provision of finishes which are intended to protect more vulnerable components. These finishes, of which an obvious example would be paint, are recognised as having a life shorter than that required of the building and it is accepted that they will need to be renewed at regular intervals throughout the life of the building.

Estimating the expected life of a building component is often a complex and difficult task which will be considered further in following pages. However, any discussion on the causes of deterioration of buildings and their components needs to recognise that in the expected life of a building, some items (for example paint finishes or a central heating boiler) will be considerably less than that of the whole building.

The primary purpose of this book is to provide some understanding and guidance on unexpected building defects. These can usually be attributed to either one or a combination of the following factors:

- The inappropriate use of a component or material.
- The use of a material adjacent to or in combination with another that adversely affects it.
- A lack of knowledge by the designer regarding the potential deterioration of a material.
- The building being subject to forces or agents not considered in the design.
- Inaccurate information from manufacturers.
- Poor manufacturing quality.
- Poor workmanship.
- The failure to carry out necessary routine maintenance at the appropriate time.

The eight factors listed above provide the mechanism for the failure; the manifestation can usually be analysed to have occurred through the action of one, or more commonly through the actions of a combination of external forces or agents. These have been classified by ISO 624 into five major groups (see table). A brief description of the salient features of these five groups follows.

Mechanical agents

These impose a physical force on a building. They may be static and permanent such as ground pressure or static and temporary such as a snow load. Alternatively, the force can be dynamic such as wind or vibration. The design of structural items should take account of the action of most predictable mechanical agents, though occasionally failures do happen, for example because the user of a building has imposed loads greatly in excess of design loads. However, it is sometimes important to remember that non-structural components, particularly plastics, may also be subject to creep and deflection due to self weight.

Other mechanical forces which act on and within components do manifest themselves as building defects. In this latter category the most common occurrences are probably:

- Frost action through the formation of ice crystals within components already saturated with water, in very cold weather. The wetting may be due to rainfall or piped water, but melt water from a partial thaw or snow or ice is a particularly common precursor to damage, as it is likely to produce saturation by water only just above freezing point. Recent research suggests that the severity of the frost (lowest air temperature), speed of freezing (rate of fall of air temperature) and change of air temperature during the period of frost, were all more critical to the risk of frost damage than either air temperature at the time of thawing, or the duration of the frost.
- Wind action can have one of two results: The action of driving rain and of other wind borne pollutants may cause penetration into the building and the

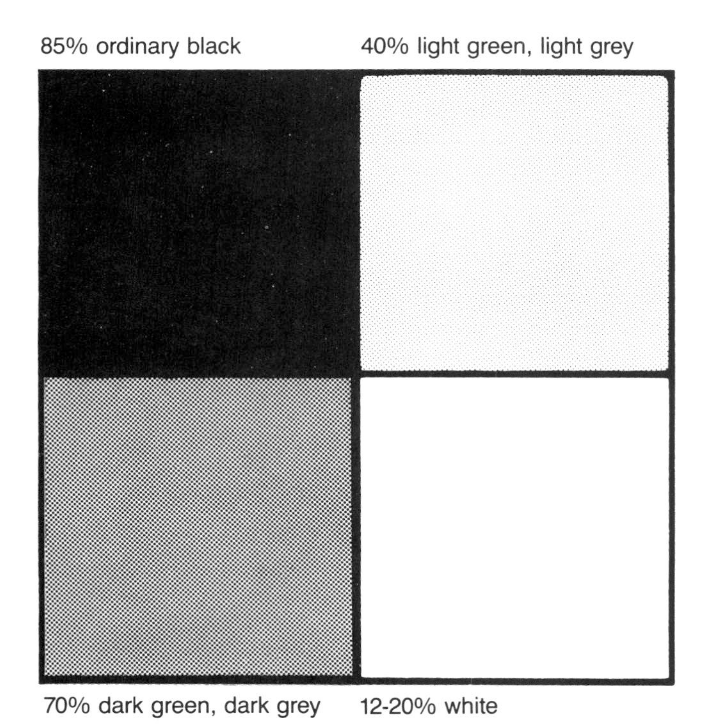

Figure 1 Absorbtivity of infra red radiation (700 – 1000 nm) solar reflective paint will reduce the temperature rise of a black ashphalt roof by 20 – 35%.

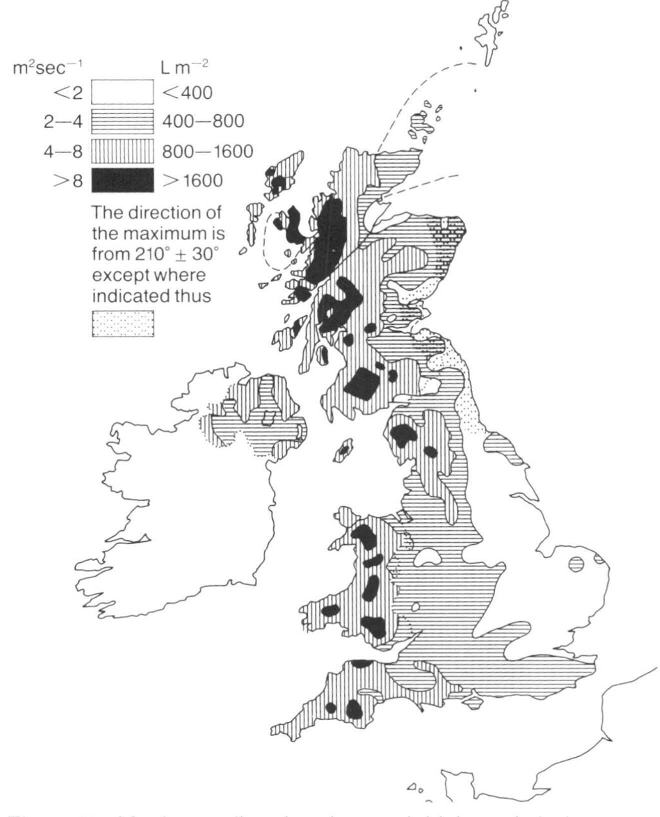

Figure 2 Maximum directional annual driving rain index map for the UK – more detailed local maps are available.

Figure 3 Isotherms of minimum shade air temperatures (deg C) for the UK.

Figure 4 Isotherms of maximum shade air temperatures (deg C) for the UK.

CHAPTER 1 CAUSES OF DETERIORATION

continued

erosion of the external fabric. The creation of differential pressures in localised positions on a structure resulting in deflections in components, or loosening of fixings.

- The unforeseen thermal expansion of components.
- Wear, particularly to such items as floor finishes.

Electromagnetic agents

As far as the durability of building materials is concerned the most important agent in this group is radiation.

Solar Radiation

The way it affects buildings can conveniently be split into three forms, each of which are discussed below. Although the actions of the different bands is very different, there is no readily available information on the incidence of the individual bands. Most published information concerns total solar radiation measured as bright sunshine and total radiation. Peak solar radiation is likely to occur around noon inland and during the late morning on the coast. In approximate terms, the radiation received on an overcast day is about one third of that received had it been cloudless. On very dull days the proportion would be much less. Predicted mean values of hourly irradiation (direct and diffuse) for both vertical and inclined surfaces at various orientations are given in the Department of Energy Handbook *Climate in the UK: A Handbook of Solar Radiation, Temperature and Other Data for Thirteen Principal Cities and Towns.*

Ultraviolet Radiation

A large proportion of this band of radiation (290nm – 400nm where a nm is a nanometre or one thousand millionth of a metre) is absorbed by the earths atmosphere and so has no effect. That which does penetrate the atmosphere can cause deterioration of organic materials, though as the penetrating powers are not great, the action tends to be confined to surface layers. For example, many organic dyes are degraded by ultraviolet light, as are bituminous materials and some synthetic polymers such as those used in sealants. Fading, yellowing and embrittlement may all be the result of ultraviolet action. Ultraviolet action can also be important in initiating a degradation, which is then accentuated by other agents such as temperature and moisture. An example of this is the yellowing and surface delamination of glass fibre reinforced polyester sheets. A further factor to be aware of when considering damage caused by ultra violet radiation is that the actual mechanism of the degradation may vary according to temperature. In uPVC components photo-oxidation occurs at lower temperatures and chlorination at higher ones. Both phenomena can be observed in the same component if it is subject to a range of temperatures.

The Visible Waveband

This spectrum of radiation (400nm – 700nm) is primarily experienced as heat (see also thermal agents below). The total radiation received will depend on cloud cover, the proportion being the same as for solar radiation; the season of the year; and the local topography – surfaces normal to the sun such as roofs and those that receive substantial reflected radiation experience the highest temperatures.

Infra-red radiation

This band of radiation (700nm – 1000nm) is absorbed by all forms of matter, causing an increase in temperature such that the surface temperature will be greater than the surrounding air temperature. For a given surface texture, the colour of the surface considerably affects the absorptivity (figure 1) – For example in clear sunny weather in the UK, black mastic asphalt roofing or black bituminous roofing may reach temperatures of 50° to 80°C. A reflective coating can reduce this temperature rise by 20% – 35%.

Thermal agents

Temperature is particularly relevant to components that are exposed to an unobstructed sky, for example roofing, cladding and external structural members. The actual temperatures reached can lead to either temporary or even permanent changes in physical or chemical properties such as embrittlement at low temperature and accelerated oxidation at high temperatures. Changes of temperature are also relevant when assessing the consequences of thermal expansion and contraction – such as stresses within materials when changes of size are restrained and strains imposed on jointing materials when components are free to change size.

Thermal radiation can also act from within a component. On cold clear winter nights, roofing

CHAPTER 1 CAUSES OF DETERIORATION

continued

surfaces have been found to drop in temperature to as low as −20°C with the temperature of the roofing membrane being more than 5°C lower than the ambient air temperature due to the radiation from the roof. This phenomena has been called supercooling.

The effects of thermal radiation may vary within a component according to location. Low bright winter sun following a frosty night, has caused glazing to crack because of the thermal stresses. The centre of a pane will warm at a faster rate than the edges because of the cooling effect of surrounding masonry which does not respond to the warming of the sun as fast as the glass.

Another thermal reaction worth attention is the cooling effect of a summer shower, particularly with metal roofing and cladding. The drop in temperature caused by the rain can be sufficient to cause the relative air pressure within the roof void to fall below that outside, and for the pressure differential to be such that the rain is drawn into the building from the outside through joints between the sheets.

Chemical agents

Water

The chemical agent that is most prevalent is water. It is probably also the agent with greatest influence on the properties of materials, particularly when it is combined with extremes of temperature. In many instances the presence of moisture is a necessary prerequisite to enable physical, chemical or biological reactions to take place. Examples are:

- the effect of sulphate attack on Portland cement products
- corrosion of iron and steel products
- electrolytic corrosion between metals
- fungal attack of wood products

Most materials absorb moisture to some degree. The direct affect of water alone on a material can be:

- a volumetric change
- a change in mechanical properties, for example ordinary chipboard looses its strength and can disintegrate when it becomes saturated
- the development of twisting and turning forces such as happens to some unrestrained timber boards
- a change in electrical properties
- a change in thermal properties, many insulants lose their performance if they become wet
- a change in appearance

Water, in relation to buildings appears in three major forms.

The simplest manifestation is as water in the ground – all UK soil can be considered as permanently damp. Building failures caused by damp from the ground are likely to be caused either by inadequate detailing in the initial design or by faulty workmanship and materials.

The second manifestation of water is as precipitation which can come as snow, hail, rain or dew. The major action of snow is in the form of structural loading on horizontal or near horizontal surfaces. It tends to be only drifting snow that causes problems with walls. Snow being blown into buildings through small holes, particularly roofspaces, can also cause problems. There is very little information available on the occurrence of snowfalls as meteorologists measure snowfall as equivalent rainfall.

The prime occurrence of water outside buildings is as rain. All exposed surfaces must expect to be wetted to some extent by this phenomenom. The Meteorological Office maintain records of rainfall over the whole of the United Kingdom. As far as buildings are concerned, the most useful information is probably the Driving Rain Index Map prepared by BRE. It has been found that the amount of water arriving on the vertical faces of buildings is related more to the combined effect of wind and rain than to the actual quantities of rain on its own (see Figure 5). Hail precipitation in the UK tends to be very short lived and with the exception of occasional impact problems, may be regarded as being very similar to rainfall in its effects on buildings.

The third form of water to affect buildings is water vapour. Externally the climate of the UK is considered as being unusually humid at all times of the year, particularly when compared to continental Europe. A wide range of vapour pressures may be experienced in any one month, and there is also a large overlap of

CHAPTER 1 CAUSES OF DETERIORATION

continued

summer and winter ranges. Coastal regions usually maintain higher relative humidities than inland areas especially in summer when the major factor controlling relative humidities is temperature. Internally the major manifestation of water vapour is as condensation which is generated from within the building, often from steam producing activities such as bathing or cooking, but even in some cases simply from the presence of a number of persons exhaling.

Dew is formed by moisture present in the ambient air condensing onto relatively cooler surfaces. It is a phenomenom of the early morning, and may be regarded as having a similar effect on buildings as a light rainfall.

Oxygen

As far as buildings are concerned the second most important chemical agent is probably oxygen. It is the most reactive of the gasses and present in the air in large volumes and leads to the corrosion of metals, as well as the oxidation of paints, plastics, sealants and bituminous materials.

Sulphates

Sulphates are salts that are naturally present in industrial wastes, gypsum plaster, clay bricks (particularly those fired at lower temperature), flue condensates and as a solution in the ground water in some areas. In persistently wet conditions, sulphates react slowly with tricalcium aluminate (a constituent of Portland cement and hydraulic lime) forming a compound called sulphoaluminate. This reaction causes the cement mortar or render to expand and eventually disintegrate. When this action happens to walls behind renders the initial reaction is likely to occur in the mortar which will then cause a second reaction within the render itself. Frequently it is the presence of the render which will be responsible for the initiation of the attack, as once water has penetrated a crack the render itself will shield the wall and prevent it from drying out. Identical bricks left unrendered may well remain trouble free under normal exposure conditions.

Other chemical agents

These include gasses such as carbon dioxide, sulphur dioxide and nitrogen that are all present in the atmosphere. In the presence of moisture they contribute to the formation of acids that attack susceptible materials such as unprotected metals, concrete, other cementitious products and some building stone. It is important to note that it is not only the gasses themselves that can cause damage but also the products of the reactions may themselves be reactive towards other materials.

In industrial areas, micro-droplets of either acid or alkaline solutions of low volatility hydrocarbons or oils may be in the atmosphere and be responsible for causing changes of the properties of building materials. The nature of the action is often very specific to the pollutant, and further specialist advice should be sought.

It should be remembered that in coastal areas sea spray is corrosive, and with the influence of wind it may be transported inland for several miles.

The hydrocarbon emissions from power stations and some other industrial atmospheric pollutants may travel considerable distances in the upper atmospheres, and be returned to earth at great distance from the emission point dissolved in precipitation. This phenomenom is sometimes referred to as acid rain.

The best available guide to potential atmospheric pollution, is the Ministry of Agriculture Average Atmospheric Corrosivity Map (see figure 6). It is based on zinc reference data and provides a corrosivity rate for every 10 km grid square of the whole of the UK. The map also gives details for using the zinc index to calculate corresponding corrosivity rate for other metals – steel, aluminium, copper and brass.

Biological agents

Biological agents can be broken down into four categories:

Surface growths

These include bacteria, fungi, algae, lichen and mosses. They do not necessarily do harm but some release acidic metabolic products which are corrosive, and others invade the surface of substrata and cause deterioration.

Insect vermin

The predominant damage or deterioration to

Figure 6 Atmospheric corrosivity values (1986) darker (and striped) areas have the most corrosive atmosphere.

CHAPTER 1 CAUSES OF DETERIORATION

continued

construction materials caused by insects occurs with timber and timber based products. Although the moisture content of timber is the predominant factor when considering its durability with regard to both fungal and insect attack (over 20% is the crucial threshold), temperature is critical too. Infestations are inhibited by low temperatures, accelerated by warmth and are occasionally destroyed by hot conditions.

BS 5268 Part 5 gives hazard categories for risk of decay or insect attack of timber and tabulates these for different situations in buildings.

In the UK the only other building material known to have been subject to insect damage is masonry. BRE have published case histories of damage caused by mason bees and it has also been reported that brickwork has been damaged by ants.

Animal vermin

The pests most likely to cause damage to building materials are rats and mice, who gnaw timber, other organic substances and uPVC casings to electric cables. Insulated profiled metal cladding and roofing has been damaged by birds pecking at unprotected insulation on cut sheets, and damage has also been caused by birds dropping debris.

Plant agents

Probably the major damage to buildings from plants occurs from tree roots disrupting foundations and penetrating underground drains. Considerable damage has occurred to buildings with shallow foundations or faulty infilling that have been erected on shrinkable clay subsoils and with trees and large shrubs nearby. Damage to buildings from plants has also been caused by plants growing in gutters and blocking them.

The Building User

Although not classified as an agent by ISO 6421 the building occupant can be a very important factor to a building defect. Leaving aside deliberate misuse such as vandalism, defects do occur that are the result of action by the building occupants. Condensation caused by the use of propane gas and paraffin heaters is a well known example.

In conclusion

When considering the durability of materials the following two points need to be born in mind at all times. Agents do not act uniformly – individual agents have differing effects depending on the materials being acted upon. For example most plastics are impervious to damage from fungal and insect attack but are susceptible to damage from ultraviolet radiation, whereas timber is susceptible to attack from both. The effect of an agent is often governed by the presence or absence of another. For example, an increase of 10°C can double the rate of chemical reaction, and fungal attacks on building fabric will only take place when moisture is present.

The extent of any deterioration is likely to vary according to the nature of the attack. For damage caused by actions like corrosion and wear, the duration of the action is the dominant factor. Whereas when radiation thermal agents or moisture attack a building, the intensity is usually the critical factor – beyond a certain point there is generally a qualitative change. For other agents like driving rain the frequency of the action will usually determine the extent of the damage.

CHAPTER 2 DURABILITY OF MATERIALS

CHAPTER 2 DURABILITY OF MATERIALS

The durability of building materials is a massive subject which this one chapter cannot hope to cover exhaustively. Rather it sets out to look at those features of a selection of common building materials that seem to recur in building failures.

A material failure of a building component is usually due to one of the following three factors:

- An inherent weakness in the material due to faults in its manufacture.
- Poor design, where the material has been used in a situation for which it is not suited.
- The degradation of the material through the attack of an agent.

This chapter does not propose to discuss the first of these factors. Generally such defects are of a highly technical and specialised nature and it is hoped that the increasing awareness of the need for quality control and the extent to which manufacturers have been and are submitting themselves to the rigours of BS 5850 should help to reduce faulty manufacturing failures in the future. The majority of the defects detailed in this book are the result of one or both of the latter two of these factors.

Bituminous materials

Bitumen products are primarily used in buildings for waterproofing purposes. The three major forms of these products are:

- Bituminous liquid coatings.
- Bituminous felts – in which bituminous compounds are manufactured into a sheet product often onto a base of another material such as a polyester fleece.
- Asphalt – in which the bitumen is mixed with inert mineral matter.

Bitumen products are inert to moisture and are primarily specified as waterproofing products – particularly for roofing and for the protection of buildings against ground water. A major weakness of bituminous products is their susceptibility to damage by ultraviolet radiation, and/or atmospheric oxidation. This damage usually takes the form of a hardening of the product accompanied by shrinkage and slight surface crazing. Because bituminous products are black or dark grey in colour, if they are not given solar protection, their absorption of infra-red radiation can cause high temperatures to build up within the material. In some circumstances this can cause softening and even flow characteristics. Conversely on cold clear nights, thermal radiation from the same materials has been found to result in such materials falling to temperature below that of the ambient air temperature. The most critical factor to whether such temperature changes cause damage, is the rate of temperature change. Slow changes will be accommodated without internal damage, but sharp changes, particularly at low temperatures, can cause cracking and weakening. Although water does not affect these materials if it becomes trapped either immediately below a bituminous membrane or between layers of a finish, solar radiation can cause the moisture to vaporize and expand which will manifest itself as blisters on the finish. Other than being unsightly, such blisters may remain for periods of several years without any failure of the membrane occurring, however they do represent a weakening of the membrane, and they are more susceptible to mechanical damage.

Both asphalt and traditional roofing felt are also susceptible to damage through differential movement of the substrate, particularly where isolating membranes have not been provided. New modified polymer bituminous membranes have been formulated to provide better elasticity, though improved qualities do not absolve the architect from proper detailing.

Bituminous products are not normally affected by biological or chemical agencies other than oils.

Bricks and tiles

Both bricks and tiles are made of clay and concrete, whilst in addition bricks can also be made of calcium silicate (sand-lime bricks).

Clay bricks and tiles

These are manufactured in a generally similar manner, from prepared clay which is shaped either by hand or mechanically and then dried and fired. Different clays will give different qualities and strength to the products, and the temperature at which the product is fired will also affect the same. Generally these clay products are

CHAPTER 2 DURABILITY OF MATERIALS

continued

extremely durable, with the two major causes of deterioration being frost action and the crystallization of soluble salts, though there have also been problems with expansion when bricks have been used too fresh from the kiln.

Frost damage occurs when masonry or tiles have been saturated. The pore structure of the product is more critical than the absorptivity of the material. It has been found that, as a rough rule of thumb, the larger the pore size, the more susceptible the item will be to damage. In the case of brickwork, location is also important; with chimneys, parapet and free standing walls being at far greater risk than walls between the eaves and DPC level. Increasing insulation on the insides of walls or in the cavity will cause exposed external walls to stay colder in deep winter weather, and may thus make them more susceptible to frost damage. With roof tiles the lower the pitch the greater the danger of moisture being absorbed and retained, and the consequent possibility of frost damage. Ventilation beneath roof tiles, such as that created by counter battens, helps tiles to dry out and will lessen the risk of frost damage.

The action of soluble salts is much more prevalent with bricks than tiles. Small amounts of salts are present in a wide range of bricks. A spell of dry weather following wet weather may bring these to the surface as efflorescence (see 5.3.3). This phenomenon is usually harmless though unsightly. In some exceptional instances, these salts may cause a crumbling on the face of the brick, but this usually only occurs with underfired bricks. Occasionally, when magnesium sulphate salts are present in the bricks, the salts can migrate inwards and crystalise between the bricks and the plaster, causing the plaster to detach.

The most serious problem with salts in bricks occurs when they react in solution with Portland cement or hydraulic lime mortars and renderings with the result that a sulphate attack occurs. Occasionally these salts may be present in the mortar but in the more serious cases the salts are usually present in the bricks.

Calcium silicate bricks

These bricks are made from hydrated lime and silica, usually either sand or crushed flint, the two materials being mixed and shaped and then the bricks being hardened by reacting the lime under pressure in a steam autoclave. The pressure of the steam and its duration determines the strength of the brick. Calcium silicate bricks show a marked correlation between strength and frost resistance; stronger classes should always be specified in more exposed situations.

These bricks are free from the salts which cause efflorescence and are less susceptible to sulphate attack. However they are liable to attack from sulphur dioxide which causes decomposition of the calcium silicate bonding, and when in a very polluted atmosphere may cause general weakening and blistering. These bricks should also be avoided in coastal locations, as sea spray causes erosion of the surface and will reduce resistance to frost attack.

Calcium silicate bricks vary considerably in drying shrinkage and moisture movement, so the type of mortar used is critical. The weaker the mortar the less likely the chance of damage. Strong mortars should only be used with 'special class' bricks. Movement joints should also be incorporated in all walls over a certain length (see 5.1.6).

Concrete bricks and tiles

These are manufactured from a mixture of cement and selected aggregates, which is moulded into shape and then cured under controlled conditions. They are durable materials and inert to most biological and chemical pollutants. Occasionally they are prone to efflorescence. All cement products have a high initial shrinkage which should be allowed to take place before the products are used. Concrete bricks in use have drying shrinkage limits that exceed those of calcium silicate bricks, so the need to avoid strong mortars is essential.

Concrete

Concrete is a mixture of cement, water and aggregates which takes the shape of a mould and when cured forms a solid mass. This mass may or may not incorporate another material such as steel for reinforcement, and other special additives to give the material special working characteristics or qualities.

Concrete technology is very specialised, and constantly developing. Concrete is perhaps better viewed as a family of products rather than a single entity, because the nature of the basic cement and aggregate used, how they are mixed together, whether

CHAPTER 2 DURABILITY OF MATERIALS

continued

or not additives are incorporated, the method of placement, and the control maintained over curing, can all vary and differing end products result.

Concrete is generally looked on as a very durable and intrinsically strong material. However, it is also a porous product so that any concrete that is exposed will be subject to penetration by atmospheric gasses, and to wetting and drying. The take up of water and resultant expansion being a factor that will vary according to both the nature and density of the mix, and the properties of the aggregate used. This wetting is always more pronounced on the most exposed areas. Many of the problems of concrete durability occur because of this water penetration. Probably the simplest defect is that of frost action on saturated material. Susceptibility to attack is usually related to poor quality of the concrete. Frost resistance can be increased through the judicious use of additives to incorporate fine rather than large pores.

The wetting of concrete can also cause reactions with other constituents of the mix, or between the concrete and an outside agent, the corrosion of reinforcement probably being the commonest. Steel within concrete is usually inhibited from rusting by the high alkalinity of the surrounding concrete. However, both carbon dioxide and sulphur dioxide from the atmosphere reduce this alkalinity by carbonating the alkalis, and increase the vulnerability of the steel to corrosion and rusting. Inadequate cover and poor concrete exacerbate the carbonation problem. The result is cracking as the rusting steel expands followed by spalling. Rust staining is also often pressent.

Sulphate attack can also occur to concrete in contact with sulphate contaminated ground. London, Oxford and Kimmeridge Clays, the Lower Lias and Keuper Marl soils are all suspect. Sulphate resisting Portland cement must be used instead of ordinary Portland cement below or at ground level.

Concrete is also vulnerable to chloride attack. In marine atmospheres even unreinforced concrete has been known to disintegrate if heavily attacked by sea salts. However, chloride attack has also occurred through the use of calcium chloride admixtures, or the presence of chlorides in aggregate. Although the problem of chloride attack on roads and bridges through the use of chlorides as de-icing agents is outside the scope of this book, the resultant introduction of chlorides onto hardstandings and garage floors from car tyres should be remembered.

Some forms of silica found in aggregate can react with alkalis in the cement in the presence of water, and cause what is called an alkali/aggregate reaction resulting in the production of a gel, expansion and subsequent cracking and damage. The actual mechanism is complex and dependant on a number of factors being present together. As far as the UK is concerned damage seems to have been confined to cases involving reactive minerals in aggregate from the South Coast, the Bristol Channel and the Thames Estuary. This reaction needs specialist testing to diagnose its presence.

Another problem that should be mentioned when considering defects in concrete is the developing weakness that occurs when high alumina cement has been used. This type of cement forms a concrete with a rapid rate of strength development, and a marked resistance to sulphate attack. However, depending on ambient temperature and humidity, this form of concrete has been found to 'convert' from an initial metastable form to one which possesses higher porosity and lower strength. Highly 'converted' high alumina cement is vulnerable to chemical attack in the presence of water such as might be caused by damp gypsum plaster, and reduces the concrete further in strength.

Having described in a very generalised way some of the defect mechanisms that occur with concrete, it cannot be too strongly emphasised that in any situation where a structure is at risk, specialist advice is critical.

There are a wide range of tests available. A growing number of them are non invasive, others require samples to determine not only the nature, but also the extent of any failure. Although immediate action may be needed to secure the safety of a structure, it is often found that concrete distress progresses relatively slowly, so the task is not so much one of taking immediate action, but of understanding the mechanism, and then devising a strategy to deal with it effectively and economically. Both the specification and the actual process of repairing concrete is extremely sophisticated, and should only be contracted

CHAPTER 2 DURABILITY OF MATERIALS

continued

to knowledgeable personnel that specialise in this particular field.

Metals

These days, iron, steel, copper, aluminium, lead and zinc are the metals most commonly used in buildings. The use of cast iron and to a certain extent copper has declined in recent years, in many instances however, these have been replaced by other materials such as plastics in plumbing.

When considering the durability of metals, corrosion can be regarded as the major problem. In nearly all cases where metals are specified for buildings, their strength characteristics are an important factor, and corrosion if left unchecked will compromise this. Corrosion itself is a complex electro-chemical reaction, that in almost all circumstances requires the presence of moisture. Although dampness and oxygen is all that is needed for rusting and some other forms of corrosion, however corrosion to most non ferrous metals usually requires the presence of another material – generally another metal, but sometimes something different such as brick, plaster or timber, and in a number of instances other atmospheric pollutants. In anaerobic conditions where suitable nutrients are present, bacterial activity can induce decay of metals. BS PD 6484:1979: Commentary on corrosion of bimetallic contacts and its alleviation, gives comprehensive details.

The other factors to be considered in the context of the durability of metals are deformation due to structural stresses which can result in fatigue and creep. The latter is more common, as all metals subject to a steady load will gradually deform. Generally this is a factor considered in design and will not cause serious problems, though such things as insufficiently supported flashings are prone to failure from this cause. Metals subject to repeated movement, particularly when components are large and tightly restrained may suffer stress fractures and fatigue.

Ferrous metals

All iron and steel materials are classed as ferrous. The major factor which distinguishes this group of metals from others is their propensity to rust in the presence of air and moisture so that protective measures need to be taken. This can be achieved by painting or the application of other applied coatings, galvanising, or by alloying steel with other metals such as copper or chromium to produce 'stainless steel'. The term 'stainless' needs to be regarded with care; there are a wide variety of specifications for alloys offering different degrees of resistance to corrosion; the particular alloy chosen needs to be carefully matched to its environment.

When a protective coating fails and rusting occurs, the rust not only degrades the component but also causes it to expand at the point of rusting. This expansion can cause failure in other components (see also stone).

Aluminium

Aluminium is widely used for cladding, window frames and flashings. Generally components are factory fabricated by either casting, rolling or extruding. Although cutting and bending are readily undertaken on site, welding is difficult because of the low melting temperature of the aluminium, and is therefore usually done in a workshop. Because of its molecular structure, aluminium is usually only worked in one plane. Great care and skill are needed to work aluminium in three dimensions, without cracking occurring.

Mill finish aluminium will develop a protective layer of aluminium oxide upon exposure to normal atmosphere, but these days most aluminium used in construction has been anodised in the factory for improved protection. Aluminium can suffer electrolytic corrosion in damp or wet conditions. Contact with copper, copper alloys such as brass, bare mild steel, copper based preservative treated timber and oak should all be avoided. Damage has also been recorded with rainwater running off lead. Contact with zinc and stainless steel is usually safe in non marine conditions. Damage from copper can happen at a distance, for example rainwater that has run off a copper roof will attack an aluminium window below. The anodised layer on aluminium is also attacked by splashes of cement and lime, and care must always be taken when mortar is used near aluminium components.

Copper

Copper is primarily used in construction in sheet form for roofing and cladding, and in factory fabricated components for water supply and services distribution. It is also used in the form of wire for electrical distribution.

CHAPTER 2 DURABILITY OF MATERIALS

continued

Copper is generally very resistant to corrosion, particularly to sea water, but it is attacked by strong mineral acids and ammonia. When used externally a green patina will develop which acts as a protective skin, but soldered joints in external situations may cause corrosion. Electrolytic action will take place between copper and iron or copper and zinc in the presence of moisture, so these combinations should never be allowed in any plumbing system. Copper is also attacked by flue gasses containing sulphur dioxide so great care should be taken with any chimney design near copper (see 8.2.1).

Lead

Lead is used mainly in sheet form for roofing, flashings and DPCs. Externally it is highly resistant to corrosion through the formation of a dense protective film of lead sulphate caused through reaction with atmospheric pollutants. It is however attacked by organic acids released from damp timber, particularly Oak, Douglas Fir and Western Red Cedar, and should never be allowed to come into direct contact with these. It is also attacked by the free alkali present in rich cement/sand mortars. Corrosion of unprotected DPCs has been known to occur within 10 years. Protection is usually provided by a bitumen coating.

With the advent of better insulated buildings, there have been several recent failures of lead roof caused by condensation forming on the underside of the lead leading to its decomposition into lead carbonate (see 8.2.3). This can be prevented by providing ventilation directly under the lead to keep it dry.

Zinc

In modern times, zinc as a material on its own, is used primarily in buildings for roofing, flashings and occasionally rainwater goods. However zinc is also used extensively for protective coatings on steel – ie. galvanising, sherardizing and zinc spraying.

As with lead, exposure to normal atmospheric conditions will cause the formation of a protective layer consisting principally of zinc carbonate. Although zinc has good resistance to normal – ie average urban conditions, and marine atmospheres, it is very susceptible to attack from sulphur compounds and from carbon dioxide. It has been estimated that the rate of corrosion of zinc in unpolluted dry conditions is very slow and of no practical significance. In damp unpolluted conditions the rate may be four times as great, whereas in damp and polluted conditions the rate will be ten times as high. The Ministry of Agriculture Average Atmospheric Corrosivity Map, is based on zinc reference data and provides a corrosivity rate for every 10 km grid square of the whole of the UK.

Zinc is unaffected by Portland cement or lime in mortars once they have set, however if soluble salts such as chlorides are present a protective coating of bitumen paint is recommended.

Contact corrosion from lead, tin solder, iron or aluminium is unlikely but all contact with copper and copper alloys should be avoided. As with aluminium, run off from a copper roof can damage zinc at some distance from the actual copper material.

Zinc, like lead is affected by organic acids released from damp timber, particularly Oak, Douglas Fir and Western Red Cedar, and should never be allowed to come into direct contact with these.

Plastics

A wide and increasing range of plastics are used in construction, particularly for cladding, rainwater goods, damp-proof membranes, drainage, insulation, and glazing. The properties of these materials, differ markedly according to their formulation, so it is important to identify which particular material is being used before attempting any analysis of a problem. Detailed information on some individual defects is given in the appropriate pages. As most plastics components are of relatively recent development (compared to materials such as bricks or timber), knowledge on long term life is relatively uncertain, though Agrément certificates have been issued on some products estimating lives of 30 years and more.

The three major problems associated with plastics components have been either to do with expansion, creep or with degradation under ultraviolet light. uPVC products and polycarbonates in particular have high thermal expansion that needs to be taken account of in design. Ultraviolet degradation is usually greatest in exposed locations such as coastal areas. Additives in the material formulation can lessen such damage, though the addition of some fire retardant additives has been found to make the item more susceptible.

Stone

Stone is usually classified into one of three main groups – igneous, sedimentary or metamorphic.

Igneous Stone

Igneous stone is formed by the solidification of a moulten magna. The only igneous rock used extensively in construction in the UK as a building stone is granite, though others such as dolemite, basalt and pumice are used in crushed form as aggregates. Granite is highly resistant to most agents and is virtually impermeable to water.

Sedimentary Stone

Sedimentary stone, as the name suggests, is formed by the consolidation of particles of material cemented together to varying degrees by minerals and consolidated by pressure from the mass of material as it builds up.

The bulk of building stone used in the UK is sedimentary, and from one of the two principal forms:

- Limestone – which consists essentially of calcium carbonate, though in some forms a proportion of magnesium carbonate may also be present.
- Sandstone – which consists of grains of quartz cemented together by silica, calcium carbonate, or sometimes by both calcium carbonate and magnesium carbonate.

One of the main agents of deterioration in stonework is sulphur gasses present in atmospheric pollution. The sulphur reacts with calcium carbonate and forms calcium sulphate which is slightly soluble, so the action of rainwater will slowly erode the surface of stones with this constituent. Generally sandstones are much less soluble than limestones. Sulphur gasses are also one of the most likely causes of salt crystal formation within stone, though salts can also be introduced by cement mortar – lime mortar reputedly being less dangerous. The effect of salt crystallisation is that a hard and impermeable surface skin forms which subsequently blisters and flakes off.

Because all sedimentary stone tends to be formed in layers it always has a natural bed though it may not be obvious on newly dressed stone. However when soft layers are present, these are likely to weather at a faster rate and disfigurement results. Stones should always be laid with the natural bed horizontal. If they are laid otherwise, flaking of the face commonly results.

As with bricks the size of pore within a stone is probably one of the most critical factors as to whether stone suffers frost damage. As the proportion of fine pores increases generally so does resistance to damage.

Metamorphic Stone

These are formed from sedimentary stone through the action of great heat and pressure which radically changes their structure. The two metamorphic stones used in construction in the UK are slate – derived from clay, and marble – derived from limestone. Both are very durable but can be affected by sulphur gasses. BS 680 slates should be resistant to attack as they undergo a sulphuric acid immersion test which should detect any calcium carbonate present.

Damage by Fixings

Along with sulphur attack described above, a common cause of damage to stone has been caused by the corrosion of metals. The most common occurrence (which can affect any stone) is the rusting of iron or steel fixing dowels and cramps. The force exerted by expanion of the rust will fracture the stone and often cause chunks of stone to break away. These days all stonework should be fixed by a non rusting material such as bronze or stainless steel. The other effect of the corrosion of metals is to cause unsightly staining, particularly from rust and verdigris.

Timber

Timber is a natural material, and therefore there are wide variations in form and quality. Timber is usually classified into two types:

- Softwood – The timber of coniferous species. Generally softwood is specified as a general purpose utility timber for structural members and joinery. It is less common for softwood to be specified by species, though this may happen with the more expensive ones such as Douglas Fir.

- Hardwood – The timber of non coniferous species. Generally hardwoods are considered as being more durable than softwoods, though this not necessarily so. Although some of the cheaper mahogany type species are specified generically, for example for sills, it is more usual for hardwoods to be specified by species, particularly when appearance is critical or special qualities are required.

Generally the moisture content is the predominant factor when considering its durability. The amount of moisture absorbed by wood varies according to species. In general terms it can be said that wood will swell by up to 25% (of its volume) as it takes up moisture. After 25% absorption more moisture will still be absorbed but the wood will remain dimensionally stable. Sapwood can take up its own weight in water. All joinery should be designed to allow for small dimensional changes due to the absorption and drying cycle that will inevitably take place. The three major environmental factors affecting the moisture content of timber are:

- Rainfall or driving rain affects all exposed timber whether or not it is painted.
- Timber within the weatherproof skin of a building is affected by changes in atmospheric humidity. Although, provided the building remains weathertight, and appropriate ventilation has been provided to cope with moisture generating activities, it will generally remain outside the decay range.
- Enclosed structural timber and wood based materials may be placed at risk by interstitial condensation. The extent of risk will depend on the differences between internal and external temperatures and humidities, the integrity of any vapour barriers, and the adequacy of any ventilation provided.

The take up of moisture by timber is not always uniform. The face adjacent to the source of moisture may hold more than other faces. For example where timber floors are laid on damp substrates the differential moisture content can cause the timber to deform (see 7.1.11). To prevent undue movement occurring to timber components, it is always advisable to 'condition' the wood to suit its eventual location. Internal joinery should always be protected from the weather, and if it is needed, to keep components in storage before installation, it should be in conditions that match as closely as possible those of the product when installed.

Moisture is also a critical factor when considering the durability of timber with regard to both fungi and insect attack. Timbers kept with a moisture content of less than 20% are not normally subject to fungal attack. BS 5268: Part 5 gives hazard categories for risk of decay or of insect attack and tabulates these for different situations. Temperature is also important. Infestations are generally inhibited by low temperatures, accelerated by warmth, and sometimes destroyed by hot conditions.

Treatment of timber with chemical preservatives preferably through pressure impregnation will protect it from both insect and fungal attack.

When timber is used externally it must be remembered that most species are susceptible to damage from prolonged exposure to sunlight. The ultraviolet radiation causes degradation of the surface, particularly of softwood species, which will affect the adhesion of paint and other finishes.

CHAPTER 3 PRINCIPLES OF DIAGNOSIS

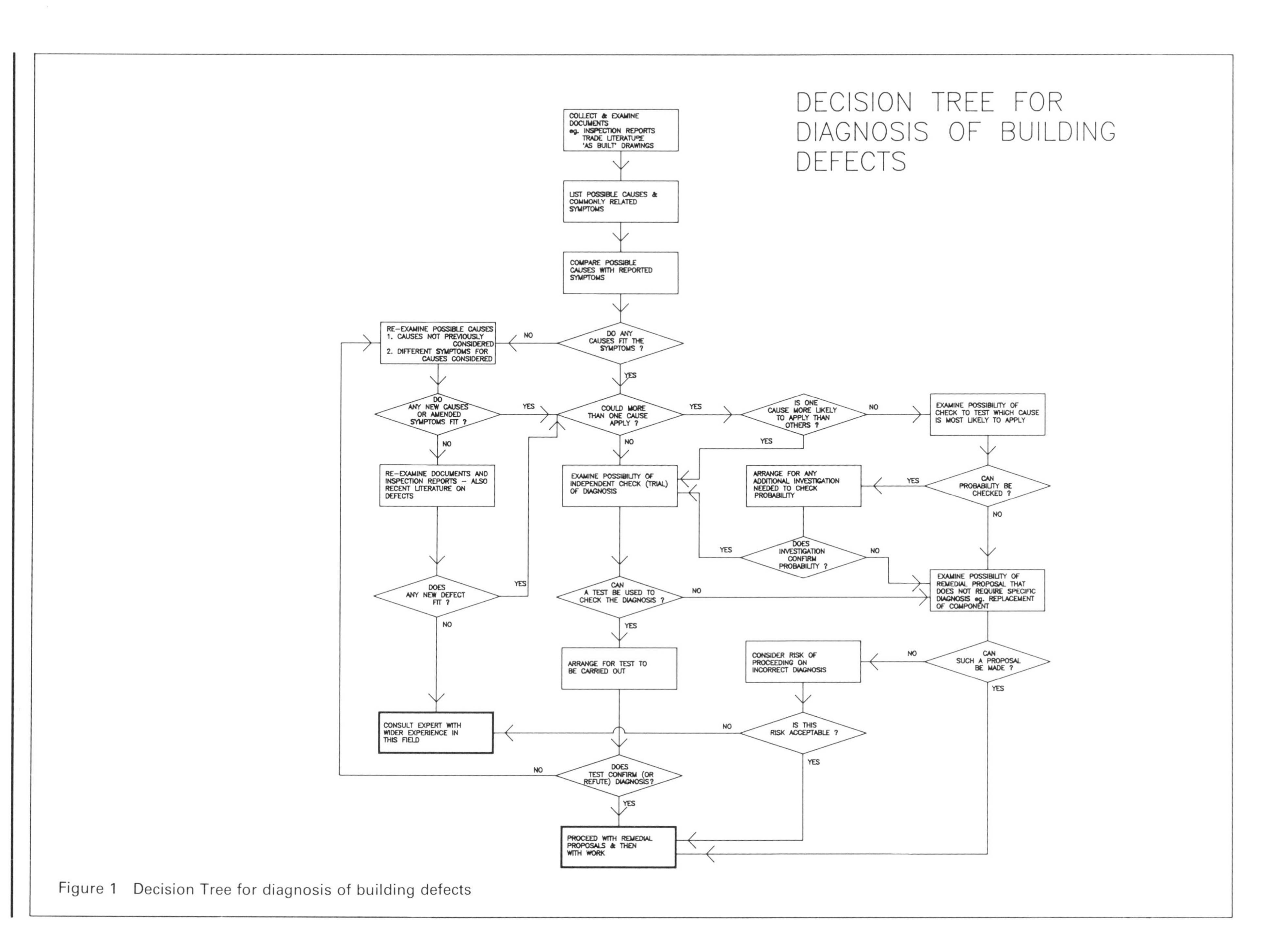

Figure 1 Decision Tree for diagnosis of building defects

CHAPTER 3 DIAGNOSIS OF DEFECTS

The diagnosis of defects has been compared with crime detection. Investigation produces evidence, then alternative hypotheses are examined in turn to assess which best matches the facts. The most obvious causes must not be accepted if any one of the clues points in a different direction. Even in cases where all the clues seem to point in the same direction, it is still important not to jump to conclusions without doing everything practicable to test the hypothesis. New, unprecedented defects occur from time to time, and data from investigations can be unreliable or misleading. A book cannot provide a complete guide to diagnosis; it can only suggest some of the possible causes for different defects. The actual investigation and diagnosis must be a patient, case by case procedure. A diagram is given on page 22 illustrating, in a general way, how to work through the diagnosis of the causes of a specific building defect. Most of the steps shown on the diagram are simply common sense, but there are some points to watch for in particular.

If an attempt is made at the outset to list the potential causes of a defect, then the list provides a reference point to return to when unexplained symptoms are discovered, or a diagnosis is challenged. If the liability for a defect is disputed it is useful to have assembled facts that support the rejection of a potential cause, as well as those that led to the diagnosis.

The emphasis must be on methodical step by step procedures, but it is essential to recognise that diagnosis can never be entirely mechanical. The symptoms that are recorded may not give a full picture of the extent of the defect. The investigation is unlikely to reveal either the precise conditions to which the construction has been subjected, or the variations that may have occured in the manufacture of the components and their assembly. As explained in the introduction, our knowledge of defects is continually evolving. New, previously unknown defects occur quite frequently. Because the symptoms, the results of the investigation and the list of potential defects may all be incomplete, there is a place for scepticism, hunches and the application of general experience. It is as well to remember that even the most obvious diagnosis may still be the wrong solution. If the cure relies on the diagnosis being correct, possibly a small trial of the remedial solution is prudent before proceeding with a full blown remedy.

Diagnosis need not be geared to discovering a single cause for a defect. In reality, defects are often caused by a combination of factors, none of which, on their own, would require remedial work, but where they are working in combination, some action has to be taken. In such cases it may only be necessary to restore the building's performance to an acceptable level. If the contribution of the different causes can be analysed, it may be possible to cure the defect by selecting the one that costs least to eliminate. For example, in some cases condensation may be cured by additional ventilation, rather than by a different heating system or provision of more thermal insulation.

Sometimes it is only possible to say from the available information that a certain cause is the most likely reason for the occurrence of a particular defect. Other possible causes may be known, and it may be necessary to carry out further investigations before even a tentative diagnosis can be made of the full range of potential causes. In such cases an assessment of the cost or further investigation may show that, whatever the result of the investigation, the cost is not justified as it cannot lead to a large enough reduction in the cost of remedial work, or to a sufficiently large increase in the damages to be paid by whoever is shown to be liable for the defect.

When a diagnosis is complete and doubts remain about the cause of a defect, it is essential to recognise that there are still doubts. Remedial action designed to cure one cause may not be effective if the diagnosis turns out to have been wrong. A more comprehensive approach to the remedial work, such as complete replacement, may be the only way to ensure that the remedial work is successful.

It is as well to be aware of some of the common pitfalls that lead to wrong diagnosis of building defects. Even a conscientious investigator can be misled by wrong information that has been provided by others. The records available after completion of a building may not be adequate – information may be missing or simply incorrect. What is shown on the drawings as a cavity wall may have been built as a solid wall and vice versa. Whenever some aspect of a defect is hard to explain, it is worth going back to the investigation of the construction to see whether the site investigation has confirmed what was shown on the drawings or in the specification. Additional investigation may be

HEADINGS FOR TYPICAL DEFECT REPORT

(note that more space will be required under most of the headings for the majority of defects)

prepared by .... date ....
Address ....
....
Position in building ....
Description ....
....

INVESTIGATION

This report is based on the attached site investigation dated .... and on the following findings and additional information. eg. advisory publications, trade literature and interviews or correspondence with occupants or specialists

CAUSE OF DEFECT

The following possible causes of the defect were considered.

Cause (or contributory causes) of defect and reason for this conclusion ....
....
Causes rejected, for the reasons stated ....
....

FURTHER INVESTIGATION

It is recommended that the following further investigation or study should be undertaken for the reasons stated ....
....
....
Time required for further investigation ....
approximate cost .... effect of investigation on use of the building ....

REMEDIAL WORK

The following action is proposed to prevent further damage occurring ....
....
Time required for work .... approximate cost effect of work on use of building ....
....
The following permanent work is proposed to correct the defect ....
....
Time required for work .... approximate cost effect of work on use of building ....
....

Diagram referred to in first paragraph

Figure 2 Format for internal office report on a typical defect

necessary to confirm the form of construction. There will also be occasions when the "as built" drawings are perfectly correct but, because alterations and repairs have been done, the construction is no longer as it was when first completed. Records of repairs are even more likely to be incomplete than records of initial construction. There may be no record at all of patch repairs and, if claddings have been replaced over the repairs, there may be no visible sign of the repair either. The only clue in such cases could be a change in the way the defect shows itself.

Records of how a defect first became apparent, and how it has developed since it was first noticed, are important for the diagnosis of causes. They can sometimes be put together with information on changes in the environment and use of the building to help in the selection of most likely causes. For example, wind direction is likely to affect rain penetration through porous masonry, and condensation may build up over a period if conditions do not allow it to disperse in the intervals between conditions that promoted its formation. Use of a building comes into the diagnosis of defects too. Records of when uses changed, or even of what uses a building has been put to, are often incomplete or non-existent but, if such records do exist, they can give useful clues as to the causes of a defect. The danger is that such records are wrong, or at least unreliable, and that they lead to wrong conclusions.

As has been already said, this book deals only with defects that have been described in publications up to the date of its preparation. It leaves out three classes of defect; those that are so common and obvious that no one has thought it worth while to write about them – for example damage caused by slamming doors; also inevitably omitted are those that have come to light since this book was prepared; a third category are those for which expert advice is needed because those who are aware of their existence have not made their knowledge public. There are many bodies that can give expert advice in specific areas, and other professional organisations that have lists of members with specialist knowledge. Some of these sources of expert advice are given below. In most cases advice has to be paid for, but one or two trade organisations give free advice.

Aluminium Windows	Aluminium Window Association Tel; 01 637 3572
Clay Bricks	Brick Development Association Tel: Winkfield Row (0344) 885651
Clay Roof Tiles	Clay Roofing Tile Council Tel: Stoke on Trent (0782) 747256
Concrete & Render	British Cement Association Tel: Fulmer (028 16) 2727
Drains	Clay Pipe Development Association Tel: 01 388 0025
Flat Roofs	British Flat Roofing Council Tel: Haywards Heath (0444) 416 681
Floor Finishes	Contract Flooring Association Tel: 01 286 4499
Flues & Fires	Solid Fuel Advisory Service Tel: 01 235 2020
General	Building Research Advisory Service Tel: Watford (0923) 676 612 East Kilbride (03552) 330 01
General	Royal Incorporation of Architects in Scotland Tel: Edinburgh (031) 229 7205
General	Royal Institute of British Architects Tel: 01 580 5533
General	Royal Institution of Chartered Surveyors Tel: 01 222 7000
Glass	Glass and Glazing Federation Tel: 01 409 0545
Historic Buildings	Society for the Protection of Ancient Buildings Tel: 01 377 1644
House Maintenance	Building Conservation Trust Tel: 01 943 2277
Lead	Lead Development Association Tel: 01 499 8422
Mostly Structure	Construction Industry Research and Information Association Tel: 01 222 8891
Paint	Paint Research Association Tel: 01 977 4427

Particle Board	Chipboard Promotion Association Tel: Marlow (06284) 3022
Plastics Windows	Plastics Windows Association Tel: Birmingham (021) 643 6181
Rubber & Plastics	RAPRA Technology Tel: Shrewsbury (0939) 250 383
Services	Building Services Research and Information Association Tel: Bracknell (0344) 426 511
Services (Mostly Piped)	Chartered Institution of Building Services Engineers Tel: 01 675 5211
Services Electrical	Institution of Electrical Engineers Tel: 01 240 1871
Sports Buildings	Technical Unit For Sport Tel: 01 388 1277
Steel Windows	Steel Window Association Tel: 01 637 3571
Stone	Stone Federation Tel: 01 580 5588
Structures	Institution of Structural Engineers Tel: 01 235 4535
Tiles	British Ceramic Tile Council Tel: Stoke on Trent (0782) 747 147
Tiles & Bricks & Sanitary ware	British Ceramic Research Ltd Tel: Stoke on Trent (0782) 45431
Timber & Joinery	Timber Research and Development Association Tel: Naphill (024024) 3091
Zinc	Zinc Development Association Tel: 01 499 6636

Manufacturers have (or should have) an interest in advising on causes of defects where their products are involved. The disadvantage of such advice is that the manufacturer may mislead enquirers in order to shelter a flawed product. The advantage of going to the manufacturer for information, is that he is likely to have the most up to date and comprehensive information if the defect associated with that product has occurred elsewhere.

A diagnosis is rarely a private opinion. It has to be presented to others and will have to withstand scrutiny, and possibly criticism, particularly if costly remedial work is proposed. The presentation of the diagnosis is important. Others must be able to follow the detailed argument that leads to the conclusions and reasons must be given for rejecting alternatives. The illustration below gives one format that could be adopted for an internal office report of a typical defect. Other presentations may also be used, for example to explain the diagnosis to a layman or when offering evidence in legal cases. It should always be clear in a defects report whether the information is an opinion or a fact. If it is an opinion, who gave it and what is it based on? If it is a fact, where was it obtained? It is helpful if the report on a defect concludes either with recommendations on how the defect can best be cured, or with proposals for further investigations, or studies to be carried out in order that firm recommendations can be made for a cure.

If the recommendation is for remedial action to cure the defect, it must be for a cure which will succeed. Where doubts as to the cause remain, or finance is not available to ensure that the defect is put right in all respects, the recommendation may be for a trial solution. In such cases it must be made absolutely clear that the recommendation is only for a trial, and that all concerned accept the possibility of a recurrence of the defect, with more extensive remedial work in the future. As a general rule, it is even more important to avoid recurrence of a defect than it is to avoid their occurrence in the first place. Recommendations for curing defects should be on the safe side. Owners and users of buildings which have suffered defects will not want to repeat the experience.

CHAPTER 4 INVESTIGATION TECHNIQUES

Photo 1 Using a CM5 covermeter to establish the position of the steel reinforcement on a tall concrete facade: Protovale (Oxford) Ltd.

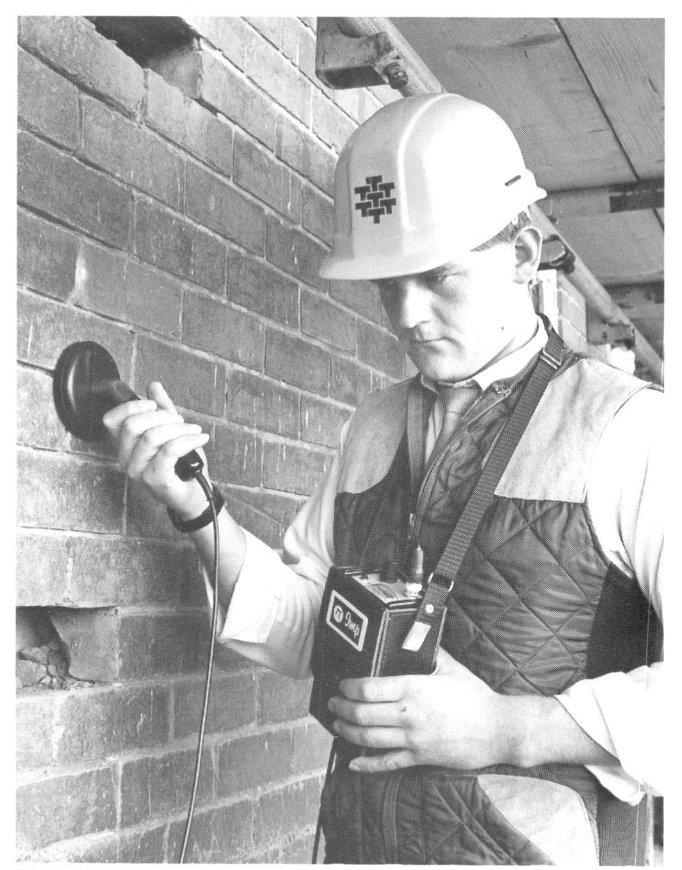

Photo 2 Using an Imp wall tie locator to check the presence and location of ties in cavity brickwork: Protovale (Oxford) Ltd.

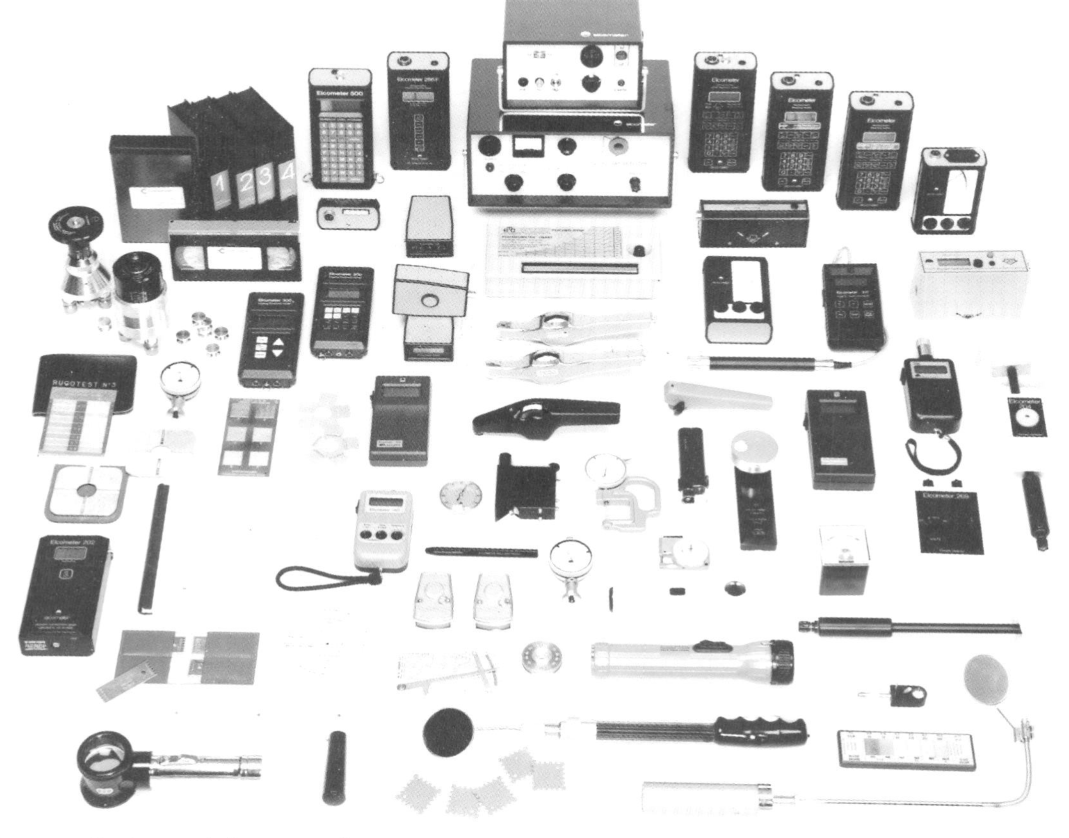

Photo 3 A range of instruments for measuring coatings on different substrates; Elcometer Instruments Ltd.

CHAPTER 4 METHODS OF INVESTIGATION

This chapter deals with some of the changes that have occurred in recent years in the methods of investigating defects. The Standard equipment that a surveyor would use to make a preliminary inspection of a building is still relevant and is listed below. In addition to this basic equipment there are now more specialised and sophisticated items available that provide: more information; more accurate information; information in a more convenient form or simply speed up the inspection process.

List of basic equipment for investigating defects

Equipment for measuring of dimensions	*Example of use*
Measuring rod	to measure height above a floor level.
Measuring tape	to measure horizontal dimensions and longer diagonal or vertical dimensions.
Ruler	for indicating the size of a defect in photographic evidence.
Vernier Calipers	for measuring small outside and inside dimensions e.g. pipes, joints, etc.
Feeler gauges	for measuring fine gaps and cracks.
Plumb bob & line	for checking bulging or leaning of walls.
Spirit level	for checking slope of roofs and floors.
Straight edge	for checking dents or bumps on flat surfaces.
Compass	to establish the orientation of the building.
For access and inspection	
Folding ladder	for gaining access to defects that are just out of reach.
Binoculars	for visual inspection of construction that is well out of reach.
Hand mirror	to inspect such things as external window putties from inside.
Torch	for roofspaces, understairs cupboards, cellars etc.
Drain keys	to lift inspection covers.
Stopcock keys (crutch and fork pattern)	to turn off water supply.
For testing and sampling	
Small Screwdriver	to probe suspect timber for rot.
Penknife	to scrape off samples of loose materials.
Hammer & Bolster	to cut out samples of plaster, masonry, screed or render.
Cordless drill	to obtain mortar samples.
Hacksaw	to cut off samples of hard materials.
Plug top circuit tester	to test whether sockets are correctly wired.
Electrical resistance meter	to check moisture content of timber and plaster.
Unused plastic bags and containers (must be sealable type)	to collect samples.
For recording investigation	
SLR camera	to record appearance at time of investigation.
Chalk (white and light colour)	to mark areas for further investigation.
A3 clipboad and tracing paper	as a firm base for inspection of drawings and for making overlay sketches.
Spiral bound A5 pad	for notes.
Adhesive labels	to label samples.
For protection	
Face mask	to reduce breathing in of dust or fibre.
Goggles	to protect eyes from dust.
Plastic gloves	to handle samples.
Hard hat	to wear when headroom is restricted.

Measuring and checking dimensions

Traditional measuring and surveying equipment is now supplemented by a whole range of devices that can be used to obtain accurate information on the dimensions of buildings and any deformation that has occurred. There are small, handy, electronic instruments with digital displays that can give a reasonably accurate reading of the distance from a solid object. They measure the distance by the time a reflected ultrasonic pulse takes to return to the instrument. The use of these instruments is limited because they cannot easily give running dimensions of features such as openings along a flat wall surface but they are suited to measuring room dimensions or distances between buildings that are within 30 m. The accuracy of different models varies considerably as it depends on the quality of the built in temperature compensator and on the width of the beam. A narrow beam is less likely to be reflected from adjacent objects. Some of these instruments can give readings of areas and volumes and can store a series of measurements in their 'memory'.

Low cost electronic levels are also available giving a very accurate digital display of the angle to the horizontal of a flat side of the instrument. These are useful for a check on verticality or horizontality of surfaces, for example for measuring the falls of paving or flat roofs, but traditional surveying instruments, brought up to date with electronic extras are more suitable for accurate measurement of levels over longer distances, for example, to establish the difference of levels at the corners of a building, or the extent of a bulge in a wall's façade. Traditional surveying levels are now fitted with digital displays and with memories to record long series of measurements. They can also be used with lasers that help to aim the level at a point or, by rotating, provide a reference plane from which measurements can be taken. Periodic measurement from the same base point may be needed to check on the movement of a building, but the small movements that cause cracking are usually measured by fixing "tell tales" across cracks. Traditionally tell tales were cement patches with the date scratched onto them or strips of glass fixed firmly to the wall on either side of the crack. Purpose made tell tales marked with a vernier calibration that allows small dimensions to be read accurately are now available. They are screwed to the wall so that they bridge the crack. The markings allow the extent of any movement to be read off. Another method that gives accurate readings is to fit projecting non-rusting screws close to the crack on either side and take measurements of the distance between the screws with a vernier caliper.

Access and inspection

To reach parts of the outside of a building that are not accessible from the ground or from raking ladders the traditional solutions are to use chairs, cradles, scaffolding or steeplejacks Yorkshire and Lancashire type sectional ladders. More flexible methods of access are now used to inspect inaccessible external walls. A range of hydraulic platforms can be used by a non specialist surveyor up to about 15 m above the ground. Some of the platforms are self propelled, others have to be towed into position. Specialist teams of surveyors are available to reach higher levels by absailing on ropes down the face of the building. This method has been used successfully for the systematic inspection of joints in large panel concrete buildings.

Defects diagnosis sometimes relies on information that is hidden in the voids and cavities of a building. Opening up for inspection and making good after the inspection is an expensive business and can disrupt the use of the building. Endoscopes (also known by the trade name Boroscopes) are used to inspect and photograph cavities and voids through small holes drilled in the enclosing materials. It is now common practice to examine wall ties in masonry with an endoscope. The head of the endoscope and a beam of light is swivelled round inside the cavity and the instrument can be focussed on any feature within range of the lens. Other probes are used to test conditions in enclosed spaces such as heating ducts – more detail is given below under moisture and thermal investigation.

Endoscopes can be linked to closed circuit television (CCTV) to record the view through the instrument as the lens is rotated. There are other applications of CCTV for defects inspection. The most common is for drain inspections where the camera passes along the drain continuously recording its condition. When the tape is replayed it is possible to pin point the exact position in the drain run where a particular defect was recorded. However, there may well be disagreements on the diagnosis of the defect shown on the tape or a dispute over whether it is a blemish that can be left or a defect that needs putting right.

It is sometimes important to discover the line and depth of buried or embedded services, either to avoid damage in the course of remedial works, or as part of a diagnosis where defects are related to services, for example, where pipes leak or where service trenches destabilise foundations. Radio detection equipment can help in three ways.

1. Where the hidden service carries electrical power a radio detector will pick up a clear signal from up to 3 m below the ground.
2. Signals can also be picked up from buried metal pipes or services that are not being used as electrical conductors.
3. Non metallic or deeply buried drains can be tracked up to a depth of 18 m by running a radio transmitter through the drain and picking up its position with a radio detector on the surface.

Similar instruments are used to track reinforcement, these are described under testing of materials below.

Leaks in soil and waste pipes can also sometimes be traced by using dyes to colour the water in suspect pipes. Each pipe in turn is filled with coloured water until the colour appears in the water that leaks to the outside. New techniques are available for on site soil investigation using sound or radio waves to detect changes in the composition of the soil concealed below the surface. Voids, obstructions, and changes in the depth of soil strata can be guaged without excavation. Instruments are also available to assess the bearing capability, etc., of the soil. CIRIA (Tel: 01 222 8891) have published books on the 'core penetrometer' and the 'pressuremeter'.

Moisture and thermal investigation

As a result of problems with condensation in buildings, particularly in dwellings, techniques have developed rapidly in this area. The investigation of moisture, including humidity, and of thermal conditions in a building has to be very precise if it is to be used as a basis for a confident diagnosis. It may also have to be continued over a period of weeks or even months to see how the symptoms vary with changes in the weather or the use or the internal environment of the building.

The basic 'moisture' meter is an electrical resistance meter that measures the reduced resistance between two probes when they are embedded in damp materials. Its original use was to check on the moisture content of softwood joinery to avoid problems of rot, movement of the timber after installation and poor adhesion of paint. It can also be used to trace areas of walls affected by rising damp or rain penetration. In these applications the true cause of the defect is sometimes difficult to establish because surface and interstitial condensation can produce some of the same symptoms. Electrical resistance meters have become more refined to deal with this problem, and other instruments are used with the meters to provide simultaneous records of related data. Meters are now available with a range of probes and sensors to hammer into the wall, or to attach to the surface of the wall, so that surface dampness can be distinguished from dampness within the wall. Other instruments can be used to measure the temperature of the air and wall surfaces and the relative humidity of the room, with a further set of sensors collecting similar data on the outside of the building. When the outside and inside conditions are related to the occurrence of dampness in a wall it should become clearer, over a period, whether the problem is straight forward condensation, water from some other source, or a combination of both. Measurements will need to be taken at regular intervals, or better still, continuously. Instruments have to be set up in positions where they will not be damaged. The practical problems of this sort of investigation are sometimes greater than the technical ones, particularly where the use of the building also has to be monitored. The probes are very sensitive and easily damaged.

Other instruments used to detect moisture can sense the presence of water through an impervious covering, such as a waterproof roof membrane; 'nuclear' detectors react to the presence of hydrogen in H_2O, temperature sensing devices rely on the cooling effect of entrapped water on the temperature of the surface, electronic scanners detect water by the increase it produces in the electrical conductance of the construction. There is a range of instruments from small hand held moisture detectors, to trolleys that can be wheeled across a roof to 'map' the the areas where dampness below the surface suggests that leaks have occurred.

Thermal and moisture sensors are also used to check thermal comfort, particularly of air conditioning systems. They are then used in conjunction with anemometers to record the speed of air movement and thermometers that measure radiant temperature in different directions. These, more sophisticated, instruments are only likely to be used by specialists.

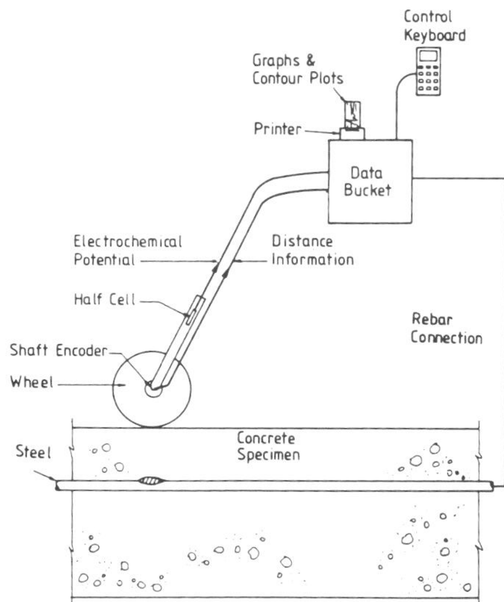

Photo 4 & Figure 1 The potential wheel and data bucket used to determine the probability of reinforcement having corroded within concrete. CNS Electronics Ltd.

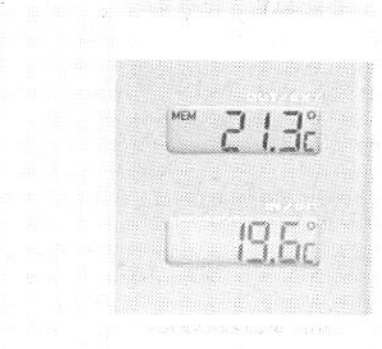

Photo 13 Indoor/outdoor digital thermometer ST-200: Solex International.

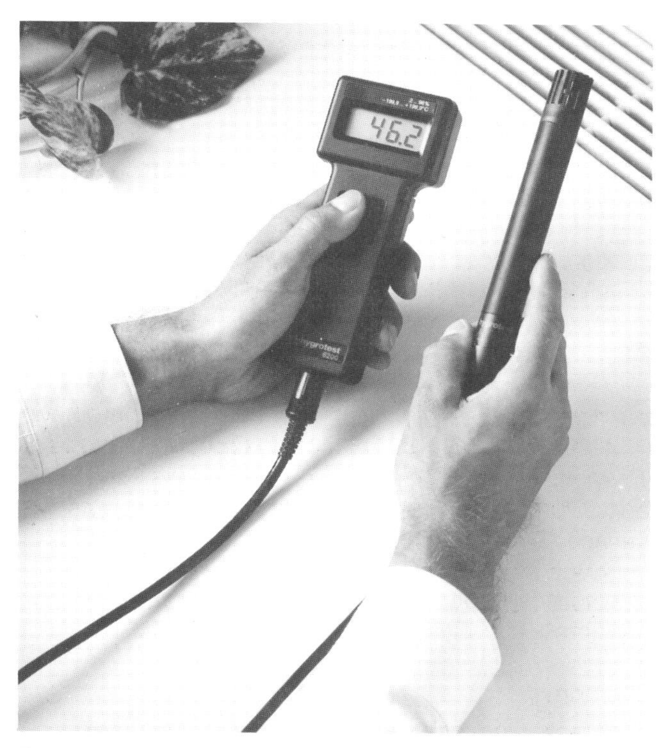

Photo 6 The Hygrotest 6400 for temperature and humidity measurement: Testoterm Ltd.

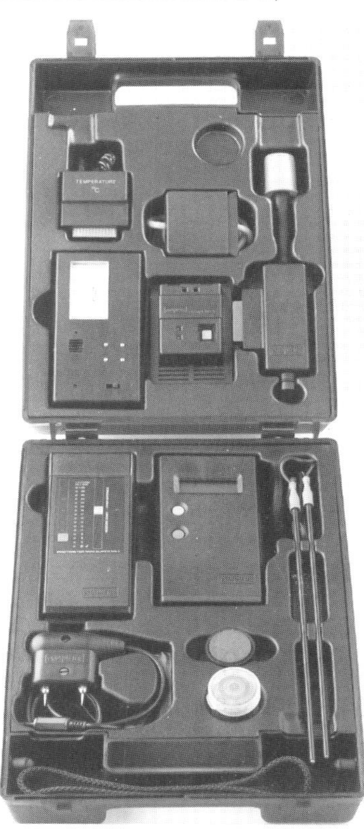

Photo 5 The compleat dampness kit that includes surface thermometer, thermal hygrometer, salts detector, on site salts analysis kit, electronic tell tale and moisture meter: Protimeter Instruments Ltd.

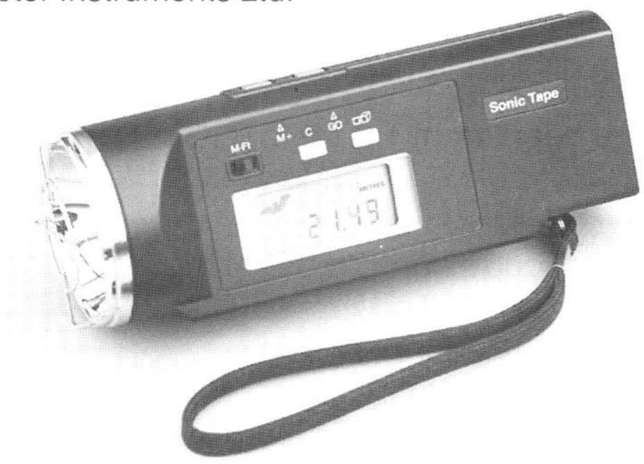

Photo 7 A sonic "tape measure" that can be used to display distance, area or volume in either metric or imperial units: Blue Chip Products Ltd.

Each of the methods described and the instruments employed has potential limitations which may render the results suspect. It is important that the investigator fully understands the equipment, is skilled in the interpretation of the results and uses these methods as diagnostic aid rather than crediting them with the ability to unfailingly provide definitive conclusions.

For example, an electrical resistance, or moisture meter, will give very different results in similar materials. Salts or metals, such as those found in concrete blocks made from ash, will lower electrical resistance so that the meter will indicate that even dry materials have a high moisture content.

Sound testing

Where construction defects are suspected of leading to unacceptably low levels of sound insulation, tests may be required to establish the scale of the problem in the first place and the effectiveness of remedial measures once the work is completed. Equipment is available to measure the sound received in one room from a standard sound source in another and to produce a print out covering each octave of the audible range.

Material testing

Most materials can be subjected to a whole range of tests to establish their composition and properties. Many of the tests relevant to building defect diagnosis can be best carried out in the laboratory, however, some materials can also be tested on site. Site tests that are used in defects diagnosis are described here and the names are given of some of the organisations capable of carrying out the more commonly required forms of laboratory tests.

Concrete

There have been many recent developments in the non-destructive testing of concrete, particularly of reinforced concrete structures for the investigation of reinforcement corrosion. Where reinforcement corrodes as a result of carbonation a visual inspection may find rust stains, or pieces of concrete starting to spall away. Samples of the affected concrete can then be sent to a laboratory to check the extent of carbonisation. Alternatively samples can be brushed with a liquid that changes colour in areas where the alkalinity is reduced by carbonation. In the early stages there may be no sign of carbonation on the surface of the concrete but systematic tapping of the surface of the concrete over the reinforcement will discover hollow sounding patches where the reinforcement has corroded. The line and depth of reinforcement in concrete can be found with an electromagnetic covermeter (a hand held instrument that is passed over the surface of the concrete giving a reading of the closeness of the steel to the surface). Where corrosion is due to chlorides, from salt sea spray, de-icing chemicals used in the building or faulty manufacture of the concrete in the first place, the presence of the corrosion can be predicted by instruments that measure the electrical potential of the reinforcement and compare it with a reference potential level derived from a half cell. The areas with the greatest negative potential are the most likely to be corroded. Arrays of half cells are mounted on frames or wheels which pass over the surface of the concrete taking a set of readings from which the likely areas of corrosion can be plotted. With this method, an electrical cable has to be connected to some part of the reinforcement.

Testing the composition and strength of concrete and cement based materials is usually done in the laboratory using sample cores drilled out on site. BRE has developed a simple instrument for in-situ testing of screeds (BRE Screed Tester I.P.11/84). It measures the indentation caused by a weight being dropped from a fixed height onto the screed surface. There is also a recommended BRE method for determining the dryness of a sub-floor using a hydrometer set on the floor surface to establish whether the floor is sufficiently dry for flooring to be laid.

BS 1881 Part 201 is a guide to the use of non-destructive methods of test for hardened concrete. The table below gives information on the principle methods from this guide that are relevant to defects diagnosis. The British Cement Association's materials service department can offer an investigation and analysis service (Tel: 028 16 2727). For laboratory tests of concrete the British Ready Mixed Concrete Association publishes a register of test houses (Tel: 01 381 6582).

Masonry

In defects diagnosis, the most common feature of masonry that requires testing is the presence of chemicals, particularly sulphates that cause

Photo 8 Photographic equipment available with boroscopes (endoscopes) for recording the condition of cavities in buildings Key Med (Medical and Industrial Equipment) Ltd.

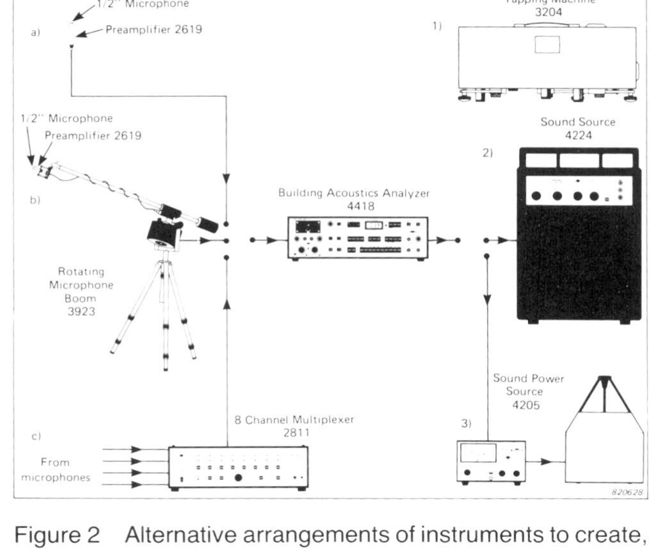

Figure 2 Alternative arrangements of instruments to create, record and analyse sound pressure levels Brüel and Kjear.

Photo 9 Using a boroscope (endoscope) to examine the inside of a hollow construction: Key Med (Medical and Industrial Equipment) Ltd.

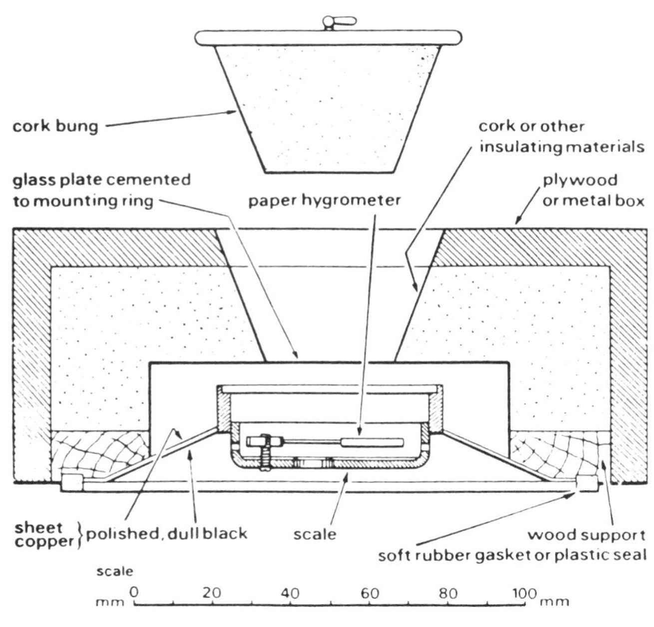

Figure 3 Method of determining the dryness of a sub floor: BRE Digest 18.

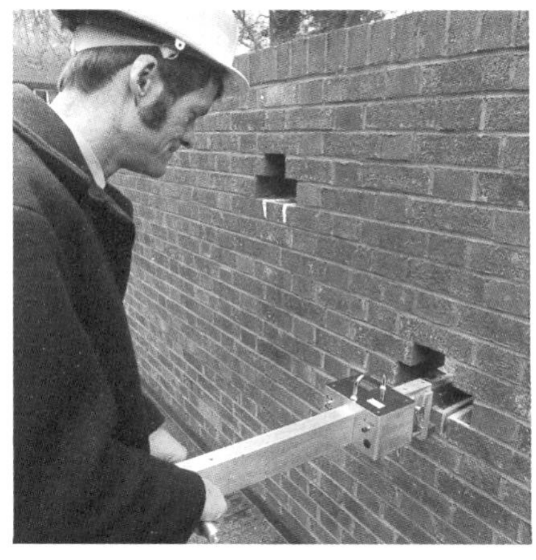

Photo 10 The "Brench" used to test the strength of existing masonry.
Building Research Station.

Photo 11 The Dec Scanner roof moisture detector used to indicate the presence of moisture below a built up roof membrane: Tranex Ltd.

Photo 12 The Screed Tester used to assess the soundness of floor screeds (see chapter 7).

deterioration of mortar, and sodium salts that accelerate the breakdown of stone masonry. The cement content of the mortar is also frequently tested. All these tests are normally carried out in the laboratory from samples taken on site. Normally, samples used to establish structural properties are drilled out from the central depth of the wall as material on the surface may not be typical of the body of the wall.

BRE has developed an instrument for measuring the flexural strength of masonry on site. It can help to establish whether existing damaged walls can be repaired rather than rebuilt.

British Cement Research Ltd have laboratories specialising in analysis of clay products (Tel: 0782 45431).

Asphalt and bituminous products

Laboratory tests are necessary to establish the composition of asphalt, its bitumen content and hardness and of materials, such as oil or fat that has damaged it. The private sector laboratories that specialise in bitumen products are: Stanger Ltd (Tel: 01 207 3191), Sandbergs (Tel: 01 730 3461) and Manchester Building and Testing Laboratories (Tel: 020 488 667).

Timber

Apart from the visual inspections that are described under the defects listed in the following chapters, many investigations of the causes of defects in timber require tests to be carried out in a laboratory. There are a few areas however, where site investigation methods are suited to use on site:

Moisture content: Electrical resistance meters are regularly used on site to check conformity with specifications for new timber, to check susceptibility of existing timbers to rot and to check whether timber is dry enough to accept paint systems. Decisions should not be made on the basis of one measurement showing a high moisture content. A set of measurements should be taken from different timber sections to establish the general moisture level.

The extent of rot: Probing and drilling timber is the normal method of establishing the extent of rot when it is not visible on the surface, but other instruments using sound waves, which were originally developed for checking telegraph poles, are being used experimentally on large structural timbers. Trials are also being made with instruments that are sensitive to the odours given out by fungus to establish whether there is any rot present in a structure.

Extent of treatment: With colourless preservatives, it is often not possible to tell by a visual inspection whether a particular piece of timber has been treated. The preservative manufacturers will supply chemicals that can be painted or sprayed onto timber to confirm – by changing colour – that the timber has been treated. On untreated timber the chemical does not change colour. If it is also necessary to establish the extent and concentration of preservative treatment, samples must be tested in a laboratory. The on site test kits can give misleading results where timber has been in contact with a variety of chemicals that are used on building sites.

Several laboratories are capable of carrying out tests on timber. The Timber Research and Development Association (Tel: Naphill 024 024 3091) is the only one in the country specialising in timber. They will identify rot and infestation, investigate problems with paints and preservatives and assess the condition and strength of existing timbers.

Metals

Testing the composition of metals and analysing the causes of corrosion or fracture is normally done in a laboratory. Many laboratories are capable of such analysis. The Lead Development Association recommend – Cookson Group plc (Tel: 01 997 5635). The Copper Development Association use – The British Non-Ferrous Metals Laboratories (Tel: 0235 772992). The Stainless Steel Fabricators Association of Great Britain suggest – The Fulmer Research Institute (Tel: Fulmer 02816 2181). The Zinc Development Association deal with – Sandbergs (Tel: 01 730 3461).

Instruments have been developed for in-situ magnetic testing of the thickness of coatings on porous metals which, when applied to the surface of the metal, can give a measurement of the thickness on a digital display without marking the coating. Other instruments can measure the thickness of coatings on non-ferrous metals. Coatings on non-conductive materials such as plastic, cement, concrete and wood have to be cut or drilled to allow the thickness of the various coats to be measured. Special cutting and measuring instruments are available for the job.

Plastics and glass

A wide variety of plastics are produced and it is usually necessary to identify the type of plastics in order to diagnose the cause of a related defect. BRE has produced a kit to help with identification which includes instructions for simple tests and samples for comparison. The Rubber and Plastics Advisory Service (Tel: 01 235 9888) can recommend experienced consultants for specific investigations and these, in turn, will propose test houses for the somewhat complex laboratory tests required to diagnose faults in plastics products.

The British Glass Manufacturers Confederation (Tel: 0742 686201) have their own Research and Development Laboratories that inspect and carry out laboratory tests on such things as unexplained cracking, failure of adhesives in double glazing and surface discolouration.

Defects in sealants using polysulphide polymers can be investigated by Thiokol Chemicals (Tel: Coventry 0203 416632) who manufacture the raw materials for this type of sealant.

Other materials

Laboratories acquire experience and specialist knowledge in analysing the composition and properties of different materials. For example the Industrial Research Laboratories in Birmingham (Tel: 021 235 3204) have specialised in fibre cement products such as man made slates. NATLAS – The National Testing Laboratory Association Scheme (Tel: 01 977 3222) has published a useful concise directory which lists laboratories under a number of headings that are relevant to construction.

Defects may affect a combination of materials as well as materials in isolation. BRE has pioneered many of the test methods that can be used on site to assess the performance of buildings. Tests for such things as air and water tightness can be carried out on completed structures.

They also carry out many more routine tests of materials and will send experts to investigate defects on site. The BRE Advisory Service has two contact points one at Watford (Tel: 0923 676612) and one in Scotland (Tel: 03552 33941).

New methods of investigation are regularly being developed. Some to deal with new problems that have arisen, others borrowed from other industries are simpler or more convenient than those previously used in construction. The Construction Industry Research and Information Association (Tel: 222 8891) has undertaken to produce a handbook of practical information and guidance on the use of testing and monitoring equipment.

Building and site records

The starting point for diagnosis of some defects is the record that exists in plans and other documents related to a building or to a site. When a building is completed building owners should, but rarely do, possess a full set of as-built drawings. It is often necessary to look for other sources of original drawings and of drawings or specifications that were modified during the contract. Any one of the consultancy firms or specialist contractors involved may have retained copies of documents after the contract. Subsequent alterations, routine maintenance or remedial works may also have required documentation or the logging of data on computer files that is a useful source of information. Sometimes it may be necessary to draw up the details of a construction that can be guessed from superficial inspection and confirm the detail by opening up critical points for inspection.

Where the suspected cause of a defect relates to the ground conditions, BRE digests 64 – *Soils and foundations 2* and 318 *Site investigation for low rise building: desk studies* give useful advice on soil identification and sources of information related to topography, vegetation and drainage. For example, a catalogue of abandoned mines was compiled between 1928 and 1939, and a directory of quarries and pits in 1973.

Other documentary evidence may be relevant when a building is affected by atmospheric pollution. A map showing general levels of atmospheric corrosivity values for the United Kingdom is published by the Farm Buildings Group of the Ministry of Agriculture.

Local authorities may have more detailed information affecting particular sites.

Recording investigations

A suitably detailed and systematic record should be made of any defects investigation. At the least, enough information should be put down to avoid the need to revisit the building for missing details before a diagnosis can be made. The aim is often to produce a record that can be used in comparison with a subsequent survey to establish whether there has been further deterioration. If a defects inspection can also provide enough information to specify the extent of remedial works there may, in the long run, be a saving of time and money.

Records of investigations are collected in various forms, eg. samples, photographs, measurements, sketches, descriptions, video tapes, mechanically drawn graphs and laboratory test results. In each case it is vital to record exactly what is referred to and where and when the record was made. It is too easy for the investigator to delude himself that he will never be able to forget which block or which façade was measured. A systematic method of identifying records is essential in any defects investigation.

When a series of similar investigations are being made, for example of the roofs of a housing estate, pro formas and outline drawings can be prepared before an inspection takes place so that the collection of information is speedier and more complete.

Recent developments in surveying equipment can help to provide systematic, clearly identified records. Electronic equipment of all sorts can be linked to tape recorders or computers to log a long series of measurements. Hand held computers can be programmed to accept information in a particular sequence so that no records will be accepted until the basic information on the address of the property, the position of the defect in the building and the date of the inspection have all been entered. Such programmes have been developed where maintenance of a series of similar components, for example windows, each requires an individual record. They may also be of use where there is a need for the data that is collected to be manipulated on a computer. If the record is collected directly onto the machine the inevitable mistakes made when manually entering data can be avoided.

In some inspections, covering a large area or a great number of instances, it may be necessary to develop a statistical basis for selecting a sample of defects to investigate. This is a specialists job but it has been done successfully, for example, with the joints of large panel concrete structures and the results have correlated well with the actual findings when all the joints were opened up for remedial treatment.

continued

Principal non-destructive test methods for hardened concrete – based on Table 2 BS 1881: Part 201: 1986 Section one

Method	BS 1881: pt 201: 1986 clause no.	Principal reference	Principal applications	Principal properties assessed	Surface damage	Type of equipment	Remarks
Pull-out test (drilled hole)	2.18	BS 1881: Part 207*	In situ strength measurement	Strength related	Moderate/ minor	Mechanical	Drilling difficulties on vertical surfaces or soffits Surface zone test
Internal fracture	2.17	BS 1881: Part 207*	In situ strength measurement	Strength related	Moderate/ minor	Mechanical	High test variability Surface zone test
Surface hardness	2.15	BS 1881: Part 202 (supersedes BS 4408: Part 4)	Comparative surveys	Surface hardness	Very minor	Mechanical	Greatly affected by surface texture and moisture Surface test unrepresentative on concrete more than 3 months old Strength calibration affected by mix properties
Initial surface absorption	2.8	BS 1881: Part 208* (supersedes BS 1881: Part 5)	Surface permeability assessment	Surface absorption	Minor	Hydraulic	Difficult to standardise in-situ moisture conditions and to obtain watertight seal to surface Comparative test
Surface permeability	2.9		Surface permeability assessment	Surface permeability	Minor	Hydraulic	Surface zone test Water or gas
Resistivity measurements	2.3		Durability survey	Resistivity	Minor	Electrical	Surface zone test Related to moisture content Indicates probability of reinforcement corrosion in zones of high risk
Half-cell potential measurements	2.4		Survey of reinforcement corrosion risk	Electrode potential of reinforcement	Very minor	Electro-chemical	Cannot indicate corrosion rate
Thermography	2.11		Structural integrity survey and void location	Surface temperature differences	None	Infra-red radiation detection	Extraneous temperature effects have to be excluded Temperature differentials small Shortage of data and development

* In preparation at time of writing

CHAPTER 5 WALLS

CHAPTER 5 WALLS Contents

continued

CHAPTER 5 WALLS Contents

continued

Other walling defects

CHAPTER 5 WALLS Contents

continued

INTRODUCTION

Wall defects usually show as cracking or dampness (or both together) or as defects of appearance such as stains and irregular surfaces. Other types of defect, for example those involving bulges or defective claddings, may be noticed only when cracks or signs of dampness appear.

This chapter starts with cracks – the critical symptoms being: the width and depth of the crack, its position and its direction. Other features also come into the picture, particularly changes in the dimension of the crack, the general appearance of the edges of the crack and its relationship to the materials used in the building's construction. Cracks must be looked at in the wider context of nearby trees, ground conditions, foundation design, adjacent loadings and any alterations that have been made to the building.

Defects associated with dampness follow the section on cracks – here the critical features are the extent of the dampness and when it appeared. It is most important to know precisely how the building is put together and what materials were used in the construction.

Other defects are listed under their particular symptoms. Superficially similar defects – such as falling ceramic cladding tiles and falling brick slips may have very different causes. The walling defects covered by this book are all specific to particular materials. It is often necessary to have materials tested to find the true cause of these defects.

CRACKS IN WALLS

General points

When cracks in walls are being investigated there are certain general points to bear in mind:

- Some cracking in the early life of the building may be inevitable. Cracks resulting from drying out or taking up of moisture or from the initial drying out of materials cannot always be avoided. Such cracks are invariably fine and can usually be repaired in a way that avoids recurrence.
- The structural significance of cracks is often exaggerated; this is a natural reaction of the owners or occupiers of a building. Some cracks may be an indication of instability of the structure but many others, even those that look quite serious, will have little or no effect on stability or other aspects of building performance apart from appearance.
- It is probable that thermal expansion is given as the cause in cases where it is not the true cause. A crack, whatever its origin, provides a convenient observation point. It is easy to observe changes in size where a crack opens and closes, whereas changes in the overall dimensions of a wall can be measured only with special equipment.
- In large buildings the cumulative affect of slight movements is likely to cause cracking where no cracking would occur in smaller buildings of the same construction. This can be seen where a large concrete roofslab disrupts the perimeter parapet as it expands and contracts in response to temperature variations.
- Cracks may be caused by a temporary load or lack of support that is not evident when the causes of cracking are being investigated. For example materials stacked against a wall or service entries taken under a wall may have been the initial cause of cracking but can be difficult to diagnose a year or so after the damage has been done.
- Cracks do not always take the typical form described in this book. The restraint to which the wall is subjected influences the form and position of cracks. Restraint is exercised by abutting walls and adjoining construction.

The specification of repairs to cracks in walls must take into account two problems.

- It is desirable (but not always possible) to establish whether cracking is due to movement that has occurred and will not continue or be repeated, or whether it is due to progressive or cyclical movement that could continue.
- Once a building has cracked it is difficult to repair it in a way that re-establishes the original strength of the construction. The crack can remain a weakness that will reopen in response to forces acting on the building. This sometimes leads building owners to complain of unjustified expense or inefficient work, particularly if they have not been made aware of the limitations of the repair.

Significance of cracks

Apart from their structural significance cracks can affect the performance of a building in other ways. Sound insulation between spaces is greatly reduced, particularly at the higher frequencies, if there is a crack between the spaces. Cracks in external walls are likely to increase air infiltration and heat loss and may well provide pathways for rain penetration to vulnerable materials in the construction or to the inner surface of walls.

Whenever cracks are thought to be structurally significant a structural engineer should be consulted – even in cases when cracking is slight but the cause of the cracking is not quite clear. As most buildings have some cracks it is not practicable (or necessary) to involve a structural engineer in every condition survey. Large areas of walling designed before 1978, when design criteria were changed, should be checked for structural adequacy even if they show no signs of cracking, and particularly if they are:

i less than 220 mm in overall thickness

or ii have an average height of over 9 m

or iii have a length between lateral supports of more than 6 m.

(see BRE digest 281 *Safety of large masonry walls*)

The BRE has classified visible damage to walls in six categories (see table 1). The categories relate particularly to ease of repair of plaster and brickwork or masonry – and therefore more relevant to low rise

continued

Table 1 Classification of visible damage to walls with particular reference to ease of repair of plaster and brickwork or masonry

Category of damage	Degree[1] of damage	Description of typical damages *Ease of repair in italic type*	Approximate crack width *mm*
0	Negligible	Hairline cracks of less than about 0.1 mm width are classed as negligible	Up to 0.1[2]
1	Very slight	*Fine cracks which can easily be treated during normal decoration.* Perhaps isolated slight fracturing in building. Cracks rarely visible in external brickwork.	Up to 1[2]
2	Slight	*Cracks easily filled. Re-decoration probably required. Recurrent cracks can be masked by suitable linings.* Cracks not necessarily visible externally; *some external repointing may be required to ensure weathertightness.* Doors and windows may stick slightly.	Up to 5[2]
3	Moderate	*The cracks require some opening up and can be patched by a mason. Repointing of external brickwork and possibly a small amount of brickwork to be replaced.* Doors and windows sticking. Service pipes may fracture. Weathertightness often impaired.	5 to 15[2] (or a number of cracks up to 3)
4	Severe	*Extensive repair work involving breaking-out and replacing sections of walls, especially over doors and windows.* Window and door frames distorted, floor sloping noticeably[3]. Walls leaning[3] or bulging noticeably, some loss of bearing in beams. Service pipes disrupted.	15 to 25[2] but also depends on number of cracks
5	Very severe	*This involves a major repair job involving partial or complete re-building.* Beams lose bearing, walls lean badly and require shoring. Windows broken with distortion. Danger of instability.	usually greater than 25[2] but depends on number of cracks

Notes:

1 It must be emphasised that in assessing the degree of damage account must be taken of the location in the building or structure where it occurs, and also of the function of the building or structure.

2 Crack width is one factor in assessing category of damage and should not be used on its own as direct measure of it.

3 Local deviation of slope, from the horizontal or vertical, of more than 1/100 will normally be clearly visible. Overall deviations in excess of 1/150 are undesirable.

In general 3, 4, and 5 all involve significant repairs. Category 2 sometimes involves such repairs. Categories 0 and 1 can usually be dealt with by filling and decoration. But even these categories can be significant if they indicate weaknesses that could lead, for example to detachment of claddings or collapse of free standing walls, or if they are the first sign of progressive cracking. BRE Defect Action Sheet 102 discusses when a crack indicates progressive movement. It advises that the movement of the crack should be monitored over a minimum period of six months to establish whether movement has ceased or is progressive. There is a warning that cyclic movement may well be found to be superimposed on progressive movement and a longer period of monitoring may be needed to distinguish between seasonal and progressive movement.

domestic scale buildings in masonry construction than to framed structures and large buildings.

Most progressive cracking of walls is due to changes in the support for the wall. Ground movement can affect the foundation of a wall, particularly near trees or where trees have been recently felled. Beams supporting walls can fail or if the structure moves they may lose their bearing. Alterations can overload one part of the structure until it buckles or causes foundations to subside. Progressive cracking can also be caused by cyclical movement that does not recover to the original position after each cycle. For example when a crack opens it may become packed with debris that does not allow it to close up again. In the next cycle of contraction and expansion more debris is packed into the crack. This process can progressively move the two sides of the crack further and further apart.

Where slight or even moderate cracking can be shown not to be progressive, e.g. when the cause of cracking can be removed, for example by repairing a leaking water service pipe that has saturated dry clay soil beneath a partition, there may still be an argument about what repairs are necessary – particularly if the building is to be sold. In such cases a structural engineers report should be obtained before carrying out extensive but possibly unnecessary work to reassure a potential buyer.

In framed buildings and very large buildings cracks can have different implications to superficially similar cracks in low rise masonry buildings. Cracks between panels and frames may be the result of movement in either the panel or the frame – or both. The frame itself is not likely to collapse when cracks start to show at the perimeter of a panel, but the panel, or parts of the facing materials surrounding it, could become detached and fall from the building. Types of movement that would only cause very slight cracking in small masonry buildings, can cause much more extensive cracking in large buildings because of the cumulative effect of movement over longer dimensions. In a large building, thermal movement in copings and parapets will produce greater displacement and more cracking at unrestrained corners.

Investigation of cracks

The techniques that can be used to measure and record cracks have been described in chapter 4. The following notes are given to guide the investigator in the examination of cracks to provide evidence for diagnosis.

Direction – Can conveniently be classified as vertical, horizontal or diagonal. There should also be a note of whether the crack is straight, toothed or variable and irregular. If several cracks run parallel this should also be noted.
Extent – Note the starting and finishing points of the crack; whether it extends across openings or passes round the edge of material. Does the crack pass through a DPC or is it confined to above or below the DPC?
Width – Note the width and if the crack tapers, note the variation in width, record the date, time of day and temperature (and if possible the humidity) when the width was measured.
Depth – How far through the construction does the crack extend? In a multi-layer construction – which layers are cracked? Does the crack in a cavity wall extend through to the inner leaf?
Alignment – Note should be taken of the levels of the materials on the two sides of the crack. One side sometimes being proud of the other. This can usually be done by feeling with the finger tip, passing it across the crack in both directions, any difference of level between the sides will be evident. Alignment can show whether the crack has been produced by a straight pull – as with tensile force, or by a diagonal pull – as with shear action.
Edge Sharpness – Most cracks have sharp edges but they may be rounded, roughened or have flakes spalled off by compressive forces or vibration if the edges have been brought together.
Cleanliness – Examine the cracks, preferably with a magnifying glass, to see whether the edges are bright and clean; dirt, insects, algae, or paint may give a clue to the crack's age.
Time Effects – It is not usually imperative to repair a crack as soon as it is noticed. A period for observing its behaviour is desirable to give a more accurate diagnosis of its cause. Cracks are seldom noticed at the moment they appear except when the cracking of materials is audible (even then the crack that is found after a cracking noise is heard often turns out to be an old crack unrelated to the noise).

continued

Adjacent Construction – It is essential to note the materials near the crack and their condition. For example if there is a timber lintel below a crack – is the timber in good condition? Record any distortion in the adjacent walling:–
A. By plumbing the walls at corners and at intervals away from the corners.
B. By levelling along a brick course or other suitable line (bearing in mind that the building may not have been level to start with).

Surrounding Conditions – Note anything in the environment and surroundings of the cracked construction that could have caused or accelerated the damage, e.g. proximity of trees, exposure, dampness and flooding, lightning, structural alterations, changes of use or loading, changes of temperature (frost and fire), vibration, collision, or use of the building.

1 & 2 (*above and below*) Cracking that is wider (or splits in two) at the top.

3 R.C. Beams between pads to replace inadequate foundations.

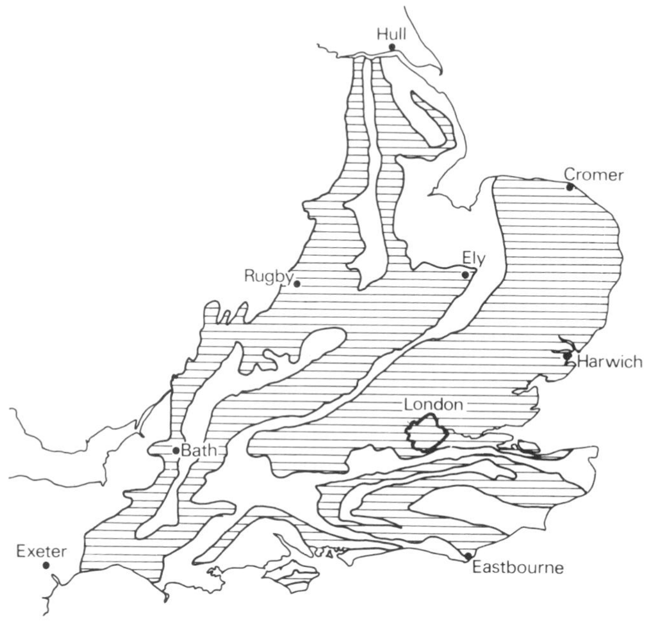

4 General distribution of firm shrinkable clays.

5.1.1 VERTICAL CRACKING OF EXTERNAL WALLS

Away from the corners and running between the foundation and the top of the building. The defect described here is in masonry walling.

Symptoms

Cracks are usually widest at the top of the building diminishing to a hair-line crack at or near foundation level. They may run through the foundations or they may only start above first floor openings. Often there will be a single crack in each of the opposite elevations of the building and they may be connected by a crack in a concrete floor or a flat roof. The roof finish of a pitched roof may be stretched or pulled apart. Diagonal cracks may also be associated with the defect (see 5.1.8)

Investigation

This defect is likely to occur only where there is a clay subsoil. Similar cracking on other subsoils is more likely to be moisture expansion of brickwork (see 5.1.3)

Find out the history of the site before the building was put up. Check the type and depth of foundations. Measure any changes in the width of the crack over a period of at least 6 months (see chapter 4).

Diagnosis and cure

This defect is a result of swelling of clay subsoil that was drier than normal when the building was put up. The clay was probably dried out by large trees that were cut down before the damage occurred or it may have been exposed during an abnormally dry period. The defect will only occur if the foundations are not deep enough to reach the subsoil beneath the dry clay. When the clay expands under the centre of the building there will be a vertical crack in the centre of the façade. If the expansion is at the corner it may affect only the corner, sometimes producing diagonal cracks (see 5.1.8).

Occasionally a building may break its back and show the symptoms described here when subsidence occurs at the end of the building, but the foundation in the centre performs as designed. An example would be when there is a cellar in the centre with foundations at a lower level than those at the ends which are subject to subsidence.

No permanent repairs to cracked walls should be carried out until the upward movement of the soil has virtually finished. The small daily fluctuations in the width of the cracks caused by thermal expansion and contraction must be ignored. Wide cracks should be repaired with compressible material or cover strips. Fine cracks can be left alone, unless they allow water to penetrate into the building, in which case they must be made watertight.

When swelling of dry clay is occurring, deepened concrete trench fill foundations used to overcome the problem may be displaced by the lateral swelling of the clay. Lateral pressure on foundations can be avoided by placing a compressible layer between the side of the foundation and the clay.

REFERENCES

BRE Digest 298: *Influence of trees on house foundations in clay soil*

BRE Defect Action Sheet 96: *Foundations on shrinkable clay: avoiding damage due to trees*

BRE Defect Action Sheet 102: *External masonry walls: assessing whether cracks indicate progresive movement*

NHBC: *Building near trees*

Tree species	Maximum tree height H (m)	Safe distance from building
Poplar	24	
Oak	23	
Willow	15	
White Beam/Rowan	12	H (see footnote)
Flowering Cherry/Plum	8	
Plane	30	
Elm	25	
Cypress	25	
Lime	24	
Maple/Sycamore	24	0.5 H
Common Ash	23	
Beech	20	
Birch	14	
Pear	12	
Apple	10	

Footnote H may be reduced to 0.5 H where the 'shrinkage potential' of the clay has been established as medium or low. Tests to determine the properties of clay soil are specified in BS 1377 and are readily available from soil testing laboratories at little cost.

Table 1 'Safe' distances of trees from buildings – see also NHBC guidance on building near trees.

Age of tree	Future growth		Solutions	Advantages	Disadvantages
Tree older than house	More growth expected		Prune tree	Cheap; tree preserved	Not a permanent solution; effect uncertain
			Deepen foundations (underpinning)	Permanent solution; tree preserved?*	Expensive; disruptive
			Remove tree	Permanent solution to shrinkage problem; cheap	Possible heave problem
	No more tree growth		Repair damage after winter recovery; perhaps prune tree	Simple; cheap	May be slight recurrence in a future dry summer if tree not pruned
Tree younger than house	More tree growth expected	Very dry summer caused cracking	Repair damage after winter recovery; perhaps prune tree	Simple; cheap	May be slight recurrence in a future dry summer if tree not pruned
		Normal summer caused cracking	'Slight' cracking**: repair cracks after winter recovery	Simple; cheap	Probable future cracking
			Worse cracking: remove tree	Cheap permanent solution	Loss of amenity
			prune tree	easy, cheap	Maintenance required
			deepen foundations	probably permanent solution	expensive, disruptive; may not give indefinite protection
	No more tree growth		Repair damage after winter recovery; perhaps prune	Simple; cheap	Possible slight recurrence in very dry summer if tree not pruned

* Most underpinning firms insist on tree removal in the terms of the guarantee
**See Digest 251

Table 2 Summary of solutions to tree root problems.

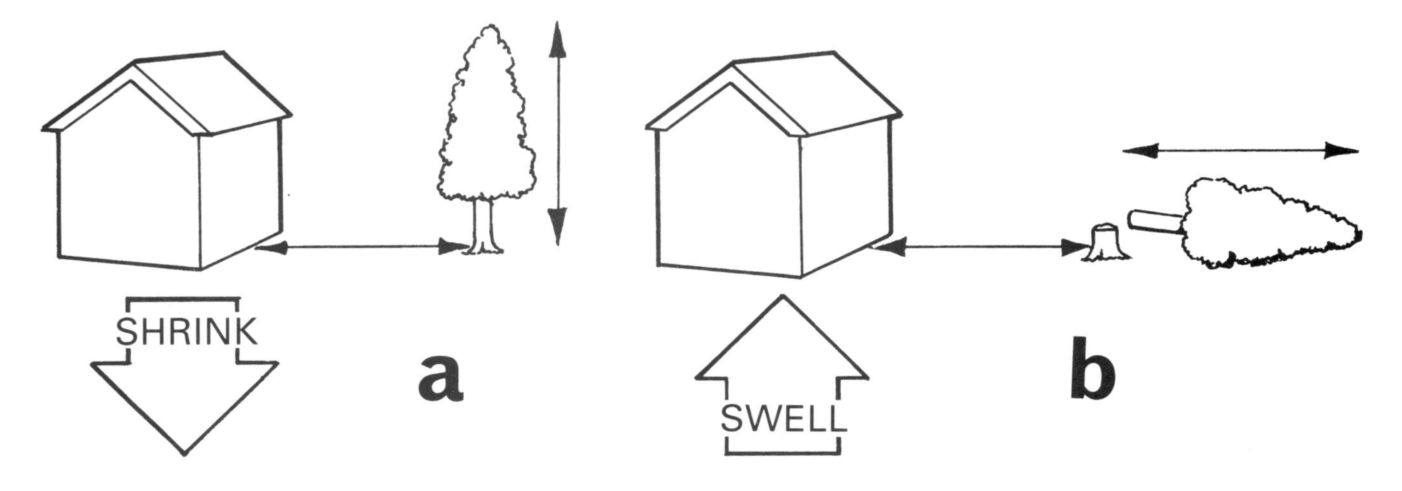

5 Compare tree height with distance from foundations – See Table 1.

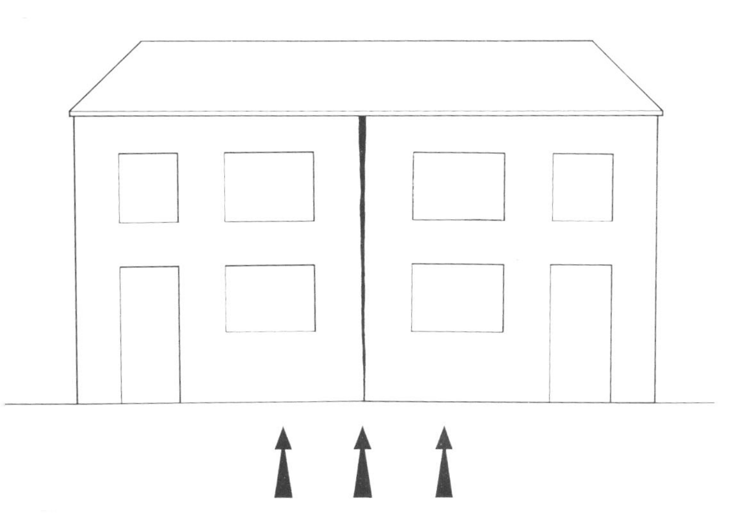

6 Effect of swelling clay under the centre of a building.

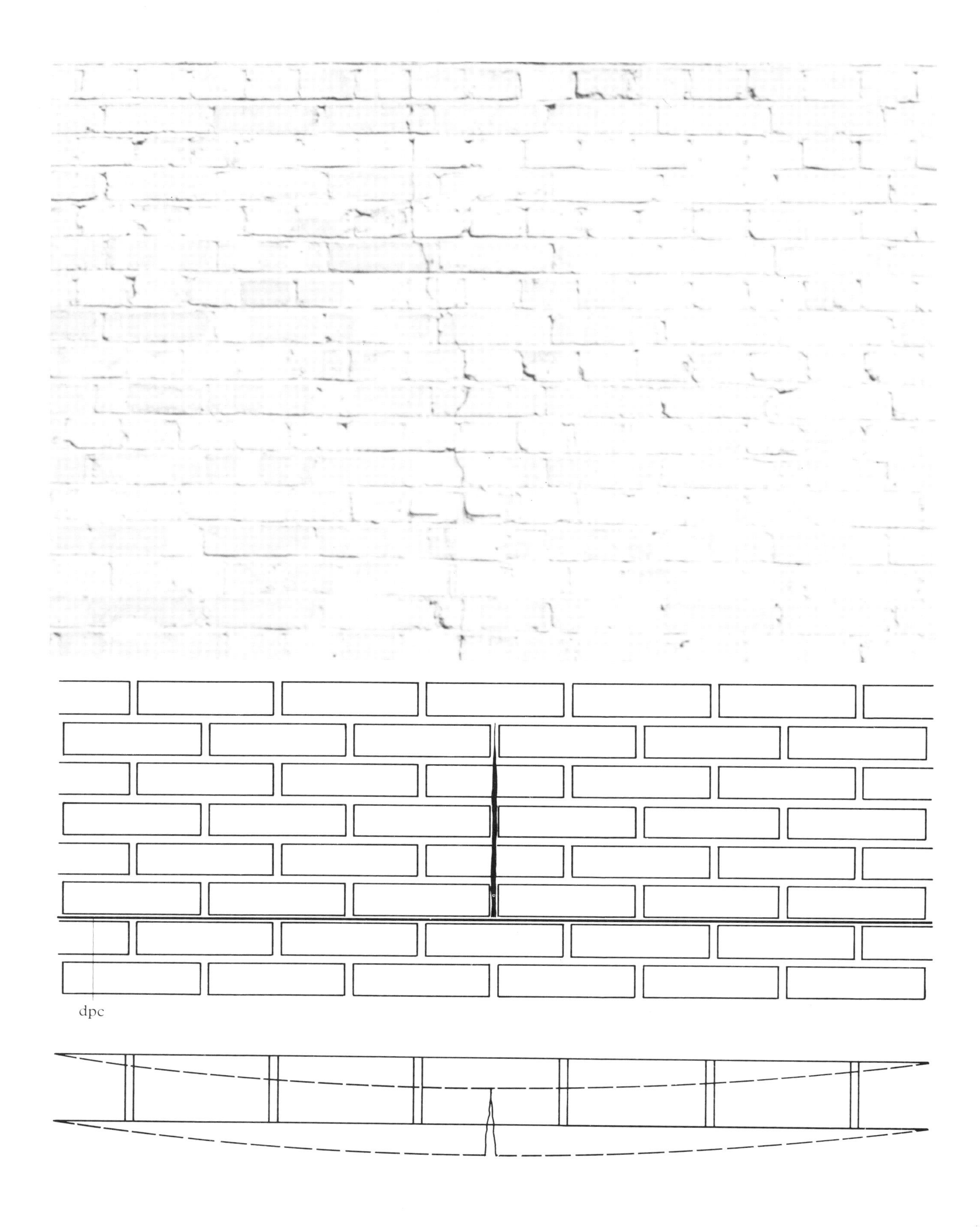

Crack combined with outward bowing of brickwork panel above DPC – may extend full height or just for a few courses. The crack is due to expansion of clay bricks particularly where a brick panel is set between concrete columns.

5.1.2 VERTICAL CRACKING OF EXTERNAL WALLS

Near the centre of a panel of cavity brickwork above DPC level. The defect described here is in clay brick masonry.

Symptoms

The crack is widest at the bottom and may narrow to a hair-line crack at the top (especially where the brickwork does not reach the top of the panel) and may extend for a few courses above the DPC (or bottom of the panel) or for the full height of the panel. Bricks on alternate courses are cracked through.

The crack is generally accompanied by a slight outward bulging of the brickwork which is greatest at DPC level where the panel may oversail the construction below.

These cracks usually occur where panels of brickwork are set between columns, especially if they are of in-situ reinforced concrete.

Investigation

Check the construction of the brickwork – this defect is confined to the external leaf of cavity construction.

Check that the crack does not appear below the DPC.

If the brickwork is supported on nibs formed on the face of a reinforced concrete structure, check that the nibs have not been broken. This will occur only if there has been appreciable movement of the brickwork.

Find out what type of brick has been used and, if possible, check the time interval between the production of the bricks and their incorporation into the building.

If the crack extends only through part of the panel height, note whether any extension occurs over a period of a few months. Check the depth of the crack. Note the amount of bowing and whether this increases with time.

Diagnosis and cure

This defect is caused by moisture expansion of clay bricks where the panel of brickwork is restrained at the ends leading to tensile cracking as the panel bows outwards. Most of a clay brick's moisture expansion occurs in the first two weeks after production, so this defect is commonly found where recently produced bricks are built into a panel between columns. It may occur during construction. Thermal movement can contribute to the expansion of the brickwork.

Once observation shows that the crack is no longer developing, it should be made good. The extent of the making good will depend on the extent of the crack and on the displacement of the brickwork. Do not use strong mortar. If cracking and bowing is extensive, introduce movement joints into the panels.

Moisture movement of clay bricks can also cause diagonal cracking (see 5.1.9) and horizontal cracking and detachment of brick slips (see 5.1.21).

REFERENCES

BRE Digest 75: *Cracking in buildings*

BRE Digest 160: *Mortars for bricklaying*

BRE Digest 164: *Clay Brickwork 1*

BRE Digest 165: *Clay Brickwork 2*

BRE Digest 200: *Repairing brickwork*

CIRIA Special Publication 44: *Movement and cracking in long masonry walls*

BDA Design note 10: *Designing for movement in brickwork*

1 & 4 Vertical cracks at corners can be caused by unrestrained expansion of clay bricks.

2 & 3 When part of the brickwork is restrained, the unrestrained section breaks away.

5.1.3 VERTICAL CRACKING OF EXTERNAL WALLS

Near corners of brick buildings above DPC level. The defect described here is in clay brick masonry.

Symptoms

A predominantly straight vertical crack occurring near the corner of the building starting at the DPC and extending upwards by varying dimensions up to a height of several metres. This type of crack usually occurs early in the life of a building but could occur at any time during the first 20 years of the building's life. The brickwork may oversail the DPC and the DPC may be slightly squeezed out.

Investigation

This type of defect only occurs in clay brick masonry and does not show below the DPC (unless the DPC is some distance above the ground). Check the age of the building in relation to when the crack was first noted.

Diagnosis and cure

This defect is caused by moisture expansion of clay bricks, sometimes in conjunction with thermal movement. The lateral expansion can produce a crack at the point of least restraint in a straight elevation. This may be at the corner of the building.

As the defect is unlikely to significantly weaken the structure of the building, repairs must simply improve appearance and prevent rain penetration. The crack can be sealed with a flexible sealant that will accommodate subsequent movement. In general, providing that the brickwork is at least three months old, it can be assumed that the initial expansion is complete and that a satisfactory and permanent repair can be made.

REFERENCES

BRE Digest 75: *Cracking in buildings*

BRE Digest 165: *Clay brickwork: 2*

BRE Digest 200: *Repairing Brickwork*

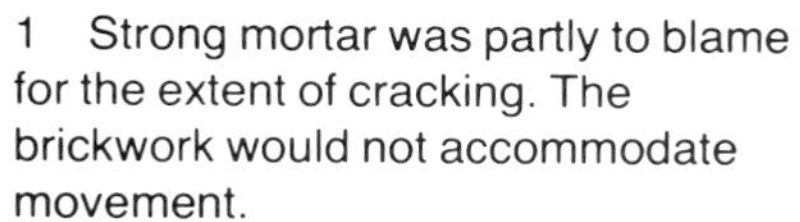

1 Strong mortar was partly to blame for the extent of cracking. The brickwork would not accommodate movement.

2 & 3 Typical cracks caused by the rotation of a short return.

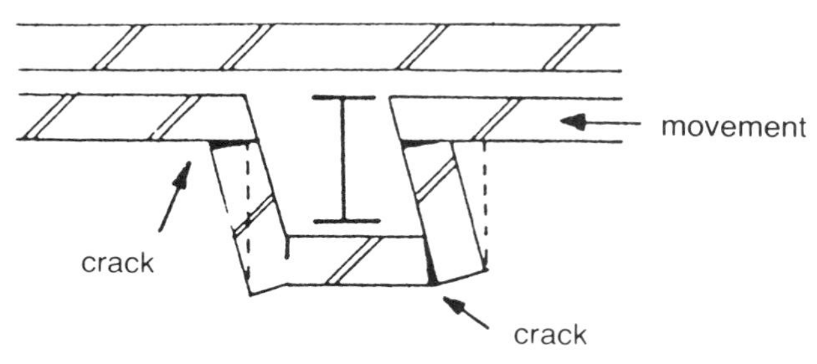

4 Movement at pier

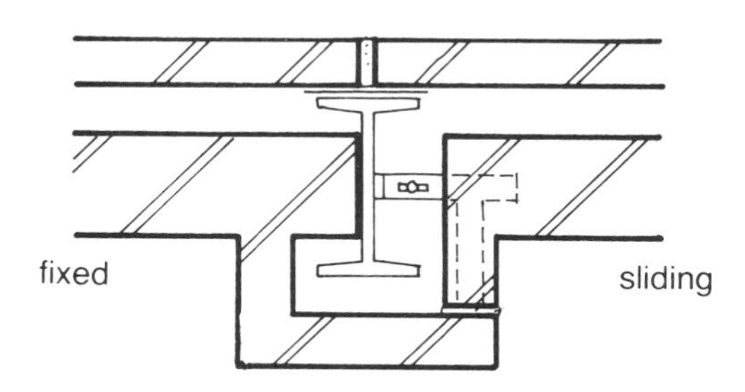

6 Allowance for movement at an intermediate stanchion

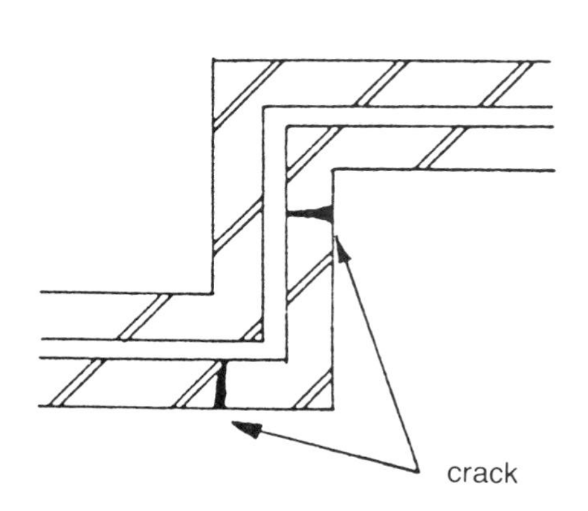

5 Movement at short return

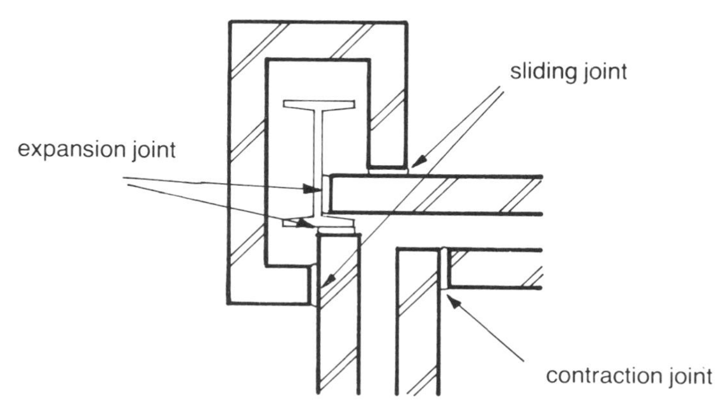

7 Allowance for movement at corner stanchion

5.1.4 VERTICAL CRACKING OF EXTERNAL WALLS

Cracking near to short return walls above DPC level. The defect described here is in clay brickwork.

Symptoms

The crack is predominantly straight and vertical, generally ½ brick away from the corner. It extends from the ground level DPC upwards for the full height of the wall. It appears during the early life of the building as a fine hair crack and may become progressively wider for a period of up to as much as 20 years, the ultimate width being about 10 mm. A slight rotation of the short return of the brickwork is usually observable if the crack is wide. It does not usually occur with longer returns (ie of more than 680 mm).

The crack does not occur in the brickwork below the DPC (unless the DPC is at least 1 m above the ground).

Investigation

Check that the crack does not appear below DPC level (except where DPC is 1 m or more above the ground). Note the height of the crack and, if it is not the full height of the wall, see whether its height increases.

Find out what type of brick was used and, if possible, how soon after being taken from the kiln the bricks were incorporated into the building.

Establish the relationship on plan between the masonry and steel or concrete columns.

Diagnosis and cure

The crack does not pass through low level DPC's and therefore should not be confused with settlement cracking. It is caused by moisture expansion of clay brickwork possibly enhanced by thermal expansion. The effect may be increased by the shrinkage of a concrete frame restraining the return brickwork. If cyclical thermal movement is involved, debris filling the crack can lead to progressive movement even if the bricks themselves have virtually completed their expansion (the initial expansion is usually complete within three months).

If the appearance is acceptable, the crack can be filled when expansion is complete. However, for appearance sake, the quoin may have to be rebuilt (which also creates problems of appearance as the new work has to be matched in). Where thermal changes are likely to reopen cracks, one or more movement joints may have to be formed by cutting into brickwork or by rebuilding. Attention should be paid to tying in sections of brickwork between movement joints. If this type of crack appears in a freestanding wall, it may become unstable. The return will need to be rebuilt or tied in to stabilise the wall.

REFERENCES

BRE Digiest 75: *Cracking in buildings*

BRE Digest 165: *Clay brickwork*

BRE Digest 200: *Repairing brickwork*

BDA & BSC: *Brick Cladding to Steel Framed Buildings (position of movement joints).*

BDA Design note 10: *Designing for movement in Brickwork*

1 Central portion of bay dropping away – in this case the canted flanks were well bonded to the main structure.

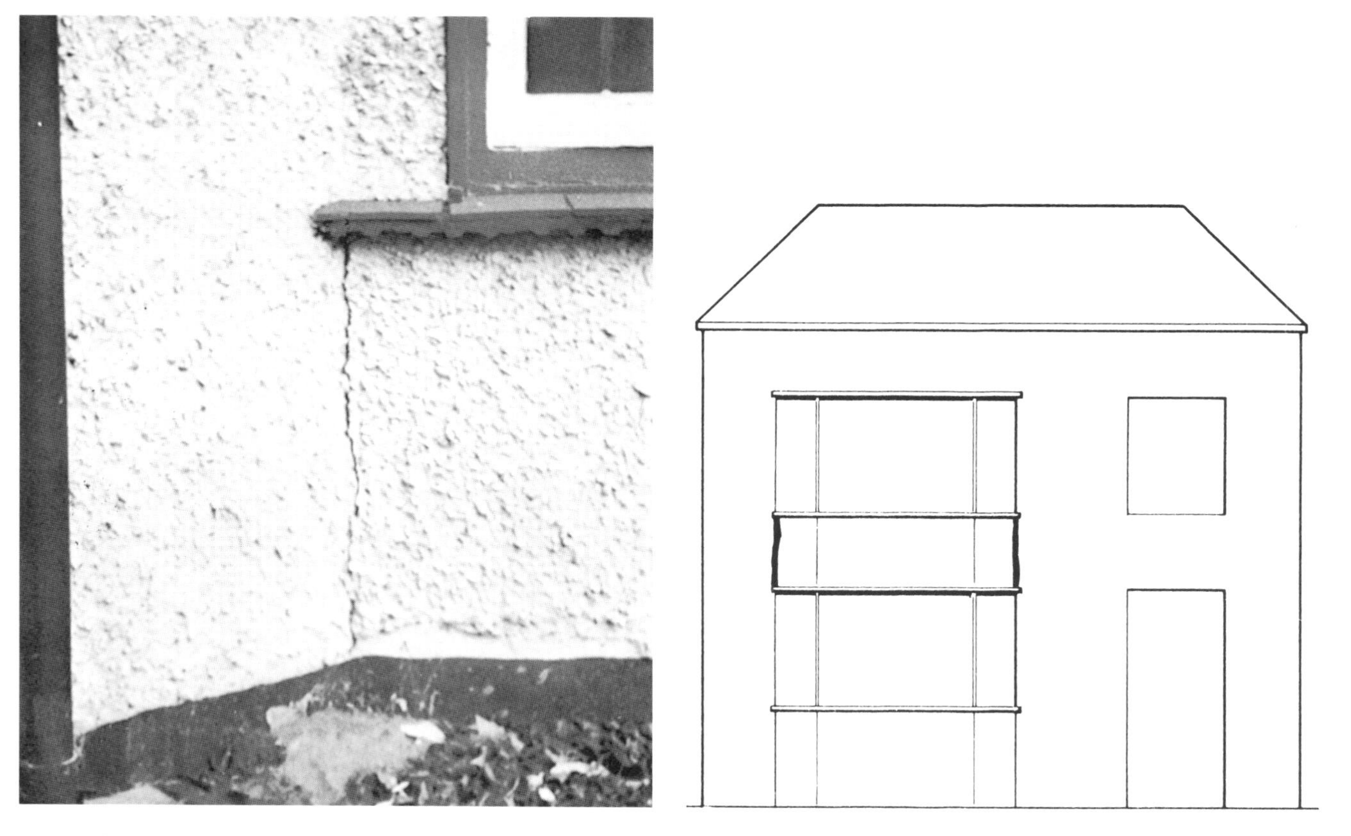

2 & 3 Cracks at the side of a bay are common.

5.1.5 VERTICAL CRACKING OF EXTERNAL WALLS

Cracking at sides of bay windows. In either masonry or timber stud construction.

Symptoms

Vertical cracks between spandrel walls of the bay and the wall on either side – may become quite wide (5–15 mm).

Investigation

Establish the construction of the spandrel walls – they may well be different at ground and first floor levels. Find out the depth of the bay foundations and whether the subsoil is shrinkable clay. Note the construction of the windows, particularly any corner posts and whether the windows are original or replacement. Look for displacement of flashings around the roof of the bay.

Diagnosis and cure

First floor bay spandrels are often of timber stud construction. The timber may shrink and might be inadequately tied into main walls or joists. Where there are cracks in timber stud spandrels at first floor level, the spandrel should be tied in to the main structure with galvanised steel straps. The crack can then either be filled to match existing finishes or, better still, concealed with a cover strip as there is always a strong possibility that it will reopen.

The ground floor level spandrel is usually masonry, and a crack at this level is due to subsidence of the bay foundation which may be shallower and more vulnerable to clay shrinkage during droughts. Underpinning or a pad and beam foundation is likely to be needed to support a ground floor bay spandrel that has cracked away. The cracked brickwork should be cut out and made good.

REFERENCES

BRE Digest 200: *Repairing brickwork*

BRE Digest 240 & 241: *Low rise buildings on shrinkable clay soils, Parts 1 & 2*

BMCIS Design/performance data, Serial No. 141: *Building owner's report: external walls. 32 Unsatisfactory replacement bay windows*

1–4 Provision for movement is especially necessary with concrete and calcium silicate bricks. Diagonal cracking below windows is fairly common.

5.1.6 VERTICAL AND DIAGONAL CRACKING OF EXTERNAL WALLS

Typically these cracks are seen in brickwork above or below openings. The defect described here only relates to cracks in concrete or calcium silicate brickwork or concrete blockwork.

Symptoms

Where the mortar is weak, the cracks will run round the bricks (giving a toothed effect). In strong mortar the bricks are more likely to break giving a straight crack. In very weak mortar there may be a number of fine cracks. The cracks usually appear within a few weeks of the wall being built and increase in width over a period of two or three years. Moisture and temperature may affect the width of the cracks which can close up temporarily in certain conditions. The defect is more likely to occur in long low panels of brickwork or blockwork.

Investigation

Check whether the bricks are calcium silicate or concrete and check the quality of bricks used and the composition of the mortar.

Check that the sides of the crack match and that they will fit together if the crack closes. Check variations in the width of the cracks during hot weather and between dry and wet weather.

Diagnosis and cure

The cracking is caused by the shrinkage of these bricks and blocks during initial drying and curing and also by moisture and thermal movement. When the cracks have been made good, movement joints can be cut into the brickwork to take up any further movement but this may not be necessary if the cracks are fine and the initial curing is complete. Bricks or blocks that have cracked across should be cut out and replaced. Movement joints in new concrete or calcium silicate brickwork should be at 7.5–9 m centres and for concrete blockwork at 6 m centres.

REFERENCES

BRE Digest 157: *Calcium silicate (sandlime, flintlime) brickwork*

BRE Digest 160: *Mortars for bricklaying*

Addleson, Lyall. *Building Failures* Architectural Press 1982.

BDA design note 10: *Designing for movement in brickwork*

1 Upper section of wall restrained by concrete floor.

2 Inadequate allowance for movement.

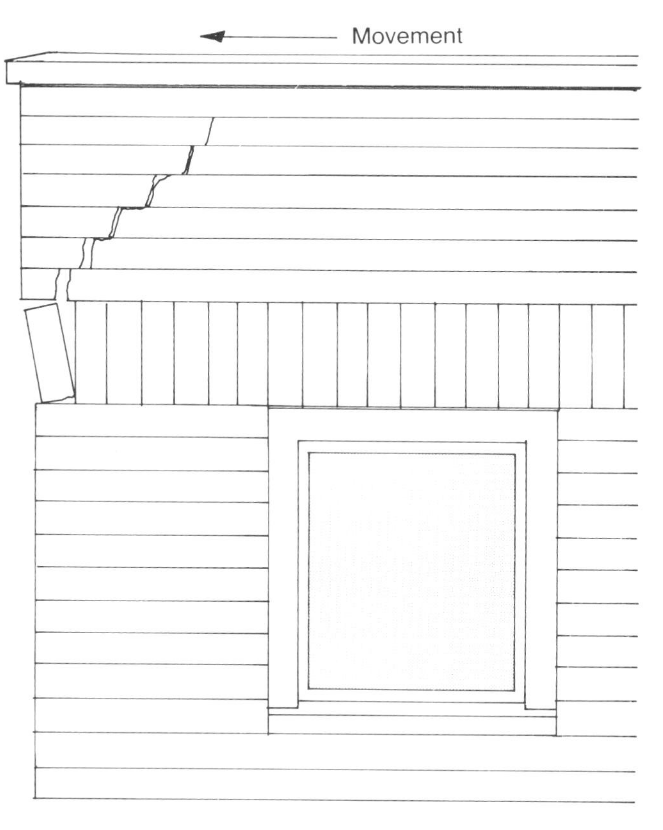

3 Diagonal crack at end of unrestrained parapet.

5.1.7 VERTICAL AND DIAGONAL CRACKING OF EXTERNAL WALLS

Isolated cracks usually near corners, running through three courses or more, that do not pass through the ground level DPC. The defect described here relates to all types of masonry.

Symptoms

Cracks may be toothed and stepped or straight. Vertical cracks are characterised by horizontal parting of the faces and are not normally wider than 5 mm.

Investigation

Examine the restraints on the brickwork that has cracked. Locate the DPC's on which the brickwork can slide. Where are abutments or restraining columns and piers? Where are the concentrated loads?

Examine the materials and decide if possible, how soon after manufacture they were incorporated into the building.

Diagnosis and cure

This defect may either be caused by moisture expansion of clay bricks (see 5.1.10) or shrinkage of calcium silicate or concrete bricks and concrete blocks (see 5.1.6). In both cases thermal movement can contribute to cracking. The cracking is due to the local restraint variations in the masonry, possibly caused by a discontinuous DPC allowing the masonry to move freely in places and restraining it in other places. Parapets are vulnerable to such movement because there is not enough dead load on parapet brickwork to restrain movement.

Once the mechanism of restraint and cracking has been discovered, movement joints (or horizontal slip joints such as DPC's) can be provided. Cracked units should be cut out and cracked joints raked back and made good.

REFERENCES

CIRIA Practice Note: Special Publication 44 *Movement and cracking in long masonry walls*

BRE Digest 200: *Repairing brickwork*

BDA Design note 10: *Designing for movement in brickwork*

4 & 5 Typical corner cracking due to subsidence.

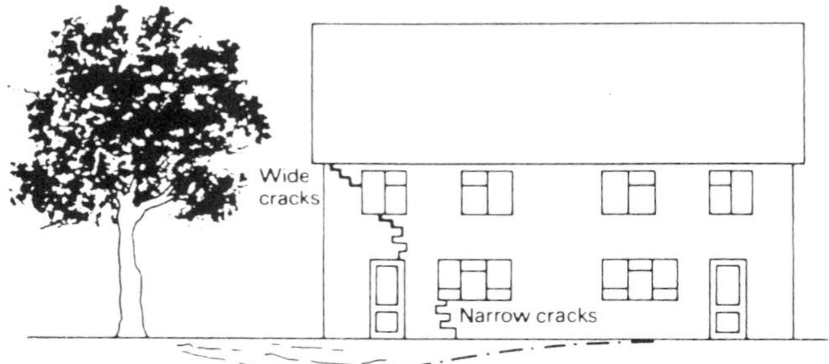

3 Crack pattern associated with a tree near the end of a building.

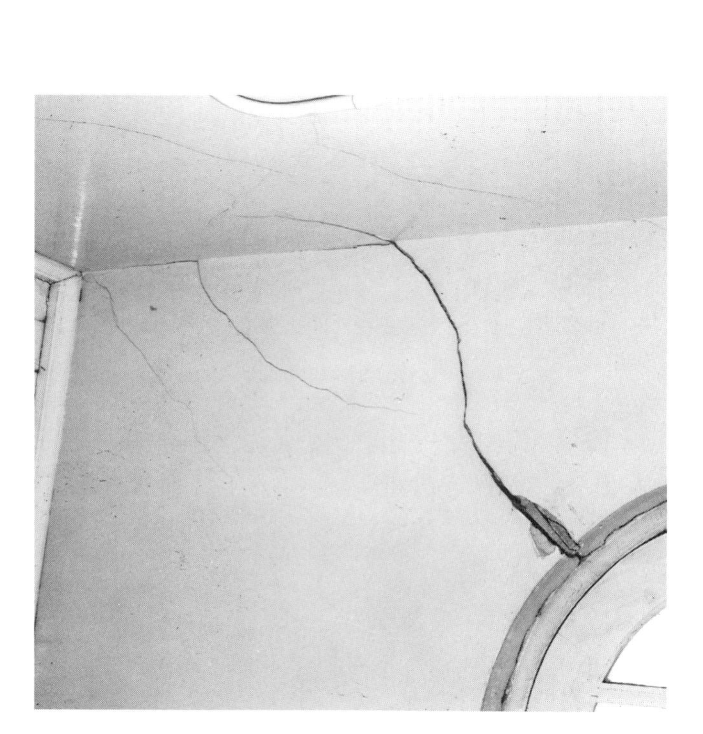

1 & 2 Inside and outside showing crack over opening.

5.1.8 EXTERNAL WALLS. DIAGONAL CRACKS

Often at the corner of a building with similar cracks on both faces. Cracks may extend from corners of openings. The defect described here could occur in any masonry or in-situ concrete construction.

Symptoms

The crack is widest at the top and can extend downwards through the DPC. In brickwork and blockwork the crack will mostly be stepped. In in-situ concrete it can be straight. These cracks may become very wide (25–50 mm) in places.

Investigation

Check the support for the wall below the crack. Is it a foundation on shrinkable clay, an overloaded pier or stone or timber that has deteriorated?

Note alterations that have been made to the building or changes of use that may have increased loads.

Note the nearest trees or large bushes to the cracked wall.

Note changes in the dimensions of the crack (at least four observations over a period of six months).

Note the time of year and recent weather conditions.

Note position of drainage trenches recently excavated near foundations.

Note the condition of adjacent buildings – are they also showing cracks?

Check the verticality of walls and relative levels around the building.

Plot all adjacent internal and external cracks onto a drawing noting their width and direction of taper.

Diagnosis and cure

Typically these are subsidence cracks, often caused by tree roots extracting moisture from shrinkable clays under shallow footings, but they may appear whenever the support for a wall is removed or lowered, for example when a lintel or its bearing is damaged. The vital issue is whether the cracking is progressive. If the cause of cracking can be removed, the wall can either be repaired and stitched together or, if this proves impracticable, displaced walling can be rebuilt. In extreme cases, buildings may have to be demolished to avoid excessively expensive repairs.

Permanent repairs should be postponed until it has been established that the cracking is not progressive – or likely to start again as soon as there is a spell of dry weather. If it is necessary to remove trees, they should be reduced in size progressively over several years (depending on the size and species of tree). This will help to avoid ground heave when a clay subsoil, that has been dried out by a tree, expands as it starts to take up water again (see 5.1.9).

REFERENCES

BRE Defect Action Sheet 96: *Foundations on shrinkable clays: avoiding damage by trees*

BRE Defect Action Sheet 102: *External masonry walls assessing whether cracks indicate progresive movement*

BRE Digest 75: *Cracking in buildings*

BRE Digest 200: *Repairing brickwork*

BRE Digest 240: *Low rise buildings on shrinkable clay soils*

BRE Digest 251: *Assessment of damage in low use buildings*

BRE Digest 298: *The influence of trees on house foundations in clay soils*

NHBC Practice Note 3 1985: *Building near trees*

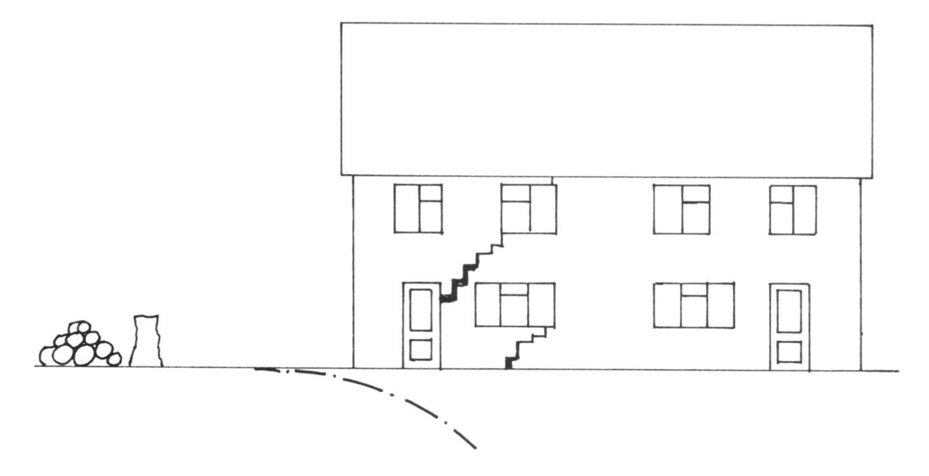

1 Ground can swell where tree has been felled. Cracks are widest at bottom.

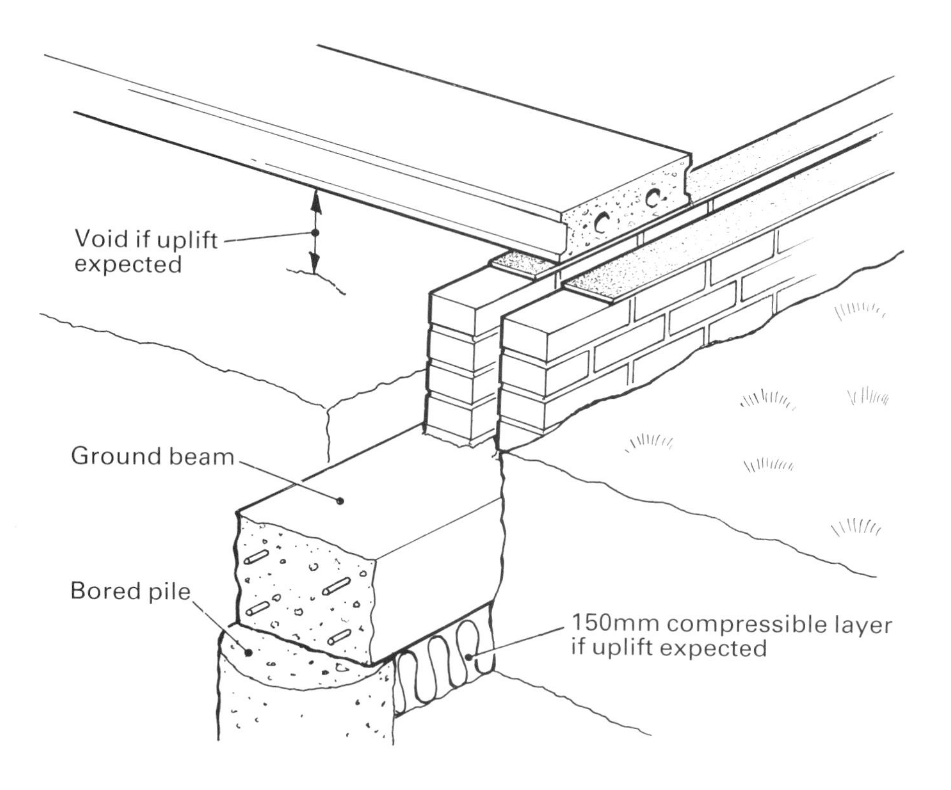

2 Piles or pads deep enough to reach stable ground can support ground beams below a wall where uplift has occurred, or is expected.

5.1.9 EXTERNAL AND BOUNDARY WALLS. DIAGONAL CRACKS

Cracks extending from below the DPC. These cracks are usually near to the corner of the building at the bottom. The defect described here relates to masonry walls and to in-situ concrete.

Symptoms

The cracks are widest at the bottom and may affect one or more corners of the building. Opposite sides of the building may show similar cracks. There may be cracks in a concrete ground floor slab related to the position of cracks in the external wall. Internal partitions running parallel with cracked external walls may also show cracks.

Investigation

This defect only applies where the subsoil is shrinkable clay.

Note whether the bricks are clay, calcium silicate or concrete and whether concrete blocks have been used.

Record the position, extent and width of the cracks and whether there is any flaking of materials at the lower end. See whether the crack runs through the foundations.

Check the depth of the foundations.

Check the history of the site immediately before the crack appeared. Were any trees cut down? Was there any leaking or flooding? Has there been any excavation that would have dried out the soil?

Diagnosis and cure

This defect occurs where shallow foundations bear on shrinkable subsoil that is drier than normal. Subsequent wetting of the subsoil expands the clay which exerts an upward pressure on the foundations. If the expansion takes place at the corners or ends of the building, it will create diagonal cracks, if at the centre of the building there will be a vertical crack (see 5.1.1). The wetting of the subsoil is likely to have been caused by removal of trees which were previously drawing water from the soil and keeping it dry.

Frost or tree root heave (see 5.1.11) can cause similar damage in dwarf walls on very shallow foundations, usually away from completed buildings.

REFERENCES

BRE Digest 75: *Cracking in buildings*

BRE Digest 200: *Repairing brickwork*

1 Horizontal displacement above a DPC.

2 Oversailing of DPC.

3 Horizontal displacement in an external wall.

4 Movement in a boundary wall.

5.1.10 EXTERNAL AND BOUNDARY WALLS. DIAGONAL CRACKS

Narrow cracks starting from DPC level (unless the DPC is high up the wall). The defect described here relates to clay bricks.

Symptoms

The cracks often run through to the jamb of an opening where some oversailing of brickwork can be seen. In other cases the diagonal cracks in the wall run first to one side and then to the other forming a horizontal vee. The cracks are stepped and show signs of horizontal rather than vertical displacement.

Investigation

Check that the bricks are clay (if they are calcium silicate or concrete, the defect is likely to be 5.1.6).

Note the age of the building and the date when the crack was first seen.

Check that the crack does not pass through the foundations. (If it does see 5.1.9.)

Diagnosis and cure

This defect is likely to be caused by moisture expansion of new brickwork (see also 5.1.2, 3 & 4) coupled with thermal expansion and contraction. When the brickwork expands, it pushes out towards an opening or to the end of the wall, then, on contraction, it pulls back and leaves a crack. Once the brickwork at the end has been displaced by the expansion, the tensile strength of the brickwork alone is not sufficient to pull it back into position.

Having cracked in this way, the wall is likely to move again at the same spot. Repairs are best done using a flexible material to replace the mortar joint. Garden walls are most likely to show this cracking where the mortar has been damaged by sulphate attack or frost. If this is the case, the wall will probably have to be rebuilt. Movement joints in new clay brickwork should be at 12 m centres.

REFERENCES

BRE Digest 75: *Cracking in buildings*

BRE Digest 165: *Clay brickwork 2*

BRE Digest 200: *Repairing brickwork*

1 Roots will raise boundary walls and light structures.

2 Trees will push walls sideways (and disrupt pavings).

5.1.11 EXTERNAL AND BOUNDARY WALLS. DIAGONAL CRACKS

Cracks near large trees.

Symptoms

Diagonal cracks wider at the top and extending through to the foundations often affecting the alignment of the wall.

Investigation

Note the position and girth of trees (or signs of large trees having been removed recently).

Diagnosis and cure

The wall has been cracked by the actions of a large tree root which as it grows, exerts an upward pressure on the footing or base of the wall.

The tree root needs to be growing near the surface of the ground so that it exerts the upward pressure as it grows. The extent of the cracking will depend upon the weight of the wall (and on any loads superimposed on the wall). A wall carrying a heavier load is less likely to be cracked in this way. For this reason the wall of a building will very seldom be affected – though it may be claimed that this is the case. The defect must not be confused with the action of tree roots drying out shrinkable clay sub-soils.

1 The damper section of wall below the window has been more affected by corrosion of the ties causing cracks at every 4th course.

2 Cracks in brickwork showing through rendering – in this case probably every 6th course. Movement is cumulative and usually more obvious towards the tops of walls.

3 Vertical twist tie showing how corrosion can expand the end embedded in the outer leaf.

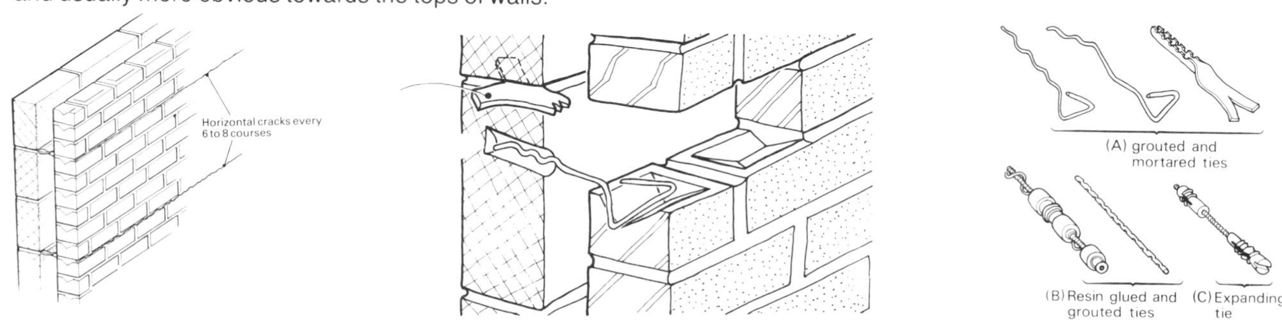

4 & 5 & 6 Symptom and repair methods. It is essential to cut back and remove corroded ends of ties from the outer leaf to prevent continuing movement.

5.1.12 EXTERNAL WALLS. HORIZONTAL CRACKING

Cracks at every fourth course in cavity walling. They may be continuous or seen only at intervals along the joint. The defect relates to cavity walls in masonry.

Symptoms

Distinguished from sulphate attack (see 5.1.13) by the irregularity of the cracks. Normally the cracks are only in the external wall but occasionally similar cracks may be seen in internal plaster. The interval between the cracks will depend on the spacing of all the ties. The cracks may be seen in external rendering over cavity masonry. The wall may bulge or the outer leaf may push up at eaves level and disrupt roof finishes. Similar cracks have occurred where joist hangers bear on the external leaf.

Investigation

Check the position of the cracks, particularly the lowest crack. Is it four courses above the DPC?

See whether there are any related cracks on the internal leaf.

Check the type of mortar used and the type of wall ties. Is there a high salt content in the air (ie. coastal areas, chemical plant)? If the wall has not been filled with foam insulation, the cavity can be inspected by using an endoscope to examine the position and condition of the ties. Alternatively a brick can be removed from the return to allow a view down the cavity. The position of the wall ties can be established with a metal detector (see Chapter 4). The wall can be opened up at the widest point of the crack to remove a wall tie for examination. A mortar sample, taken from the centre rather than the face of the wall, should be sent away for analysis.

Diagnosis and cure

This defect is caused by corrosion of ferrous wall ties, particularly in black ash mortar or where there is a high salt content in the air. The corrosion expands the ties and opens up horizontal cracks in the wall. A long term cure may only be possible by removing the outer masonry leaf and rebuilding it with new ties (new ties made for the purpose can be fixed to resin filled holes drilled in the inner leaf).

Alternatively consider using plastics wall ties, thus avoiding the risk of repetition of the defect. Shorter term solutions are possible (their adoption will depend on the structural significance of the wall), eg. the outer leaf can be secured by drilling and fixing through it to the inner leaf with special fixing devices (existing ties should be cut out as they will continue to corrode). Cladding or rendering the wall is not recommended because, although dampness could be excluded and the rate of corrosion reduced, it would be more difficult to follow the rate of any continued deterioration. Tying the external leaf to the internal leaf with a heavy duty polyurethane foam is not recommended – (see 5.2.6 which discusses the problems that can be associated with foam filled cavities).

REFERENCES

BRE Digest 329: *Installation of wall ties in existing construction*

BRE Digest 200: *Repairing brickwork*

BRE Defect Action Sheet 21: *External masonry cavity walls: wall tie replacement*

BRE Defect Action Sheet 70: *External masonry walls: eroding mortars – repoint or build?*

Addleson, Lyall. *Building Failures* Architectural Press, 1982.

Hollis, Malcolm RA. *Diagnosis and treatment of defective cavity wall ties* Structural Survey Vol 4

5 Expansion and disintegration of mortar due to sulphate attack.

1 & 2 Expansion of mortar joints shows through rendering. The cumulative effect of movement is more evident towards the top of the wall.

4 The first sign will be hairline cracks in mortar joints.

3 Brick sub-sills are particularly at risk from sulphate attack.

5.1.13 EXTERNAL WALLS. HORIZONTAL CRACKING

Horizontal cracks in the mortar of bed joints. It will usually be found in several joints at least – and may well be in every bed joint. The defect described here relates to clay brickwork.

Symptoms

Most commonly occurring in parapets or garden walls and walls of buildings exposed to regular or continuous dampness. The cracking may be associated with an overall expansion of the brickwork, which can be seen as an oversailing of the corners of the building or as an increase in the height of the wall. In the case of a cavity wall, it is generally only the outer leaf which expands, but in doing so it often causes some horizontal cracking of the inner leaf which may be limited to one or two major cracks near eaves level.

Mortar may become friable and break away, leaving recessed joints. Horizontal cracks – externally in rendering (matching the bed joint of the wall behind) is a typical symptom of this defect. The faces of softer more friable bricks may spall away.

Investigation

Examine the mortar in the bed joints. Send a mortar sample for analysis of sulphate content in the bricks. (The sample should be taken from the centre rather than the face of the wall.)

Find out whether the wall could have been regularly or continuously soaked with water. If it is not in an exposed position, has it been wetted by overflows or broken guttering or by roof or sill discharge?

Diagnosis and cure

If the cracking only affects one or two bed joints it is unlikely to be the defect described here.

The defect is sulphate action on portland cement or semi hydraulic lime. The bricks contain soluble sulphate salts which cause the mortar to expand in the presence of a plentiful supply of water. Most clay bricks contain some sulphate salts (though less in special quality bricks than in other types).

The expansion can be considerable – as much as 50 mm in the external wall of a two storey house.

Dense rendering which has cracked may allow water in behind the face and hence promote sulphate action in the walling behind.

The spalling of friable bricks is due to added stresses across the face of the bricks caused by expansion of the mortar exposed to the weather.

If the damage is limited and the source of dampness can be eliminated, for example by a wider coping or greater sill projection, it may be enough to rake out and repoint the section of wall (using sulphate resisting cement mortar), but if the wall is badly damaged and likely to become unstable it will be necessary to have it rebuilt.

REFERENCES

BRE Defect Action Sheet 70: *External masonry walls: eroding mortars – repoint or rebuild?*

BRE Defect Action Sheet 72: *External masonry walls: repointing*

BRE Digest 75: *Cracking in buildings*

BRE Digest 89: *Sulphate attack on brickwork*

BRE Digest 165: *Clay brickwork 2*

1 Top of wall oversailing due to lateral thrust of roof.

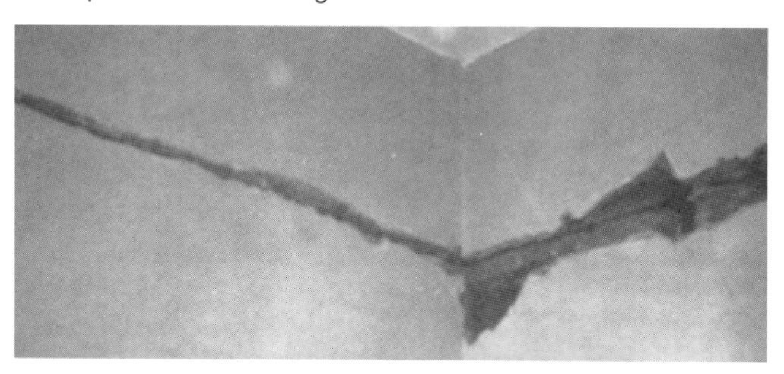

3 The thick bed joint cracks has been exposed inside the room where plaster had cracked between the solid wall above and the cavity wall below.

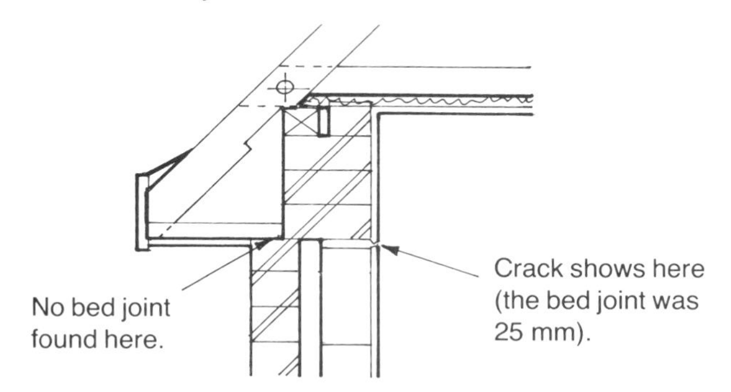

Change in wall construction and difference in levels of the two leaves of the cavity wall has created a weakness that is easier to disguise than remove.

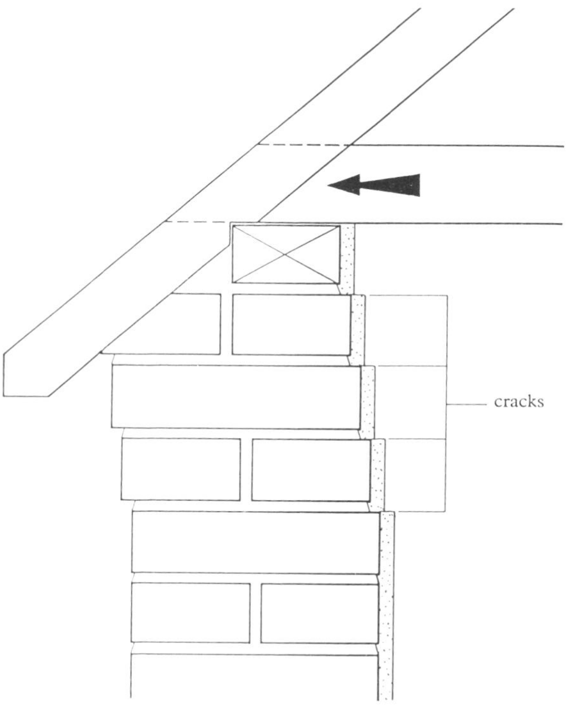

2 The cracks shown in this diagram may be concealed behind fascias and cornices.

5.1.14 EXTERNAL WALLS. HORIZONTAL CRACKING

Near pitched roof eaves level on exernal and internal leaves accompanied by outward movement of top courses under the eaves. The defect relates to masonry.

Symptoms

Horizontal cracks in one or more of the top courses of brickwork or blockwork, often more apparent on the inside of the building.

Investigation

Check whether movement is still taking place. Examine the wall externally to estimate the amount of outward movement.

Obtain information about any change in the roof covering (e.g. a heavier covering). Check on the construction of the top courses of brickwork. How is the cavity closed? Do both leaves of the cavity continue right up to eaves level?

Examine the roof timbers for signs of beetle or fungal attack. Look particularly for fungal attack at the points where the timber is in contact with the walls, and for beetle attack at timber joints.

Diagnosis and cure

The normal cause of this defect is the spreading of the timber roof which causes the top few courses, where not restrained by separating (or abuting) walls, to move outward. This movement generally affects three or four courses externally but sometimes only showing as one crack on the inside wall. Older buildings in weaker mortar are most subject to roof spread of this type.

There is an instance of a similar defect (showing on only one course near the top of the internal leaf) where the cause was the change in construction from a cavity wall below to a whole brick wall above and the unsatisfactory horizontal joint between the two.

Minor movement in the shelter of the eaves may be tolerated but where courses have moved out by more than 25 mm (in total) they should be rebuilt in sections and tied back to a member that can resist horizontal forces. Trusses and purlins should be examined for sag or joints that have opened and they should be then strengthened as necessary.

Where the crack is due to a change from cavity to solid construction the solution, short of rebuilding the top courses of the wall as a cavity wall, is to conceal the crack behind a horizontal picture rail, as the crack is likely to reopen after any making good.

REFERENCES

Benson, Evans, Colomb & Jones – *The Housing Rehabilitation Handbook* Architectural Press 1980.

Construction Feedback Digest 52, Autumn 1985 *Internal Cracking of Walls at eaves level* B&M Publications (London) Ltd

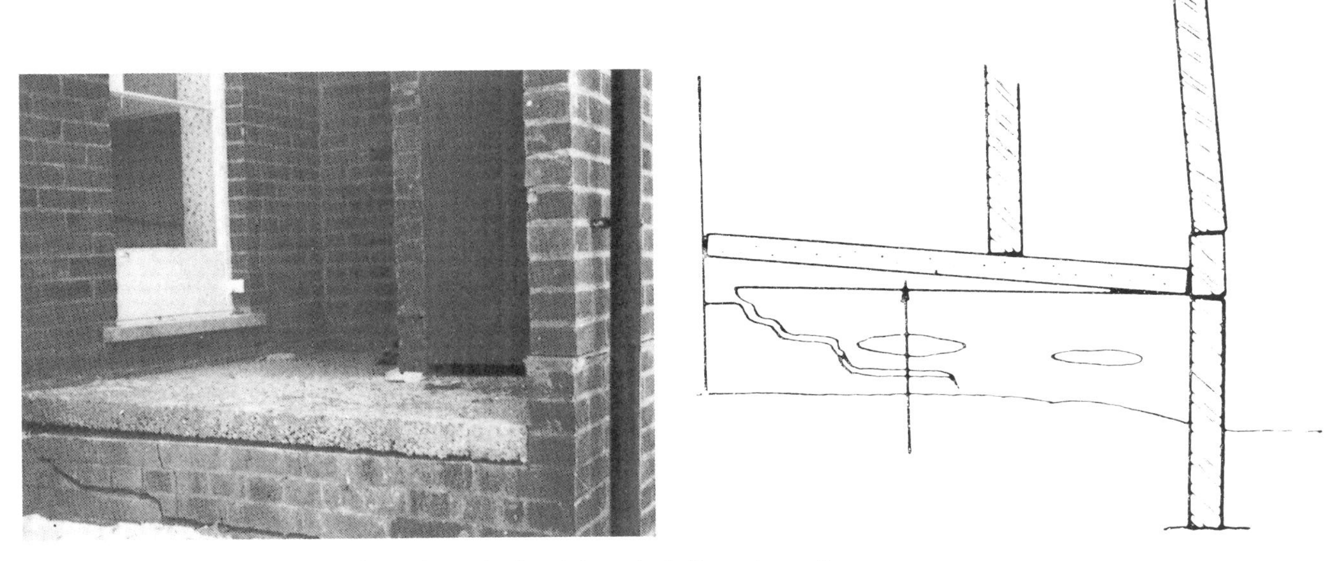

1 & 4 Frost heave under a porch produces horizontal cracks in the outer wall.

2 Sulphate attack on cement in a concrete floor can also cause horizontal cracking.

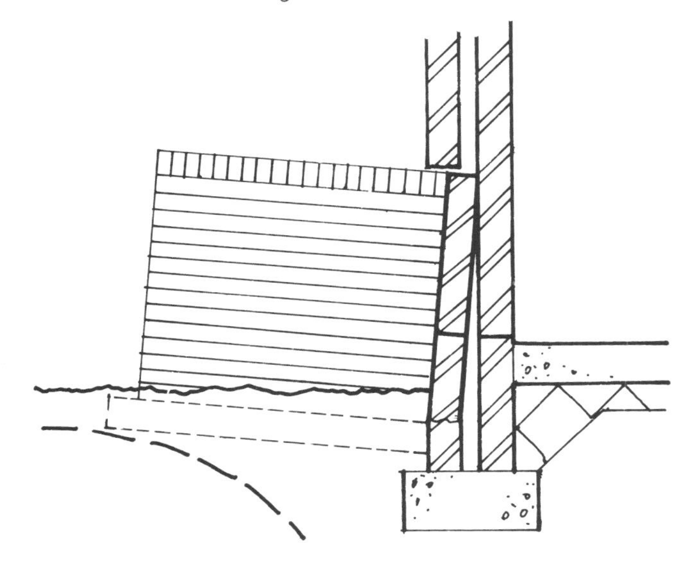

3 Frost (or moisture expansion of clay soil) lifts an abutting construction, such as a boundary wall and damages the main structure.

5.1.15 EXTERNAL WALLS. HORIZONTAL CRACKING

Cracks where rigid construction abuts a wall at an intermediate level. This defect relates to masonry construction.

Symptoms

Horizontal crack or cracks close to or just above the top of the abutting construction (a wall, a slab or a building) usually accompanied by tilting of the abutting construction.

Investigation

Examine the foundations of the abutting construction. Are they on shrinkable clay; are they shallow and on chalk or other soil that might be susceptible to frost heave?

Check the verticality of the damaged wall face. Is there any distortion in the region of the crack?

Diagnosis and cure

The cause of this defect is movement of the abutting construction (or possibly of the building) where no allowance has been made for differential movement. Typically shrinkage or heave caused by the drying out or wetting of soil beneath a screen wall with a shallow foundation, or frost heave under a raised porch slab pulls or pushes the adjacent walls and cracks them. Sulphate action expanding the concrete of a ground floor slab can produce these symptoms. The swelling of timber floors in damp conditions (steam pipe leaks) has been known to produce similar symptoms.

The defect will probably only be noticed if the damage to the wall has gone beyond simple making good. It may well be first noticed through the tilting of the abutting construction. It is likely that the damaged section of wall will have to be rebuilt. The abutting construction should be cut back or rebuilt to allow a movement joint between it and the wall. If the abutting construction is a screen wall which depends on the main wall for stability, sleeved ties could be provided through the movement joint.

REFERENCES

BRE Digest 75: *Cracking in buildings*

Bickerdike, Allen, Rich and Partners in association with Turlogh O'Brien. *Design Failures in Buildings second series* George Goodwin Ltd 1974.

1 & 2 & 3 Expansion of a concrete roof slab can disrupt the surrounding parapet.

5.1.16 EXTERNAL WALLS. HORIZONTAL CRACKING (OR VERTICAL OR DIAGONAL)

Typically below parapets at the level of the underside or the top of a flat roof or at window heads or sills below a flat roof. The defect described here relates to masonry and concrete construction.

Symptoms

The parapet masonry is displaced and oversails the masonry below. When the movement only affects one elevation and the displaced parapet is broken away from the return, there may be diagonal cracking at the corners and diagonal cracking running upwards and outwards from the corners of the windows (the masonry over the window having moved with the parapet above). Vertical cracking below the corner of the slab can be related to the same cause. The vertical crack will be wider at the top than at the bottom.

Investigation

Examine the construction at the perimeter of the concrete roof slab. Is the facing masonry built tight up against the edge of the slab?

Examine the finishes and insulation of the slab. Is the slab insulated above or below? Is the finish dark or light?

Note any internal cracking at ceiling level under the roof.

Note the surfacing of the roof. Are there paving slabs built close up to the inside of the parapet?

Under what condition did the defect first appear? Was it after very hot weather? Do the cracks open and close in response to changes in temperature?

Diagnosis and cure

The defect is caused by the thermal expansion of the concrete roof slab or, exceptionally, by the expansion of a concrete paving finish on the roof slab. In the case described above there may be cracks at the top of partitions or along separating walls inside the building.

The cure is to reduce the temperature variation within the concrete so that there is less expansion and contraction. This can be done by providing insulation above the concrete and by covering the roof with a layer of light coloured chippings or solar reflective paint. If the problem relates to concrete pavings on the roof these can be taken up around the edge and cut back to leave a perimeter margin for expansion.

REFERENCES

BMCIS. Design performance Data Serial No.141 *Building owners reports: 1 External Walls* June 1985.

Bickerdike, Allen, Rich & Partners in association with Turlogh O'Brien: *Design Failures in Buildings second series* George Godwin 1974.

3 The balcony parapet has been cracked by corrosion expansion of the rail support.

1 Corrosion has split concrete coping.

4 In this case a movement joint was pointed up with mortar and the top courses have lifted.

2 Corrosion expansion has caused the brick coping to lift.

5.1.17 EXTERNAL AND BOUNDARY WALLS. HORIZONTAL OR VERTICAL CRACKING

The top courses of a freestanding parapet or garden wall are displaced or broken, while the wall below is unaffected. The defect relates to masonry which may have ferrous metal railings or cramps built into the joints.

Symptoms

The coping may arch upwards leaving gaps below. Coping stones may be broken or missing. Brick copings may be displaced sideways or at the ends and the coping may oversail the wall below.

Investigation

Find out whether there is ferrous metal in the coping joints, and if so is the metal corroded? Is the damage confined to the courses in which the metal is embedded?

Note whether there is a DPC on the line of the horizontal crack.

Note the condition of the wall below and the type of bricks or blocks used for both the coping and the wall.

Diagnosis and cure

The corrosion of metal in mortar joints expands the joint and lengthens the top course (or courses) of masonry. If the ends are restrained the top course can rise up. If one end is unrestrained the course will oversail. If the coping stones are large and each one is restrained from movement by its bedding the corrosion will split the ends of the coping stones.

Similar, but less extensive cracking can develop locally where movement joints are not carried up through copings.

It is likely that the coping will require rebuilding with new bricks or stones. Both new metalwork and reconditioned metalwork must be well protected from corrosion before being built into a reconstructed coping.

REFERENCES

BDA Design Note 10: *Designing for movement in brickwork*

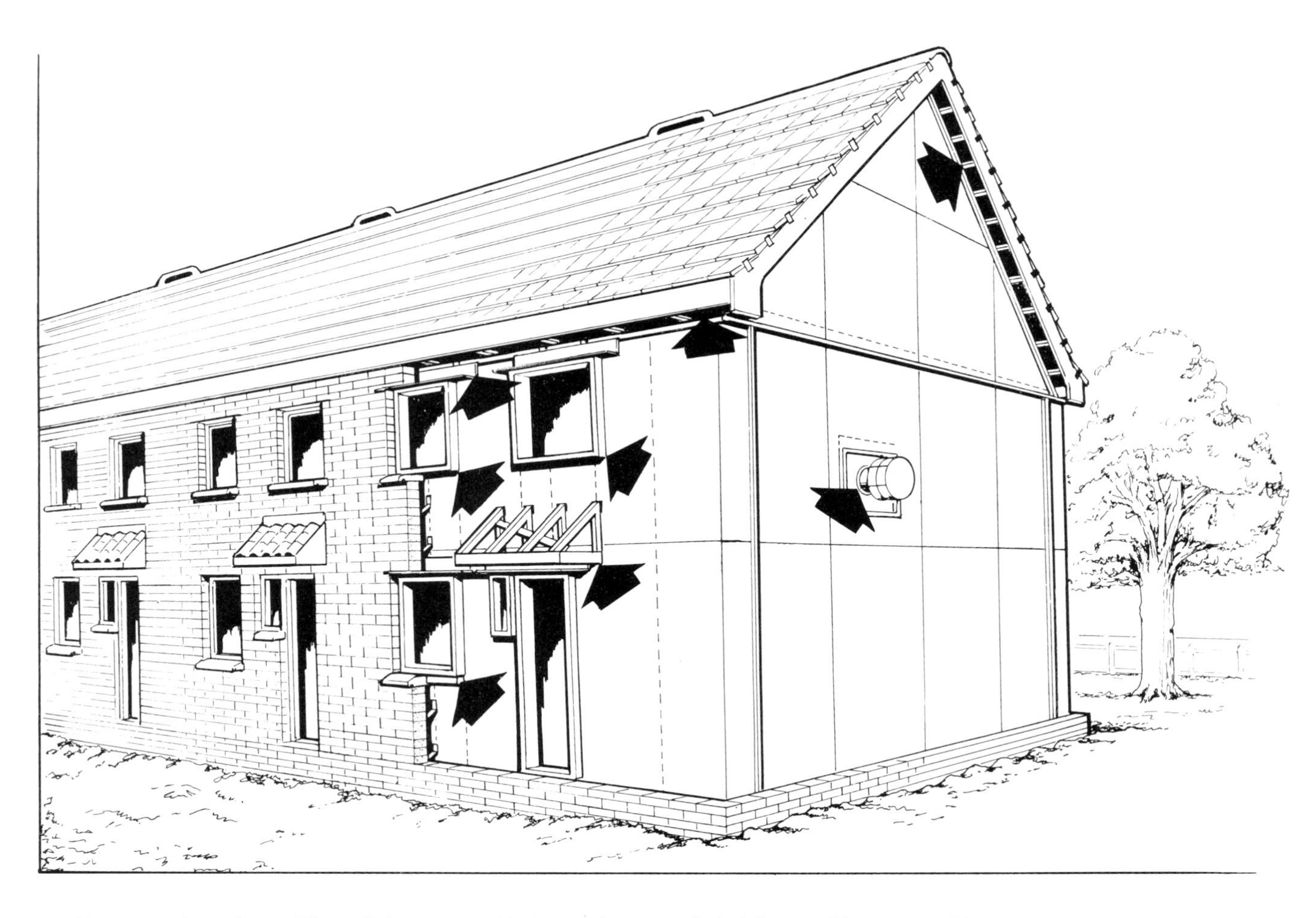

3 Danger points where differential movement between frame and cladding could cause problems.

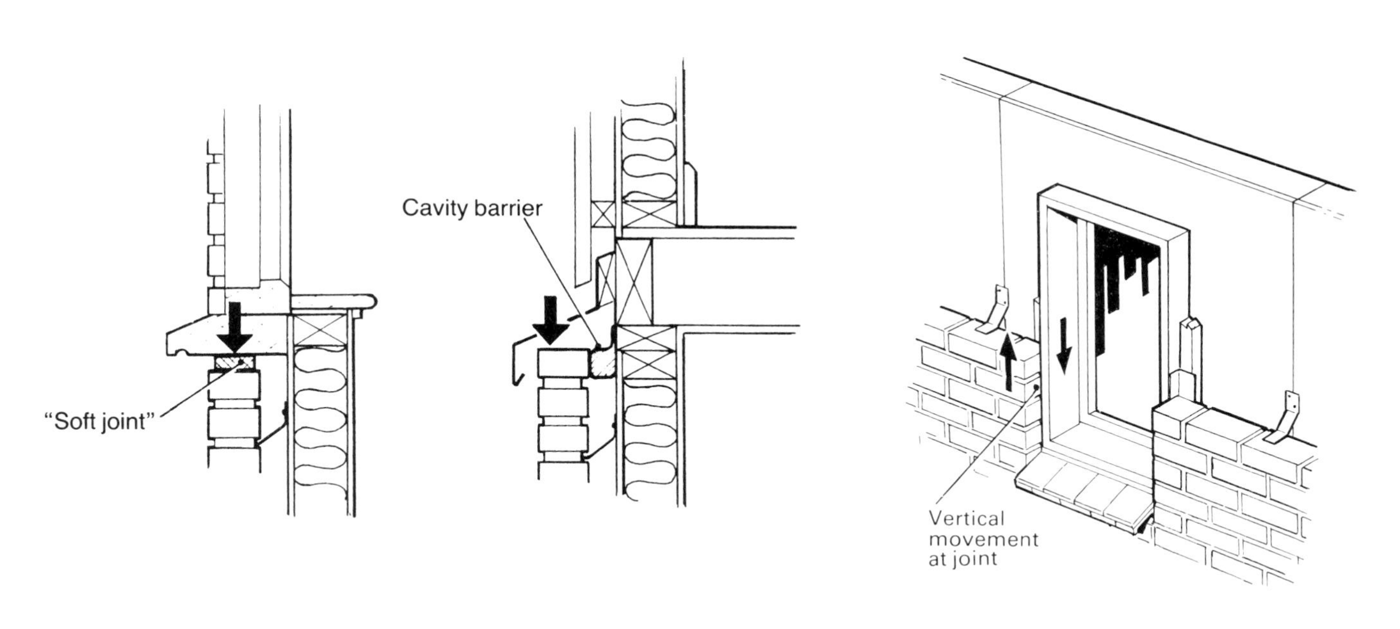

1 & 2 Ways of allowing for movement below a sill or flashing that is connected to the timber frame.

4 Vertical movement beside windows is also likely to take place.

5.1.18 EXTERNAL WALLS. HORIZONTAL CRACKS

Crack below tiled sill. Applies to a building with clay brick cladding to timber framed structure.

Symptoms

The sill is displaced, and may even slope back towards the building. There may be some bowing out of the brickwork below the windows. The rafters may rest on the external leaf of brick cladding rather than on the top member of the timber frame, gable ladders and porch canopies may be displaced.

Investigation

Examine the construction. Is the structure of timber framing? If so, has adequate allowance been made for shrinkage of the framing?

Are there cracks on the inside? (i.e. between the ceiling under the roof and the wall).

Examine the construction and finishes for damage along the top of all brickwork.

Examine the wall ties if bowing of the cladding has occurred. Do they hold firmly into the brickwork? Are they of a type currently recommended?

Diagnosis and cure

This defect is caused by shrinkage of the timber frame. Cross sections of timber making up the height (i.e. the platform floor and the head and sill rails) lose moisture after the building is enclosed and inhabited. The damage can easily be repaired if brick cladding has not bowed out.

Sills will have to be reset and the top course of bricks will have to be cut down to leave a gap above to accommodate any further shrinkage of the timber. NHBC recommend that a gap of 18–21 mm should be left above the brickwork under eaves and verges in three storey timber frame construction and 9–12 mm above tiled sills at first floor level.

REFERENCES

BRE Defect Action Sheet 75: *External walls: brick cladding to timber frame – the need to design for differential movement*

NHBC practice note 5: *Timber framed dwellings*

1 Crack finds a point of weakness at change of level.

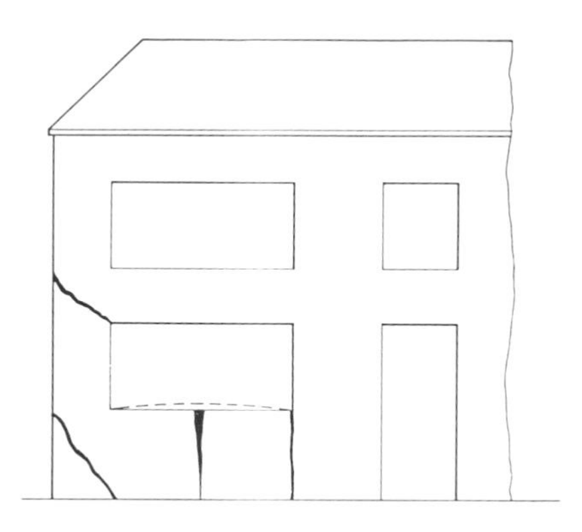

2 Expansion of foundations due to sulphate attack.

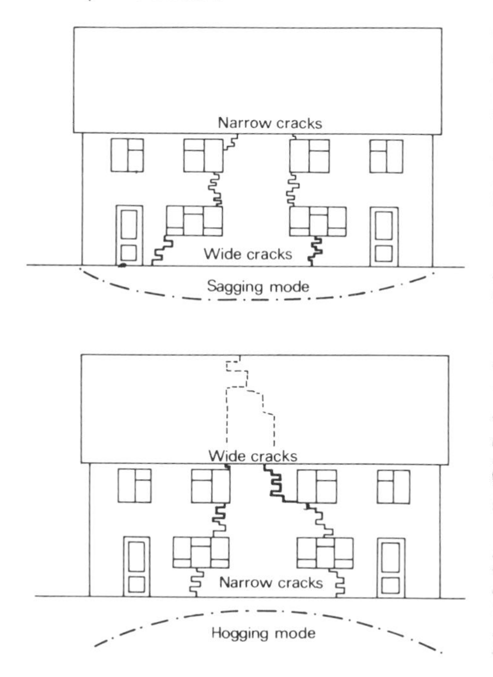

3 Typical crack patterns resulting from hogging and sagging.

4 Sagging away from corner.

5 Corners of openings are often affected.

5.1.19 EXTERNAL AND INTERNAL WALLS. CRACKS – VARIABLE IN DIRECTION

Variable cracks not usually confined to one wall or corner of the building. Typically cracks run from corners of openings. The defect relates to masonry.

Symptoms

The cracks vary in width from fine hair-line cracks to as much as 25 mm wide. Solid ground floors and foundations may also show cracks and walls may be out of plumb. The cracks sometimes appear suddenly (accompanied by some noise). They can appear long after the building was put up. There may be some outward movement of the walls at the DPC.

Investigation

If the foundation bears on shrinkable clay and is close to trees or to where trees have been removed, the possibility of damage from tree roots should be examined first (see 5.1.8 & 5.1.9).

Note the nature of the ground under the building. Is it made-up ground? Is there mining subsidence or a geological fault in the area? Much information can be obtained from an examination of the near-by buildings and of records held by the local and other authorities.

Examine the mortar in the brickwork and concrete in foundations and slabs for sulphates. If they appear, make a similar analysis of the sub-soil, ground water, hardcore and bricks.

As much information as possible should be obtained about the time the cracks occurred and any increase in the number or width.

Diagnosis and cure

This defect can result from two quite different factors. One is the movement of the ground on which the building is founded (A) and the other is the effect of sulphate salts on the foundations (B).

A There are several causes of ground movements, the more important being as follows:

1 Mine working, especially coal and salt.

2 Geological faults

3 Made-up ground

4 Vibrations from traffic, machinery, etc. (rarely responsible in spite of many claims)

5 Seismic tremors and earthquakes

6 Shrinkage of clay sub-soil

There should be little difficulty in obtaining enough information to decide which is responsible.

Cracking of buildings on made-up ground is likely to be a more gradual process than such causes as mine workings and geological faults.

B Sulphate attack of the foundations does not usually produce such dramatic or disastrous effects as ground movements. Sulphate attack of the Portland cement in the concrete or in the mortar joints of the brickwork in the foundations causes the concrete and or the the mortar to expand and hence to lift up the walls. The effect of this may be variable. If a solid ground floor of concrete has no damp-proof membrane underneath it, there may also be cracking and lifting of the floor (7.1.4) and there may be disturbances of the brickwork immediately above the DPC.

The cracking does not appear for some time after completion of the building, often several years, since sulphate attack is a relatively slow process.

Demolition may be the only course of action if the cracks are numerous and wide or if the walls are out of plumb, but slight cracking may be treated by normal methods. Whatever the cause, it will be necessary to take into account the possibility of continuing movements and structural engineering advice should be obtained.

REFERENCES

BRE Digests 63, 64 & 67: *Soils and foundations*

BRE Digest 75: *Cracking in buildings*

BRE Digest 250: *Concrete in sulphate bearing soils & groundwaters*

BRE Digest 276: *Hardcore*

BRE Digest 278: *Vibrations: buildings and human response*

1 Crack at foot of partition indicates deflection of floor.

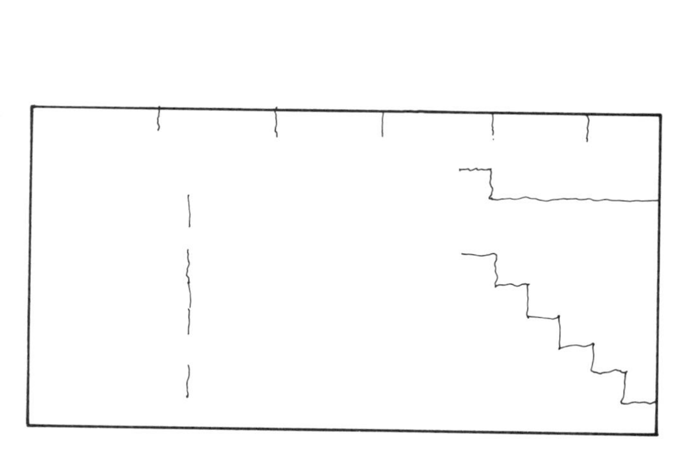

2 The pattern of cracks in blockwork is related to the way the blockwork is restrained by surrounding construction.

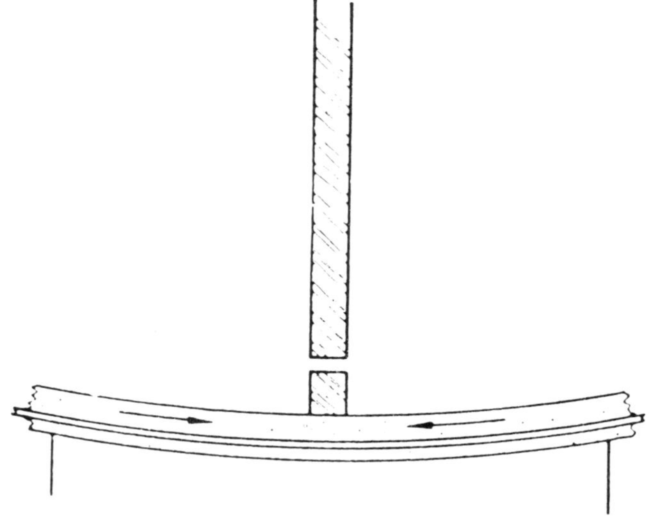

3 How floor deflection causes a crack at the foot of a partition.

5.1.20 CRACKS IN INTERNAL WALLS

Cracks in plaster on internal partitions corresponding to joints in blockwork. The defect relates to concrete masonry units (also calcium silicate bricks and in-situ concrete partitions).

Symptoms

Cracks may be vertical or horizontal and follow joints in blockwork behind. They do not affect the outside faces of external walls. They may outline the shape of the concrete blocks used to construct the partition. These cracks are not normally more than 5 mm wide.

Investigation

Establish what materials were used in the construction and, if possible, how soon after manufacture the bricks or blocks were built into the wall.

Record the position of cracks and any variation in width from day to day.

Note whether the partition has been subjected to wetting or a wide range of temperature.

Check the level of the floor below the partition. Does it sag? Is there a gap under the partition?

Check the walls at the ends of the partition. Do they lean or bulge?

Check the verticality of the partition itself.

Check the depth of the crack. Is it only plaster depth or does it carry through into the blockwork behind?

Diagnosis and cure

The most probable diagnosis is moisture or thermal movement of concrete blocks, particularly if they have been built into the wall within two weeks of manufacture. The direction and position of such cracks will depend on the restraint given to the partition by surrounding construction (or by reinforcement within the wall). If strong mortar (with a high cement content) has been used, the crack is likely to be vertical in the centre of the wall or at one end. In-situ concrete partitions can show similar shrinkage cracks.

Horizontal cracks above skirting level can be caused by deflection of the floor slab, and plaster depth horizontal cracks above the skirting can be caused by contraction of a floor finish that is bonded to a coved skirting. If the partition is fixed to a solid floor at the top and bottom, the horizontal crack may show in the centre.

These cracks can be made good by cutting out and filling the plaster (the groove should be undercut to provide a key). There is always a risk that the crack will continue to open and close with variations in moisture and temperature and that the filling will show a hair-line crack. If the crack is in a suitable position, a cover strip can be used or a finish can be applied that will disguise the crack.

REFERENCES

BRE Digest 35: *Shrinkage of natural aggregates in concrete*

BRE Digest 75: *Cracking in buildings*

BRE Digests 227–229 *Estimation of thermal and moisture movements and stresses*

1 Spalling bricks covering a concrete frame. Where brick slips have been used they are likely to become detached.

2 The clay brick spandrels exert a horizontal pressure on the cladding to the vertical frame member.

3 Expansion of the clay bricks (and shrinking of the concrete frame) have snapped the nib supporting the brickwork on the flank wall.

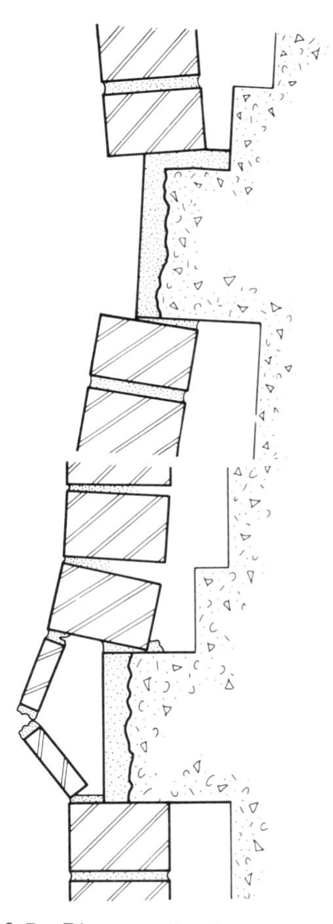

4 & 5 Diagram showing tendency to disrupt brickwork if no allowance is made for movement between concrete frame and clay brick cladding.

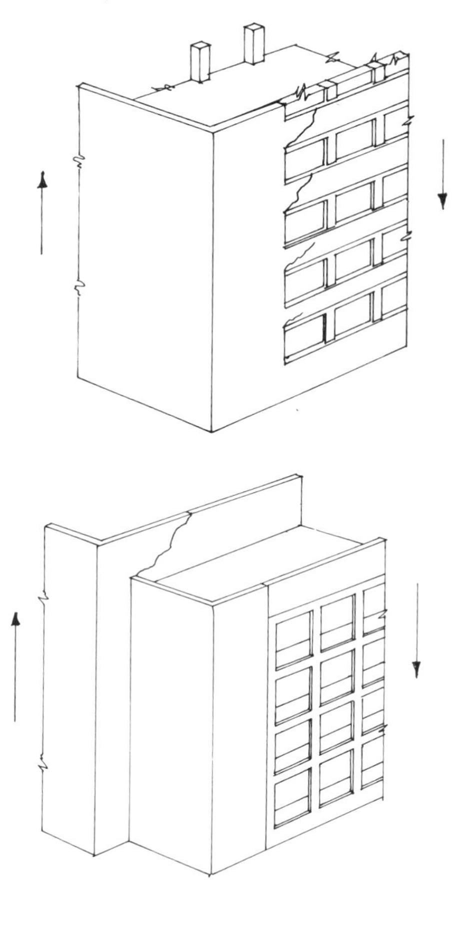

6 & 7 Other situations where cracks may occur as a result of this differential movement.

5.1.21 EXTERNAL BRICK CLADDING. CRACKING ON CONCRETE FRAME STRUCTURE

Horizontal cracks at floor levels or diagonal cracks adjacent to tall brick façades. This defect relates to clay brick (and sometimes to stone) cladding of reinforced concrete framed structures.

Symptoms

Displacement of brick slips on concrete edge beams supporting storey height brick cladding (often first seen when one or two brick slips fall from the building). The horizontal cracks may be less obvious than the bulging of the slips (or sometimes the cladding). Where tall uninterrupted areas of brick cladding meet spandrels or roof structures supported on the concrete structure, diagonal cracks may develop. The defect is commonly associated with tall buildings. Sometimes concrete nibs supporting brickwork are seen to be broken away from the concrete frame.

Investigation

Establish the form of construction and specific materials used for the brick (or stone) cladding and its connections to the concrete frame. Has allowance been made for differential movement between frame and cladding?

Examine brickwork at or adjacent to the top of any tall section of brick cladding and below high level concrete projections from the frame.

Examine brick courses adjacent to cracks for spalled edges.

If possible establish how much time elapsed between the erection of the concrete frame and the fixing of the cladding to the frame.

Diagnosis and cure

This defect is caused by the shrinkage of the concrete frame over a number of years. It is accentuated by irreversible moisture movement of clay bricks. As the bricks expand and the frame shrinks, stresses are set up that can displace and crack the brickwork, as well as spalling of the brick slips covering the nibs supporting the brickwork (and sometimes spalling of the nibs as well).

If there is any doubt about the stability of the brickwork, structural engineering advice should be obtained.

If there is a possibility of masonry falling, protection and fencing should be provided below the cladding.

Isolated diagonal cracks can be made good as they do not usually involve safety. (Unsightly cracks may have to be rebuilt with new movement joints.)

Cracks associated with the horizontal supports for the brick cladding will need to allow for further movement and structural engineering advice should be sought.

As this is a common defect in tall brick clad blocks of flats, many repair solutions have been tried. They range from: raking out a movement joint at the top of the brick panel and individually fixing each brick slip back to the concrete nib, to recladding the building with the brickwork supported on purpose made metal rails. Solutions using rendered bands or precast units below brick panels have also been tried.

REFERENCES

BRE Defect Action Sheet 2: *Reinforced concrete framed flats: repair of disrupted brick cladding*

BRE Digest 75: *Cracking in buldings*

BDA Design note 10: *Designing for movement in brickwork*

1 Crack in reinforced concrete structure showing salts brought to the surface by dampness.

5.1.22 CRACKING OF CONCRETE

Isolated hair-line cracks at midspan of beams, near junctions of frame members and from corners of openings. The defect applies to in-situ concrete walls and frames.

Symptoms

The crack follows a line of weakness (a line of maximum stress). If it is exposed to dampness, salts may show along the line of the crack. The crack may only show on one side of a beam or column. Cracks normally appear in the first year after construction.

Investigation

If the cracks are wider than hair-line cracks or in cantilevered construction or extend in a continuous line across several bays of the building, structural engineering advice must be sought.

Note the position of the cracks and any changes in their dimensions.

Note any flaking or spalling concrete related to the cracks.

Establish the time that the cracks first appeared.

Note any rust stains at cracks.

Investigate the possibility of foundation failure (on shrinkable clays and made ground) or alterations, changes of use and accidental impact that could have led to overloading.

Diagnosis and cure

Some cracking of concrete structures is fairly common due to shrinkage of the concrete. It does not necessarily imply a defect that requires attention. Progressive cracking does require attention, as does cracking that exposes steel to moisture and corrosion. If cracks are made good or sealed over, a flexible material should be used to maintain the seal as the crack opens and closes in response to variations in temperature and moisture.

REFERENCES

Turton, C P: *Non-structural cracking of concrete* British Cement Society: Building Technical File No.5 April 1984

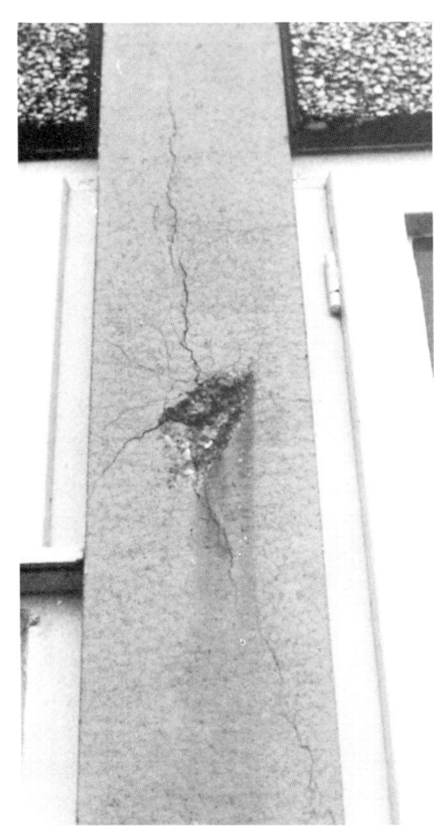

2 Corrosion of reinforcement soon after construction may be caused by inadequate cover.

3 Reinforcement corrosion can lead to disintegration of concrete members.

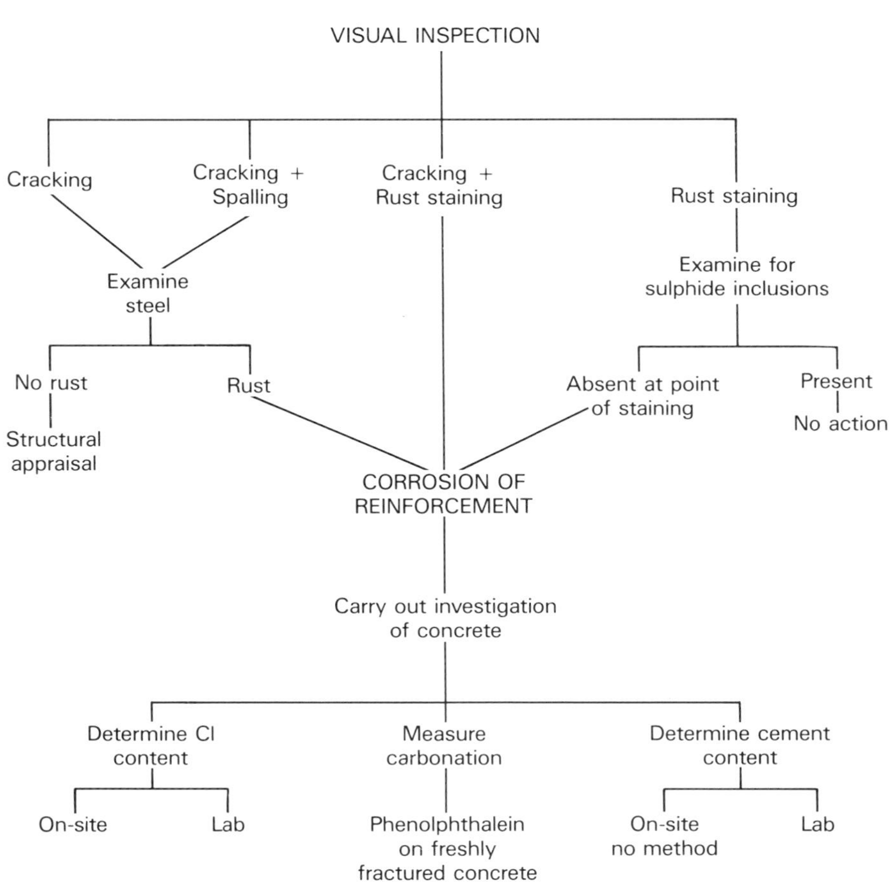

1 Flowchart for inspection of corroded steel in concrete.

Nominal cover to reinforcement

Condition of exposure	Nominal cover *mm*				
	Concrete grade				
	20	**25**	**30**	**40**	**50 and over**
Mild: e.g. completely protected against weather or aggressive conditions, except for brief period of exposure to normal weather conditions during construction	25	20	15	15	15
Moderate: e.g. sheltered from severe rain and against freezing whilst saturated with water. Buried concrete and concrete continuously under water	–	40	30	25	20
Severe: e.g. exposed to driving rain, alternate wetting and drying and to freezing whilst wet. Subject to heavy condensation or corrosive fumes	–	50	40	30	25
Very severe: e.g. exposed to sea water or moorland water and with abrasion	–	–	–	60	50
Subject to salt used for de-icing	–	–	50*	40*	25

*Applicable only if the concrete has entrained air

4 Nominal cover required for different exposure conditions and grades of concrete.

5.1.23 CRACKING OF CONCRETE

Cracking and spalling concrete along the lines of steel reinforcement or encased steel sections. The defect applies to in-situ and precast concrete.

Symptoms

The cracks follow the line of the reinforcement where it is near the surface. Sometimes the concrete is stained with rust, particularly where pieces of concrete have spalled away. The defect occurs where concrete is exposed to intermittent wetting. It is less likely to occur where concrete has been painted and is sealed from the atmosphere.

Investigation

Where these symptoms are present on structural concrete, a structural engineers advice must always be sought.

Establish whether the steel near the surface has rusted. A piece of the concrete can be cut away to expose the reinforcement and tapping the surface of the concrete with a hammer will help to show how extensive the rust is. Well developed rust just below the surface of the concrete will sound hollow.

If large pieces of concrete appear to be loose or if there is a possibility that they may come away and fall on people below, arrange for fencing and protection.

Test for carbonation in the vicinity of rusted steel reinforcement. Check on the chloride content of the concrete – particularly in the case of precast concrete. On site or laboratory tests can be used. The distribution of chloride in the original mix can be an important factor.

Determine the cement content of the concrete – a sample has to be tested in the laboratory.

Test kits are available for the on site testing of acidity and alkalinity of concrete using dyes that change colour to indicate acidity (see chapter 4). The position and depth of reinforcement near to the surface of reinforced concrete members can be tested with special equipment (see chapter 4).

If structural members are affected seek the advice of a structural engineer.

Diagnosis and cure

With in-situ concrete the defect is usually caused by carbonation of the surface layer of concrete, (shrinkage cracks in recently constructed concrete structures can also follow the lines of reinforcement). The carbonation process is the result of the action of carbon dioxide in the atmosphere and proceeds more rapidly in porous concrete or concrete with a low cement content. The action lowers the Ph level of the concrete – making it less alkaline. Where moisture is present, the lower Ph level allows the steel reinforcement to rust. With precast concrete, and sometimes in-situ concrete, calcium chloride was used to speed up the hardening of the concrete. As the calcium chloride was in the original mix its action is not a gradual process confined to the surface layers exposed to the atmosphere.

Treatment will depend on the extent of the corrosion and on whether it is practicable to seal the concrete from water and carbon dioxide. Several specialist firms offer survey and remedial services that involve cutting back carbonised concrete around reinforcement, cleaning and coating the reinforcement and replacing the concrete – sometimes with a resin based filler which is then sealed over with a protective coating. This can be a costly and disruptive process but is often the only alternative to demolition. If a concrete structure is found to have a high chloride content (over 0.5%) and is exposed to moisture it will probably have to be replaced. There are a number of chloride test kits on the market that can be used for site testing.

REFERENCES

BRE Information paper 6/81: *Carbonation of concrete made from dense natural/aggregates*

BRE Information paper 21/86: *Determination of Chloride and cement contents of hardened concrete*

BRE Digests 263/264 & 265: *The durability of steel in concrete*

Construction Feedback Digest 44 Autumn '83 *Calcium chloride accelerates beam deterioration* B&M Publications (London) Ltd

Hollis, Michael *Surveying Buildings* Surveyors Publication 1983.

5 Corroding reinforcement exposed.

6 Ring beam damage due to high chloride content of concrete. The two most prominent horizontal rust stains correspond with the flanges of a rusting steel beam.

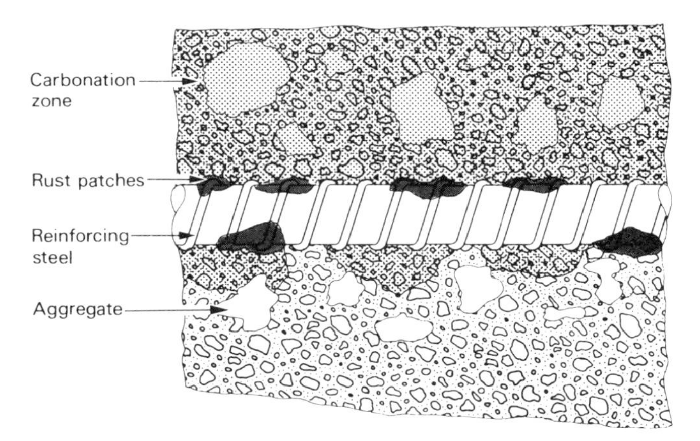

Diagramatic view of steel corroding in carbonated concrete

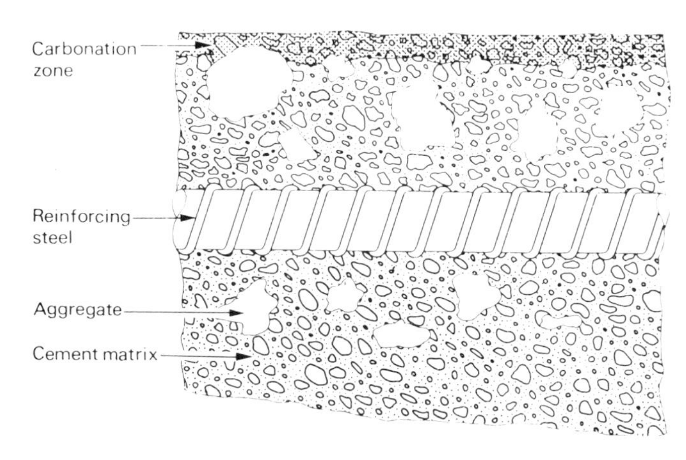

Diagramatic view of steel protected from corrosion in partially carbonated concrete

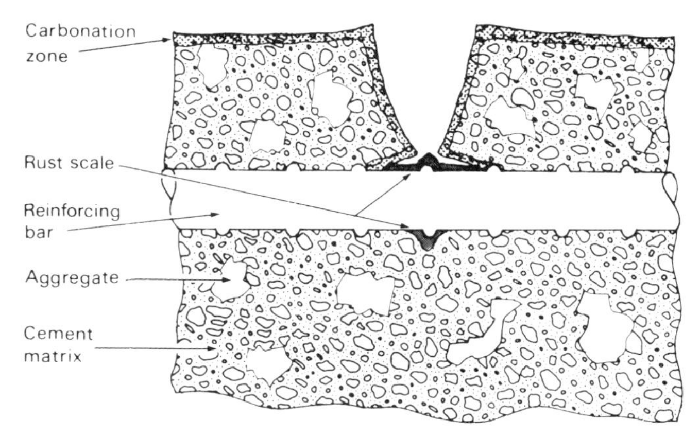

Diagramatic view of steel corroding in cracked concrete

7 & 8 & 9 Diagrammatic view of steel corroding in cracked concrete.

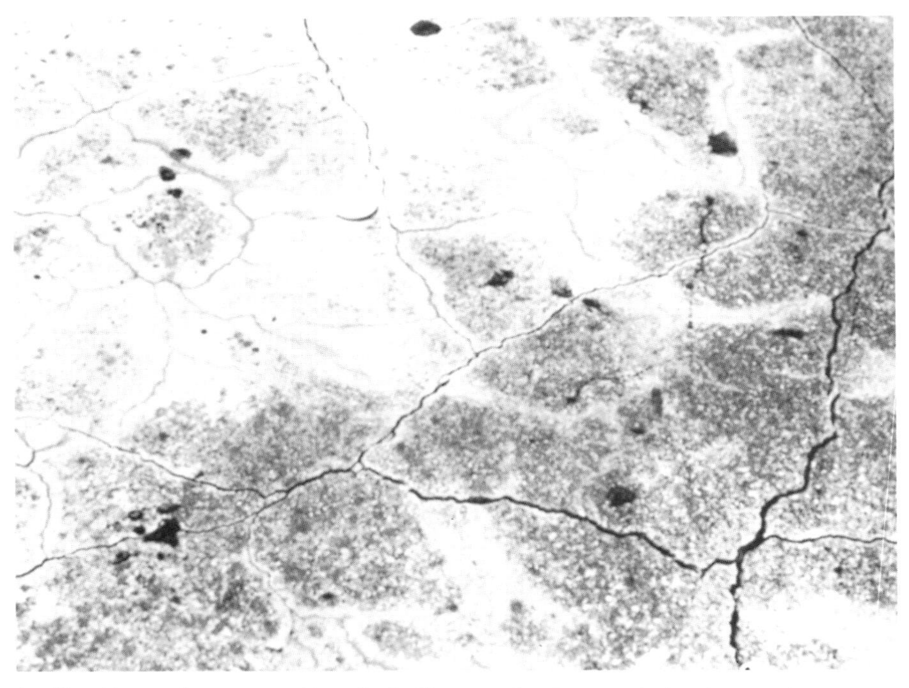

1 Cracking pattern caused by alkali-silica reaction in unrestrained concrete.

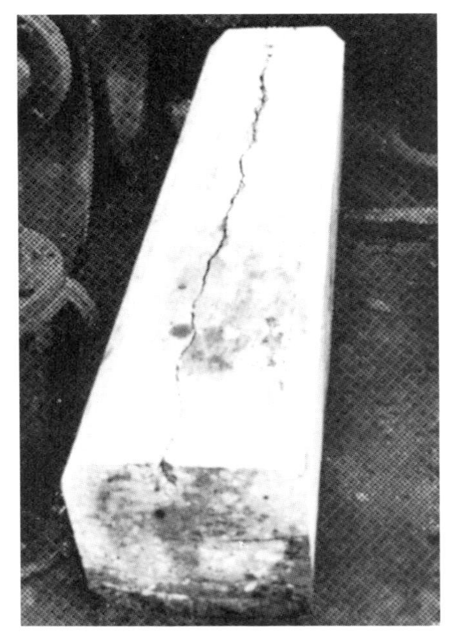

2 Where expansive forces are restrained by reinforcement directional cracking can result.

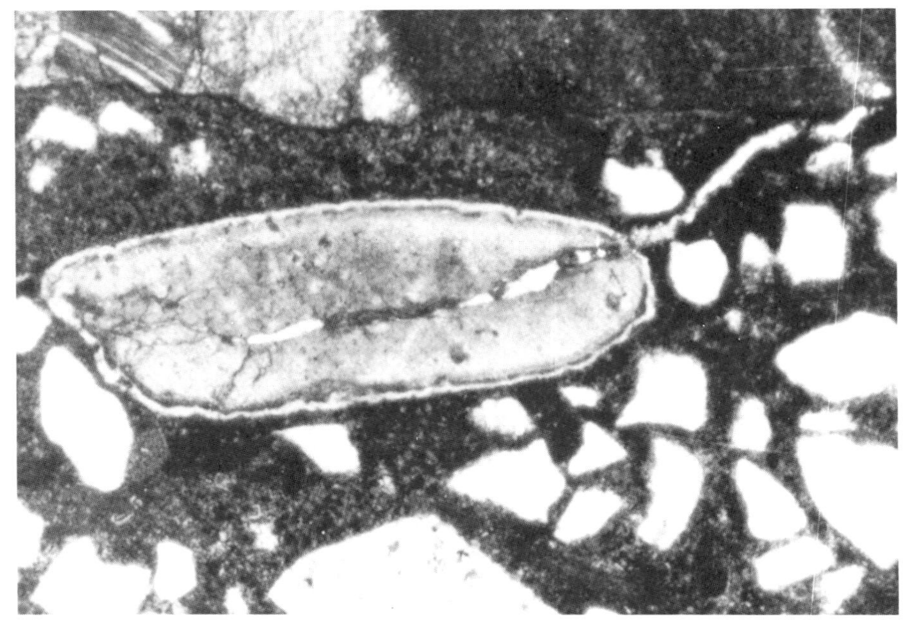

3 Gel showing as a line passing through aggregate particles when a section of affected concrete is examined under the microscope.

5.1.24 CRACKING OF CONCRETE

Typically random 'map' cracking over the concrete surface appearing some time after concrete has been cast. Sometimes cracks running parallel to main reinforcement bars. Applies both to in-situ and to precast concrete.

Symptoms

This type of cracking usually appears in structures that are over 5 years old. Most of the cracks will be very fine – sometimes only visible as darker lines when the concrete surface is drying out. Some cracks will be wider, particularly where the concrete is restrained by reinforcement. Sections of concrete surrounded by cracks may fall out.

Investigation

When these symptoms are present structural engineering advice must always be sought.

Note the areas of concrete affected by cracking (fine cracks can be seen if the concrete surface is brushed clean and wetted. The cracks will show as the surface dries out).

Note the weather conditions when the cracks were first noticed – similar cracking can be caused by frost action.

Establish the time interval from construction to first evidence of cracking – similar cracking caused by shrinkage is likely to appear early in the life of the structure.

Determine, if possible, the source of aggregate and make enquiries about concrete structures of a similar age using aggregate from the same source. Arrange for core samples to be taken from cracked areas for microscopic examination in the laboratory.

Note over a period of time whether steelwork corrosion or frost action is affecting areas where cracking has occurred.

Note signs of gel (light coloured coating) to edges of cracks and aggregate particles on the concrete surface.

Further evidence can be obtained from measurements of core samples stored in warm, water-saturated atmosphere.

Diagnosis and cure

The cracking is caused by alkali-silica reaction between the silica in the aggregate and the cement. It occurs only with certain aggregates and was originally thought to be rare in the U.K. but more instances are discovered every year. The reaction produces a gel that imbibes water and expands. The expansion disrupts the concrete. When the expansion is near the surface and unrestrained it produces the typical 'map' cracking. When it is restrained by reinforcement it can produce cracks running parallel to the main lines of reinforcement.

The reaction develops fairly slowly and is likely to be noticed on visible concrete members before it affects the stability of the structure. However it may not be seen before lumps of concrete start to fall from concrete at high levels. The compressive strength of the concrete will be little affected at the early stages, but tensile strength and elasticity will be reduced. The expansion of the gel can cause bowing and displacement of concrete units.

No method is known of reversing the alkali silica reaction. However, its affects can be reduced by keeping the concrete as dry as possible (with coatings or claddings) and by reinforcing the building where the reaction threatens its stability. The assessment of affected structures must always be made by a structural engineer.

REFERENCES

BRE Digest 258: *Alkali aggregate reactions in concrete*

1 & 2 Two examples of chimneys affected by expansion of mortar joints on one side.

5.1.25 CHIMNEYS. HORIZONTAL & VERTICAL CRACKING

Freestanding chimneys are distorted and show cracks. The defect applies to masonry and in-situ concrete.

Symptoms

In chimney stacks that have been in use for several years cracks appear, sometimes along the mortar joints and sometimes as a vertical split. The stack may bend or twist. The chimney pot may be displaced. Damp patches are often associated with this defect either on the faces of the stack or lower down on chimney breasts within the building.

Investigation

Examine the materials used in the construction of the stack. Is there a flue liner? Is the liner correctly installed? Is the stack damp?

Note the type of appliance discharging into the stack – is it a boiler or slow burning fire? Check that the stack is stable. If there is any doubt, have it taken down to a safe level.

Note the extent of the cracks and reinspect over 6 month to 1 year to see whether cracking is progressive.

Note whether the stack is reinforced.

Have a laboratory test carried out on the concrete and mortar in the stack to check for sulphates and to check the cement content.

Examine the flashings at roof level.

Diagnosis and cure

The usual cause of the defect is condensation of the water vapour in the flue gasses as they pass up the exposed section of the chimney. The water may contain sulphates which cause the mortar in the stack to expand. If driving rain is predominantly from one direction the expansion of the mortar will be greatest on that side. The stack will bend away to the opposite side. Clay flue liners are not damaged by condensates containing sulphates (but if the liners are fitted upside down the sulphates will run into the surrounding structure). In-situ concrete containing colliery shale could also be the source of sulphates which cause damage in the damp environment of an exposed stack.

The cure is to rebuild the chimney in good quality materials with correctly fitted flue liners.

REFERENCES

BRE Digest 200: *Repairing brickwork/concrete chimney*

Bickerdike, Allen, Rich & Co in association with Turlogh O'Brien. *Second series design failures in building* George Godwin & Co. 1973.

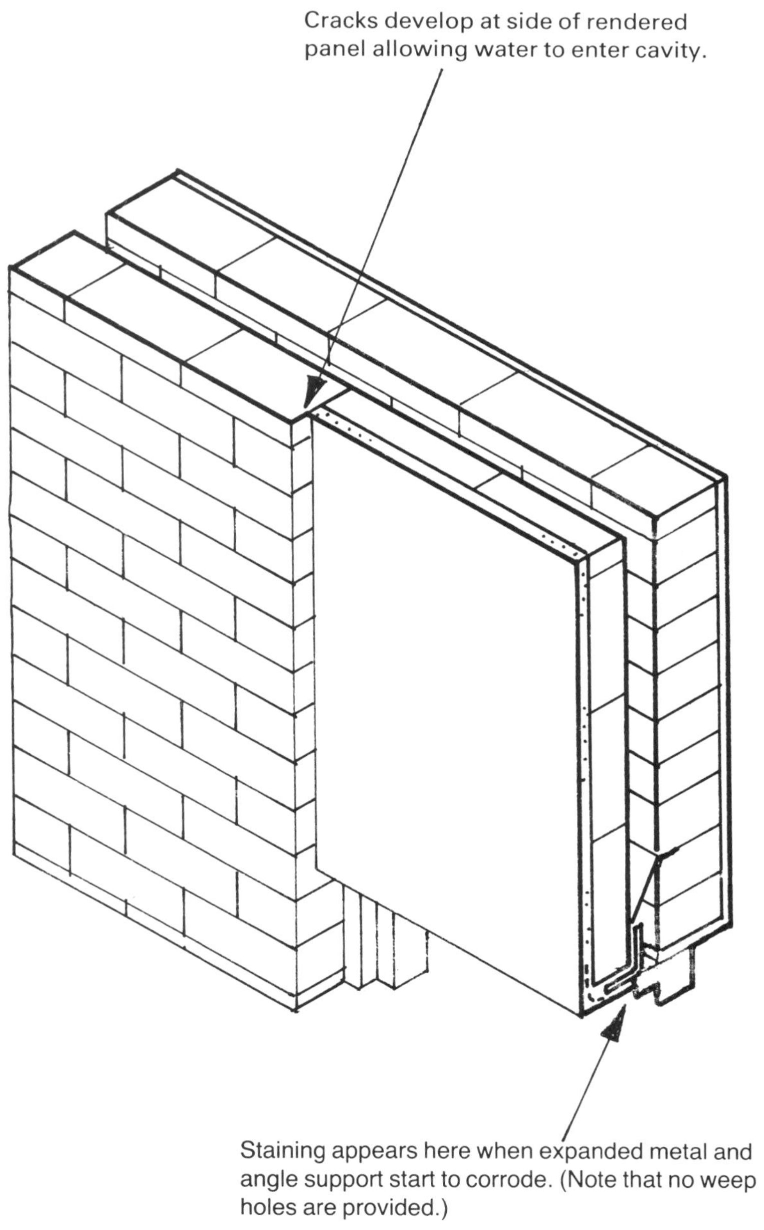

Cracks develop at side of rendered panel allowing water to enter cavity.

Staining appears here when expanded metal and angle support start to corrode. (Note that no weep holes are provided.)

Typical rendered panel against brickwork showing where defects first appear.

5.1.26 EXTERNAL CAVITY WALL. CRACKING OF RENDERED PANEL

This defect refers to rendered panels with a backing of precast concrete blocks set in the outer clay brick leaf of a cavity wall.

Symptoms

Cracking of render, particularly at the sides and bottom of the panel. If the cracks are extensive, sections of the render may become detached and fall from the building. It may be accompanied by brown rust stains.

Investigation

Establish the construction of the wall. Investigate any signs of leakage visible internally.

If there are brown rust stains, find the source of the rust. It may be from a lintel, from wall ties or from expanded metal used to reinforce the render.

Examine the cracks under different conditions of humidity and temperature and note any change in their width. If there is a cavity wall behind the render, note whether weepholes have been provided to drain it.

Diagnosis and cure

The cause of this defect is the thermal and moisture movement of the rendered blockwork relative to the surrounding clay brickwork. The situation is made worse by rain entering any cracks that form at the edges of the panels and saturating the blockwork or, if there are no weepholes, by collecting in the cavity behind the blockwork. The moisture in the wall will rust any mild steel components that are not effectively protected. It could also make the wall vulnerable to frost damage.

In this case the cure must either prevent water entering the wall – a lightweight cladding could be fixed to the panel; or the wall must be replaced with one that is less susceptible to thermal and moisture movement. For example the concrete block backing to the render could be replaced by hollow burned clay blocks with more or less the same coefficients of expansion as the surrounding brickwork. Weepholes should be provided at the bottom of the panel and steel components should either be totally avoided or otherwise adequately protected from corrosion.

REFERENCES

Bickerdike, Allen, Rich and partners, in association with Turlogh O'Brien. Design failures in buildings second series George Godwin Ltd. 1974

DAMPNESS IN WALLS

General points

Dampness in walls has taken on added significance in recent years. When buildings are sold, the purchaser's surveyors invariably check for dampness (which is now more easily done with electrical resistance meters). If even quite low levels of dampness are found, the value of the building can be affected. In domestic buildings dampness is increasingly seen as a health hazard. Many successful cases have been brought against landlords whose property was shown to be damp – even though the dampness was at least partly the result of improved heating standards. A third factor has been the increase in multi-storey buildings in which practically all spaces have external walls with comparatively cold surfaces, and a potential for rain penetration, while relatively few spaces are vulnerable to roof leaks or rising damp.

When dampness in walls is being investigated the following aspects should be considered.

- The pressure which moves moisture through spaces and structure has wide fluctuations and several types of dampness will only occur when pressure is greatest.

  ie when wind driven rain or snow is from a certain direction and the wind is above a certain speed.

  when humidity within the building is very much higher than humidity outside.

  when the ground water table is unusually high.

  when a gutter overflows in exceptionally heavy rain.

  The conditions that produce dampness may only occur as isolated incidents. Observation over a period of time is often necessary to see whether the occurrence is repeated.

- The action of many chemical and biological agents can only take place in damp conditions. Although this type of damage to buildings is often seen as breakdown of materials, rot, worm infestation, etc., these defects have only occurred because the construction in a particular spot was damp. If other parts of the building had been damp, the defect might well have occurred there too. The underlying defect can be dampness while the symptoms are material failures or biological attack.

Sources of dampness

Dampness in buildings originates from seven sources.

1. *Rain and snow* – precipitation can be wind driven so that it penetrates joints that remain watertight in normal weather conditions. It can also collect and cause problems when gutters overflow or drifting snow piles up against walls.
2. *Condensation* – dampness may result from humid air condensing on cooler surfaces or within, or between, building materials (interstitial condensation). Air can become humid in several ways – not always as a result of the occupants generating water vapour.
3. *Rising damp and flooding* – buildings may be directly in contact with ground water or flood water, or the ground water may be absorbed by the walls and then transported up the wall by capillary action.
4. *Services leaks* – not just from pipes and tanks, but also the overflowing of condensation forming within ventilation systems.
5. *Construction processes* – processes that use water – generally to form mixtures that dry out before the building is used, but sometimes by retaining moisture (sealed in by impermeable finishes) that shows and causes problems in the completed building.
6. *Use of the building* – including cleaning of the building, spills, apparatus leaking, etc.
7. *Moisture in the air* – as distinct from condensation. Hygroscopic salts can extract moisture from the air in conditions that would not allow that moisture to form condensation.

To cure defects involving dampness, it is essential that the source of dampness is correctly diagnosed. Even when a cure seems to have been achieved, and the wall drieds out, remedial work that is based on a wrong diagnosis will allow the dampness to return. However identifying the source of dampness is not always straightforward. More than one source can contribute to dampness in a wall and there is an interaction between sources of dampness that sometimes makes it difficult to decide what remedial action would be most effective – particularly when condensation is involved. Largely as a result of these problems, many new techniques for measuring and analysing dampness have been developed recently, together with methods of calculating the probability of dampness occurring.

Dampness comes and goes with changing conditions. Sometimes it leaves stains or traces of mould and lichens and mosses. Sometimes no traces remain. Continuous monitoring may be the only way to establish the cause of dampness. If monitoring by the building's users cannot be arranged, mechanical recorders are available.

The examples given in this part of the book illustrate typical instances of dampness in walls from different sources (and sometimes from more than one source). Do not assume, because of a superficial similarity, that the causes of dampness in a building will be as described in the examples. The full range of potential sources of dampness should always be considered so that the diagnosis is made by a process of elimination.

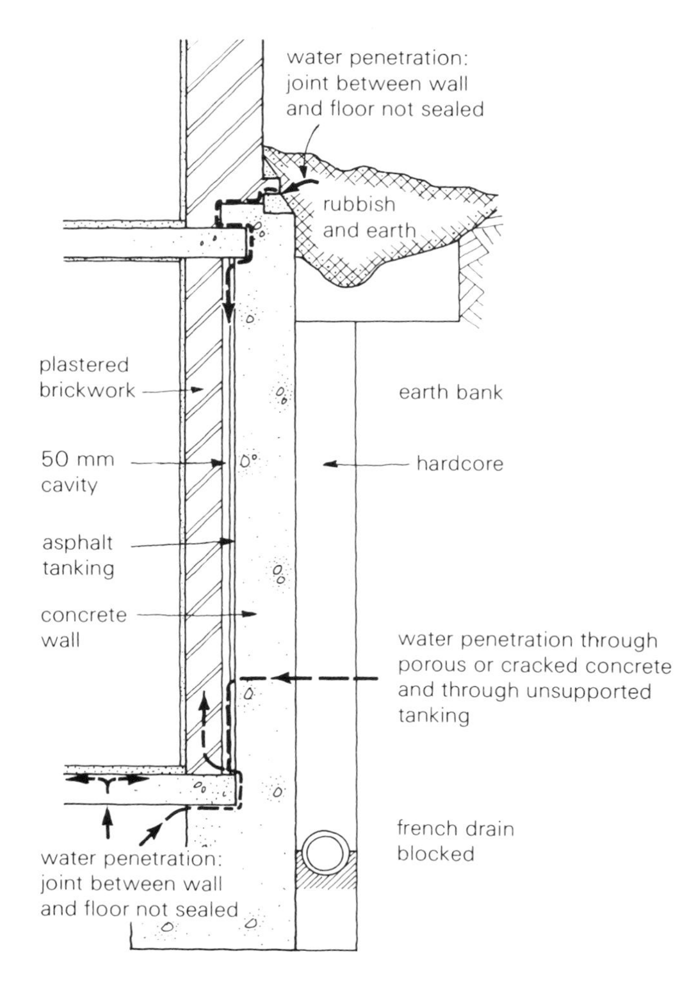

Figure 1 Existing basement construction showing water penetration.

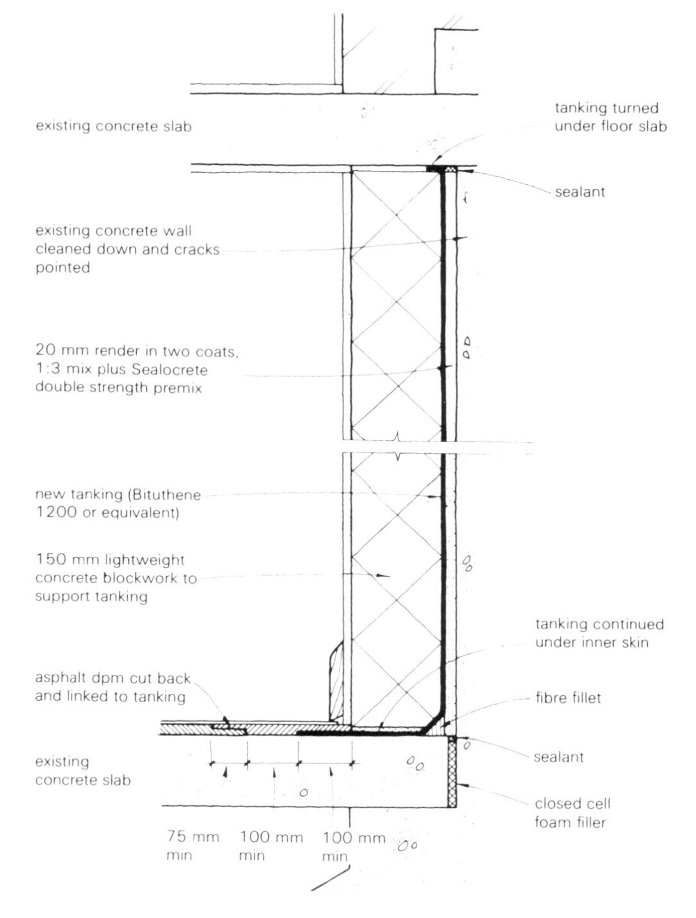

Figure 2 Remedial work to provide continuous, supported tanking. Note that the removal of rubbish and the provision of manholes, that allow the french drain to be properly cleared, were also included in the remedial work.

5.2.1 BASEMENT PERIMETER WALLS. DAMPNESS IN ANY LOCATION

Where the wall is below ground level and the outside of the wall is in contact with the ground. This defect relates to in-situ concrete and masonry construction.

Symptoms

Visual signs of dampness, or high electrical resistance meter readings occur near service entry points. Structural cracks and daywork joints in concrete or along the bottom of the perimeter wall (and in partitions abutting the perimeter wall) open up. The floor of the basement may be flooded. Symptoms may only be noticed when occupation or heating is intermittent.

Investigation

Has excavation or drainage work recently been done near the outside of the wall?

Record the area of dampness and any changes in the area affected. Is the source of ingress of water obvious? If not, wipe the wall surface dry and watch whether the water spreads over the surface from any particular place.

Establish the form of construction. Is there a cavity? Is there a layer of tanking? How is the junction of the basement floor and wall waterproofed?

Investigate structural movement – has it damaged the basement waterproofing?

Check on the position of service entries and any other below ground perforations in the perimeter walls.

Establish the water level in ground adjacent to the basement.

Does the area or degree of dampness fluctuate with variations in the ground water level?

Examine drain runs near the building. Apply pressure test to drain runs. Are there signs that manholes have been flooded recently?

Diagnosis and cure

The likely causes are:

1 Ground water under pressure, leaking through faults in the waterproofing system. The water pressure may have increased as a result of the blocking of a normal water channel or from the ground being saturated by additional water from blocked drains or gulleys. New service entries, structural cracks or construction defects can allow water through the waterproofing system.

2 Damp penetration by moisture from the soil. The moisture in the soil does not have to be under pressure to find its way through permeable materials if there is any fault in the waterproofing system.

3 Condensation on the cold inside surfaces of basement walls. This is more likely to occur if heating is intermittent and the use of the basement generates moisture vapour (e.g. washing, sauna, meetings, sports). Moisture entering the basement from 1. and 2. may either evaporate and condense elsewhere or back on the original damp patch which will be cooler than the dryer areas of wall. Condensation can make other dampness problems worse rather than being the original cause of the dampness.

If the dampness is all at low level it may be running down the inside of a wall cavity or behind a facing material – or it may be a fault in the floor to wall junction – but it may also be that the basement has been flooded, possibly during cleaning.

Dampness at higher level, and in one place, makes a specific fault in the waterproofing system the most likely cause. An alternative reason is water from a leak in the superstructure above, which only shows when it reaches the differently constructed basement wall. If the leak allows rain penetration, the damp area is likely to be more marked in wet weather.

Various specialist treatments are available to plug holes in basement walls – even where water is pouring or squirting through under pressure. An alternative option when dampness is more widespread is to construct an inner leaf to the wall and drain away the water from the cavity behind. Drainage channels, a sump and a submersible electric pump can be used if there is no surface water drain at basement level.

REFERENCES

PSA TP E01 703 *Technical Guidance Basements* (Deals with new basements rather than repair)

RIBA product selector (13) for proprietory waterproofing services, RIBA services.

Building Technical File October 1985 *Penetrating damp problems*

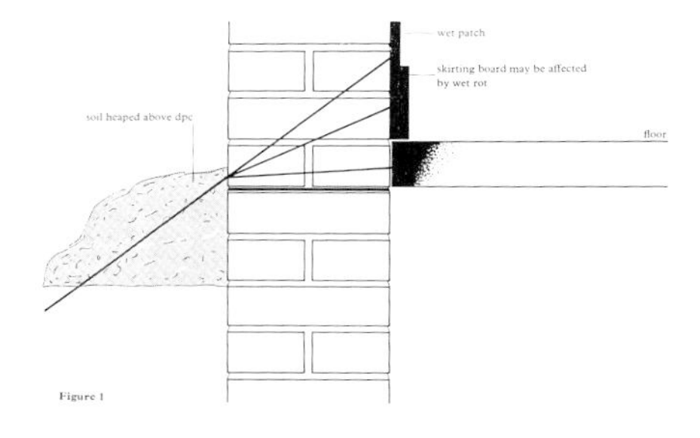

Figure 2 Soil heaped against the external wall allows moisture to pass into the wall above the dpc.

Photo 1 Staining above skirting level.

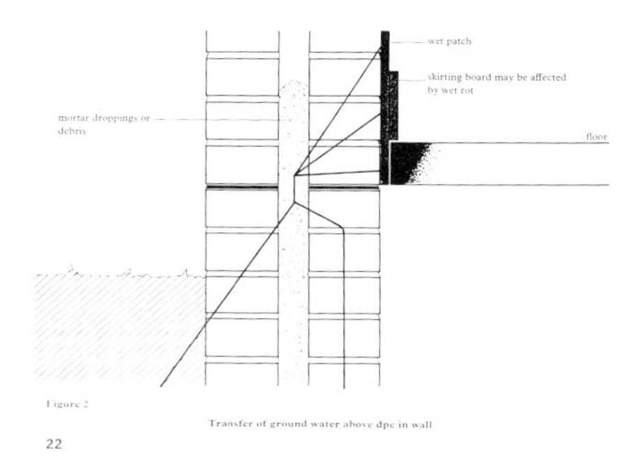

Figure 1 Mortar and debris in cavity provided a route for moisture that bypasses the dpc.

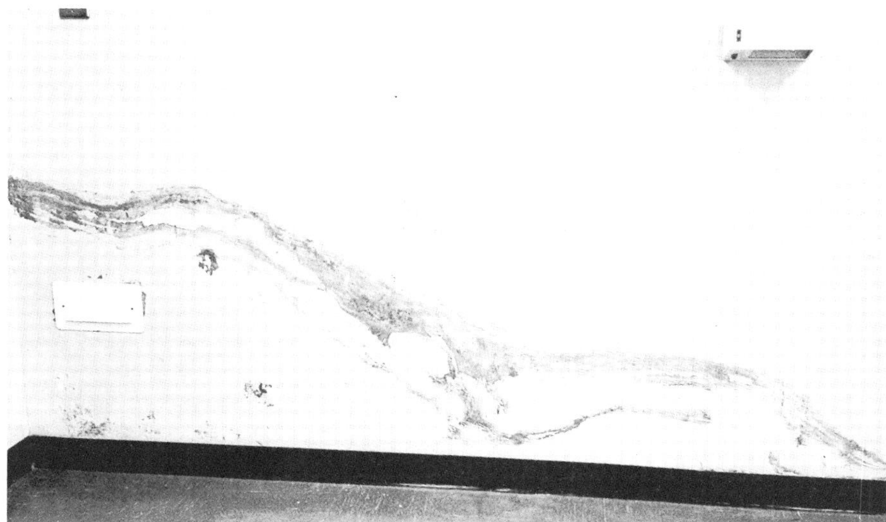

Photo 2 Tidemarks as damp advances and recedes.

Photo 4 Timber frames are at risk when walls remain damp.

Photo 3 Level of dampness varies.

5.2.2 EXTERNAL AND INTERNAL WALLS. DAMPNESS AT LOW LEVEL

Dampness showing most clearly on internal wall surfaces from floor level up to about 1 m. This defect relates to masonry and to porous stone claddings.

Symptoms

The interior of the wall is damp, fluffy crystalline salts are brought to the surface forming 'tide marks' around areas of damp. Timber skirtings, door linings and the ends of timber joists may be rotted. Decorations may be stained or blistered. The dampness may affect the edge of a floor screed. Externally there may be discolouration and stains on the masonry or cladding. Pointing and rendering at low level may be broken and loose.

Investigation

Establish the form of construction. Is there a DPC and if so what is its condition? Is there a cavity and if so is it clear at DPC level? Are there porous surface materials bridging the DPC? (such as Carlite plaster or weak render). Is the DPC being bridged by soil or built up paving against the outside of the building or by a raised solid floor inside? Do internal partitions bear on solid floors or do they bear on their own foundations below? Are there any other sources of water such as dripping gutters, or overflows, or water used for floor cleaning? In recently completed construction, has the wet screed been physically isolated from the absorbent wall finish?

Note the extent of dampness and record whether the extent of dampness on the surface differs from that in the substrate (using electrical resistance meter with deep probes, see chapter 4). Does the area of dampness vary with rainfall, humidity, or level of water in the ground or with the use of the building? It may be necessary to keep records over a period of six months to one year, to establish sufficient variations in circumstances.

Are hygroscopic salts present on the wall surface? (see chapter 4 on electrical resistance meter with an attachment that indicates presence of salts at the surface).

Remove a section of plaster from the wall and drill out samples of brick and mortar dust from the inside of the wall. Plot the moisture content and the hygroscopic moisture content of the samples at different levels in the wall, (say at intervals of 100 mm) (see Appendix to BRE Digest 245 for method).

Diagnosis and cure

Rising damp is caused by

(a) Lack of DPC

(b) Bypassing of the DPC

(c) Failure of the DPC material

It will often be possible to establish the presence of this defect purely by examination and electrical resistance meter readings, but where rain penetration, condensation or leaks from embedded pipes are the source of dampness the symptoms may be misleading. The method recommended in BRE Digest 245 is a surer diagnosis. If the moisture content of the bricks and mortar at the base of the wall is less than 5% then the problem is unlikely to be rising damp. Where the dampness extends more than 1 m up the inside of the wall it is likely that another source of dampness is at least contributing to the problem.

Rising damp often brings hygroscopic salts to the wall surface. These salts will take up moisture from the atmosphere and make the wall damper in humid conditions. The presence of hygroscopic salts does not necessarily prove the presence of rising damp, but as they are commonly brought to the surface of the wall by water from the soil and foundations a likely indication is that there has been rising damp.

Where the cause of rising damp is the bypassing of the DPC the cure is to stop the route by which the DPC is bypassed, for example by replacing porous plaster with waterproof render. If this solution is not feasible, or if the cause is lack of, or failure of the DPC then a new DPC will need to be installed. This may be done by either cutting a DPC into the wall or by the injection of a chemical DPC. Specification and workmanship are critical to the success of injected chemical DPC's (see references). Where there is a possibility of alternative sources of dampness contributing to this defect, an effort should be made to eliminate these sources before treating the rising damp. A new DPC is expensive to install (where plaster and decorations have to be replaced). If the dampness proves to be from another source the DPC may not be necessary, or a less extensive treatment may be possible.

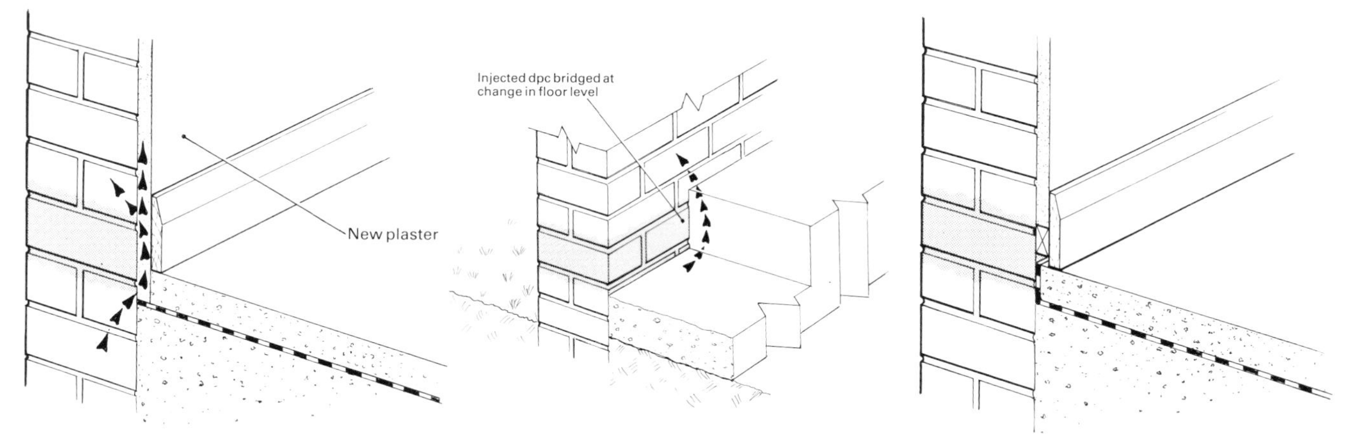

Figures 3 & 5 Routes that bypass injected dpc's

Figure 4 Correctly injected dpc.

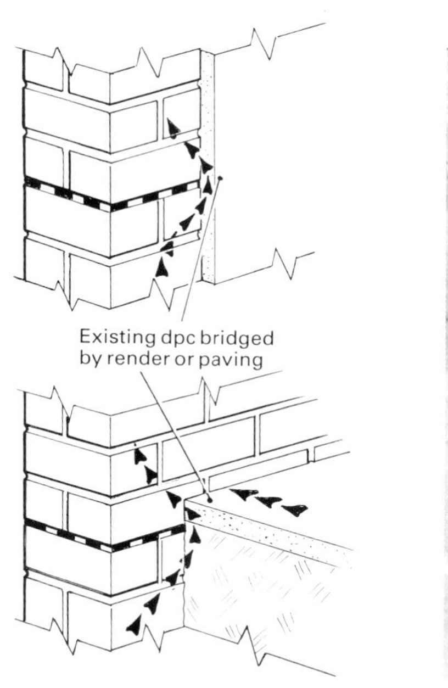

Figure 8 Render and paving bridging the dpc.

Photo 5 Damage from poor drilling permits loss of dpc fluid during injection.

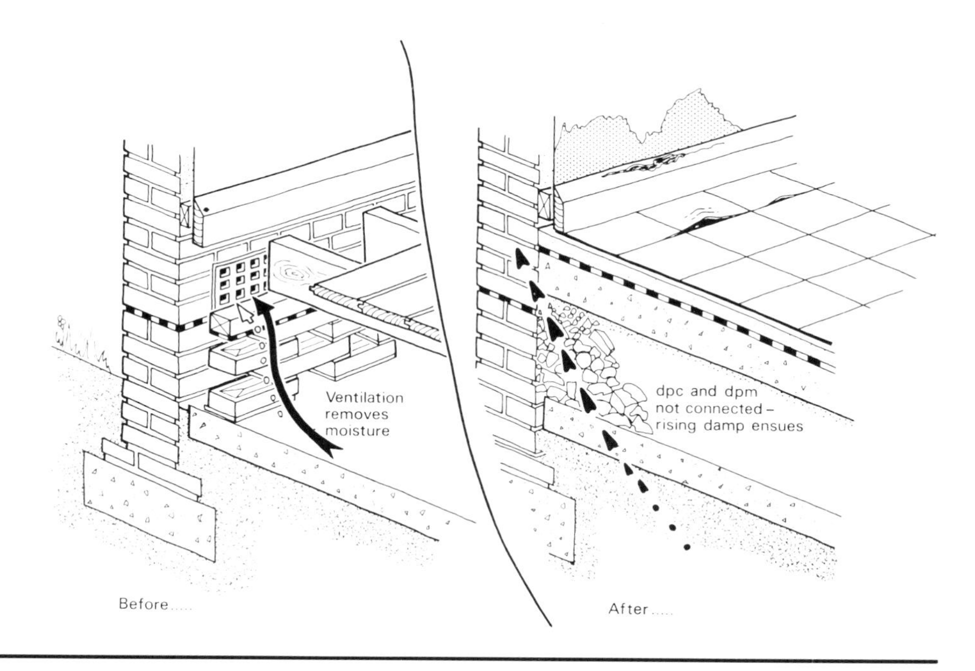

Figures 6 & 7 Replacement of a suspended timber floor by a concrete floor allowing dampness to bridge the dpc.

continued

REFERENCES

BRE Defect Action Sheet 22: *Ground floors: replacing suspended timber with solid concrete – DPC's and DPM's*

BRE Defect Action Sheet 26: *Substructure DPC's and DPM's: installation*

BRE Defect Action Sheet 35: *Substructure DPC's and DPM's: specification*

BRE Defect Action Sheet 85: *Brick walls: injected DPC's*

BRE Defect Action Sheet 86: *Brick walls: replastering following DPC injection*

BRE Digest 245: *Rising damp in walls: diagnosis and treatment*

BS 6576: 1985 *Code of practice for installation of chemical damp proof courses*

BMCIS Design/Performance Data/External Walls *Ineffective DPC between stone wall facing and pavement*

Building Technical File 3.10.83 *Penetrating damp problems*

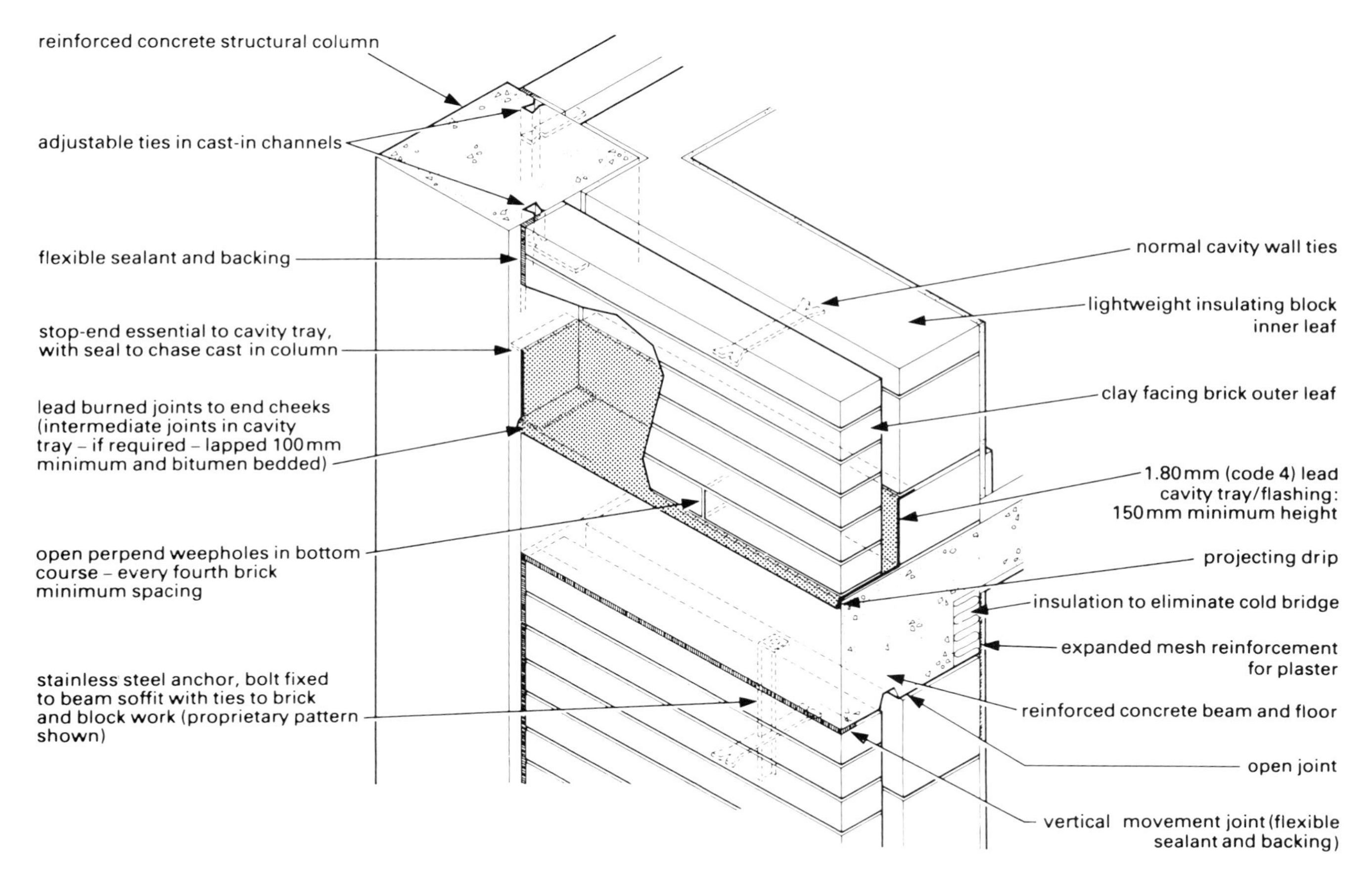

Figure 1 Detail recommended by PSA to deal with movement and damp penetration problems associated with brick infill to concrete frames – note the stop end to the dpc.

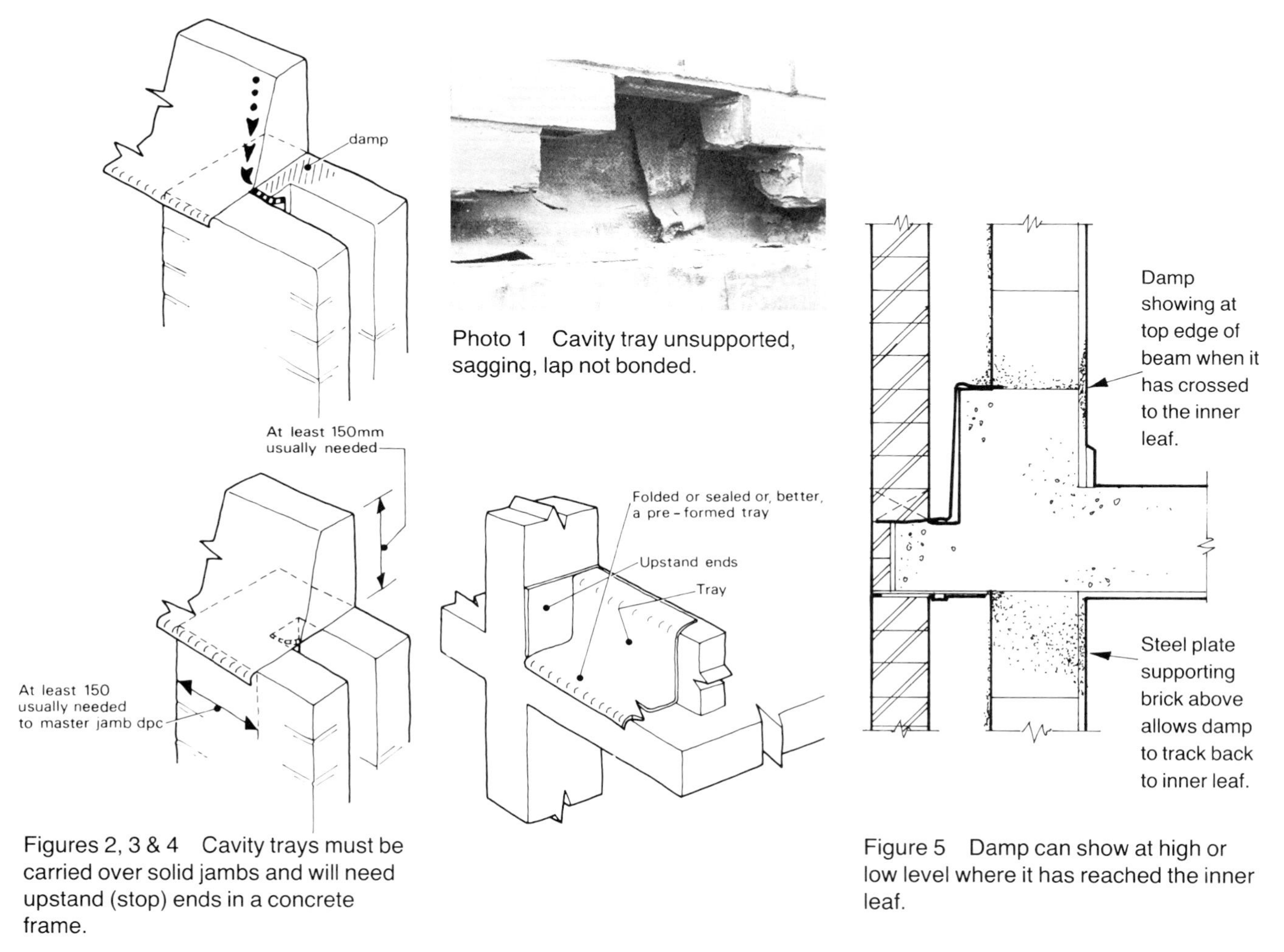

Photo 1 Cavity tray unsupported, sagging, lap not bonded.

Figures 2, 3 & 4 Cavity trays must be carried over solid jambs and will need upstand (stop) ends in a concrete frame.

Figure 5 Damp can show at high or low level where it has reached the inner leaf.

5.2.3 EXTERNAL WALLS. DAMPNESS AT LOW LEVEL

Dampness associated with rainfall showing on inside of wall near floor level. This defect relates to structures with concrete floors and cavity or curtain walls.

Symptoms

Damp stains extending along the inside of external walls just above skirting level, particularly on the inside of walls exposed to driving rain. Dampness will be most pronounced after a period of rain that has been driven onto the wall. There may also be dampness at the ceiling to wall junction on the floor below. The dampness may show as a horizontal line up to 450 mm above the floor level (i.e. above a concrete edge beam).

Investigation

Establish the construction of the external wall – if necessary by removing part of the inner or outer leaf or a panel of the curtain walling. Does the concrete floor cross the cavity or project into the cavity?

Record the extent of the dampness – Does it occur on every floor? Does it vary over a period? Is it more or less obvious at any structural columns in the external wall?

Examine the cavity with an endoscope (see chapter 4).

Is the bottom of the cavity full of mortar droppings?

Diagnosis and cure

The defect is caused by ineffective waterproofing of the joint between masonry or curtain walling and the edge of the floor. Rain that penetrates the outer leaf of the wall, or the curtain walling, reaches the inside walls where the floor either bridges the cavity or protrudes into the cavity. It runs back on the surface of the floor (or the top of an upstanding edge beam) and shows itself on the inside face of the wall.

The junction must be designed to let water drain to the outside. A number of faults can cause this defect.

1. DPC trays can be badly formed (or absent). Where DPC's are dealing with water from above, the laps should be sealed (BS CP 102: 1973)
2. Cavities can be obstructed by mortar droppings.
3. Water may be able to cross the cavity on ties that slope down to the inside or on fixings for the curtain walling.
4. If the concrete extends to the outside face of the building, water may enter at the external joint between the concrete and the wall above and then run back over the surface of the slab (or on the top of a ring beam).

On tall or exposed buildings, water behaves in unexpected ways. Quantities of water run down the outside of a tall building. Some of it may be blown up under flashings and through small holes into cavities that would not be so vulnerable in a lower, more sheltered, building.

Remedial work should aim to stop the water passing through the external leaf of the wall, prevent it reaching the internal leaf and provide a means of draining it back to the outside before it reaches the floor slab.

Pointing, flashings, gaskets, and sealants should be inspected and, if necessary, repaired or replaced. Cavities should be cleaned. Laps in the DPC trays should be sealed, the DPC should extend out over the edge of the concrete with the edge stuck to the concrete and an upstand stuck around concrete columns. Repairing and improving the DPC may have to be done in sections, removing a short length of brickwork at a time, in order to maintain support for the wall above. Weep holes should be provided to prevent a build up of water over the cavity tray.

REFERENCES

BRE Defect Action Sheet 12: *Cavity trays in external cavity walls: preventing water penetration*

National Cavity Insulation Assocation (and other organisations) *Cavity Insulated Walls – specifiers guide*

BMCIS Design/Performance Data/External Walls *Failure of cavity wall damp proof course*

BS CP 102:1973 British Standard Code of Practice for *Protection of buildings against water from the ground*

Photo 1 Cracked rainwater drainpipe wetting the adjacent wall.

Photo 2 Mould growth resulting from condensation on cold surfaces.

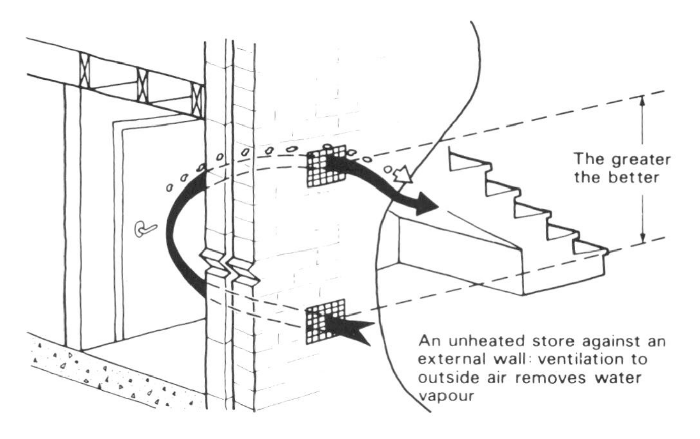

Figure 1 Ventilation at high and low level can remove water vapour from the atmosphere in an unheated space within a heated building.

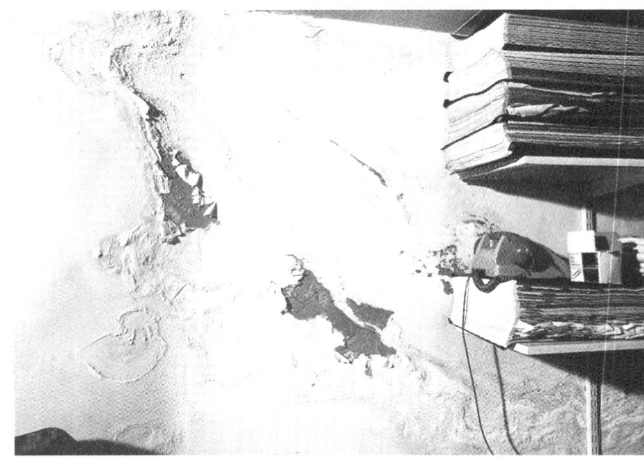

Photo 3 Rain penetration can appear as patches at any level after rain, often only after driving rain in a particular direction.

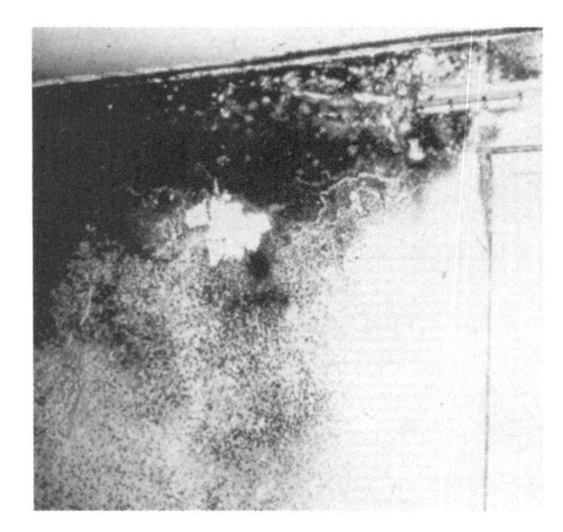

Photos 4, 5, & 6 Different levels of severity of mould growth causing "inconvenience" "discomfort" or "acute distress".

5.2.4 EXTERNAL WALLS AND ADJACENT PARTITIONS. DAMP PATCHES

The patches may occur in any position. The defect described here relates to solid masonry.

Symptoms

Damp patch or patches, usually not permanent but often recurring. Sometimes the patches run into each other and the whole wall appears damp. If there is a row of patches, are they at high or low level only? – see 5.2.12, 13, and 14 or 5.2.2 and 3. If the patches are close to windows only, see 5.2.7.

Investigation

Note extent and location of patches and whether they change over a period of 6 months.

Establish the construction of the wall and the type of finishing plaster.

Is the occurrence of dampness related to rain? If so is it rain from a particular direction? Is the dampness related to cold or humid weather?

Had the degree of ventilation (including blocking up of flues) changed when the damp was first noticed?

Examine the wall externally for sources of water discharging onto the wall (e.g. defective gutters or flashings, overflows, splashing by vehicles).

Examine the wall externally for surface defects such as cracks, broken rendering, loose or missing pointing, eroded or flaking mortar joints, damaged brickwork etc. What type of bricks were used? Is the dampness related to any junction or service penetration of the wall?

Diagnosis and cure

The dampness may be from any one of a number of sources or from more than one source. The most likely cause particularly in a solid wall, is rain penetration but condensation is also a common cause. If there is anything that tends to increase the U value of the wall where the dampness is noticed (such as a reduction in wall thickness or a dense render internally) then condensation is a probable cause. It may well be that initial rain penetration soaks the wall and increases its U value to a level where condensation forms on the inside, extending the effect of the rain penetration.

Plumbing or guttering leaks, hygroscopic salts in the plaster (see 5.2.10) and construction moisture are all possibilities to examine. The key factors that can distinguish one source from another are the timing and conditions with which the dampness is associated.

Rain penetration (or defective guttering) will be associated with wet weather particularly if the wind is blowing onto the external face of the wall. If the dampness is only occasional and limited in extent, it may be cured by repairing copings and verge overhangs (with effective drips and an overall projection of 50 mm). Painting the wall, or applying colourless water repellant treatment, is unlikely to give long term protection and may promote frost damage if water behind the protective layer freezes and spalls off the surface. A wall built of low porosity masonry is more likely to leak than a wall of more porous units, because the porous materials soak up water, whereas water streams over the surface of impervious materials and will find any routes that exist through the wall. External rendering is more likely to last and be effective when rain penetration is frequent or more general. If Carlite plaster has been used on the inside of the wall, it will absorb damp where lime plaster or cement render would not. Special types of plaster have been developed for use in renovation. Solid walls have a high thermal transmittance coefficient (U value), particularly when they are damp. If humidity in the building is high, it is likely that condensation will be at least a contributory source of dampness. The introduction of additional insulation, additional heating and additional ventilation will have to be considered. Insulation can be increased externally by rendering or cladding over an insulating layer, and internally by an insulated lining.

REFERENCES

BRE Defect Action Sheet 16: *Walls and ceilings: remedying recurrent mould growth*

BRE Defect Action Sheets 37 and 38: *External walls: rendering*

BRE Digest 177: *Decay and conservation of stone masonry*

BRE Digest 273: *Perforated clay bricks*

BRE Digest 297: *Surface condensation and mould growth in traditionally built dwellings*

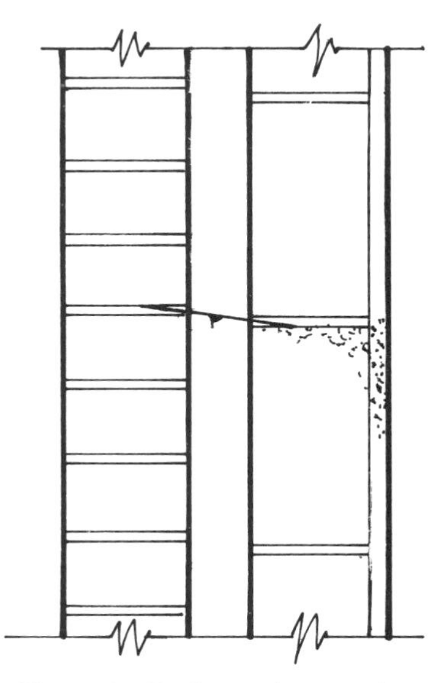

Figure 2 Bad coursing causing wallties to slope down to the inner leaf and resulting in damp patches.

Photo 2 Dampness showing on internal surface of external leaf of a cavity wall – in exposed conditions this surface is likely to be regularly wetted.

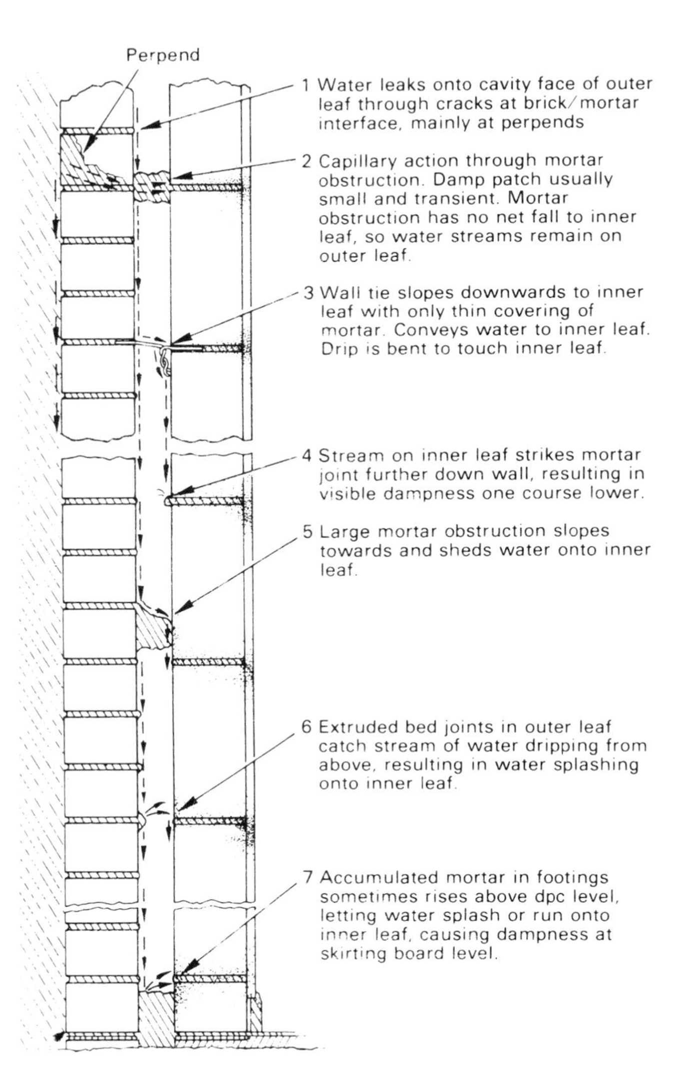

Figure 1 Routes for water penetration across an unfilled cavity.

Photo 1 Local dampness on internal surface of unplastered cavity wall.

5.2.5 EXTERNAL WALLS AND ADJACENT PARTITIONS. DAMP PATCHES (AWAY FROM WINDOWS)

The dampness may be in any location. The defect described here relates to masonry cavity walls (with no insulation in the cavity).

Symptoms

Damp patches and damp areas showing on internal surfaces. Usually associated with wet, or cold and damp weather. May be isolated patches or a pattern of dampness, or dampness showing at a particular level.

Investigation

Note the extent and location of the patches and whether they change over a period of 6 months. Establish the construction of the wall. What width is the cavity? Is the internal finish solid plaster, or a dry lining?

Note the exposure and whether the wall is vulnerable to driving rain from a prevailing direction. Is the dampness associated with cold or wet weather?

Examine the cavity with an endoscope (see chapter 4) and note the type and condition of the wall ties. Are there any obstructions in the cavity? (e.g. services, masonry nibs, projecting columns, or columns within the cavity, fire stops, cement bridging or rubbish).

Note the details at the top of the wall cavity.

Diagnosis and cure

The most likely cause is rain penetration combined with condensation and most of what has been said about similar defects in solid walls applies here also (see 5.2.4). Unlike a solid masonry wall, a cavity wall should only show dampness if there is a specific fault in its construction (conditions of severe exposure or high internal humidity are an exception to this rule) or if the principal source of dampness is neither rain penetration nor condensation. Because there is a cavity, and with dry lining there are two cavities, water can run down inside the cavities and appear some distance away from where it originally entered. Leaking parapets, plumbing leaks in hollow concrete floors and rain penetration in upper storeys can show as damp patches below. Whether its appearance and development is related to wet weather or not, this type of defect is invariably caused by something bridging the cavity. Cavity ties can have been set to slope towards the inside leaf so that water penetrating the outside leaf tracks across the ties to the inside of the building. Mortar droppings may have lodged on the tie to form a bridge for water across the cavity. Routes for water to cross the cavity may also be created by a build up of mortar droppings at the bottom of the cavity (see 5.2.4) or by rubbish left in the cavity (yes – donkey jackets, whole bricks and loose lengths of DPC have all been found in cavities).

Where parts of the wall have an increased U value, damp patches may form. Columns or nibs that cross the cavity and upstand or downstand edgebeams are vulnerable to surface condensation and may show as vertical or horizontal lines of dampness. Additional insulation, ventilation and heating may all be needed to cure patches of condensation – see 5.2.4.

Damp patches near windows are discussed in 5.2.4.

Damp patches near abutting walls and roofs are discussed in 5.2.13.

REFERENCES

BRE Defect Action Sheet 12: *Cavity trays in external cavity walls*

BRE Digest 297: *Surface condensation and mould growth in traditionally built dwellings*

Photo 1 External leaf of cavity removed showing gaps in foam filling that can collect and channel water.

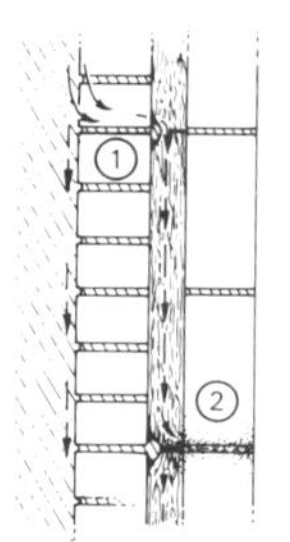

1 Water directed into the batts by mortar extruded from facework

2 Mortar obstruction at batt joint compresses batt towards inner leaf. As it flows between laminations, water is pushed closer to the inner leaf. At next batt joint, even slight ponding will allow water to reach inner leaf and cause dampness

Some routes for water penetration across a cavity filled with mineral fibre insulation batts

Figure 1 Mortar extruded from facework, and mortar obstructions that compress insulation batts, allow water to reach inner leaf.

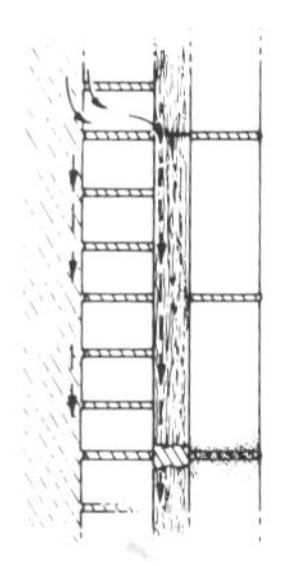

Capillary action through mortar obstructions

Figure 2 Capillary action through mortar obstructions in filled cavity.

1 Water leaks through outer leaf into cavity, mainly at perpends.

2 Wall tie drip sheds water into the cavity.

3 Displaced insulation board projects into the cavity and catches drips from above. Water runs to the inner leaf and causes a succession of damp patches.

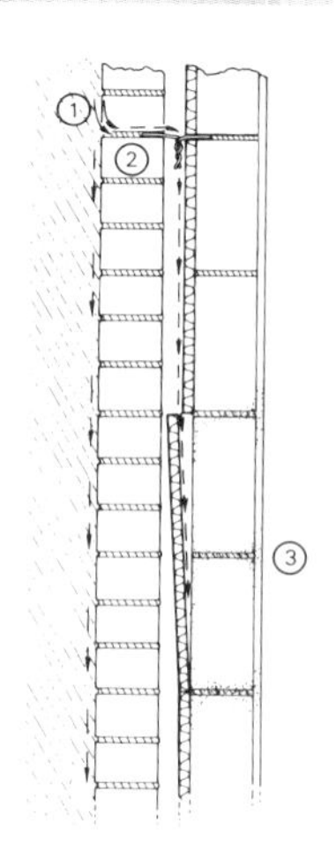

Figure 3 Displaced insulation board providing a water penetration route.

5.2.6 EXTERNAL WALLS AND ADJACENT PARTITIONS. DAMP PATCHES (AWAY FROM WINDOWS)

The patches may appear in any position (the defects described here relate to cavity walls with thermal insulation in the cavity).

Symptoms

Damp patches showing on the inside of external walls and sometimes extending back along abutting internal walls. The patches may run into each other giving the impression of an overall damp wall. They are more pronounced after wet weather. If the patches are near windows see 5.2.7.

Investigation

Establish the wall construction and type of cavity fill by cutting out one brick from the external leaf (the cavity will be more accessible if a corner brick is removed).

Note the location and extent of the damp patches and whether they change over a period of 6 months.

Note the dimension of the residual cavity with partial fill construction.

Note how the cavity insulation is fixed and jointed, whether the joints between cavity batts are clean of mortar snots and whether the batts are butted at corners. Note also whether there are cut sheets or batts and, if so, how these are fitted.

Assess the exposure of the wall as recommended in BS 5618.

Note, if possible, any fissures or unfilled areas in foam filled cavities.

Check whether patches correspond with cold bridges across the cavity e.g. lintels, downstand and upstand beams etc. – if they do see 5.2.3 and 5. Further investigation is likely to be necessary where dampness shows inside. If possible, a section of the damp inner leaf should be removed. (A 450 × 450 mm hole can normally be taken out of the inner leaf without endangering the structure, provided it is not in a critical position e.g. under a beam or close to an opening.) Monitor the presence of water on exposed surfaces in the cavity.

Diagnosis and cure

As a general rule, faults in cavity fill are due to faulty workmanship. Even where materials have failed the fault can be due to using the wrong mix of foam or filling the wall in unsuitable conditions. PSA restrict the use or type of cavity insulation in prescribed circumstances. There have been many failures and when faults occur they are sometimes difficult to put right with the confidence that the problem will not reappear.

These faults are all versions of the same defect. In one way or another the cavity insulation provides a bridge for water to cross the cavity to the inner leaf. If it is an isolated fault, for example, where a partial fill insulation board has become displaced and is touching the other side of a narrow residual cavity, it may be possible to refix the displaced board and cure the defect. But when the problem is more widespread, for example where urethane foam fill has developed fissures that act as water channels, the cavity insulation may have to be completely removed. This can either be done with a chemical spray that dissolves the foam or by breaking up the foam into small pieces with special equipment inserted between the rows of wall ties and then clearing the pieces through holes at the bottom of the wall. In order to restore the U value of the wall after removal of the insulation, external insulation, under a render coating or an insulated lining is likely to be more satisfactory than replacing the foam filling in the cavity.

REFERENCES

BRE Defect Action Sheet 17: *External masonry walls insulated with mineral fibre cavity width batts: resisting rain penetration*

BRE Defect Action Sheet 79: *External masonry walls: partial cavity fill insulation*

BRE Digest 236: *Cavity insulation*

BRE Digest 277: *Built-in cavity wall insulation for housing*

National Cavity Insulation Association and others *Cavity Insulated Walls – specifiers guide*

BS 5618:1978 Code of practice for *thermal insulation of cavity walls (with masonry inner and outer leaves) by filling with ureaformaldehyde foam.*

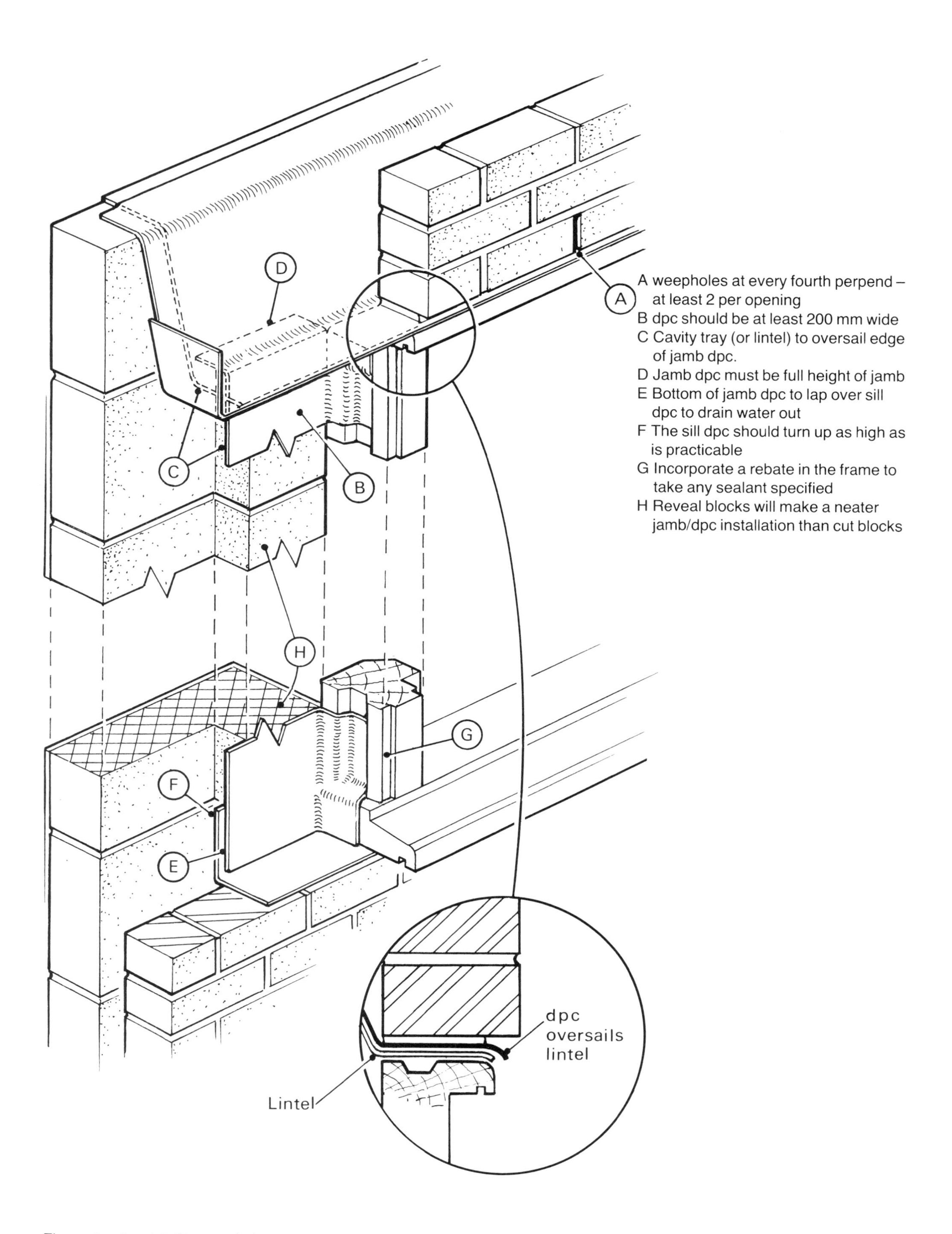

Figure 1 dpc detailing at window surround.

5.2.7 EXTERNAL WALLS. DAMP PATCHES CLOSE TO OPENINGS

Dampness showing on the inner face of an external wall. The defect described relates principally to masonry cavity walling but faults similar to some of those described here occur in solid masonry walls, large precast concrete panel walls and timber framed walls.

Symptoms

Dampness within 300 mm of a window opening. The damp may spread beyond 300 mm, but if it does it is also likely to extend up to the window. The dampness is associated with rain or with cold and wet weather. The defect may be due to a number of different causes, or to a combination of causes. In each case the symptoms are slightly different and are discussed with the alternative diagnoses below.

Investigation

Record the location and extent of the dampness and, if emergency action is not required, note any changes over a period of 6 months.

Establish the form of wall construction including the position of damp proof courses, insulation, damp proof trays, cavity closers, solid columns, fire stops and lintels.

Note the material and construction of the windows (and external doors). Are the frames hollow and are the insides of hollow sections drained externally? Where does condensation on the glass drain away? Are the windows double glazed? Are they draught proofed?

Diagnosis and cure

The defect is either caused by rain penetration through faulty construction of the window, or it is caused by condensation on the window surround. In either case the position and the spread of dampness should give a clue to its source. Common faults and their symptoms are described below.

- Faults in the window surround showing as damp patches, developing and spreading quickly in rainy weather. They are generally related to displaced, damaged or missing DPC's and to sills that do not protect the wall below, i.e. too little projection, ineffective drip, poor seal between window and sub-sill or below the sill. The most common faults with damp proof courses are:

  DPC's missing at jambs or under sills

  Jamb DPC not lapped over wall DPC

  Jamb DPC's displaced or only extending for part of the window height

  DPC tray over lintel obstructed by mortar, not provided with weepholes or not extending beyond cavity closer below or perhaps not provided with stop ends in an insulated cavity.

- Window leaks will almost certainly show on the surface of the internal window sill if rain has been driven onto the window. They may also show as damp patches below the sill where the leak has soaked into the plaster.
- Condensation on the window or frame shows in a similar way to window leaks (dampness on the internal sill and damp patches below), but it does not develop as quickly or in the same conditions. It is invariably associated with cold wet weather. Condensation on metal frames can show as a stained margin where a plaster surface meets the metal frame.
- Cold bridges occur where a section of external wall has a higher thermal conductivity than the adjacent or surrounding walls. Lintels, columns, cills solid underwindow panels etc. can all form cold bridges. In cold weather the building loses heat more rapidly through the cold bridge than through the rest of the wall. Consequently, the internal wall surface against the cold bridge is cooler than the wall surfaces in general, and condensation is more likely to form on this cooler surface. A cold bridge is likely to be revealed as a damp wall surface and as spots of mould that develop fairly slowly in cold weather. This will occur when there are humid conditions in the rooms and condensation is forming on windows.

Condensation will often contribute additional dampness where the initial dampness was caused by a leak or a poorly detailed DPC.

Isolated faults in the window surround will require rebuilding. Sections of wall can be opened up and DPC's inserted or repositioned to shed water externally. Where the faults are more widespread and can be seen at most of the windows, a more general solution such as externally applied cladding may be a cheaper solution.

Leaks in the window itself can be dealt with in a number of ways depending on the specific fault, see chapter 6.

5.2.7 EXTERNAL WALLS. DAMP PATCHES CLOSE TO OPENINGS

continued

Condensation may be curable by improvements in insulation, ventilation and heating. It may be feasible to drain condensation on the window through weepholes in the frame or by fitting specially shaped extruded aluminium sections along the bottom of each glazed light. Condensation on the inside of windows is often much worse during the first year or so of a building's use because the wet materials used in construction are still drying out.

REFERENCES

BRE Defect Action Sheet 12: *Cavity trays in external cavity walls: preventing water penetration*

BRE Defect Action Sheet 98: *Windows – resisting rain penetration at perimeter joints*

National Cavity Insulation Association & others *Cavity insulated walls: specifiers guide*

BMCIS Design/Performance Data External Walls 1. 14 *Damp penetration between concrete cladding panels.* 15 *Inappropriate sealant to wall panel joint*

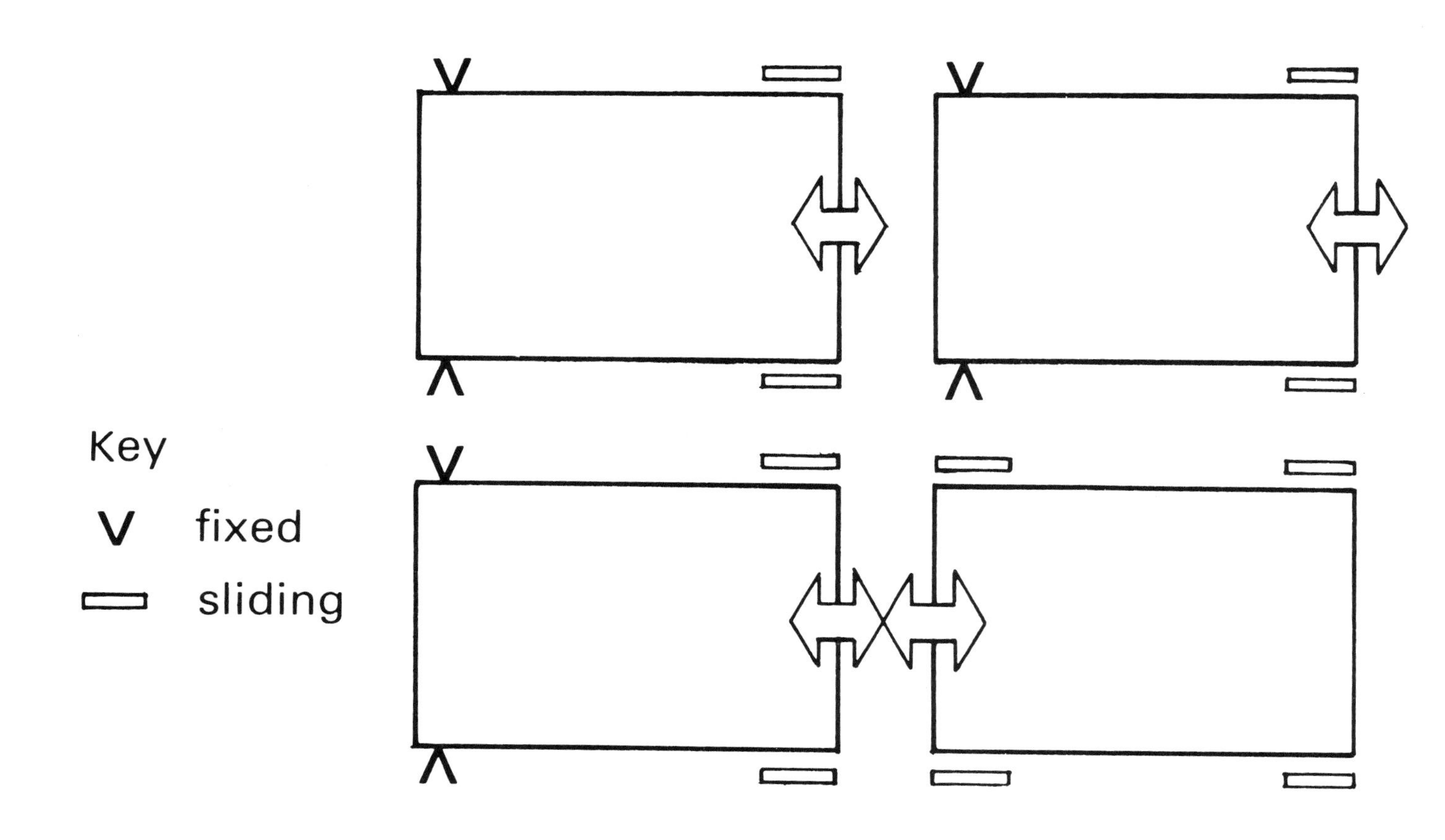

1. Movement of two adjacent panels may be concentrated at one joint and then lead to the failure of the sealant (see also 5.3.23 on sealant failure in general).

5.2.8 EXTERNAL WALLS AND ADJACENT PARTITIONS. DAMP PATCHES

The patches may occur in any position. The defect described here relates to large concrete panel walls.

Symptoms

Damp patches showing on inside surfaces and extending to the maximum after rain, then drying back gradually in good weather. They are more pronounced on the side of the building that faces driving rain. They may also show as a line of dampness above skirting level (see 5.2.3).

Investigation

Record the extent of the dampness and any changes in the area affected over a period.

Establish the form of construction. Where are the cavities in the construction? What is the profile of the joints? How are the joints sealed? Where is the vapour barrier?

Examine the concrete panels for signs of damage or displacement. Is there any cracking or spalling of the panel edges? Are the panels aligned: along horizontal joints; along vertical joints; with the surface in line with the external surface of the wall? Do the joints taper? What is the width of the joints? What variation is there in the joint width?

Examine the seals. Are they: missing; broken; cracked; coming away from panels; showing signs of distortion?

Look into the history of repairs to the building. Has any joint of the façade been repaired or strengthened since it was constructed? Have the seals been replaced?

Diagnosis and cure

The defect is rain penetration through the concrete cladding, possibly with a contribution of damp from condensation (see 5.2.2). The investigation may reveal obvious routes for rain penetration, in which case repair should be comparatively easy (see references for advice on sealants and concrete repair). If there are isolated patches with no obvious entry point, the water may be running down cavities within the construction and will have to be traced by removing sections of the inner skin or by endoscope examination (see chapter 4). Widespread damp patches after rain in multistorey construction may have to be tackled by overcladding (or overcoating) the whole building, as scaffolding costs prohibit frequent minor repairs.

Where panels are damaged or displaced, or damp penetration is extensive, structural engineering advice should be sought and the attachment of the panels must be checked (including corrosion of the fixings).

REFERENCES

BRE Defect Action Sheet 97: *Large concrete panel external walls: re-sealing butt joints*

BRE Digest 217: *Wall cladding defects and their diagnosis*

BRE Digest 235: *Fixings for non-load bearing precast concrete cladding panels*

BRE: *Overcladding external walls of large panel system dwellings*

BRE information paper 9/87: *Joint primers and sealants: performance between porous claddings*

BRE information papers 8, 9 & 10/86: *Weatherproof joints in lartge panel systems, identification and typical defects, Remedial measures, investigation and diagnosis of failures*

B S 6213: 1982 *Guide to selection of constructional sealants*

Photo 1 In a curtain walling system water may run along concealed channels before showing on internal surfaces.

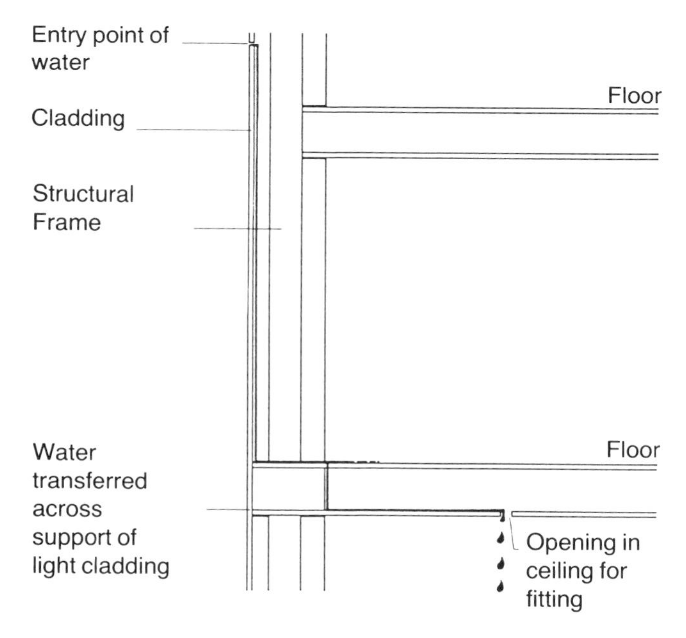

Figure 1 Section showing distance from entry point to where the defect is apparent.

Joint not filled, adhesion of sealant poor, no side packing. In exposed situations the wind 'pumping' the glass can help the water to pass round the end of the glass.

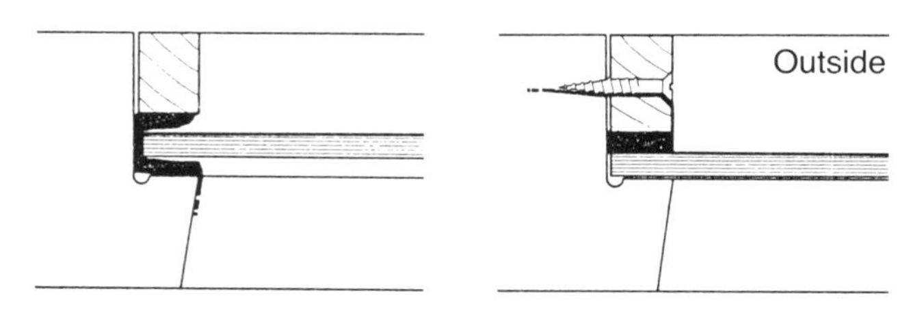

Figure 2 Water finds its way down bead fixing screw holes and round poorly fitted glazing

5.2.9 EXTERNAL WALLS AND ADJACENT PARTITIONS. DAMP PATCHES

The patches are in any position and are associated with rainfall. The defect described here relates to curtain walls.

Symptoms

Damp patches in any position but often at low level on the inside faces of external walls and abutting partitions. The area of dampness is greatest after rain but sometimes not for some hours after it has rained. It shows on the side of the building exposed to driving rain. It may be more marked on the elevation most exposed to sunlight. It may spread along at low level (see 5.2.3) and may appear some distance from the point of entry, even several floors below.

Investigation

Record the extent of the damp patches and how they vary over a period.

Establish the form of construction. What materials were used to seal the glazing and panels to the metal surround? How are the solid panels constructed? Is the glazing double or single? How is the framework fixed? Where are the firestops and how are they formed? Where are the cavities? Is there an effective vapour barrier? Inspect the glazing and panels – are there any cracks? Are there signs that the beads, gaskets or sealants around the glazing and panels are deteriorating?

Inspect the joints between metal frames and inspect any joints provided to take up movement either horizontally or vertically. What type of sealant has been used? What is its condition?

Remove a section of the internal lining in a damp area and examine the fixings of the curtain walling for signs of corrosion.

The suspect section of wall can be sprayed with a hose to discover where the water is getting in or an elaborate in-situ test can be used to put a section of wall under pressure.

Diagnosis and cure

The defect is caused by rain penetration through the joints or through cracked panels in the curtain walling. It may be exaggerated by surface or interstitial condensation – (see 5.2.2). The most likely source of penetration of water is through ineffective and deteriorating gaskets or sealants around glazing and panels. Both gaskets and sealants are likely to deteriorate faster than the frame or the glazing and infill panels – particularly in strong sunlight or where movement puts them under stress. Plastics used in gaskets and sealants may be incompatible with adjacent materials such as bituminous DPC's (see reference). Different curtain walling systems are subject to different defects. The manufacturers may be aware of problems that have developed in systems that they have sold. A record of the investigation could be sent to the manufacturer for comment.

Strong winds can cause flexing of glazing or panels which opens up perimeter joints.

Wholesale replacement of the sealants or of the whole curtain walling system may be the only solution once leaks start to appear because close periodic inspection may not be feasible.

REFERENCES

BMCIS Design/Performance Data: 1 *External walls*, 6 *Reaction between bitumen DPC and polysulphide sealant*, 10 *Curtain wall/floor slab detail leak*, 16 *Wind pressure forcing moisture through cladding*, 25 *Failure of cladding panel joints*, 41 *Stress cracking of glass wall panels*

BS 6213: 1982 *Guide to selection of constructional sealants*

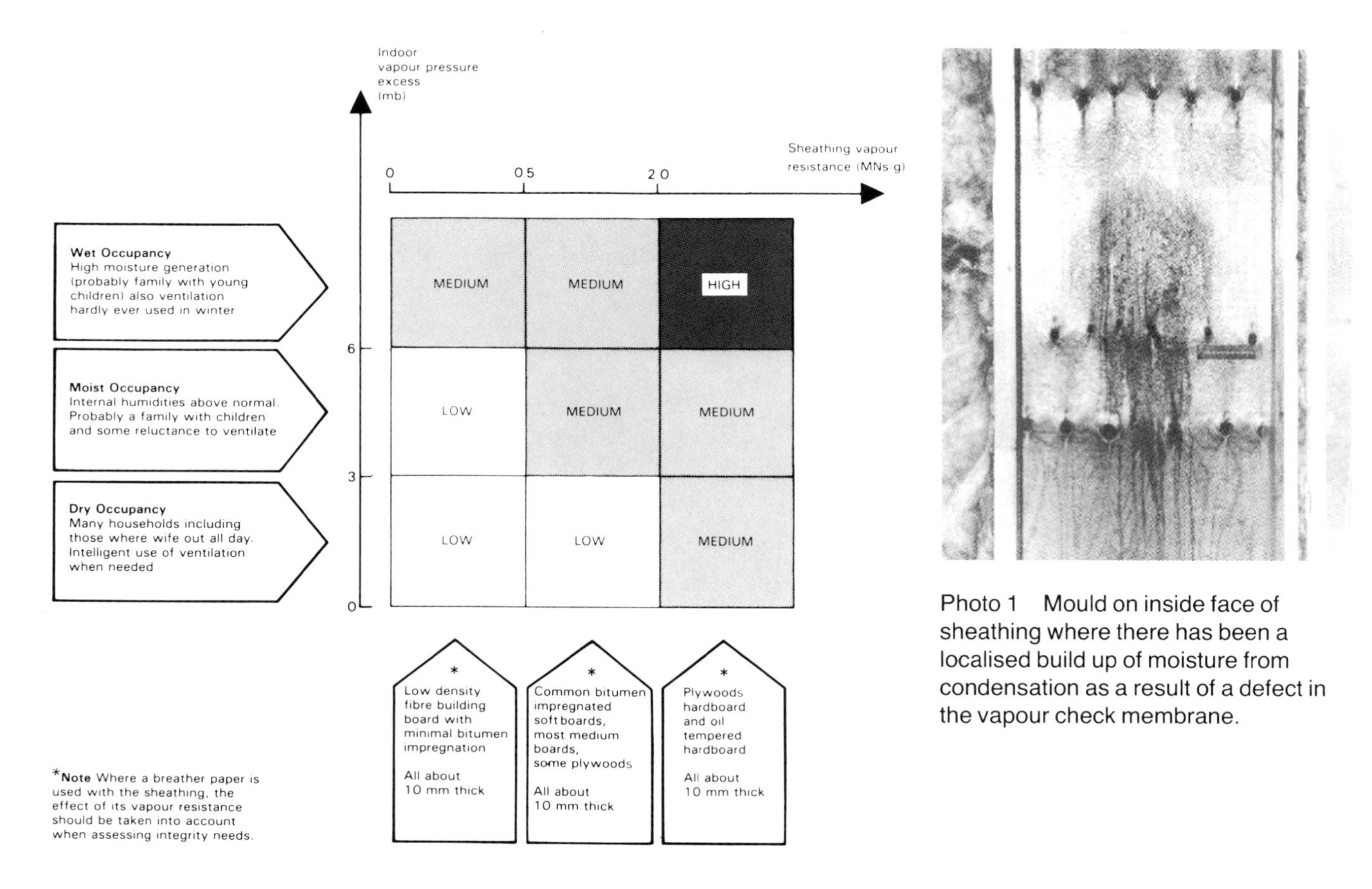

Photo 1 Mould on inside face of sheathing where there has been a localised build up of moisture from condensation as a result of a defect in the vapour check membrane.

Figure 1 Requirements for sheathing vapour resistance related to occupancy and materials.

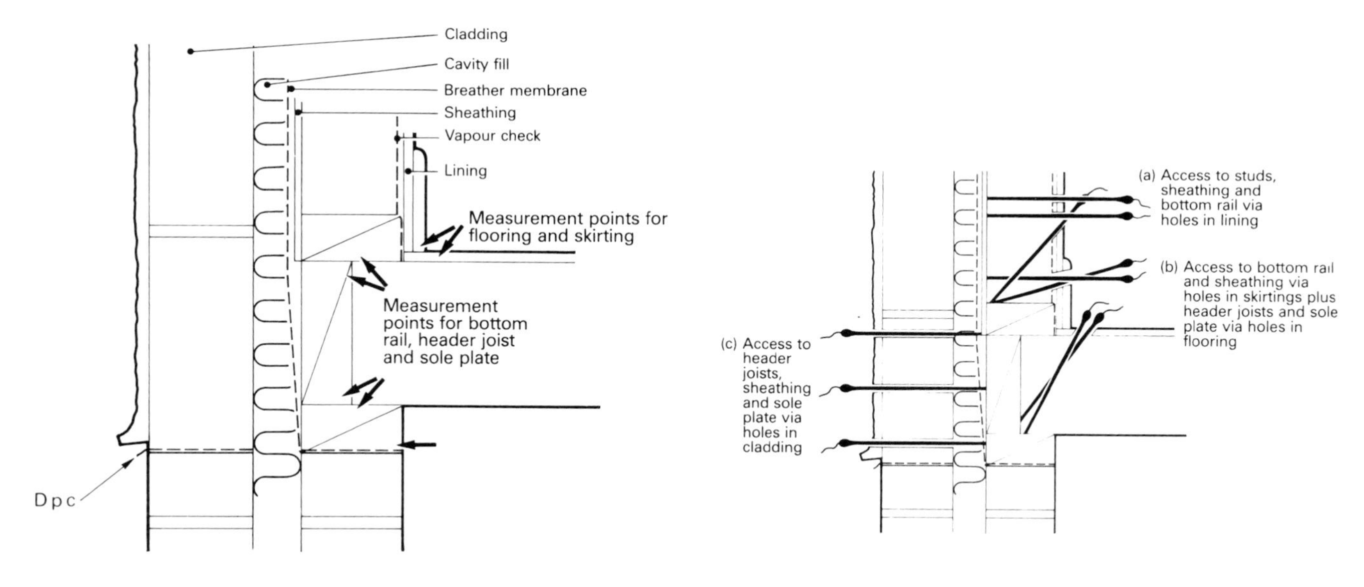

Figure 2 Measurement points to be surveyed when investigating moisture content of timber framed dwellings.

Figure 3 Position of holes to be drilled to take measurements at ground or first floor levels.

5.2.10 DAMPNESS IN FRAME MEMBERS OF EXTERNAL WALLS

The dampness may show as stained or mouldy patches, or it may only be discovered during a condition survey. The defect relates to timber framed external walls.

Symptoms

Dampness in timber frame members (over 20% moisture content in January and February or over 18% at other times of the year). The dampness may be in one or two places only or generally throughout the framework of the external wall.

Investigation

Record any visible signs of dampness and use an electrical resistance meter to establish whether other areas have high readings.

Establish the construction of the external wall. Is there a vapour check and if so what is it? The same applies to the sheathing and the breather membrane. Where is the insulation? Is the space between the studs filled with insulation? Has a cavity foam fill been used behind an external brick leaf?

If possible, remove a floor board where a timber floor abuts the external wall, and take readings of the moisture content of the exposed external frame members. Remove lengths of skirting and take readings of the moisture content at the bottom of vertical members where they abut the horizontal sole plates. The aim is to discover the dampest spot within the timber frame.

An alternative method of investigation, that reduces disruption of the buildings use, is to insert the probes of an electrical resistance meter designed to measure moisture content at the back of a cavity. The long probes are inserted through pairs of holes drilled in the internal lining of the timber frame (see chapter 4).

Examine the exterior of the wall for cracks, displaced or missing flashings, the presence of a DPC and anywhere that might allow water to enter the frame – see 5.2.3 and 5 for rising damp and cavity bridging.

Diagnosis and cure

This defect needs to be dealt with, even if it hasn't reached a stage that affects the building's performance, because it is potentially serious and could cause rapid deterioration of the structure. Localised dampness is likely to be due to rain penetration through the external wall due to a fault in the cladding, the flashings, the breather membrane or the windows. If the dampness is more general, it is probably caused by interstitial condensation due to holes in the vapour check, and to a sheathing that is relatively impermeable to vapour. Where there is a large hole in the vapour check, signs of dampness may show on the inner face of the sheathing facing the hole.

Dampness is likely to be most highly concentrated in sills and where the vertical members meet the horizontal members, because dampness runs down to this junction. The investigator should measure the moisture content at these points.

Isolated areas of dampness can be cured by repairs to the external cladding to stop rain penetration, or by sealing up any holes that have been discovered in the vapour check. If the vapour check appears to be the most likely cause of more general dampness, it may be necessary to provide additional heating and ventilation to dry out the house, and to maintain a dryer atmosphere within it. If there is dampness, and the cavity between the sheathing and the external cladding has been filled with insulation foam, rain penetration through defects in the foam is the most likely cause. Where dampness is widespread in this type of construction, the cavity insulation may have to be removed – see 5.20.6.

REFERENCES

BRE Information paper 21/82: *Moisture relations in timber framed walls*

BRE Information paper 1/85: *Surveying the moisture content of cavity filled timber framed dwellings*

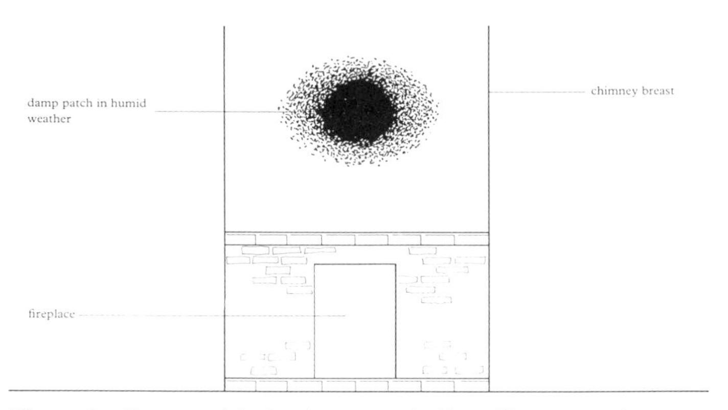

Figure 1 Damp patch due to concentration of hygroscopic salts derived from chimney soot.

5.2.11 CHIMNEY BREASTS AND WALLS. DAMP PATCHES

Dampness that appears on internal walls and internal faces of external walls. Typically it is seen in older buildings near flues, or at low level. The defect relates to masonry walls plastered direct.

Symptoms

The dampness may appear in any position on a chimney breast particularly in humid, but not necessarily cold weather. The same conditions will make dampness show more prominently at low level, or in rooms that have been flooded with sea water, or in rooms that have been used for storage of salty goods.

Investigation

Record the extent of the dampness and how it varies over a period of not less than six months. Is it related to rain or snow?

Investigate the previous use of the building. Has it been used for storage of salty goods? Has it been flooded by the sea?

Arrange for laboratory tests on a sample of the plaster from the damp patch to establish whether it contains an unusually high concentration of hygroscopic salts.

Diagnosis and cure

The defect is caused by concentrations of hygroscopic salts which have the ability to absorb moisture from the atmosphere. The investigation should reveal the source of the salts. On chimney breasts it is likely to be a build up of salts derived from soot in the flue that have migrated from the flue to the internal plaster surface. At low level it may be a symptom of rising damp (see 5.2.2) or of salt water flooding, or of salts from goods that have been stored against the wall. If the dampness appears only on a chimney breast after rainfall, it is more likely to be connected with rain penetration and faulty roof finishes. The laboratory test of the plaster sample should establish whether the problem is from salts or rain penetration.

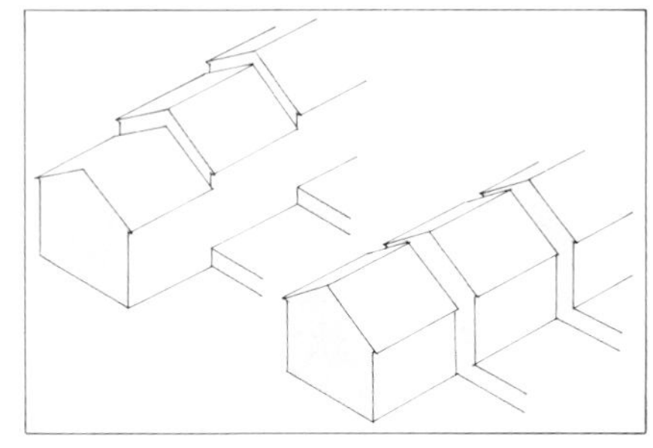

Figure 1 Roofs abutting cavity walls need careful detailing to prevent water in the exposed leaf of the cavity wall above from reaching the internal wall below.

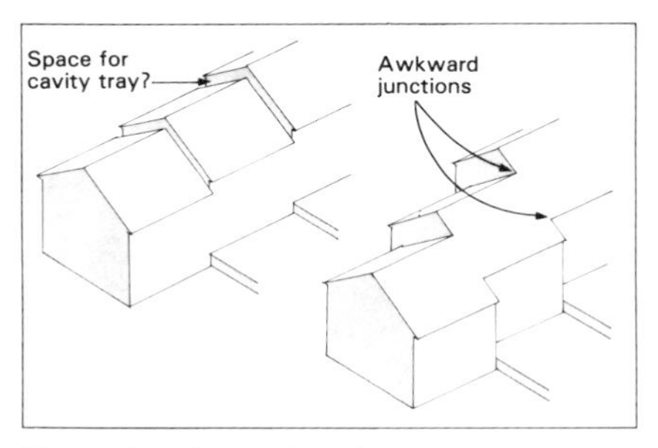

Figure 2 Some junctions are particularly difficult to waterproof for example if the roof step is less than 6 courses.

Photo 1 Dampness from brickwork showing through newly finished wall below a faulty flashing see BRE Defect Action Sheet 114

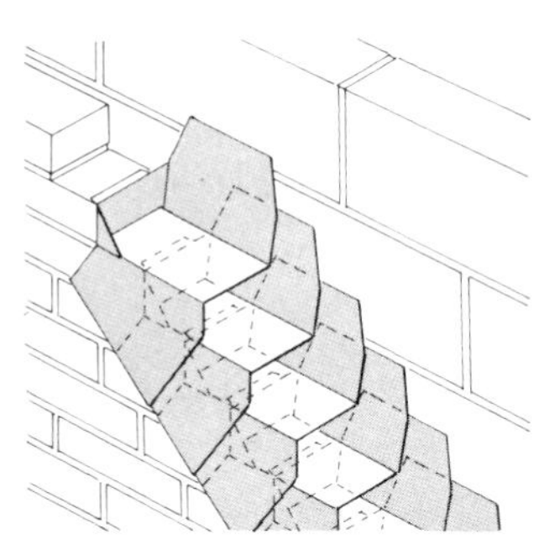

Figure 3 Trays may not need to be built into the inner leaf. They may have lead skirts that are bent up to allow the soakers to be fitted below them.

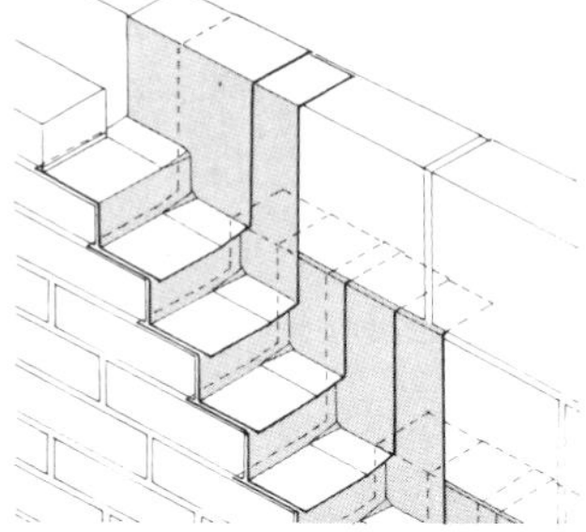

Figure 4 Other trays are built into the inner leaf and a stepped cover flashing must be tucked in below cavity tray.

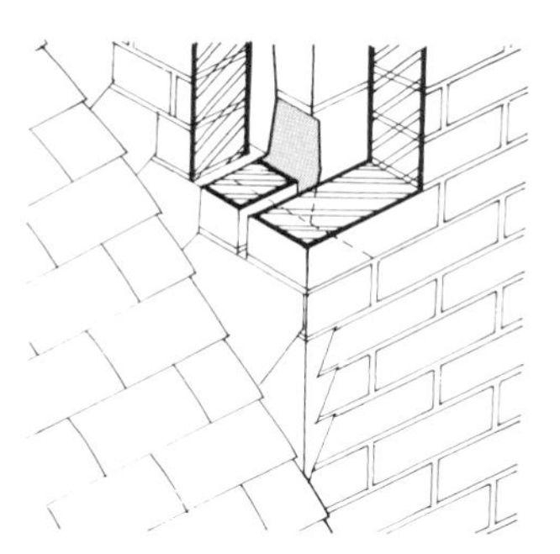

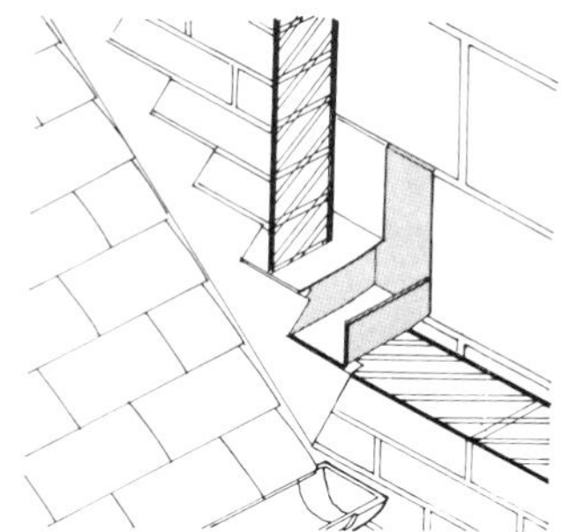

Figures 5 & 6 Weep holes are needed to drain the bottom of the cavity tray.

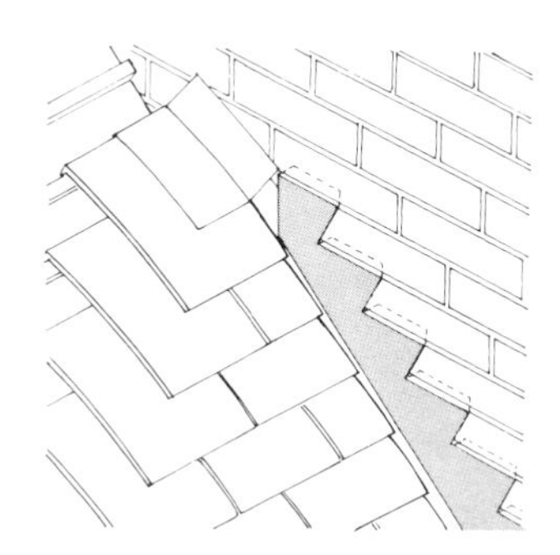

Figures 7 & 8 Lead soakers should have a minimum 75 mm upstand. Stepped flashings should be set at least 25 mm into the course joints.

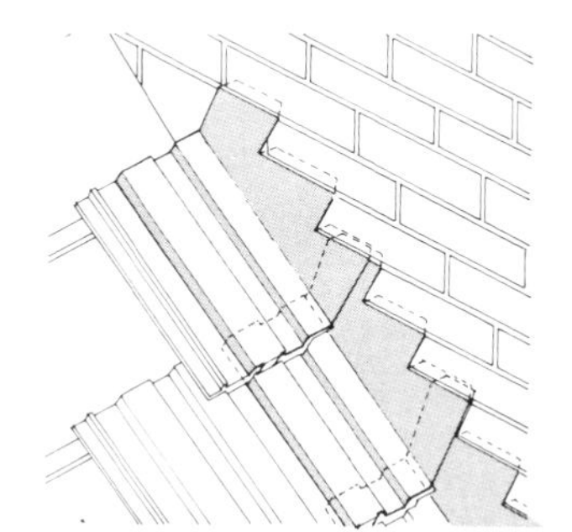

Figure 9 Where interlocking tiles will not accept soakers a wider stepped flashing should be dressed over the tiles.

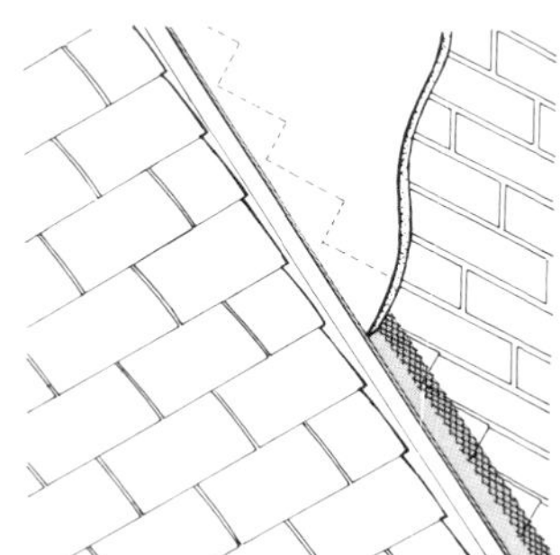

Figure 10 Below rendered walls a raking rendered stop should be specified over the flashing.

5.2.12 EXTERNAL WALL. DAMP PATCHES CLOSE TO ROOF ABUTMENTS

Dampness showing on inside face of wall at or below roof abutments. The defect relates to masonry cavity walls.

Symptoms

The damp patches appear after rain or long periods of wet weather. They may show inside the roof space of the lower roof or on the inside walls of the spaces below the abutment on either side. They may only show after driving rain (or drifting snow) has occurred from a particular direction. The same defect can cause dampness at the base of the wall below the roof abutment.

Investigation

Record the extent of the dampness and note any changes over a minimum period of 6 months.

Establish the construction of the wall and abutting roofs, and the precise relationship of the dampness to the profile of the abutment. Does the cavity extend down to the base of the wall? Is there cavity insulation, and if so, how far does it extend? What material has been used for cavity trays and flashings? Are the joints in horizontal cavity trays lapped and sealed? Do the stepped flashings extend through to the inner leaf? How high is the flashing at its lowest point above the abutting roof?

Note the condition of the roof. If it is a flat roof, are there signs of deterioration of the roof? Check for springiness or sponginess.

Diagnosis and cure

The most likely cause of this defect is attributable to faulty installation of the DPC or flashing at the abutment. Condensation may be the cause, particularly if the dampness shows on the inside of a section of wall above the abutting roof. Rising damp and faulty tanking may be the cause if the dampness shows near the base of the wall, particularly if the ground floor levels of the abutting buildings are different and the dampness is more marked in the lower building. The same symptoms may have been produced by overflowing appliances or plumbing leaks in the higher of the abutting buildings. A further possibility is rain penetration through the exposed section of wall above the abutment (see 5.2.5, 5.2.6 and 5.2.13).

Some clues as to the source of water may be gleaned from examination, records and an endoscope examination of any unfilled cavity. However, if the defect relates to faulty flashings it is likely that a section of the inner leaf will have to be removed to examine the detailing and workmanship of the DPC's and flashings. Examination from outside is feasible but, if the examination is from the inside, a hose can be played on the wall outside to see where the water gets in.

Correcting the detailing of the flashing can be very expensive. It may be easier to fix a waterproof external cladding to the section of wall above the abutment.

REFERENCES

BRE Defect Action Sheet 12: *Cavity trays in external cavity walls: preventing water penetration*

BRE Defect Action Sheet 34: *Flat roofs: built up bitumen felt – remedying rain penetration at abutments and upstands*

BRE Defect Action Sheet 35: *Substructure DPC's and DPM's*

BRE Digest 77: *Damp proof courses*

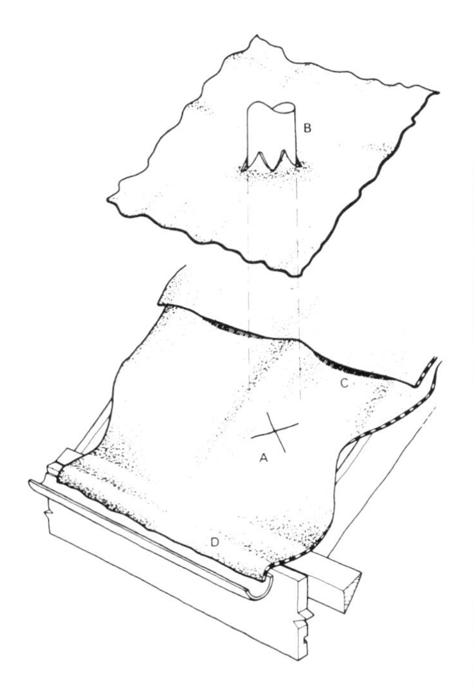

Photo 3 & Figure 1 Sarking felt defects can cause dampness in walls. The felt must be neatly fitted round pipes and into gutters so that water is not directed into the building.

Photo 1 Gutter leak showing externally.

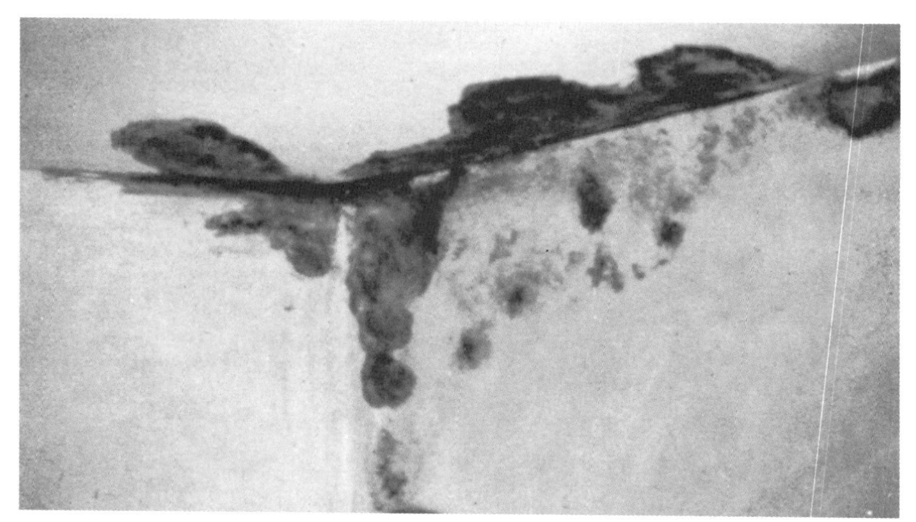

Photo 2 Internal dampness at high level may be from a number of causes – see text.

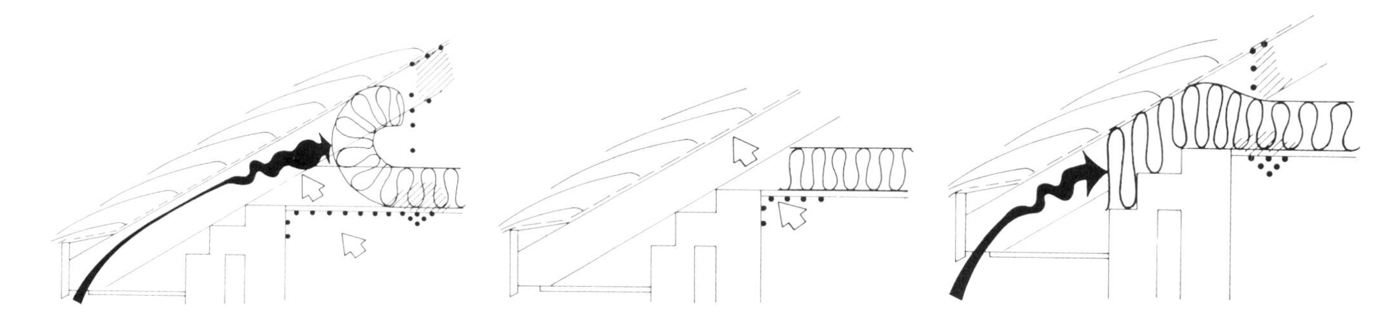

Figures 2, 3, & 4 Faulty fitting of insulation in a pitched roof eaves may cause dampness to show at the junction of the wall and the ceiling.

5.2.13 EXTERNAL WALLS. DAMP PATCHES AT HIGH LEVEL BENEATH A PITCHED ROOF

The damp patches appear at high level in rooms below a pitched roof. The defects described here relate to all materials normally used for wall or roof construction.

Symptoms

Dampness that appears after rain or in cold damp weather. It may also show on the ceiling near a wall or on the inside of gable walls within a roofspace. Where there is an internal gutter and a parapet wall see 5.2.14.

Investigation

Record the extent of the dampness and how it varies over a period of 6 months to 1 year.

Establish the construction of the roof at the perimeter. Is there insulation? Is there sarking? Is there adequate ventilation to the roofspace?

Note the type of roof covering and the pitch of the roof. Is the sarking felt carried out over the verge and over the gutter? Are there any visible defects in the roof coverings or flashings (use binoculars for multi-storey buildings)?

Establish the construction of the wall. Are both masonry units and mortar in good condition? Is there a cavity (and if so, is it filled)? Are there any signs of cracks or displaced masonry? Does the dampness show on the outside of the wall?

Diagnosis and cure

The most likely causes of dampness are:

- Condensation within the roof space running down to the eaves and soaking the top of the wall. This is likely to be seasonal (worse during the cold months) and there will be signs of water runs, drips in the roofspace or dampness in the roof insulation. The damp may also show on the ceiling. The cure here is to improve the natural ventilation of the roofspace.
- Leaks and rain penetration at the roof to wall junction, or in the wall just below the junction. Overflowing gutters, sagging or damaged sarking, split or cut felt, short and disturbed roof coverings or torching, defective flashings round chimney stacks. These defects will develop after heavy or continuous rain. The cure will not be difficult to specify once the fault has been found.
- Condensation within loose fill cavity insulation at the top of the cavity. Where warm moist air from rooms at lower levels can get into the cavity but cannot find its way past an effective seal at the top of the cavity, and where the walls just under the roof are cold enough to promote condensation. Investigation should show whether conditions are likely to produce this phenomenon. Insulation in the cavity will be damp. The cure is to ventilate the top of the cavity to the outside (peripheral vents just below eaves or verge level). It may also be possible to block the routes by which the warm damp air enters the cavity e.g. more effective sleeves for air bricks and ventilating fan outlets where they pass through the external wall.
- Surface condensation where the construction creates a cold bridge e.g.: solid walling above a cavity wall and behind a box eaves; on the ceiling perimeter (and top of the wall), where roof insulation does not extend to the perimeter; on steel and concrete lintels or beams that are exposed to cold conditions outside. The cure may be additional insulation on its own or combined with other anti-condensation measures such as improved heating and ventilation.

Plumbing leaks should be checked when dampness is not clearly rain related. In this position a damaged or displaced overflow from the storage tank can drip onto the top of the wall and create damp patches below.

REFERENCES

BRE Defect Action Sheet 4: *Pitched roofs: thermal insulation near eaves*

BRE Defect Action Sheet 6: *External walls: reducing the risk from interstitial condensation*

BRE Defect Action Sheet 9: *Pitched roofs: sarking felt underlay – watertightness*

BRE Defect Action Sheet 95: *Masonry chimneys DPC's and flashings – installation*

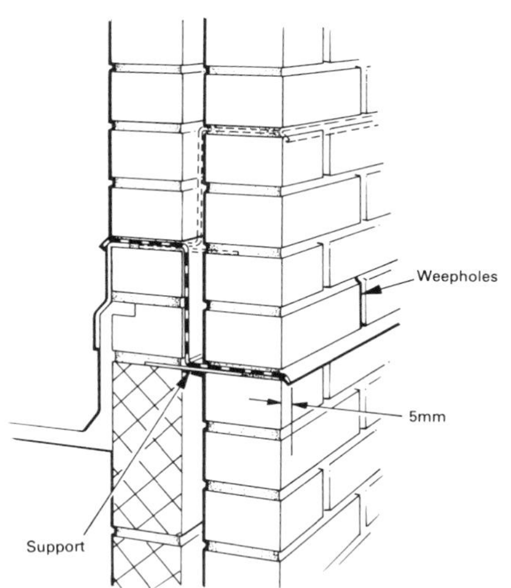

Figure 1 Method of forming an effective dpc tray (the joints must overlap at least 150 mm and be sealed). An alternative position for the dpc is shown that will avoid the risk of efflorescence on the outside face.

Photo 1 Damp staining from a defective parapet above (the vinyl wallpaper covers the dampness on the wall).

Photo 2 Defective asphalt upstand allowing water to penetrate wall below.

Photo 3 When a parapet gutter with a blocked outlet overflows into the wall extensive damage results.

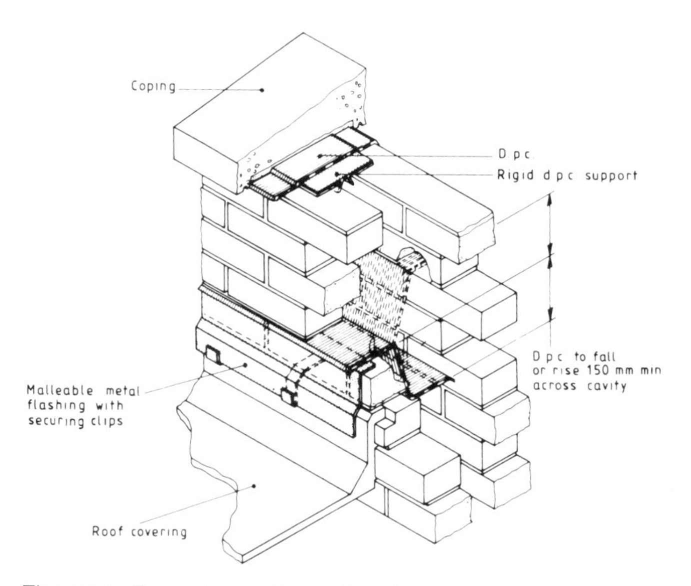

Figure 2 Properly constructed cavity parapet.

5.2.14 EXTERNAL WALLS. DAMP PATCHES BENEATH FLAT ROOFS AND PARAPETS

The patches appear on the inside of the external wall at high level, and may extend onto the perimeter of the ceiling. The defect described here relates to most forms of flat roof and to pitched roofs with internal gutter and parapet.

Symptoms

The dampness appears after rain or in cold weather. It shows as mould or stains on the plastered wall surface. These defects here relate to the perimeter details of roofs and to parapet walls. Roof leaks near the perimeter can produce similar symptoms – see chapter 8.

Investigation

Record the extent of the dampness and how it varies over a period of 6 months.

Establish the form of construction of the roof and the wall. Is there an internal gutter and if so is it insulated? Could it overflow into the building? Is there a parapet, and if so, does it have a cavity, DPC's or built in flashings? Is there a coping which overhangs and drips? What is the construction of the wall? Is there an insulated cavity? How is the cavity closed?

Does the occurrence of dampness relate to rainfall or to cold weather?

An endoscope examination of the roof space or cavities may give some clue as to the course of water (see chapter 4).

If gutter or roof leaks are suspected, sections of gutter and roofs can be dammed and flooded until a leak is found.

If damp proof courses and flashings are suspected, the wall can be hosed for several minutes and a section of the coping can be removed to examine the DPC's and masonry walls within the cavity.

Check the wall and parapet for visible defects e.g. cracks, loose or soft pointing, and any cracks or tears in the waterproofing of skirting upstands.

If there is a pitched roof with an internal gutter, check the ventilation of the internal roof space. Instruments are available to measure relative humidity within the roof space and at outlets (see chapter 4).

Diagnosis and cure

This can be a difficult defect to diagnose correctly. It is an exposed part of the building that is vulnerable to condensation and to rain penetration, as well as to cracking and thermal movement. In many cases more than one source of water contributes to the dampness. Isolating the principle source will be a problem. Testing sections of wall or roof with a hose (and dams) and watching the effects of these tests, or of heavy rain on the extent of the dampness, can give a clue to its origin. The parapet is particularly vulnerable with three faces exposed to temperature variation and to rain. There is also the possibility of condensation within the cavity of a parapet wall (see 5.2.13). In the long run it may be more economical to rebuild a section of parapet with the correct DPC's, copings, weepholes and cavity ventilation rather than trying to find faults which may be almost untraceable or only evident in rare weather conditions.

For insulation of flat roofs and gutters see chapter 8. For cracking of parapets see 5.1.16.

REFERENCES

BRE Defect Action Sheet 34: *Flat roofs: built-up bitumen felt – remedying rain penetration at abutments and upstands*

BRE Defect Action Sheet 71: *External walls: repointing – specification*

BRE Defect Action Sheets 106 and 107: *Cavity parapets – avoiding rain penetration*

National Cavity Insulation Association and others *Cavity insulated walls: specifiers guide*

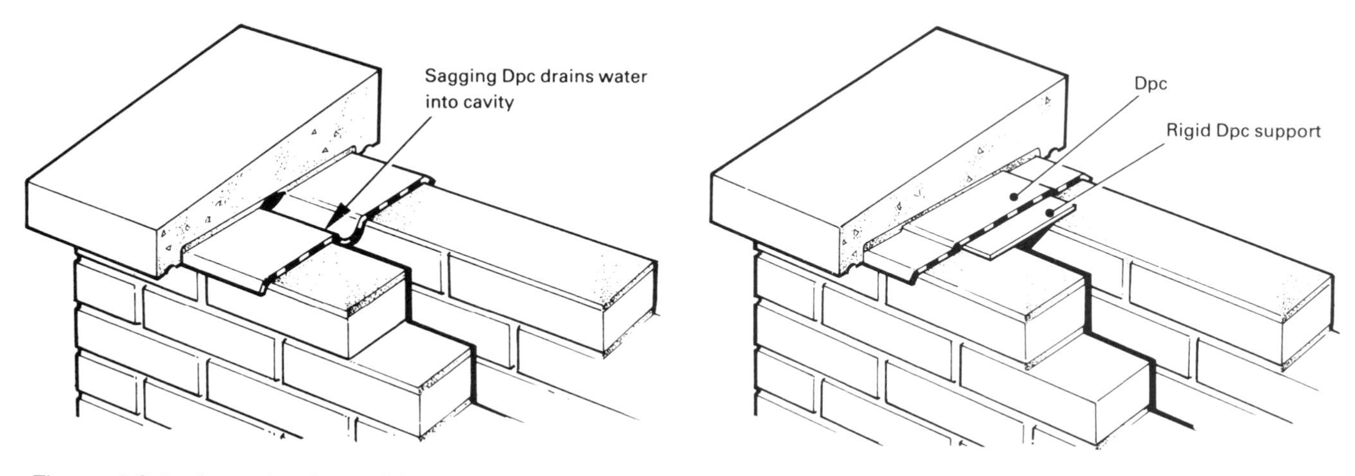

Figures 3 & 4 A sagging dpc and the remedy.

Photo 4 The dpc is unsupported and not wide enough, the cover flashing to the upstand is loose.

Figures 6 & 7 Faulty parapet edge detail to timber roof and correct construction.

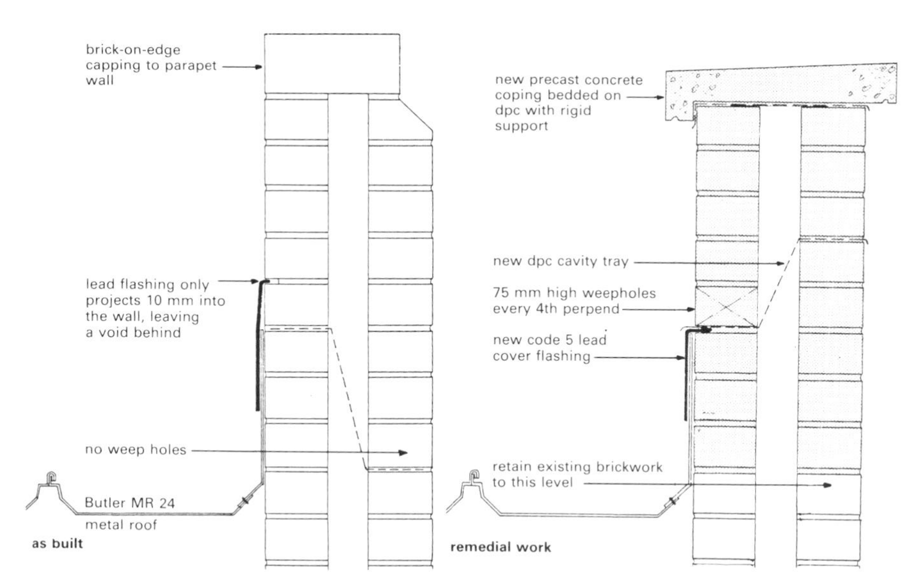

Figure 5 A parapet wall that failed and the remedial solution adopted. The remedial work also included vertical movement joints.

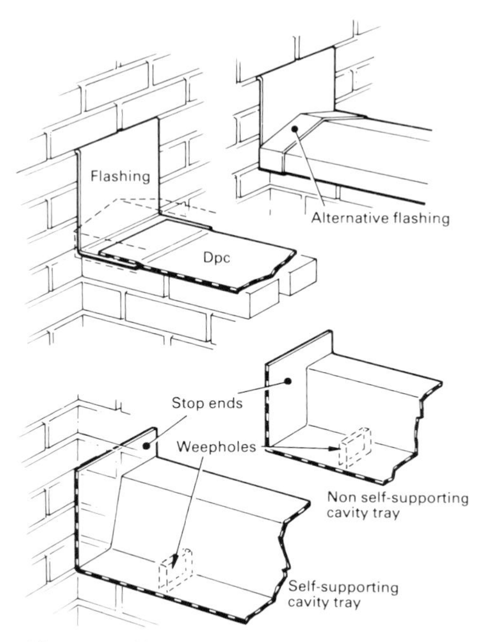

Figure 1 Parapet abutments showing stop ends to dpc's and cavity trays and the position of weepholes.

5.2.15 EXTERNAL WALLS. ISOLATED DAMP PATCH

The patch shows on the inside face where a freestanding wall abuts the external wall: the defect described here relates to masonry.

Symptoms

Where the end of a garden wall or parapet wall abuts an external wall there is a damp patch inside, and sometimes signs of dampness in the external masonry (e.g. efflorescence or discolouration). In most cases where this occurs, the freestanding wall is bonded into the external wall. The defect may show at any level from the top of the abutment to the floor.

Investigation

Record the location and extent of the dampness and monitor its development over a period of 6 months.

Establish the construction of the walls, and note the position of the damp proof courses. Is there a cavity, and if so, is it insulated?

Note the condition of the walls externally and any sources of dampness (such as a leaking gutter or dripping downpipe). Is the pointing at the abutment in good condition?

Diagnosis and cure

This defect occurs where water falling on the abutting wall can find its way through the external wall to the inside face. It is quite common where the freestanding wall is constructed using absorbant bricks with a brick coping, and is bonded into an external wall of solid construction. It can also occur with external cavity walls because rain splashes, and dampness from the saturated freestanding wall, will find faults in the cavity construction that would not show as defects under normal exposure. For investigation of damp penetration through cavity walls see 5.2.5 and 5.2.6.

This defect can normally be cured by improving the effectiveness of the coping and the flashing at the end of the coping. A surer cure is to also insert a vertical DPC at the end of the freestanding wall.

REFERENCES

BRE Defect Action Sheet 106: August 1987 *Cavity parapets – avoiding rain penetration*

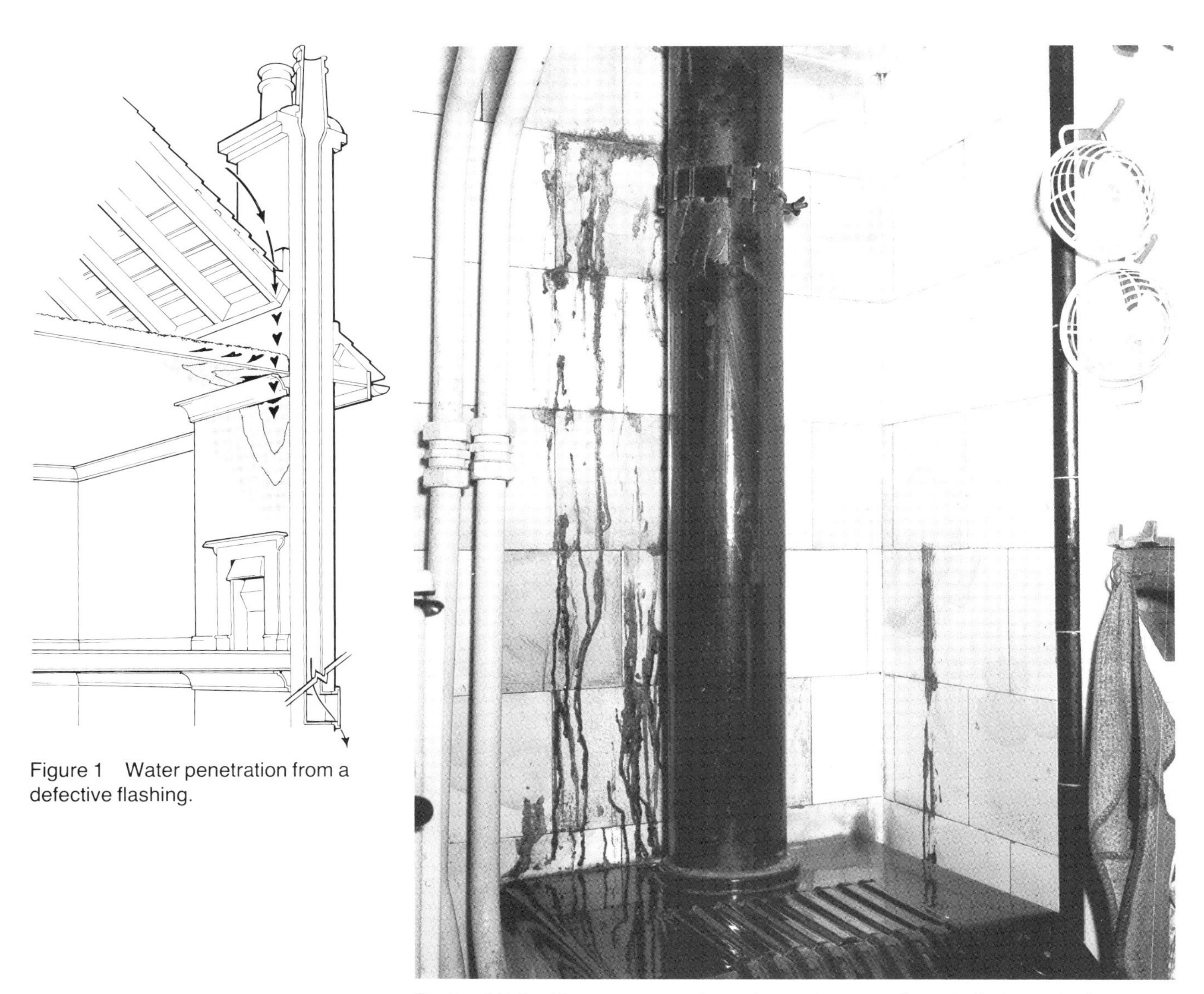

Figure 1 Water penetration from a defective flashing.

Photos 1 & 2 Flue gasses can leave brown tarry condensate that eventually appears on the wall surface.

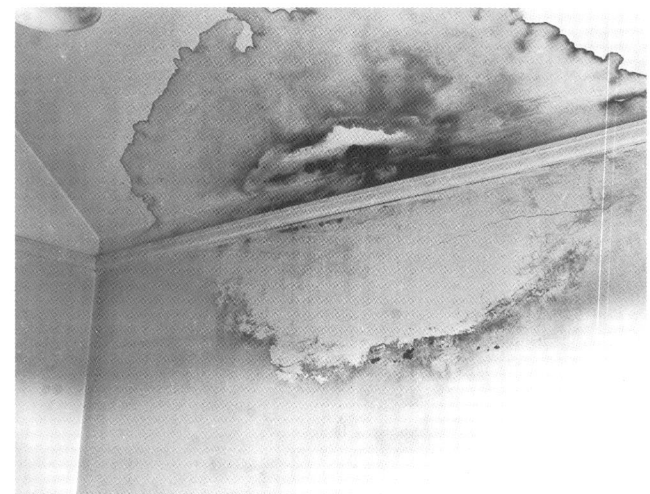

Photo 3 Chimney leaks spread the stain down the walls and across the ceilings.

5.2.16 CHIMNEY BREAST & WALLS AT HIGH LEVEL. DAMPNESS AND STAINING

This defect is usually associated with wet weather and, unlike the damp patches described in 5.2.11, it is likely to be just below roof level. It frequently affects the ceiling as well as the walls.

Symptoms

Dampness, often combined with light brown or dark tarry stains, showing internally just below eaves level where a chimney serving a fossil fuel appliance projects through the roof. The dampness may be related to either wet or cold weather.

Investigation

Establish the construction of the flue and its use.

Investigate potential sources of dampness such as gutters, haunching of flue terminal, plumbing leaks etc.

Examine the condition of the flue, flashings, DPC's, pointing and adjacent roof coverings.

Find out whether the flue has been used recently, and what the weather conditions were like before the dampness was first seen.

If the flue has been capped, check that appropriate ventilation has been provided.

Diagnosis and cure

The dampness may be caused by:

- faulty or displaced flashings
- rain penetrating the masonry of the chimney, particularly if the pointing is bad
- condensation inside the flue where it is cooled by external air. Flue gasses can leave a brown tarry condensate.

If the dampness always corresponds with rainfall, it is likely that the cause is rain penetration. If it only occurs when the flue is in use during cold weather, the cause is likely to be condensation.

A properly constructed new chimney should have a lining and damp proof courses or trays at three levels. (Under the capping and at a high and low level where brickwork is exposed as it passes through the roof.) Many old chimneys have no DPC and still function perfectly well. However, they do require flashings wherever water from the roof can flow or splash onto the masonry faces of the flue. The brickwork of the flue must be reasonably watertight, particularly if it is exposed to driving rain. (Repointing is dealt with in 5.2.4.)

Unused flues on external walls can be taken down to just above or just below roof level, and fitted with ventilators at top and bottom. Unused internal flues may be terminated below roof level. They should not be ventilated, but they should be capped, as there is a danger of their drawing warm humid air into the roofspace, and thereby promoting condensation.

REFERENCES

BRE Defect Action Sheet 93: Feb 1987 *Chimney stacks: taking out of service*

BRE Defect Action Sheet 94: Feb 1987 *Masonry chimneys: DPC's and flashings – location*

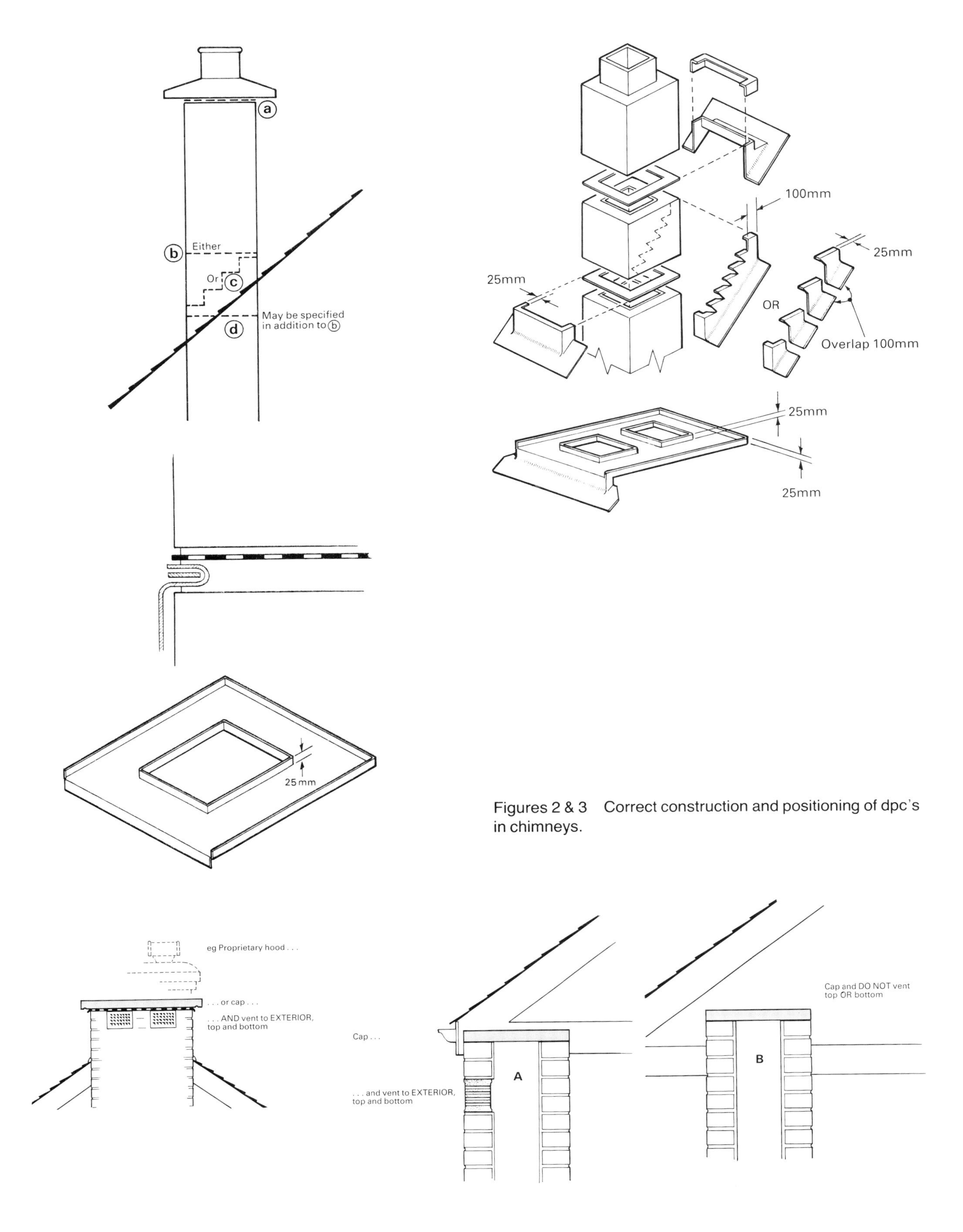

Figures 2 & 3 Correct construction and positioning of dpc's in chimneys.

Figures 4, 5 & 6 Capping of unused chimneys. Where there is an external face to the chimney it should be ventilated at top and bottom to prevent condensation or rain penetration accumulating.

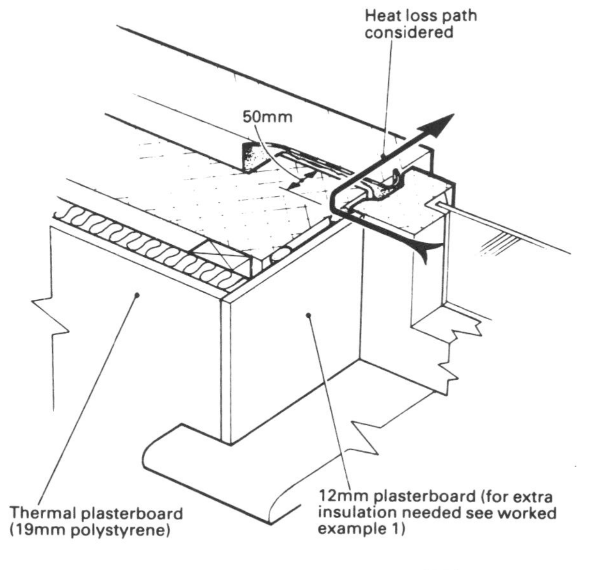

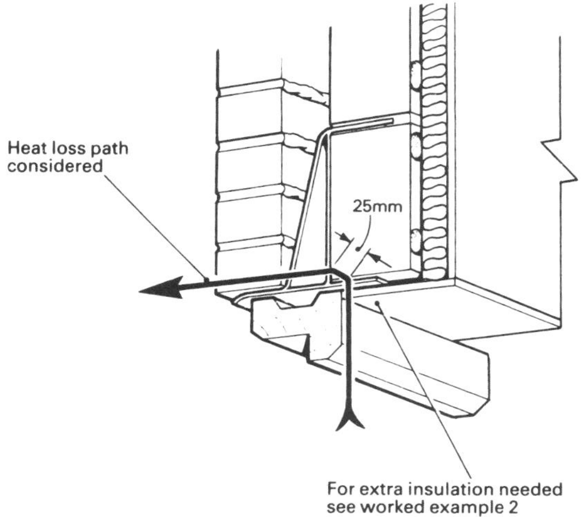

Figure 1 The general rule is that a U value of 1.2 or less is needed for surrounds to openings in housing to prevent condensation problems.

Example 1 – heat loss at the reveal

		Resistance
12 mm plasterboard		0.08
Inner leaf block (density 1100 kg/m^3), for 50 mm path length	50/100 × 0.29 =	0.15
Outer leaf brick		0.14
Sum of resistances		0.37
Total resistance needed for U-value of 1.2		0.65
'Resistance deficit'	0.65–0.37 =	0.28

Approximately 11 mm of EPS or 8 mm of polyurethane board would be needed to raise the total resistance of the construction of the reveal to 0.65 and give a U-value of 1.2

Example 2 – heat loss at window head

Sum the resistances as in Example 1:

		Resistance
12 mm plasterboard		0.08
Inner leaf block (density 1100 kg/m^3), for 25 mm path length	25/100 × 0.29 =	0.07
Air space		0.18
Outer leaf brick		0.14
Sum of resistance		0.47
Total resistance needed for U-value of 1.2		0.65
'Resistance deficit'	0.65–0.47 =	0.18

Approximately 7 mm of EPS or 5 mm of polyurethane board would be needed to raise the total resistance of the construction of the reveal to 0.65 and give a U-value of 1.2

Table 1 Worked examples referred to in figure 1.

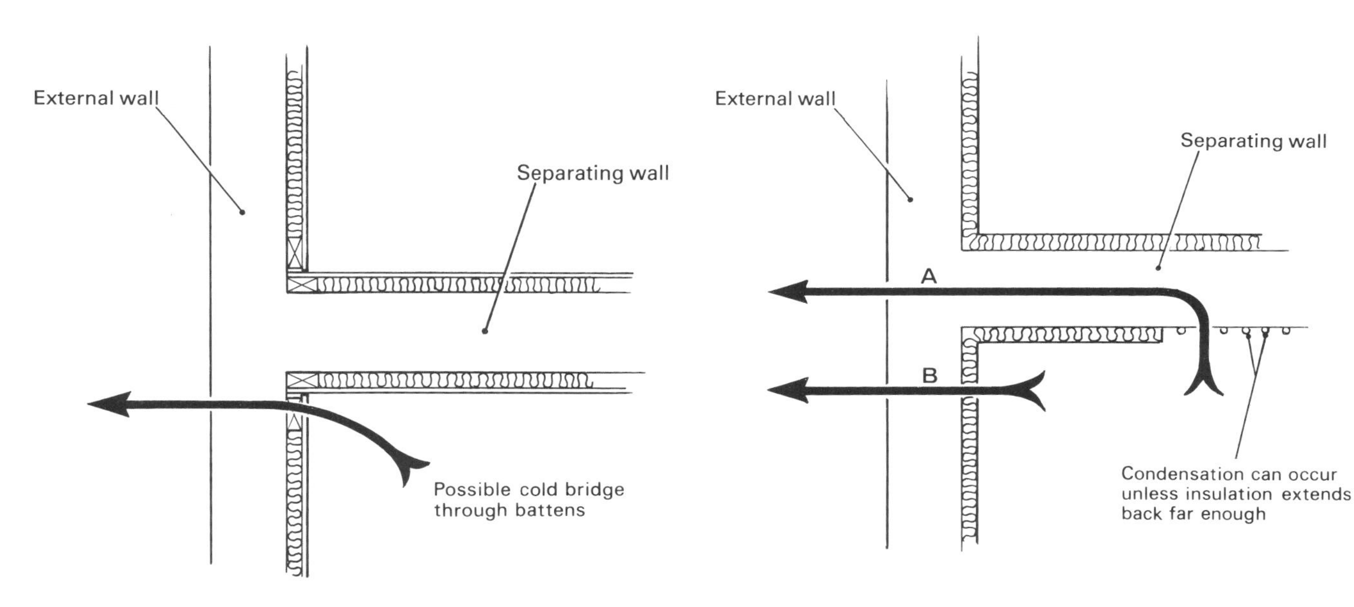

Figure 2 Fixings can create cold bridges.

Figure 3 Heat loss through A can be several times that through B.

5.2.17 EXTERNAL SOLID WALLS WITH DRY LINING. LOCALISED CONDENSATION

The defect is most commonly seen in rehabilitated property where dry lining has been applied to existing solid walls (it is not rain related).

Symptoms

Signs of cold patches on strips of wall show as pattern staining, surface dampness in cold weather or areas of mould growth.

Investigation

Establish the construction of the wall. Was insulation fitted behind the plasterboard? Does the pattern staining or dampness correspond with the positions of plasterboard fixings, or with any other features of the wall such as concealed pipes, half brick panels etc?

Examine areas where the lining ends or is omitted, thereby allowing dampness from condensation to be concentrated, e.g. at the back of paths, inside ducts, around windows and skirtings, under stairs, and where solid internal walls or partitions are tied into external walls.

Diagnosis and cure

When the heat loss through most wall surfaces in a room is reduced by the application of additional lining and insulation, any areas of wall that are comparatively cold, become vulnerable to condensation. When internal conditions are humid and it is cold outside, even plasterboard fixing battens can create cold bridges. Any plaster strips or dabs behind dry lining can increase thermal conductivity sufficiently to allow a corresponding pattern of condensation to form. The dry lining may not have been taken around concealed spaces, which may in any case be particularly vulnerable to condensation because of the lack of ventilation.

The cure is to increase the insulation in the affected area to provide a U Value of not more than 1.2, and to reduce the moisture content of the air by increased ventilation. Constant, rather than intermittent, heating will help to reduce condensation (but not pattern staining).

REFERENCES

BRE Defect Action Sheet 77: *Cavity external walls: cold bridges around windows and doors*

BRE Defect Action Sheet 78: *External walls: dry lining: avoiding cold bridges*

BS 5250: 1975 *Code of basic data for the design of buildings: the control of condensation in dwellings*

Photo 1 Rain falling on cantilevered steps dampens wall as no waterproof upstand or watershedding was provided.

Photos 2 & 3 Steps behind the balustrade wall are allowing water to enter the wall. The saturated bricks are then subject to frost damage.

5.2.18 EXTERNAL WALLS BESIDE EXPOSED STEPS. STAINING & DETERIORATION OF MASONRY

This defect occurs when the steps are not protected from rainfall by some form of cover.

Symptoms

The brickwork or stonework beside external steps becomes discoloured and eventually starts to break up. Where a wall is carried up past the level of the treads to form a balustrade, the outside of the wall may also be discoloured.

Investigation

Establish the construction of the steps and the position of any DPC's or flashings that are built into the flanking walls. Try, if possible, to find out what type of bricks or stone were used, and their frost resistance properties.

Note whether there is a skirting and its height.

Note whether an unusual quantity of water is being discharged onto the steps (e.g. from an overflow: where a downpipe is missing; from hosing down). An inspection during or just after heavy rain will show the area affected by splashing.

Examine the adjacent rooms (including space under the stair) for internal dampness and check the moisture content of walls with an electrical resistance meter.

Record the extent of staining and damage.

Diagnosis and cure

This defect is caused by water seeping into walls flanking external steps or splashing up onto the walls. It can only be cured by installing a waterproof skirting (with a minimum 150 mm girth) beside the treads, or by covering the steps so that rain does not fall on them. Where there is already a skirting, it may be possible to recover the steps and fill between the tread and the wall with a sealant.

Often, treads bridge the wall cavity, and a stepped cavity flashing with weepholes is required above the line of the treads to direct water out of the cavity above the treads. A new cavity flashing can be installed in short sections in an existing wall. Skirtings can be formed of tiling, asphalt or similar.

REFERENCES

BRE Digest 46:1964 *Design and apperance. 2*

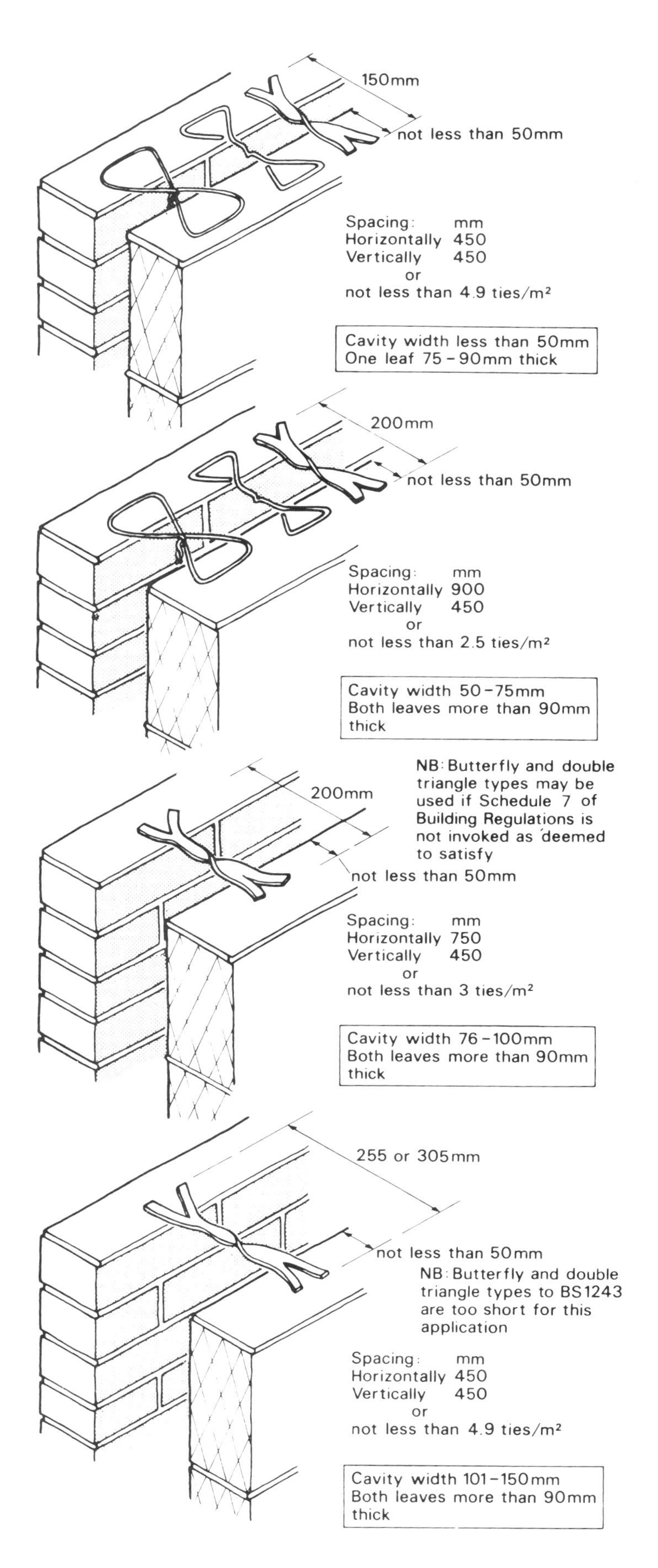

Figure 1 Types of wall tie for different cavity widths.

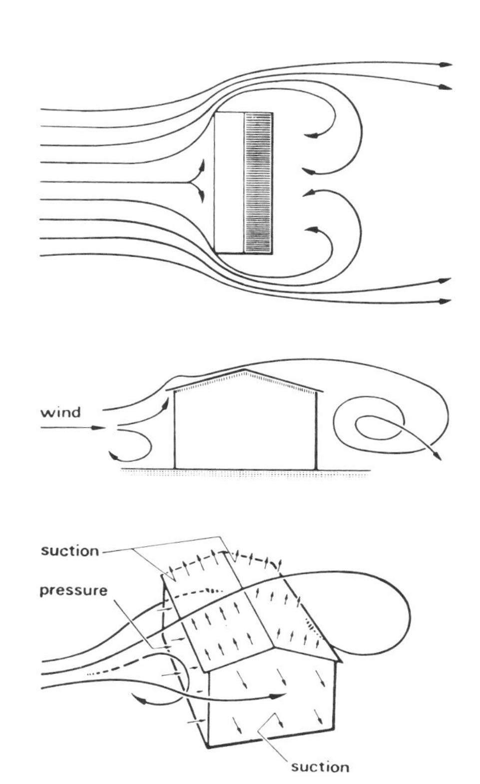

Figure 2 Wind produces pressure and suction on external walls.

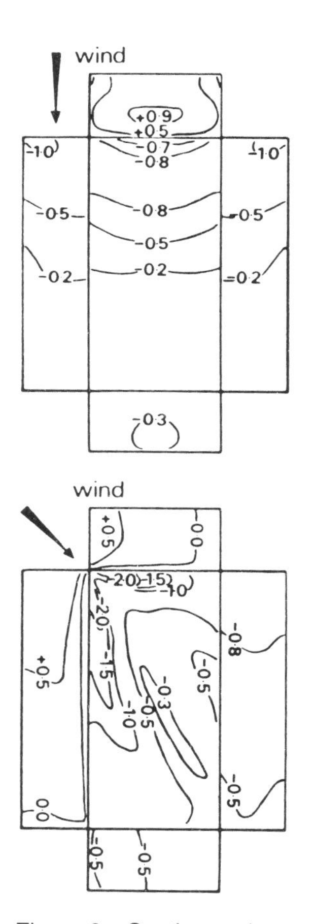

Figure 3 Suction and pressure vary with wind direction. Generally higher suction levels are produced near the edges of roofs, than on walls.

5.3.1 DETACHMENT OF BRICK CLADDING

A section of 102.5 mm thick brickwork falls from an exposed wall of a building without prior warning. This defect applies to brick cladding and to lightly loaded external leafs of cavity walls.

Symptoms

The collapse occurs in strong winds, and is most likely to affect the upper part of a large area of brickwork, or an exposed band of brick cladding not subject to significant loads from above. The horizontal cracking in 5.1.12 may result from similar causes that have not yet led to collapse. For detachment of brick slips from a concrete frame building see 5.1.21.

Investigation

Ensure that the remaining sections of wall are secure (temporary support and weatherproofing are likely to be needed).

Establish the construction of the cladding, in particular the means of tying the brick cladding to the structure. Examine the ties and record their number and position, their condition and the manner of failure (withdrawal from either leaf or failure of the tie itself).

Obtain information on the wind speed and direction at the time of the collapse.

Investigate the actual construction of the wall. Was the cavity built as specified? Were ties bent to meet course levels? Were there movement cracks in the brickwork before the collapse? Was any work done on the wall recently? (e.g. repointing or cutting out openings or movement joints). Where similar cladding occurs in other parts of the building or on adjacent buildings, particularly in exposed areas, the tying of these areas of cladding should be investigated (cavities can be inspected with an endoscope, or a metal detector, see chapter 4).

Diagnosis and cure

This defect is caused by insufficient or ineffective tying in of the cladding. The normal spacing of ties in cavity wall construction is 900 mm horizontally and 450 mm vertically (with 300 mm vertical spacing adjacent to openings). Wider cavities than 50 mm and leaf thicknesses of 90 mm or less, require closer tie spacing. Walls exposed to strong winds should have more closely spaced ties. The fixing of a brick leaf to a concrete wall requires similar spacing of fixings.

Flaws in workmanship and the specification can weaken the tying in of brick cladding. Examples are:

- Insufficient embedment of ties – at least 50 mm of the tie should be embedded in the mortar joint.
- Bending (or moving the tie in 'green' mortar) to bring it in line with a bed course of the second leaf to be built up.
- Poking ties into soft mortar when they have been forgotten as the first leaf was built up.
- Unsuitable materials used for tying in cladding e.g. light-gauge expanded metal strip, fixed to concrete backing by nails shotfired through light-gauge washers. (Strip, washers and shot firing could each be subject to failure.)

An exceptionally strong suction force may have been created by the wind blowing from a particular direction. Where the cause of this failure is not clearly attributable to defective construction, wind effects should be investigated and structural engineering advice should be sought. Before rebuilding, more of the cladding may have to be taken down, and extra fixings fitted to avoid the possibility of recurrence. When rebuilding, ensure that adequate movement joints are provided. Consideration should also be given to the support/discontinuity necessary at each storey height of multi-storey buildings faced with brick cladding. If existing ties are rusting, their complete or partial removal will be necessary as the expansion of the rust is likely to cause further damage to the wall. There are various techniques for fixing ties to existing structures, depending on the material of the backing. Expanding bolts can be used in harder materials. Wire ties with wavy ends can be pushed into resin filled holes in softer materials or materials with voids such as hollow clay blocks. See BRE Digest 329.

REFERENCES

BRE Defect Action Sheet 19: *External masonry cavity walls: wall ties – selection and specification*

BRE Digest 119: *The assessment of wind loads* (new edition 1984)

BRE Digest 329: *Installing wall ties in existing construction*

Bickerdike, Allen, Rich & Partners in association with Turlogh O'Brien. Design Failures in buildings Second series. Building Failure Sheet 8 *Brickwork cladding* George Godwin Ltd

BS 5628 Use of masonry: Part 3: 1985 *Materials and components, design and workmanship*

Photo 1 Spalling can be provoked by a relatively impervious paint coating that retains moisture in the brick and makes it vulnerable to frost.

Photo 2 Salts forming as a layer below the brick surface have forced off the faces. This defect is much less common than shown in photo 1.

5.3.2 SURFACE SPALLING OF EXTERNAL BRICKWORK

The face of the brick comes away from the body either as one piece or as flakes. The defect applies to clay brickwork.

Symptoms

Individual bricks lose their exposed face to show rough and usually lighter coloured surfaces slightly set back from the surface of surrounding bricks. Almost all the bricks in an area of walling may be affected or just one or two. The defect commonly affects exposed walls and walls below DPC level but one version of it can affect the sheltered areas of a wall under the eaves and under porches. The defect is most obvious on painted walls where it may first show as a disruption of the paint coating.

Investigation

Record the extent of the spalling and try to establish when it occurred. If possible, monitor additional spalling over a period of one year.

Establish the form of the construction, and materials used. Does the wall have cavity insulation? Has an appropriate type of brick been used?

Examine both the bricks and the mortar. Are there signs of salt crystals or white efflorescence on the spalled brick surfaces? Is the mortar eroded or porous? Is the pointing much harder than the bricks? A single damaged brick and the spalled face can be sent to a laboratory for analysis of salt content. Mortar samples (from the centre of the wall) can also be sent.

Determine what the weather conditions were when the defect first appeared. Was it wet? Was it freezing?

Diagnosis and cure

This defect has a variety of causes, the most common being frost action. Certain types of brick are intolerant of dampness and frost – they should not be used below DPC level. BS 3921:1985 has three classifications for frost resistance; 'Frost resistance (F)', 'Moderately frost resistant (M)' and 'Not frost resistant (O)'. Only frost resistant bricks (F) should be used below DPC level. Where spalling of bricks in classes (M) or (O) occurs only below DPC level, frost damage following saturation is a likely cause. Frost can also damage class (M) or (O) bricks at higher levels if freezing winds or hard frosts follow a period of driving rain that has saturated the bricks.

An alternative cause of spalling is expansion of salt deposits below the brick surface. This is less common and does not always affect the most exposed areas of brickwork. BS 3921:1985 excludes bricks with a high soluble salt content. There may well be visible signs of salts on the spalled surface or laboratory analysis may reveal high concentrations of salts.

If the wall has been pointed with cement rich mortar (i.e. with a high cement ratio of over 1:4) spalling may occur along the edges of the bricks, (particularly if relatively soft facings have been used). Absorbent bricks that have soaked up water which has then frozen, will be restricted by the mortar when the thawing action expands the ice as it melts.

If the wall has been painted, any spalling will be more obvious, and it may be that the paint coating has caused the spalling – although more commonly the paint is applied as a remedial measure after some spalling has occurred. The paint can make the situation worse, because if water penetrates the paint coat it is slower to dry out from behind the paint, and is therefore more likely to be frozen. Colourless water repellants can also cause further damage if the brickwork is not in good condition to start with. If salts become concentrated behind a water repellant film, their expansion can cause spalling of the brickface.

If an expansive particle of lime (e.g. a fossil) becomes incorporated in a brick during manufacture, it will slake in the presence of moisture and cause a 'lime blow', which spalls off a section of the brick face showing a white nodule behind. The defect affects usually not more than one or two bricks in a wall. They can be left – or if they are unsightly they can be replaced.

Where only a few bricks are affected, for example, where a small number were underburnt and therefore more vulnerable, cutting out and replacing individual bricks is the best method of repair. Where large numbers of bricks, or the majority of bricks, are affected the alternatives are rebuilding, cladding (i.e. tile hanging, sheet cladding etc) or rendering. The choice will depend on circumstances. Rendering should generally be on stainless steel expanded metal lathing, which should be fixed to the body of the brickwork as spalled bricks make a doubtful background for new render coats.

Photos 3 & 4 Gault bricks and Fletton bricks are both subject to frost action if used in positions where they remain damp.

Photo 5 Lime blows are caused by fossil nodules in clay bricks that expand as they slake. The defect usually only affects isolated bricks.

continued

REFERENCES

Hughes, Philip. *The need for old buildings to breathe* Information leaflet 4 – Society for the Protection of Ancient Buildings

Bidwell, T G. *The conservation of brick buildings* The Brick Development Association

BS 5628: Use of Masonry Part 3 1985. *Materials and components, design and workmanship*

BRE Defect Action Sheet 71: *External masonry walls: repointing – specification*

BRE Digest 89: *Sulphate attack on brickwork*

BRE Digest 196: *External rendered finishes*

BRE Digest 200: *Repairing brickwork*

Building Research Advisory Service *Painting fletton bricks* Building Technical File Number 5 – April 1984

Construction Feedback Digest 41: *Clay brickwork below DPC* B & M Publications (London) Ltd

Construction Feedback Digest 43: *Spalling brickwork hidden efflorescence* B & M Publications (London) Ltd

Photo 1 Persistent dampness can cause staining.

Photo 2 Long term effects of different dampness levels causing lightening, darkening and efflorescence.

5.3.3 PRINCIPALLY EXTERNAL WALLS. STAINING, MARKING, SOILING AND EFFLORESCENCE

Darker or lighter areas appear on the face of the wall. The defect relates to fairfaced clay brickwork.

Symptoms

A range of different types of discolouration may occur on brick surfaces giving a patchy appearance. White or yellowish crystals and dark grey or brown stains appear on the face of the bricks, and in many cases on the face of the mortar as well. (Green, red or bright yellow stains are only likely to be attributable to plants or paints.)

Analysis

Record the areas affected and, if possible, establish whether changes are taking place over a period of say 6 months.

Establish the specification used for the brickwork and whether it was followed (BS 3921 1985 includes tests for soluble salt contents of clay bricks).

Use an electrical resistance meter to establish variations in moisture content over the wall surface. (Salts will affect readings.)

Send samples for laboratory analysis if otherwise unable to identify composition of stain.

Inspect the defect during wet weather to discover if the source of dampness is rain related.

Diagnosis and cure

Staining and soiling is commonly associated with dampness, either with drying out of saturated materials or with water flowing over the surface of a wall. Occasionally it is associated with use and much less frequently with temperature differences.

Dampness – stains are usually caused by one of the following: water bringing dissolved chemicals to the surface of brickwork as a damp wall dries, for example efflorescence is often seen where sodium or magnesium sulphates from clay bricks form patches of white crystals as the wall dries. If the wall remains dry the crystals can be brushed off and will not reappear, but if it becomes damp again the efflorescence is likely to recur.

Another common example is lime leaching when lime from mortar is brought to the surface as a wall dries out. If perforated bricks are left uncovered during construction and the perforations fill with water, a band of lime bloom can sometimes be seen across the face at the level that the work had reached when the rain came. Lime can be washed off with weak hydrochloric acid (spirits of salts) masonry cleaning liquids based on hydrochloric acid are also available. Care must be taken to ensure that acidic cleaners are used safely, and are not allowed to damage historic structures. Lime can be distinguished from efflorescence because it is associated with the mortar joints rather than the bricks (although it may extend on to the bricks). Lime fizzes and dissolves in hydrochloric acid, efflorescence dissolves in water.

Other stains can be brought to the surface by dampness drying out. For example, dark brown stains from the products of condensation in flues or light coloured salts from the ground water carried up a wall with rising dampness. Such stains are not easy to remove and are best masked by paint or by matching in adjacent brickwork with a sooty solution.

Soiling of external walls can be caused by dust and grime associated with dampness, but it usually only becomes a problem when dampness causes uneven soiling. Dampness allows atmospheric dust to adhere where it does not adhere so well to adjacent dryer areas. Water also causes uneven dirtying when it flows over brickwork, washing dirt or dissolved substances from some areas, and depositing them on others. For example, some kinds of limestone can produce whitish calcium deposits where water from limestone sills runs down over brickwork. The run off from copper roofs or features can cause pale green stains on the wall below. The cure for this type of soiling is to clean the wall, and then modify projecting features to shed water clear of the wall surface. Avoid uneven dampness in the wall resulting from defective cappings, leaks and rising damp. Sometimes it may be necessary to inspect a building during prolonged heavy rain to see how the water is shed over the surface, because just a slight misalignment of a drip can direct water in a stream down one part of the façade.

Pattern staining of brickwork due to temperature difference is not so common. The underlying cause may be the difference in dampness between the warm wall and colder areas surrounding it, dirt adhering more easily to the damp areas.

Photo 3 Calcite stain from concrete can affect brickwork (see also 5.3.8).

Photos 4 & 5 Efflorescence can seriously disfigure buildings.

Photos 6 & 7 Efflorescence can pin point defective construction.

5.3.3 PRINCIPALLY EXTERNAL WALLS. STAINING, MARKING, SOILING AND EFFLORESCENCE

continued

Some staining and marking of brickwork is commonly associated with use. Ladders scrape the wall if they are raised or lowered while leaning on it. Car exhausts leave dark patches on walls surrounding parking areas. Muddy footballs, brushing tree branches, mortar splashes and scaffolding all mark brickwork in ways that may not be immediately identified.

REFERENCES

BRE Digests 45 & 46: *Design and appearance*

BRE Digest 280: *Cleaning external surfaces of buildings*

Building Technical File 4 (Jan 84) *Mortar deterioration and brickwork staining*

Ashurst, J *Cleaning of stone and brick* SPAB Technical Pamphlet 4/5

Brick Development Association Building Note 2 *Cleaning of brickwork* Harding, Smith & Brown

BS 6270 *Code of practice for cleaning and surface repair of buildings* Part 1 1982 *Natural stone, cast stone, clay and calcium silicate brick masonry*

Photo 2 Repointing has restored the weathertightness of the house on the left.

Photo 1 Frost action and incorrect mix are likely causes of this defect.

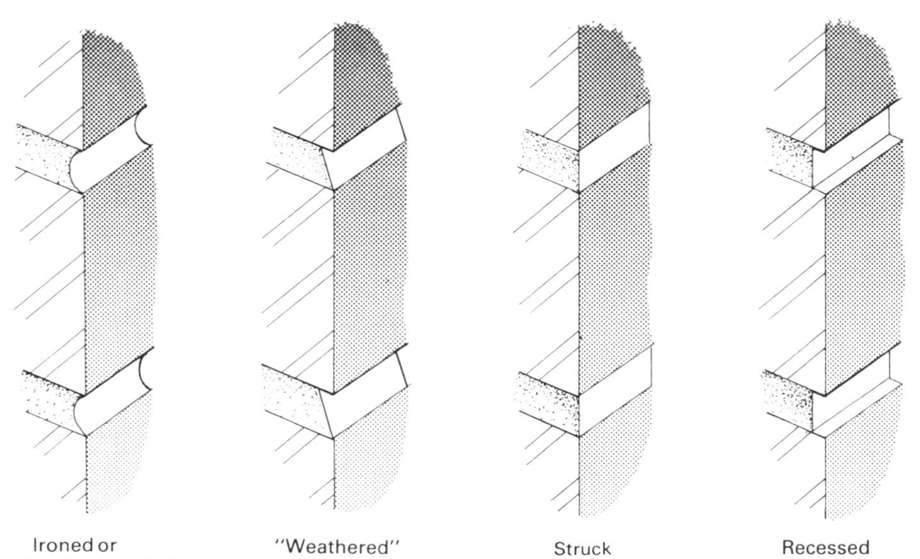

Figure 1 The joint profile affects durability and weathertightness.

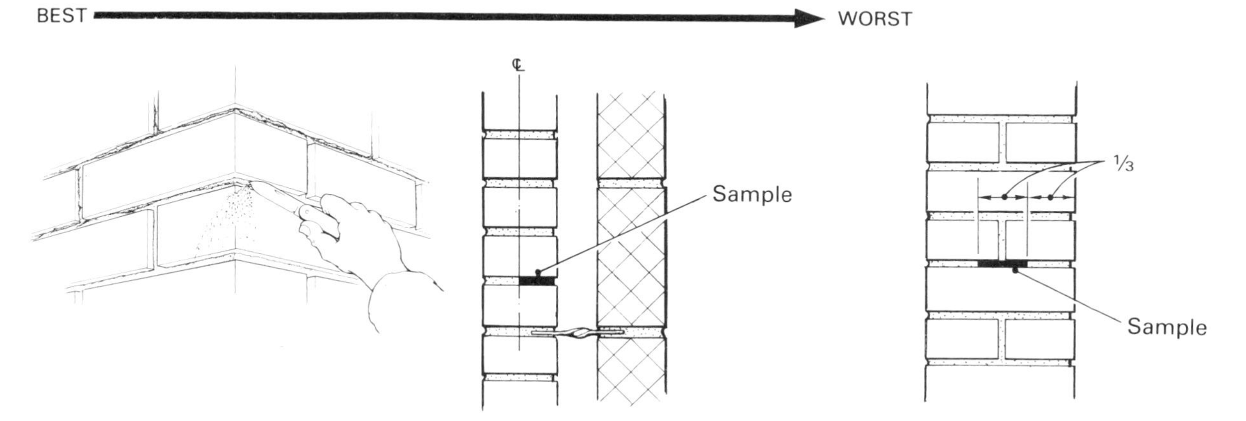

Figure 2 Mortar should not powder away as easily as this.

Figures 3 & 4 Mortar samples for analysis of the mix should be taken away from wall faces.

5.3.4 EXTERNAL WALLS. EROSION OF MORTAR

Bits of mortar flake or powder away leaving the bricks, stones or masonry units proud of the mortar face. Variations of the defect apply to any type of masonry.

Symptoms

The wall appears to have recessed joints, particularly in exposed areas. There may also be damp penetration, soiling and frost damage to the faces of the masonry units (ie bricks, stones etc.). Pointing may have fallen out over large areas. The mortar may appear soft and sandy when scraped with a knife.

Investigation

Establish the composition of materials specified for construction.

Establish the conditions and the time of year when the wall was built.

Send samples of mortar from the centre of the wall for laboratory analysis.

Investigate sources of dampness.

Note the exposure conditions and whether the brickwork is being regularly saturated with water.

Diagnosis and cure

The possible causes of this defect are:

1. Sulphate, probably from the bricks, affecting the mortar (see 5.1.13).
2. Frost action on the mortar before it had hardened. (Very weak mortar may be affected by frost after it has hardened.)
3. Poorly constituted mortar – with a low cement/lime content (less than 1:10). Note that batches of mortar used for different parts of the same wall may vary.
4. In an old wall – repointing of weak mortar joints with too shallow raking out of the joint.

A decision has to be made on the seriousness of the defect. With loadbearing or high walls (over six metres), a structural engineer should be involved. If the mortar is in very poor condition for the job the wall is required to do, the wall will have to be rebuilt. In cases where the defect is less serious, raking out and repointing in a stronger and, if necessary, sulphate resisting mortar may be satisfactory. Note that walls constructed using a weak mortar are often strong enough – a 50% reduction in mortar strength has been shown only to lead to a 15% reduction in the strength of the wall. If a sulphate problem is identified, it is particularly important to prevent the wall becoming damp.

REFERENCES

BRE Digest 89: *Sulphate attack on brickwork*

BRE Digest 160: *Mortars for bricklaying*

BRE Digest 246: *Strength of brickwork and blockwork walls: design for vertical load*

BRE Defect Action Sheet 70: *External masonry walls – eroding mortars – repoint or rebuild?*

BRE Defect Action Sheet 71: *External masonry walls: repointing – specification*

BRE Defect Action Sheet 72: *External masonry walls: repointing*

Photo 1 Spalling cladding restrained by concrete frame

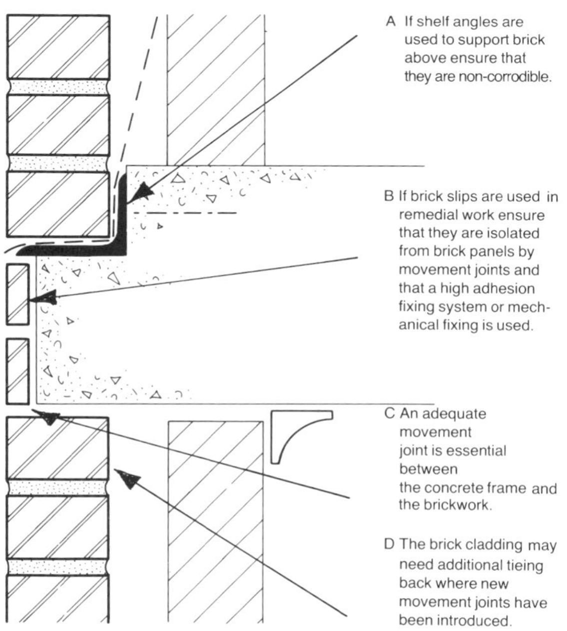

Figure 1 Precautions to be taken in remedial work.

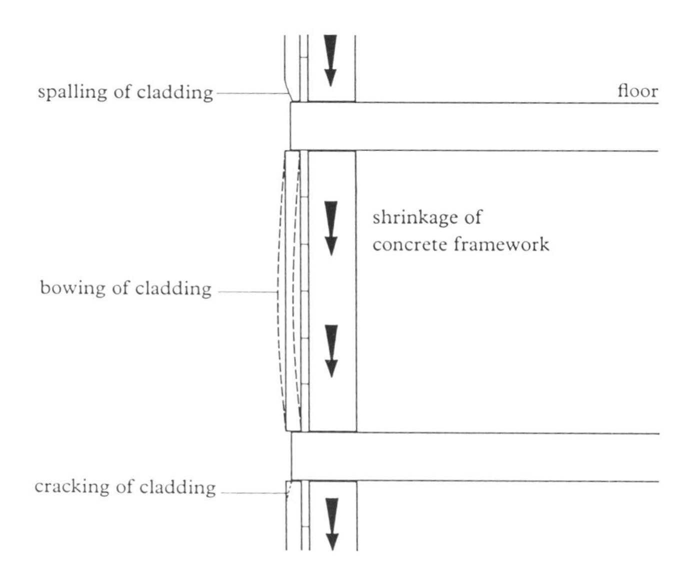

Figure 2 Effect on cladding of shrinkage of a concrete frame – the effect is increased when clay brick cladding has expanded.

5.3.5 DISRUPTION OF MASONRY PANEL EDGES AND BOWING OF PANELS

Splitting, cracking and spalling of the edges of brick or stone external wall panels within a concrete frame (the frame may be covered by brick slips or thin stone facings) see also 5.1.21.

Symptoms

Horizontal (and less frequently vertical) panel edges split away. Sometimes the edge of the concrete frame abutting the panel may spall, particularly if a mosaic or rendered finish has been applied. The panel itself may bulge or bow. The defect is most likely to occur with clay bricks but does sometimes affect stone facings.

Investigation

Establish the form of construction. Is masonry held between concrete members? Is provision made for expansion of the panel and contraction of the frame (eg compressible joint)? How is the panel supported? How is the panel tied into the structure? What type of masonry was used?

Record the extent of disruption. Does the panel bulge or bow? If there is any possibility that the panel may be, or become, unstable structural engineering advice should be sought immediately.

Establish when the building was erected and how long a period elapsed between the erection of the frame and the construction of infill panels.

Diagnosis and cure

If no allowance (ie movement joint) has been made at the panel edges for contraction of the concrete frame, the cause of the defect is likely to be the frame punching the edges of the panel as it shrinks. Clay bricks expand, particularly in the period immediately after firing. Concrete shrinks as it cures and under the influence of loads. Confirmation of the cause may come from the general appearance of the panel if it is bowed (out or in), or if cracks and fallen pieces of masonry can be seen.

If the panel is seriously out of plumb or bowing badly it will have to be rebuilt making due allowance for movement at the edges (details are given in the references). In such cases structural engineering advice should be sought if the panel does not need rebuilding.

REFERENCES

PSA Feedback: Building Technical File 2 July 1983

BRE Defect Action Sheet: 2 May 1982: *Reinforced concrete framed flats: repair of disrupted brick cladding*

Photo 1 Fallen cladding panel.

Photo 3 Close up showing faulty dpc no mortar bed, and excessive packing.

Photo 2 The wall from which the panel fell.

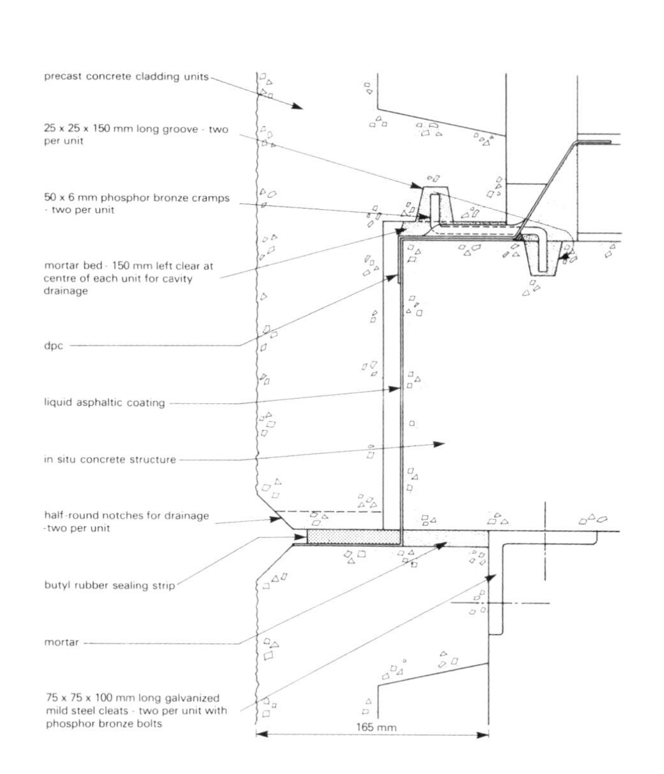

Figure 1 Fixing details as originally designed.

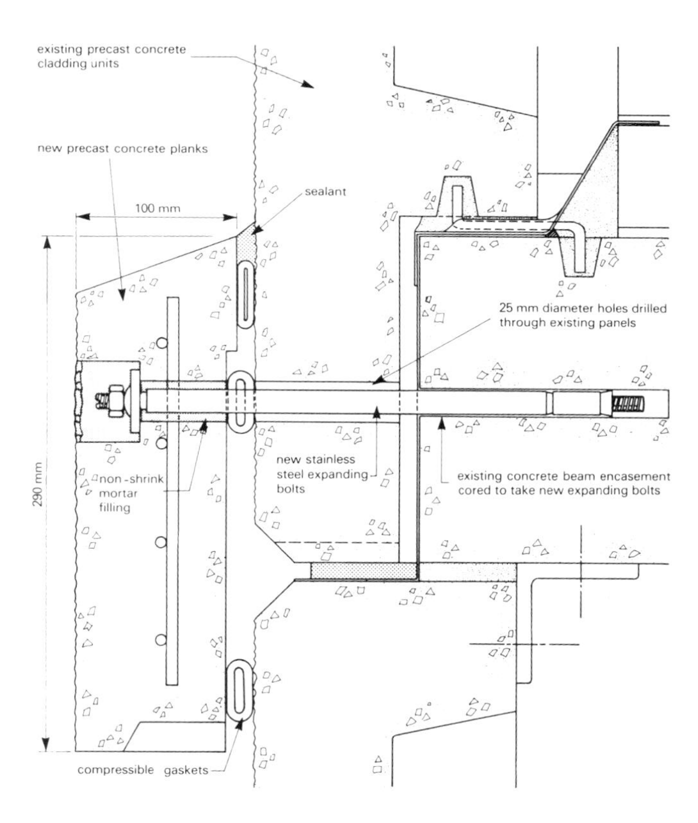

Figure 2 Remedial solution with cladding secured by concrete planks.

5.3.6 EXTERNAL WALLS. DISLODGEMENT OF NON-LOADBEARING CONCRETE CLADDING PANELS

Misalignment, rain penetration, staining or detachment of precast concrete cladding panels, sometimes involving cracking or spalling of the panels, and fracture of visible fixings.

Symptoms

The defect may be apparent because unplanned movement has occurred, or it may become apparent through associated defects – such as rain penetration. Failure of fixings on a building of similar construction may be taken as a symptom requiring investigation.

Investigation

A structural engineer should always be involved when it is suspected that concrete cladding panels may become dislodged. The investigation of this defect is more complex than most and can only be described in outline.

Where cladding panels are above ground floor level and there is a risk of detachment, precautions should be taken to protect passers-by.

Establish the form of construction and the detail of joints and fixings.

Examine the cladding for visible signs of movement such as sealant failure, particularly where horizontal joints have failed in a different manner to vertical joints. Unequal joints spalling at panel edges and misalignment of panels by more than 2–3 mm may be signs of dislodgement.

Examine the interior of the building for rain penetration through the external wall.

Look for signs of fixing failure or corrosion of fixings or missing fixings. Fixings within a cavity can sometimes be examined with an endoscope (see chapter 4). Corroded fixings may stain adjacent construction. The original production drawing may help to locate fixings and show how they were intended to work, but beware of alterations or substitutions that may have been made on site and not recorded on the drawings. For instance, bolts and brackets of different metals may have been substituted for those shown. Where there is a wide variation in joint widths, the frame of the building may have been inaccurately constructed and the designed fixings may not have fitted as originally intended.

Examine critical points in the building where the dislodgement (or potential dislodgement due to faulty fixings) is most likely to occur, ie parapets and upper levels of high buildings, and at salient angles and corners of the building.

Record the position, condition and measurements of all joints and fixings examined, and any stains or spalling that has been seen. Note the weather conditions, particularly the direction of driving rain and the temperature. Note the orientation of the wall – is it in full sun in the middle of the day? Sample fixings can be sent for laboratory analysis to establish the material used, the protective coating used and, in the case of sheared fixings, the likely mode of failure.

When preliminary investigation has found serious faults that could lead to the detachment of the cladding, a detailed survey is likely to be necessary. Some opening up of backing walls to reveal panel fixings will be required, but this can be kept to a minimum if a statistically based sample of fixings is inspected.

Diagnosis and cure

The common feature of defects in this area is that the fixings of the concrete panels have failed, but there may be many causes of this failure.

- the frame of the building may have shrunk more than the cladding panels, putting extra strain on the fixings.
- The frame may have been inaccurate in the first place, making it impossible to use the fixings as intended (extra packing with steel shims under units may cause additional stress on fixings).
- Fixings may have been omitted, or too great a reliance placed on locational fixings (which were not intended to replace mortar packing for example).
- The materials used for the fixings may have been unsuitable or incorrectly used. Leaking cladding joints, combined with inadequately protected steel, or combinations of metals promoting electrolytic action, can lead to fixing failures.
- Inadequate allowance for thermal movement of the cladding can cause the breakdown of the weatherseals on joints, and place extra strain on panel fittings.

Once the cause of failure has been diagnosed, a cure

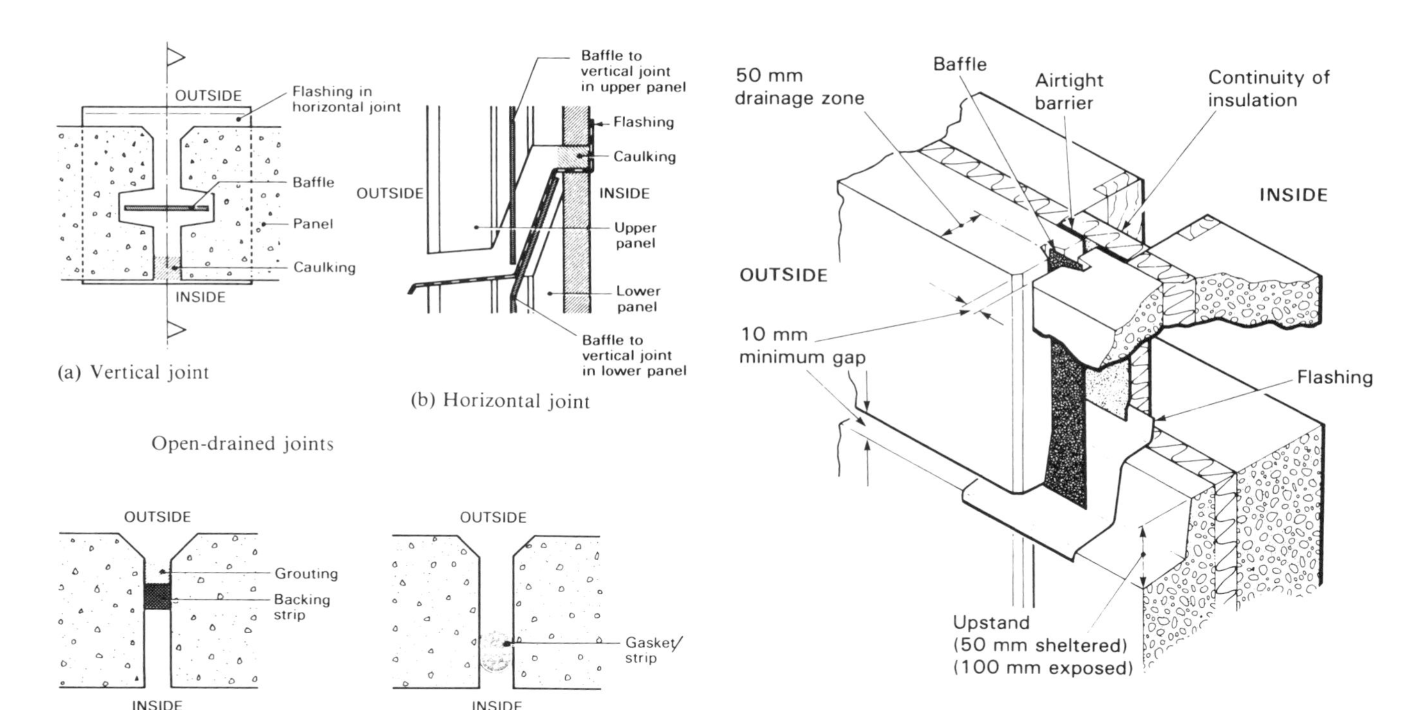

Figure 3 An open drained joint between facade panels.

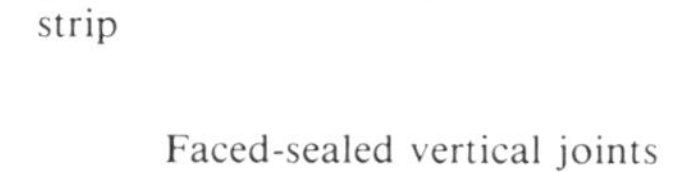

Figure 4 Types of joint.

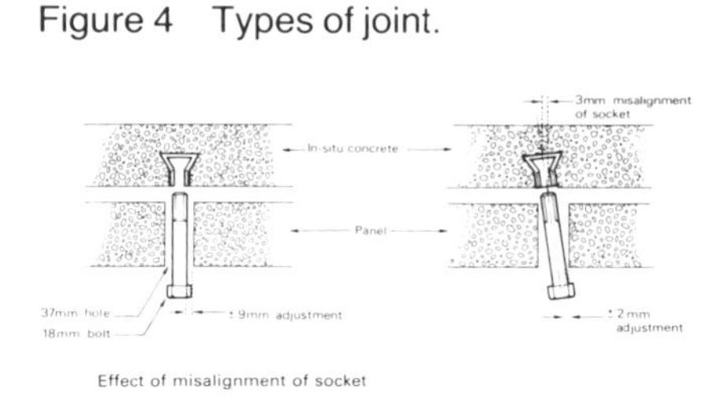

Figure 5 Effect of misalignment of socket.

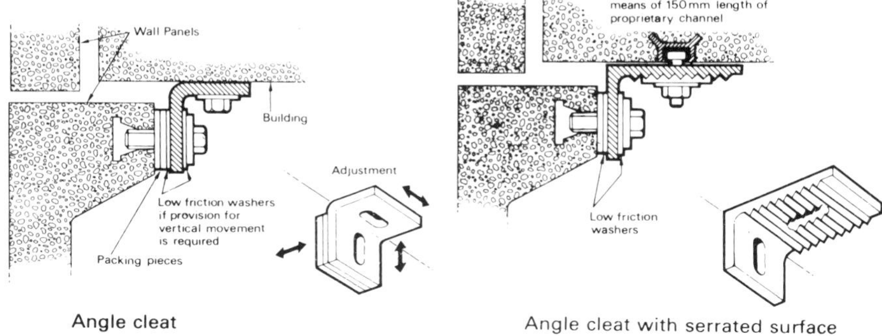

Figures 6 & 7 Angle cleats allowing and controlling movement.

Stone or pre-cast concrete cladding	
Non-loadbearing cladding formed by stone or pre-cast concrete facing units	
(a) Cracks, uniform in width or closed; at a later stage, spalling especially at edges or corners of cladding units, possibly accompanied by misalignment of faces. (b) Cracking or spalling of cladding units in a regular pattern which seems to indicate position of fixings; no evidence of compression (see a).	Cladding units under compression due to inadequate allowance for differential movement between cladding and supporting structure; more likely to occur where structure is in-situ concrete. Check provision and effectiveness of horizontal 'soft' compression joints. Check for corrosion of reinforcement. Possible corrosion of ferrous fixings: check from drawings or specification. If serious, remove unit, otherwise keep under observation.
(c) Misalignment between faces of adjoining units.	If excessive, suspect either potential compression failure (see a) or fixings absent or defective.
(d) Iron stains on surface of pre-cast concrete units, random occurrence.	If units unreinforced, likely to be iron-bearing aggregate. If reinforced, (see e).
(e) Iron stains on surface of pre-cast concrete units, occurrence suggests a pattern.	If units reinforced, likely to be corrosion of reinforcement. If adequate cover, suspect also possible use of excessive calcuim chloride in manufacture, especially if cracking or spalling evident; check whether defect occurs with particular units which may have been replacements for units damaged during construction and calcium chloride used to speed up casting of replacements.

Table 1 Latent defects indicated by pattern of occurance in stone or precast concrete cladding.

5.3.6 EXTERNAL WALLS. DISLODGEMENT OF NON-LOADBEARING CONCRETE CLADDING PANELS

continued

can be designed. Different causes will involve different cures. Often, additional fixings can be used to secure the panels. A structural engineer should always be involved in the design of supplementary or replacement fixings.

REFERENCES

Construction Feedback Digest 37, winter 1981 *Concrete cladding fixing failure.* B & M Publications (London) Ltd

BRE Digest 217: *Wall cladding defects and their diagnosis*

BRE Digest 235: *Fixings for non-loadbearing precast concrete cladding panels*

BRE Defect Action Sheet 97: *Large concrete panel external walls: re-sealing butt joints*

BS 6093: 1981 *Code of Practice for the design of joints and jointing in building construction*

BRE Information Paper 10/86. *Weatherproof joints in large panel systems: 3 investigations and diagnosis of defects*

Photo 1 Loading from brickwork above caused the cast stone units to crack below the nib.

Photo 2 Lower part of band course was restored with render on expanded metal.

Photo 3 Stone facing being forced outward.

Photo 4 Corrosion of steel behind stone facing causing the displacement.

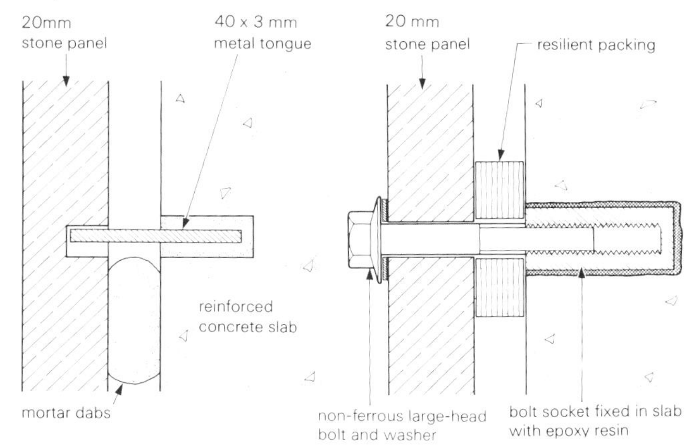

Figure 1 Original and replacement cladding fixings. The bolt and socket restrains outward movement of the cladding.

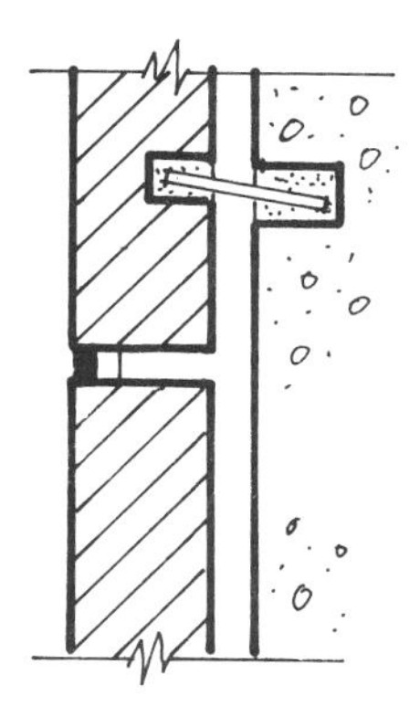

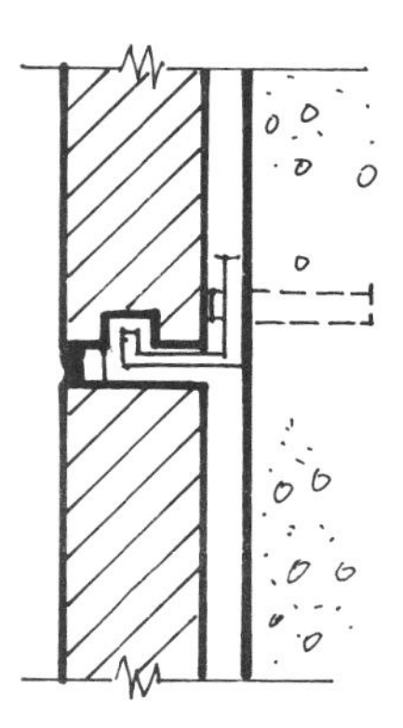

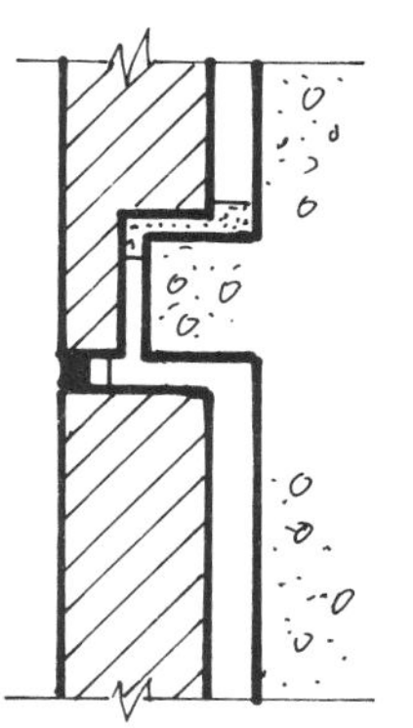

Figures 2, 3 & 4 Horizontal joints in stone cladding showing alternative methods of providing support.

5.3.7 INTERNAL AND EXTERNAL WALLS. DETACHMENT OF STONE CLADDING

Stone facing panels become dislodged and move out of alignment or fall from the building. The defect applies to stone, slate and marble facings.

Symptoms

Misalignment of facings, disrupted jointing materials, cracks in the face of the stone, rust stains, spalling or falling cladding panels (or parts of panels).

Investigation

Where claddings are above the ground floor and there is a risk of stones falling from the building, precautions should be taken to protect passers by. Structural engineering advice may be needed to assess the adequacy of fixings.

Establish the details of the cladding attachment. What type of fixing was used? What allowance was made for movement?

Check the construction and its conformity with the details. Scrutinize in particular, fixings, grouting and bedding materials, stone thickness and dimensions of chases and rebates in the stone – it may be necessary to remove a cladding unit if the backing has not been revealed by fallen sections of stone.

Record the position of any visible damage or staining and the orientation of the elevations on which the damage or staining shows. Enquire when defects were first seen.

Examine cracks, cavities, staining, broken edges etc. for signs of when defects occurred, eg soiling or accumulations of debris.

Other aspects may need to be subjected to analysis, eg sulphate content of bedding mortars, extent of microscopic cracks in stone slabs, reasons for fracture of metal fixings.

Diagnosis and cure

The first possible cause of failure to examine is thermal movement – sometimes exacerbated by shrinkage or movement of the supporting structure. There have been instances of stone claddings (particularly dark coloured claddings with a cavity behind) expanding in hot sun and bowing away from the cooler structure behind. If there is no proper provision of movement joints, thermal movement is the most likely cause of failure.

The movement of surrounding or adjacent materials may put pressure on the stone claddings and cause them to crack, spall, or break away from their fixings. Stone will only withstand small tensile forces. Sometimes extra stress in the fixings is caused by inaccurate casting in of sockets.

Failure may occur as a result of expansion of the backing materials. For example, expansion of mortar droppings behind slabs in the presence of sulphates or corrosion of metal behind the slab or, where water has penetrated behind the cladding, and its expansion after a period of freezing temperatures. A further possibility is the lack of adhesion of the stone cladding – particularly where there is a smooth surface at the back of the cladding and the bedding mortar, or where fine aggregates produce mortar with high shrinkage characteristics.

Examples of the types of failure described above will be found in the references.

The cure will depend on the extent and cause of the defect. Movement joints could possibly be cut into the cladding without refixing it. Drilling and face fixings may be an acceptable solution. It may, however, be necessary to remove the cladding and refix it. The BS code of practice CP298:1972 and the Stone Federation's Code of Practice both give recommendations on fixing details. Several proprietary fixing systems are available that allow for differential movement between the cladding and the supporting structure.

REFERENCES

Construction Feedback Digest 33: April 1980 *Horizontal cracking of a string course due to loading from brick skin above* B & M Publications (London) Ltd

Construction Feedback Digest 41: Winter 1982/3 *Inadequate fixings and corrosion behind stone balcony facings* B & M Publications (London) Ltd

Bickerdike Allen, Rich and Partners in association with Turlogh O'Brien. Sheet 5 *Sulphates causing mortar to expand behind cladding*, Sheet 16 *poor adhesion of marble to fine sand bedding*, Sheet 17 *thermal expansion of dark slate cladding*. Building Failure Sheets – second series, George Godwin Ltd.

BS CP298: 1972 *Natural Stone cladding (non-loadbearing)*

Stone Federation *Code of Practice on Natural stone cladding (non-loadbearing)*

Photo 2 The concrete edge beam is stained by water run off. The effect of tidemarks shows on the rendered wall above.

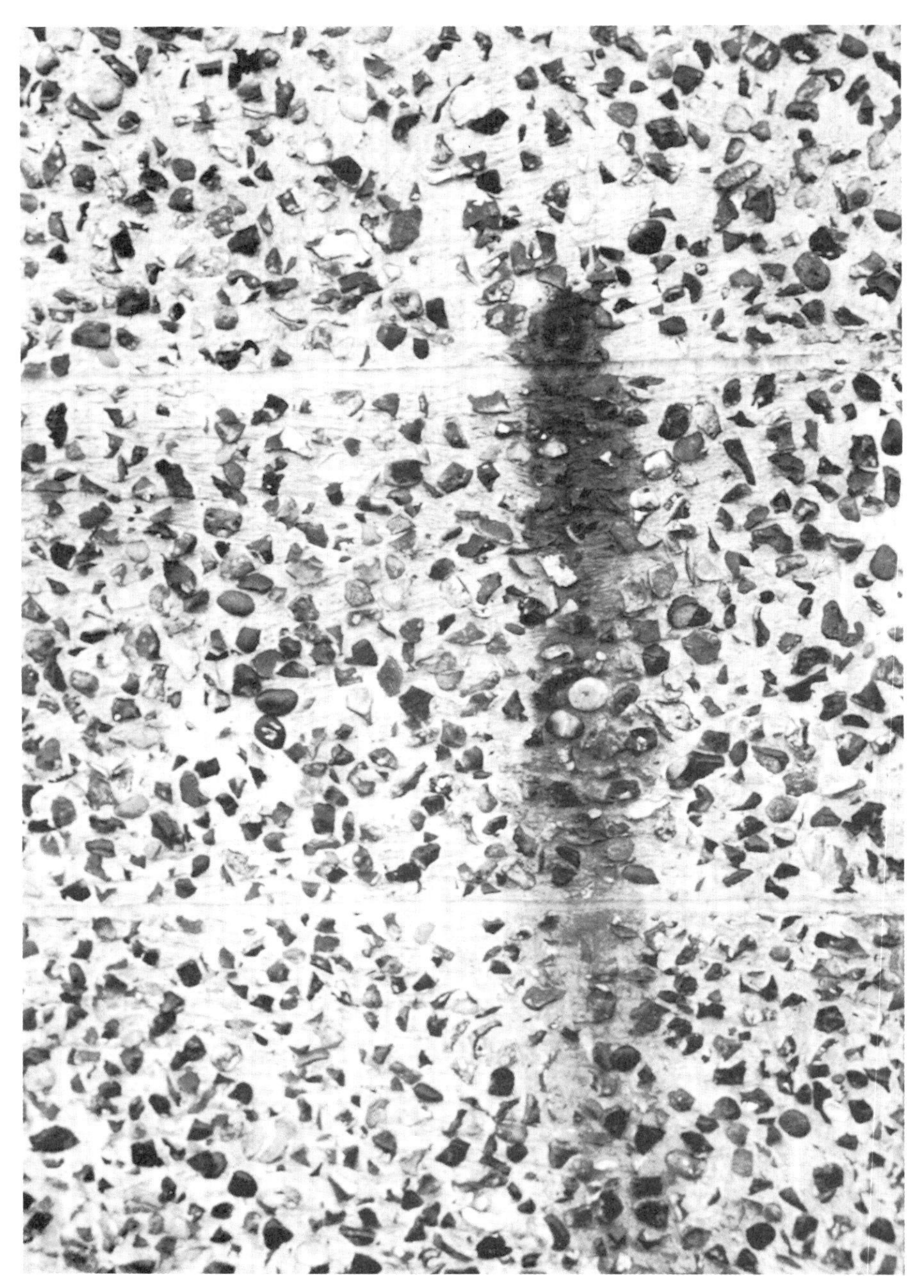

Photo 1 Isolated rusty stains can be caused by fragments of pyrite in the aggregate. They affect appearance but do not lead to general deterioration.

Photo 3 The bush hammering of this concrete parapet has not disguised what appears to be shutter marks over the arches. Isolated rust stains may either be loose tie wires from the reinforcement cage or pyrite.

Photo 4 The under surface of the canopy collects dirt while the edge is washed by rain. Cleaning would be costly.

5.3.8 EXTERNAL WALLS. STAINING AND SOILING OF UNPAINTED CONCRETE

Concrete surfaces become unacceptably shabby and stained, the marking is uneven. Some parts of the building may be more affected than others.

Symptoms

The symptoms are the change in appearance: brown or white stains; patchy, sometimes crazed surfaces; uneven soiling and tidemarks at the edge of wetted areas; organic growths combined with a build up of dirt.

Investigation

Examine the condition of the concrete. Is it sound for its job, or is it cracked? Is there any sign of corrosion of the reinforcement? Is it porous? (See 5.1.22, 23 & 24)

Record the position of the staining or soiling and how it varies across the elevations of the building. How does it relate to the flow of water over the façades in driving rain (particularly driven from the prevailing direction)? Inspection in wet weather may give a much clearer picture.

Where the cause is not immediately obvious, take samples of the surface dirt or stained concrete and send them for analysis.

Diagnosis and cure

Common types and causes of appearance defects in visible concrete are listed below:

Brown rusty stains – may be rusting reinforcement or if a very small stain, then rusting of a carelessly trimmed reinforcement tie wire. Occasional isolated rust stains are likely to be from a piece of pyrite in the aggregate. The pyrite reacts with water, oxygen and free lime from the cement to produce a rusty stain which may be spread down the façade by rainwater.

White stains or a build up of a white coating to the concrete – is likely to be calcite (calcium carbonate). It is most obvious where water carrying calcite evaporates on the surface of the concrete and the deposit is not washed off by rain (calcite can form stalactites where water drips from a soffit).

Patchiness – is probably due to variations in the way the concrete was originally formed. Even small variations in compaction, formwork or the use of release agents will affect the absorbtivity and appearance of the surface – particularly where it is smooth. Surfaces that absorb moisture will also collect dirt unless they are washed clean by rain.

Uneven soiling – the original light grey colour of smooth concrete is the colour of cement and calcium carbonate on the surface. As this is washed off, the colour of the fine aggregate shows. When in turn this is worn away, the colour of the coarse aggregate shows and the surface becomes uneven. An uneven surface dries more slowly and holds dirt more readily. Uneven soiling is often due to uneven wear and washing of the surface.

Tidemarks – when a wall surface is partially wetted by rain, the dirt on the surface is carried to the edge of the wetted area where it remains visible after the wall has dried.

Organic growth – Concrete surfaces that collect dirt and retain moisture allow the growth of plants and bacteria; these in turn retain more dirt and moisture and may add to the shabby appearance.

There may be no permanent cure for these defects. It must be remembered that exposed concrete has a limited life and that eventually some form of coating or cladding is likely to be required. Concrete can be cleaned with water spray or wet grit blasting. Dry grit blasting of concrete is not recommended because of the health hazard from silica particles. Patches can be cut out and made good to an approximate match with the original surface. The flow of water can be controlled and channelled to reduce resoiling. In the end, however, an opaque coating that is regularly reapplied is the only effective way to avoid a shabby or dirty surface. Clear water repellants have a limited life of five to ten years and may simply result in a different pattern of dirtying as more water runs further down the building. The specialist companies referred to for concrete repair in 5.1.23 can advise on surface treatments and on the risk of promoting reinforcement corrosion if wet cleaning is used.

REFERENCES

BRE Digests 45 & 46: *Design and appearance 1 & 2*

BRE Digest 280: *Cleaning external surfaces of buildings*

Hawes, F *The weathering of concrete buildings* Cement and Concrete Association 1986

BS 6270 Code of Practice for *Cleaning and surface repair of buildings*: Part 2 1985 *Concrete and precast concrete masonry*

Photo 1 The movement crack at the corner reveals lack of adhesion.

Photo 2 Fallen render from 1 – the type of brick used has poor adhesion to cement sand render.

Photo 3 Loss of adhesion allows large sections of render to fall.

Photo 4 Loss of adhesion at the bottom where no render stop was provided.

Photo 5 Loss of adhesion on the face where bricks joints were not cut back as a key.

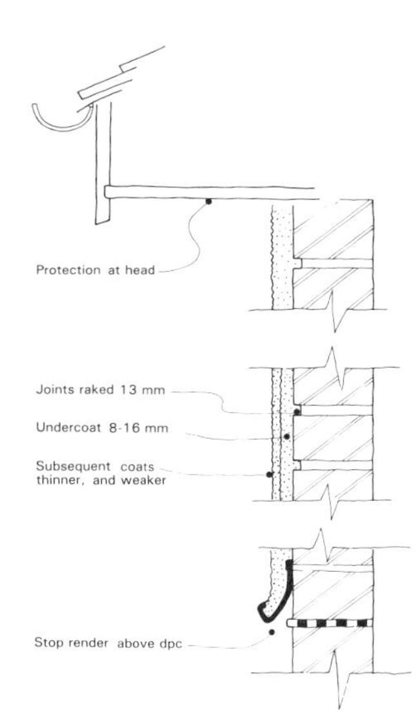

Figure 1 Points to follow with external render.

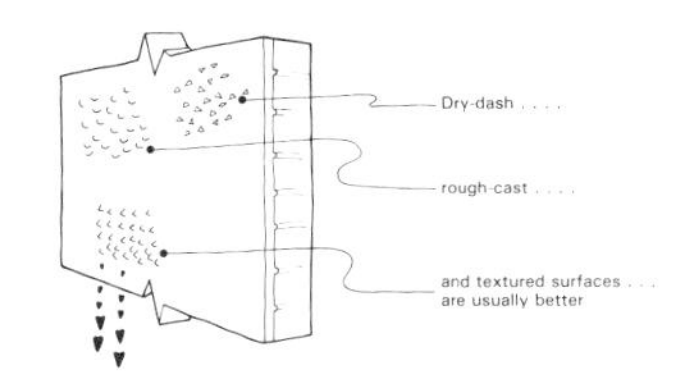

Figure 2 Surface finish affects durability.

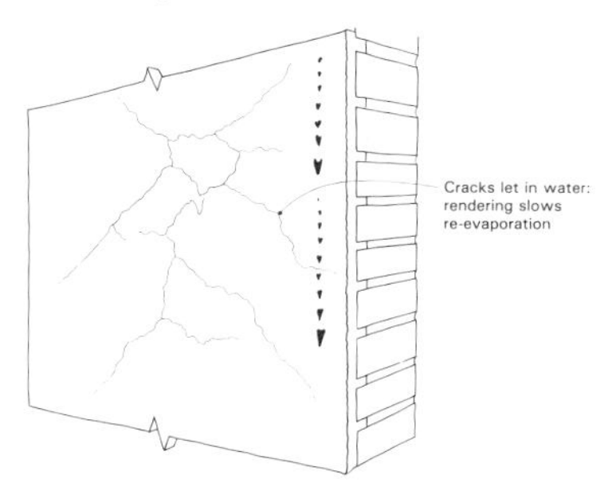

Figure 3 Cracked render can promote deterioration.

Photo 6 Render pulled off the faces of the blocks when the installation of cavity insulation allowed frost to reach the dampness behind the render.

5.3.9 EXTERNAL WALLS. CRACKING AND DETACHMENT OF CEMENT RENDER

Cracks in render that do not show on the background to the render, or sections of render that fall away from the background, or finishing coats of render that fall way from the base coats.

Symptoms

Cracks in cement render that are not related to cracks in the walls behind the render (for cracks in walls see 3.1.1–25). Areas of render or of external render coats falling from the building. Areas of render becoming 'live' (ie detached or partially detached from the background and therefore sounding hollow when tapped).

Investigation

Establish the form of construction and the number and composition of the render coats. A section of render can be sent for analysis (with each coat separately identified). Is the wall cavity insulated (if so, is the occurrence of render failure related to the time the cavity was insulated)?

Note whether there are movement cracks in the building structure; such cracks may have allowed water into backing layers, or the wall may have been wetted from another source. Subsequent freezing can then turn the water to ice. The expansion of the freezing water forces off the render coats. Water can also promote similar damage from sulphate expansion in underlying mortars or render coats, for example, where sulphates are present in brickwork, (in such cases whitish crystals may be seen on the surface of the bricks where the render has broken away).

Examine the surface of the back wall, the finish to the backing coats, and the fixing and condition of any render reinforcement (such as expanded metal lath).

Is there evidence that the bond or key has failed? Record the areas affected by cracking and detachment, and survey remaining areas for hollowness or bulges that suggest more widespread problems.

Diagnosis and cure

Rendering failures are due to a whole range of causes. It is important to identify the cause in each case so that problems do not recur after repairs have been completed – BS 5262:1976 discusses the render specifications and the problems that can arise:

- If the cause is related to application or to dampness, then the repair may be restricted to the affected area only.
- If the cause is wrong specification or unsuitable materials (for example a too fine sand that causes excessive shrinkage), then all the rendering on the building is suspect. In the long run, it may be necessary to completely replace the rendering, or to overclad it to keep the wall dry.

REFERENCES

BRE Defect Action Sheet 37: *External walls: rendering – resisting rain penetration*

BRE Digest 196: *External rendered finishes*

BS 5262:1976 *External rendered finishes*

Building Research Advisory Service: *Failure of renovated rendering* Building Technical File No.13

Construction Feedback Digest 36: *Cavity fill – A cautionary tale* B & M Publications (London) Ltd

Construction Feedback Digest 51: *Render failure (sulphate attack)* B & M Publications (London) Ltd

Photo 7 The top render coat has split away from the undercoat. A cement rich top coat may shrink and curl off a weaker coat below.

Photo 8 Random cracking is likely to be due to shrinkage of smooth trowelled cement rich render.

Photo 9 Salts from the bricks can accumulate behind the render and reduce its adhesion.

Photo 10 Cracks corresponding with masonry joints due to sulphate action on mortar when render retains moisture in clay brickwork (see 5.1.13).

Photos 1, 2, 3 & 4 Breakdown and contour scaling of calcarious sandstone attacked by acidic rainwater.

Photos 5 & 6 Combined effect of polution and frost action where less durable stone has been used.

Photo 8 Rain washing can create unplanned effects.

Photo 7 Acidic run off from timbers eating into limestone.

Photo 9 Stone facing staining by surface water that could have been avoided by detailing an impervious plinth.

5.3.10 EXTERNAL WALLS. WEATHERING AND DISCOLOURATION OF STONE MASONRY

The stone is worn away or stained, affecting its appearance in an unacceptable way, and in some cases also affecting its structural or damp resisting performance.

Symptoms

The stone spalls, wears away or changes colour. This may happen in places where the stone is exposed or in contact with other materials, or it may only affect a limited number of blocks of stone in the wall.

Analysis

Establish the type of stone and, if possible, the quarry from which it came.

Note the variation in weathering discolouration and growth of lichens, mosses and algae in different parts of the building. Where the most marked deterioration can be seen, note the adjacent materials. Look for evidence of water being channelled down the façade or splashed by rain falling on adjacent roofs or pavings.

Examine the bedding of the stones and any differences in appearance between blocks that have been affected and those that have not.

Samples of damaged stone may be sent for analysis to establish whether chemical attack has caused the damage.

Where the sources of supply of stone is known, the quarry may be able to shed light on the cause of degradation, which may be related to the quarry bed from which the stone was taken.

Diagnosis and cure

There are a number of possible causes of stone decay and discolouration – all of them related to dampness. In approximate order of the frequency with which they occur, they are:

- Staining, from uneven water flow over the surface of the building. This shows either as white crusts and tidemarks where the water has brought salts to the surface, or as dark patches where dirt has accumulated on slightly damp surfaces that are not directly washed by rain.
- Breakdown of limestone surfaces due to crystallisation of salts (usually calcium sulphate). The salts become concentrated near the surface from within the stone by evaporation of frequently wetted surfaces. Limestone with a fine porous structure is more liable to this type of breakdown than limestone with a coarser structure. Ground water and seaspray can cause crystallisation damage. Stone that has been contaminated with sodium chloride, sodium sulphate or sodium hydroxide tends to break down by crystallisation over the whole surface rather than just where the wall is wetted by rain. Dense mortar can aggravate damage caused by crystallisation of salts because, it prevents water evaporation from the joints, and concentrates the salts from evaporation on the surface of the stones rather than the joints.
- Frost damage is only likely to affect limetone with a fine porous structure, and then only when the stone is saturated, or nearly so. Frost damage shows as layers or chunks of stone cracking away rather than a wearing away of the surface. It is most common on exposed features such as parapets and areas below DPC level.
- Chemical pollution attack, rainwater containing sulphur dioxide and carbon dioxide can affect calcareous sandstone. It reacts with the calcium carbonate that joints the sand grains together and allows the grains to be washed away.
- Rainwashings from limestones, particularly from magnesian limestone, can carry calcium sulphate or magnesium sulphate down onto sandstone or brick, where it causes breakdown.
- Incorrect bedding – can result in flaking of walling stones along their bedding lines. It is usually caused by crystallisation of salts behind the faces of stones that are not laid on their natural (ie as in the quarry) bed. When the bed is parallel with the face of the wall, it creates a plane of weakness that is susceptible to crystallisation.
- Contour scaling-occurs when crystallisation due to air pollution affects the outer layer of sandstones, and wide sections of the surface crust break away.
- Staining and etching of limestone surfaces – can be caused by run off chemicals from other building materials (such as copper flashings), or water containing impurities (such as the acidic metabolic products of plants).
- Rust stains can originate from steel or cast iron fixings used in the stonework.

The cure for weathering and discolouration of stone masonry is to keep the walls dry. Damp proof courses should be provided at high and low level. Drips and flashings will be needed on projections and at abutments with horizontal surfaces, or where rainwashings from above could damage materials below. Colourless preservative treatments can be applied to prevent surfaces becoming wet, but they require periodic renewal (see BRE Digest 177). If the run off from areas of plant growth is damaging stone, the plants can be controlled with chemicals (see BRE Digest 139).

Stains can sometimes be removed from stonework with repeated application of poultices which draw moisture and the stain out of the stone. It may be necessary to patch stonework. Specialist firms offer a service which uses crushed stone and a binder to build up matching repairs. Alternatively whole stones can be replaced. It may be possible to find a more durable stone that is a reasonable match for the stones that have deteriorated.

REFERENCES

BRE Digest 45:1979 *Design and appearance 1*

BRE Digest 139:1982 *Control of lichens, moulds and similar growths*

BRE Digest 177:1984 *Decay and conservation of stone masonry* (principal reference for this defect)

BRE Digest 269:1983 *The selection of natural building stone*

BRE Digest 280:1983 *Cleaning external surfaces of buildings*

Leary, Elaine *The building limestones of the British Isles* HMSO 1986

Construction Feedback Digest 50 Spring 1985 *Calcareous stonework defects* B & M Publications (London) Ltd

BS 6270: Code of Practice for *Cleaning and surface repair of buildings* Part 1: 1982 *Natural stone, cast stone and clay and calcium silicate brick masonry*

Photos 1 & 2 Dark coloured coating failure on galvanised cladding

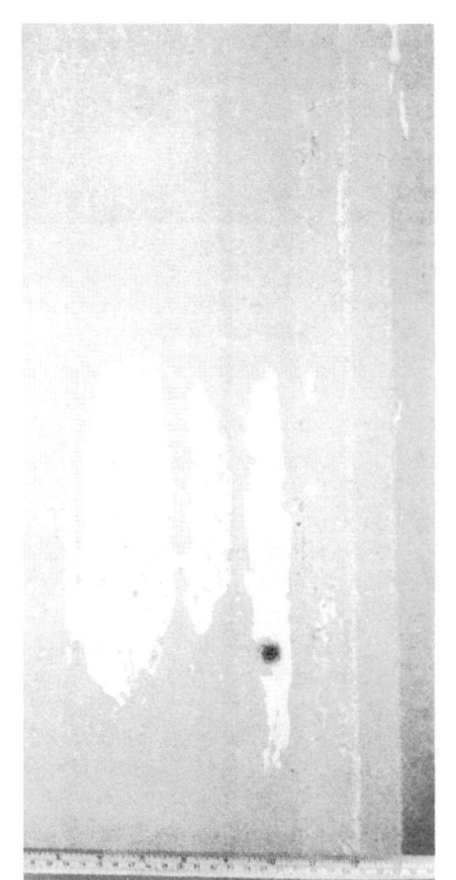

Photo 3 Typical wall cladding decorative coating failure.

Photo 4 Corrosion at bottom of wall cladding. This cladding is particularly vulnerable as it sits down onto the concrete kerb.

5.3.11 EXTERNAL PROFILED METAL SHEET CLADDING. COATING FAILURE AND DEFORMATION

This defect applies to steel or aluminium sheet cladding and to composite panels made from metal sheets bonded to a core of foamed plastics insulation.

Symptoms

Blistering and detachment of the decorative/protective finish, corrosion of the metal, bowing and deformation of the panel.

Investigation

Establish the type and, whenever possible, the manufacturer of the cladding and discuss the defect with them.

Note in particular the type of metal (steel will attract a magnet) – steel inner sheets and aluminium outer sheets are sometimes used in composite panels.

Record the position and extent of the defect. Find out when and where it first appeared, and record its development on subsequent visits.

Note the colour of the finish and whether the defect is more pronounced where the cladding is exposed to strong sunlight.

Send a sample to a laboratory to measure the thickness of the metal and of the protective coatings.

Examine the back of the panel for corrosion.

Note whether the use of the building is liable to produce high internal humidity or an aggressive atmosphere that is liable to cause, or encourage corrosion.

Diagnosis and cure

The symptoms are dealt with separately (although they may be related).

- Blistering and detachment of finish can be due to defective or unsuitable coatings, and to corrosion of the metal once it is exposed. There have been instances of dark coatings in strong sunlight failing long before either light coatings or dark coatings shaded from the sun.
- Corrosion, apart from that caused by drainage or defects in the protective coatings, is due to chafing of the sheets at joints and fixings as a result of thermal expansion and contraction, and to dampness from condensation on the back of the sheet.

  Corrosion could be quite rapid in relatively cool damp conditions and would not occur at all in hot dry conditions. However, if the coating has degraded as a result of exposure to sunlight, then the cladding could become more prone to corrosion. It is worth noting that in corrosive environments, under-sides of sheeting can become contaminated with corrosive salts. As these salts are usually able to absorb moisture from the atmosphere, corrosion may occur at an accelerated rate. Corrosion is more rapid and more marked on steel cladding that is not properly protected than it is on aluminium cladding. Corrosion of aluminium is less noticeable because of its light colour which is closer to the colour of the metal than brown rust is to steel. The back side of metal sheets is particularly vulnerable to corrosion because the sheet can act as a very cold vapour barrier when its temperature drops. To make matters worse, it has been normal practice to use a thinner protective coating on the back side of profiled steel cladding.
- Bowing and deformation of the panel can be caused by thermal movements, particularly where, in composite sandwich panels, the metal facings become delaminated from the plastics foam core insulation. Delamination will not necessarily lead to water penetration or unacceptable appearance, but it can allow corrosion to start at the back of the cladding sheet.

To repair PVC ('plastisol') coatings on steel, there is a special treatment recommended by the manufacturers. Other organic coatings may be removed with a water rinsable solvent paint remover (followed by a thorough rinsing of all traces of the paint remover). All corrosion should be removed, and rusty patches wire brushed to a bright metal finish before applying a new paint system. Suitable primers are available for application both to clean steel and to galvanised surfaces. Alternately, prime galvanised and wire brushed rusted steel areas separately, using either a special primer intended for zinc coated surfaces, or an appropriate primer for clean steel, before applying the new, decorative paint system. Etch primers may not prove effective over galvanised surfaces following the chemical removal of existing coatings. In case of difficulty, advice should be sought.

5.3.11 EXTERNAL PROFILED METAL SHEET CLADDING. COATING FAILURE AND DEFORMATION

continued

REFERENCES

GLC materials bulletin 91 Jan 1976 Item 6 and bulletin 92 Feb 1976 item 6 *An appraisal of pre-finished metal cladding materials*

A J Technical File 6 August 1986 *Element design guide external walls 5. Metal Panels* The Architectural Press Ltd

Construction 49 Feedback digest item 13 *Factory finished steel coating failure* B & M Publications (London) Ltd

Construction 52 Feedback digest item 6 *Factory finished steel coating failure* B & M Publications (London) Ltd

BSCP 143 Code of practice for sheet roof and wall coverings Part 10 1973 *Galvanised corrugated steel* para 6.2 *need for painting*

BRE Digest 217: *Wall cladding defects and their diagnosis*

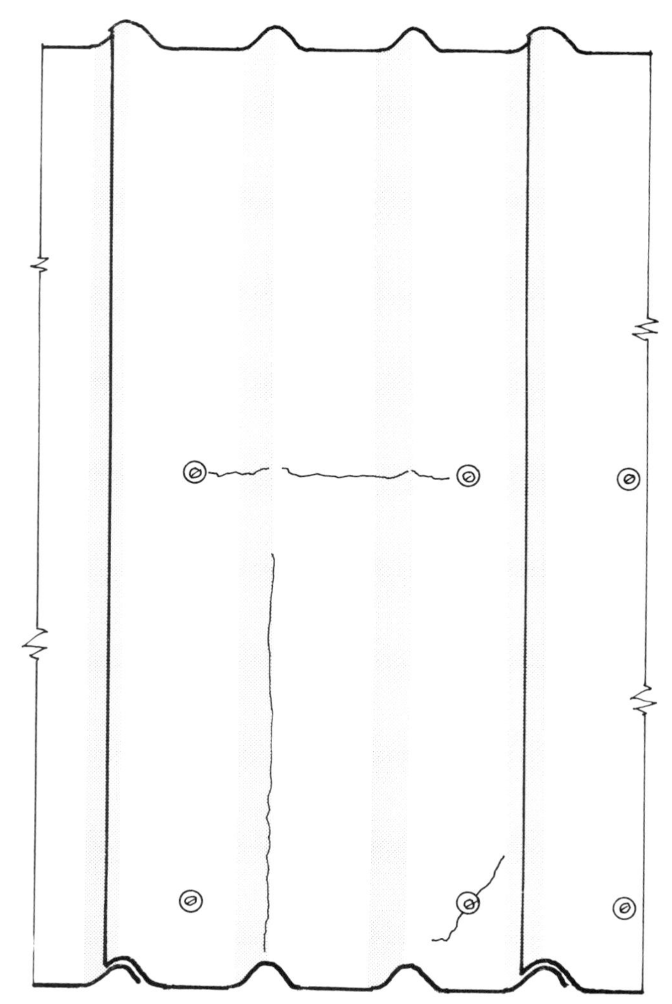

Figure 1 Splits along a ridge may be caused by bowing of the sheet. Splits related to fixings are more likely to be due to fixing faults or structural movement.

5.3.12 ASBESTOS CEMENT SHEETS. CRACKING OR SPLITTING

This defect affected asbestos cement which is now being replaced by non asbestos fibre reinforced sheeting. When the new sheeting is made with calcium silicate rather than cement, only one aspect of this defect is relevant.

Symptoms

The cracking may be in one of two forms. In one, the cracks are in the vicinity of the fixing device, and may either radiate from it, or run across the sheet to the nearest adjacent fixing point. In the other, the cracks run along the tops of the ridges of profiled sheets, or lengthwise down flat sheets between the fixing points. This latter form of cracking is more often associated with sheets that have been painted on one side only.

Investigation

Establish whether the sheets are asbestos (or non-asbestos plastic fibre) cement or of material with a calcium silicate binder – try to find out the manufacturers name and the date of supply.

Examine the fixings for corrosion and measure the size of the holes through which the fixings pass. Note whether the sheets were tightly or loosely held.

Examine the structure to see whether there has been any movement that could have put additional stresses on the sheets. Wind loads, thermal expansion, additional live loads or subsidence may have caused movement.

Note whether the sheets are painted and, if so, whether on both back and front or on one side only.

Note the type and colour of the paint and whether the sheet is exposed to strong sunlight.

Record the extent of the cracks and splits and when they were first observed.

Diagnosis and cure

Both defects are related to the clearances at the holes taking the fixings. If the sheet is held too tightly by the fixings, it may crack, either because of movement in the building, corrosion and expansion of the fixings or because the sheet used has shrunk. When an asbestos cement sheet is painted on one side only, it may bow as a result of carbonation (and consequential shrinkage) on one side only of the sheet. This type of bowing is only likely in cement based sheets. It can result in sheets that are painted on the outside bowing outwards and cracking from end to end between fixings. Large dark coloured flat sheets painted with impervious coatings are the most likely to suffer from this defect. Where cracks radiate from a fixing, it is likely that the fixing has rusted or there has been structural movement.

Temporary repairs can be carried out with waterproof tape and over-painting, but in the long term it will probably be necessary to replace the cladding sheets. Suitable precautions must be taken if asbestos cement cladding is removed, and arrangements must be made with the Local Authority for the disposal of the discarded sheets.

REFERENCES

BS CP 143 *Code of practice for sheet roof and wall coverings* Part 6 1962 *Corrugated asbestos cement* (has now been withdrawn, but is still quoted by manufacturers).

S1 1987 No 2115. HMSO *The control of asbestos at work regulations 1987*

5.3.13 DISTORTION OF uPVC CLADDING SHEETS

This defect can occur when uPVC cladding is rigidly fixed and is exposed to strong sunlight.

Symptoms

The cladding sheets deform. Some may twist, others may develop undulations. The defect is particularly marked on the South side around windows, in the vicinity of stanchions and close to the corners of the building.

Investigation

Establish the material used for the cladding, and if possible the name of the manufacturer.

Note the construction of the wall and the position of any insulation. The dimensions of the cladding sheets, the spacing of fixings, the type of fixings and the clearance of the holes around the fixings should also be noted.

Check the orientation of the affected cladding, and note its colour.

Record the extent of the distortion, and the details of edge trims and fixings in the region of the defect.

Diagnosis and cure

This defect is due to the high coefficient of thermal movement of uPVC sheets. If they are held rigidly, large sheets are bound to distort in hot sunlight. Where insulation is packed against the back of the sheet, and the sheet has a dark coloured surface, the heating-up of the cladding will be exaggerated. The thermal movement will show most clearly against fixed points such as stanchions, windows and corners of the building. The cure is to make allowance for movement in the design of the fixings, and at edges and abutments, by providing larger holes for fixings, and using flexible washers that will allow the cladding to slide relative to the fixing etc. Larger gaps and flashings around fixed windows etc. and the use of a light colour will all help to avoid the problem. uPVC claddings should not be more than 3 m long (possibly 4 m). It will be an advantage if they can be ventilated at the back, to cool them down.

REFERENCES

Bickerdike, Allen, Rich and Partners in association with Turlogh O'Brien. *Design Failures in Buildings*, Second series sheet No 18. *PVC sheet cladding*. George Godwin Ltd.

Figure 1 A bonding agent is recommended for dense concrete backgrounds.

5.3.14 INTERNAL WALLS. PLASTERWORK. BULGING; TOTAL DETACHMENT

Incorrect plastering specifications or techniques can lead to total failure of adhesion on dense concrete backgrounds. The same principles apply equally to backgrounds of high density clay or concrete bricks and blocks, ceramic tiles, firmly adhering paintwork or any other surface with a very low porosity.

Symptoms

Large areas of plasterwork may be observed to be bulging away from the concrete wall surface. There may be extensive hairline cracking to the outer surface of the plaster finish, with more defined cracking at the edges of the bulges. The plasterwork will sound hollow when tapped and will readily break free and fall away cleanly from the concrete.

Investigation

Attempt to obtain a copy of the specification for the plasterwork, or otherwise establish whether a bonding agent was applied to the concrete surface prior to the application of the base coat.

Examine the loose plaster; note the different types of material used for each layer and the bond strength between the layers. Note also whether an aggregate was added to the base layer, and if so, which type.

Inspect the surface of the exposed concrete, noting the presence of extraneous matter such as salts or dirt. A visual inspection should also be sufficient to reveal whether an attempt was made to provide a keyed finish to the concrete.

Diagnosis and cure

The reason for this defect is failure of adhesion between the dense concrete substrate and the base coat of the applied plasterwork. The cause may not be obvious as there are several factors that need to be taken into account.

Dense concrete does not provide a good base on which to apply plaster for two distinct reasons.

1 The smooth surface does not provide a good 'key'

2 The high density produces a low suction background.

Roughening the surface of the concrete by mechanical means will produce a better key for the first plaster coat. The application of a bonding agent before plastering will considerably increase the adhesion characteristics of the surface. The BS Code of Practice recommends that this should be done with a proprietory emulsion of PVA or, alternatively, a bituminous solution may be used on vertical surfaces only, where the additional benefit of providing resistance to damp penetration is required. The practice of applying a PVA emulsion bonding agent to surfaces before plastering should be checked as conforming with the PSA's own Code of Practice where necessary.

The choice of plaster type is also extremely important. On surfaces with low suction it is essential to use a bonding type gypsum plaster. Browning type plaster must not be used as it is not suitable for dense backgrounds.

When failures of this nature do occur, all loose plaster should be hacked off the wall, and the wall surface brushed clean with a stiff bristle or wire brush. The wall may then be replastered after the application of a bonding agent.

REFERENCES

BS 1191:1973 *Specification for gypsum building plasters.* Part 1. *Excluding premixed lightweight plasters.* Part 2. *Premixed lightweight plasters*

BS 5270:1976 *Specification for Polyvinyl acetate (PVAC) emulsion bonding agents for internal use with gypsum building plasters*

BS 5492:1977 *Code of practice for internal plastering* (formerly CP211)

BRE Digest 213: *Choosing specifications for plastering*

5.3.15 INTERNAL WALLS. PLASTERWORK. DETACHED TOP COAT

This defect occurs most frequently on internal walls constructed from concrete blocks. It will usually take place within two or three years of the work being completed.

Symptoms

The symptoms are not dissimilar to those of total detachment (5.3.14).

The wall surface will sound hollow when tapped over all, or parts of the surface. It may be bowing appreciably and break away readily, leaving the base coat adhered to the wall in most instances.

Alternatively, small areas of the top coat may break up and fall away in random patches.

Investigation

As much information as possible should be sought regarding the construction methods employed and the weather conditions at the time. In the case of concrete blockwork, check whether the site storage conditions for the blocks prior to being laid included adequate protection from inclement weather conditions. Attempt to establish the time lapse betwen the wall construction and the application of the plasterwork.

Examine both the intact material and that which has broken away. Note whether they have parted company cleanly or whether pieces of the base coat are still adhered to the top coat.

Identify the different types of plasterwork materials that were used.

Diagnosis and cure

If the basic wall construction is comprised of concrete blockwork, a possible cause for failure could be that the blocks were wet when laid and not allowed enough time to dry out before the wall was plastered. The blocks would shrink as they dried and the gypsum plaster would expand slightly, the differential being sufficient to break the bond between the two. Where the undercoat of plasterwork is cement based, it would shrink with the concrete blocks and remain adhered to them but debond itself from a gypsum top coat, which could then split away.

Assuming that the material shrinkage has ceased by the time the failure occurs, the wall may be replastered having first removed all the defective matter. An added precaution against future failure would be to coat the prepared surface with a PVA emulsion bonding agent before replastering. Make sure the bonding agent is not allowed to dry out, otherwise its intended effect will be reversed, and it will actually encourage separation of the coats. The practice of applying a PVA emulsion bonding agent to surfaces before plastering should be checked as conforming with the PSA's own Code of Practice where necessary.

Another cause for plaster coats debonding is incompatibility of materials. This may happen when a strong expanding finishing coat is applied to a weak undercoat or a strong shrinkable undercoat. The remedy will again involve stripping the existing finish coat. If the base coat is still sound, it may be replastered with a plaster of comparable or lesser strength than the base coat.

With the advent of reliable 'one-coat' plaster mixes this defect is becoming less common.

REFERENCES

BS 1191:1973 *Specification for gypsum building plasters.* Part 1. *Excluding premixed lightweight plasters.* Part 2. *Premixed lightweight plasters.*

BS 5270:1976 *Specification for Polyvinyl acetate (PVAC) emulsion bonding agents for internal use with gypsum building plasters*

BS 5492:1977 *Code of practice for internal plastering* (formerly CP211)

BRE Digest 213: *Choosing specifications for plastering*

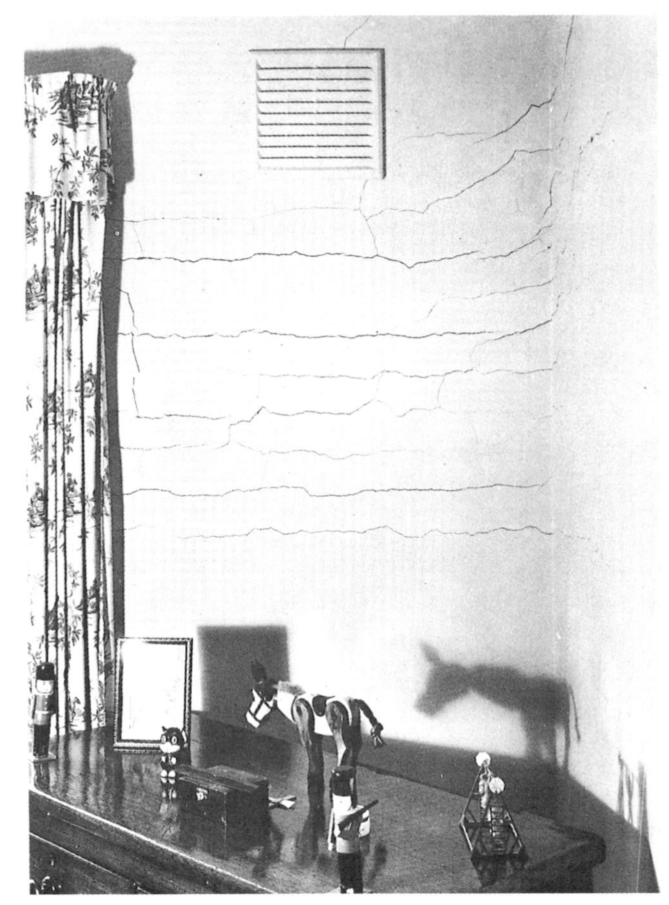

Photo 1 Cracks indicating expansion of mortar joints in the masonry background.

Photo 2 Cracks indicating movement of a wood wool slab lining.

5.3.16 INTERNAL WALLS. PLASTERWORK. CRACKING

The reasons for cracking of plaster on internal walls are numerous. Careful investigation may be required to establish the true cause.

Symptoms

Cracks in plasterwork can take several forms:–

Fine hairline cracking in a random pattern which usually covers the whole wall area.

Defined vertical and horizontal cracking in a zig-zag pattern.

At the junction of walls and ceilings (see also 7.3.2).

Vertical cracks which usually run the full height of the wall.

Diagonal cracks often adjacent to windows and door openings.

Cracking associated with surface bulging (see 5.3.14 & 5.3.15).

Investigation

Ascertain the age of the building, the construction methods employed and the materials used for building the walls.

Inspect the rest of the building for cracks, and check windows and doors for signs of binding, squareness etc. If doubts exist about the structural integrity of the building, consult a structural engineer.

Determine the depth of the cracking, except in the case of random pattern hairline cracks. It may be necessary to remove some of the plasterwork to establish whether the cracks are associated with cracking of the wall background.

In cases of dispute, especially where there are alleged damages due to external vibrations, a survey will be required to determine the amplitude and frequency of the vibrations. Bodily perceived observation of the problem will be highly inaccurate and misleading, as the human body frame is extremely sensitive to vibration. Under normal circumstances, it will require a tremor of considerable magnitude, and of enough force to cause human pain, to produce structural damage to a building.

With larger cracks that are suspected to be associated with background movements, it will be necessary to monitor the amount of movement at monthly intervals for a minimum period of six months. Traditional glass 'tell-tales' are no longer considered suitable for this purpose because, although when they break it is an indication that movement is taking place, there is no way of measuring the actual extent of movement. A more conclusive method is to measure between two non-ferrous pins or screws fixed one either side of the crack.

It should be remembered during the investigation procedure that there may be more than one cause contributing to the defect, and there may be more than one type of cracking occurring at the same time.

Diagnosis and cure

Building materials that require the addition of water to initiate a chemical setting reaction have a tendency to shrink as the water dries out. Excessive trowelling of finish plaster will draw water to the surface and cause the plaster to dry out too quickly. This, or any other reason for accelerated drying, will typically induce surface cracking which takes the form of a network of fine cracks. This kind of cracking is sometimes referred to as 'mapping'. It is quite simple to rectify using conventional decorating methods and is unlikely to recur.

Zig-zag vertical and horizontal cracks will most certainly be found to follow the mortar courses of brickwork or blockwork. It will usually be associated with shrinkage movement of the wall when it occurs at an early stage in the life of the building, but if it appears after a number of years, it could be due to structural movement. Cracks of this nature are extremely difficult to hide or disguise as the movement can be seasonal. They may be filled with a proprietory filler but should always be expected to reappear at a later date.

Long vertical or horizontal cracks are often related to joints between two materially different components such as wood and brick. Because the materials absorb moisture at unequal rates, their moisture movement is uneven causing the plaster to crack at the position of the joint between the two. This may often be observed around the outside of internal door frames. High level horizontal cracking at the joint between the wall and the ceiling may be disguised by fixing a cornice or

continued

moulding into the angle. Occasionally, vertical cracks will occur at the joints between plasterboard panels on dry-lined walls. This can be rectified by cutting out the joint area and scrim taping over the joint before applying a finish of special jointing plaster. The manufacturers application instructions are very clear and should be followed carefully.

Diagonal cracks are almost always associated with differential structural movement. The direction of movement can often be determined by the angle and the taper of the cracks. It may only be minor and associated with initial settling down of the building, but if it continues to worsen consult a structural engineer.

It is often alleged that vibrations from passing traffic and other sources are the cause of cracks in plasterwork. It is true, as with thermal movements, that vibrations may be detected and observed at a crack, but this is so very seldom the cause that it should not be seriously considered until all other possible causes have been eliminated. When considering vibrations, it must be appreciated that if there has been no other dimensional change in the construction, the ends or corners of the wall must have been moved apart to an extent equal to the width of the crack; evidence of this movement must be established before vibrations can be diagnosed as the origin of the cracks. Even so, it is far more likely that the cracking started for a different reason and was exacerbated by the vibrations.

REFERENCES

The White Book. *Technical manual of building products.* British Gypsum Ltd.

BS 5492:1977 *Code of practice for internal plastering* (formerly CP 211)

BRE Defect Action Sheet 96: *Foundations on shrinkable clay: avoiding damage due to trees*

BRE Digest 75: *Cracking in buildings*

BRE Digest 213: *Choosing specifications for plastering*

BRE Digest 278: *Vibrations: building and human response*

Photo 1 Toughened glass cladding panels that have failed.

Photo 2 Mesh fixed over cladding in case of further failures.

5.3.17 CURTAIN WALL. GLASS CLADDING. CRACKING AND BREAKING

A surprisingly common defect which has been recorded as affecting up to 1 in 40 panels on any one building.

Symptoms

Glass cladding panels are found to have broken. On occasions they may even be seen or heard to crack, particularly during periods of hot weather.

Investigation

Examine the framing system for signs of movement at the fixing points. If the frame is steel, check it for evidence of rust around the failed panels.

If the panels are of a composite sandwich construction, determine whether the layers are bonded together or whether air gaps are present. Attempt to identify the type of glass used and also the manufacturer. If this is obtainable it would be helpful to contact them and enquire as to whether they have a history of problems associated with the system.

If the panel is cracked but otherwise intact, measure the overall cut size and also the dimensions of the opening into which it is installed.

Where star-shaped fractures are occurring, especially at low level, check for evidence of stones or similar missiles that may have struck the panels.

Diagnosis and cure

Rusting steel expands and this may put excessive pressure on the edges of the cladding panel. Where the glass has not been cleanly cut it will crack at the edges quite readily (see 6.3.3).

The British Standard code of practice requires the following minimum edge clearances for glass:

GLASS TYPE	*Edge clearance for longest side*	
	Up to 2 m	*Over 2 m*
	mm	mm
Float, sheet, cast, patterned, toughened and wired glass: up to and including 12 mm nominal thickness.	3	5
Over 12 mm nominal thickness.	5	5
Laminated glass: up to and including 12 mm overall thickness.	3	5
Exceeding 12 mm but not exceeding 30 mm overall thickness.	5	5
Exceeding 30 mm overall thickness.	10	10
Insulating glass units up to and including 18 mm overall thickness.	3	5
Exceeding 18 mm overall thickness.	5	5

These clearance dimensions may have to be increased if a gasket glazing system is employed. On occasions the glass manufacturer may require slightly greater tolerances for some types of solar control glasses. Their advisory service should be consulted for their latest recommendations.

Much cracking has occurred in wired glass used as spandrel panels backed up with an inner board panel. Wired glass is particularly prone to overheating due to excessive solar radiation such as occurs when the inner panel finish is a dark colour and/or insufficient ventilation is provided between the layers. The mechanical strength of the glass is lower than that of float glass of equivalent thickness due to the wire reinforcement, which also makes it difficult to cut to a clean edge. Where sheets need to be replaced, consider using a gasket glazing system which is better able to absorb the effects of expansion and also cushion the cut edges. Care must be exercised to ensure that all the edges are either sealed or well drained to prevent water from remaining in contact with the cut ends of the wires. With a ventilated system it is recommended that a gap of not less than 50 mm be

continued

provided between the panels and, when unventilated, not less than 170 mm.

Where there is a heavy incidence of failure of toughened glass cladding panels, it may be more economical to fix fine mesh wire screens over the panels as a safety measure rather than replace all the panels. This would adequately protect any passers-by from falling panels and may be virtually invisible from the ground.

Impact damage caused by malicious attack is very much more difficult to protect against. Toughened glass has a good resistance to most impacts, but where a social problem exists, it may be cheaper and easier to replace the panels with a different and more suitable material.

REFERENCES

BS 6262:1982 *British Standard Code of practice for Glazing for buildings* (formerly CP 152)

Construction Feedback Digest 39 – Summer 1982 Item 1 – *Toughened glass cladding failure* B & M Publications (London) Ltd

Development and Materials Bulletin No. 109 (2nd series) Item No. 8 – *Failure of under-window panel glazing – Feedback.* Greater London Council

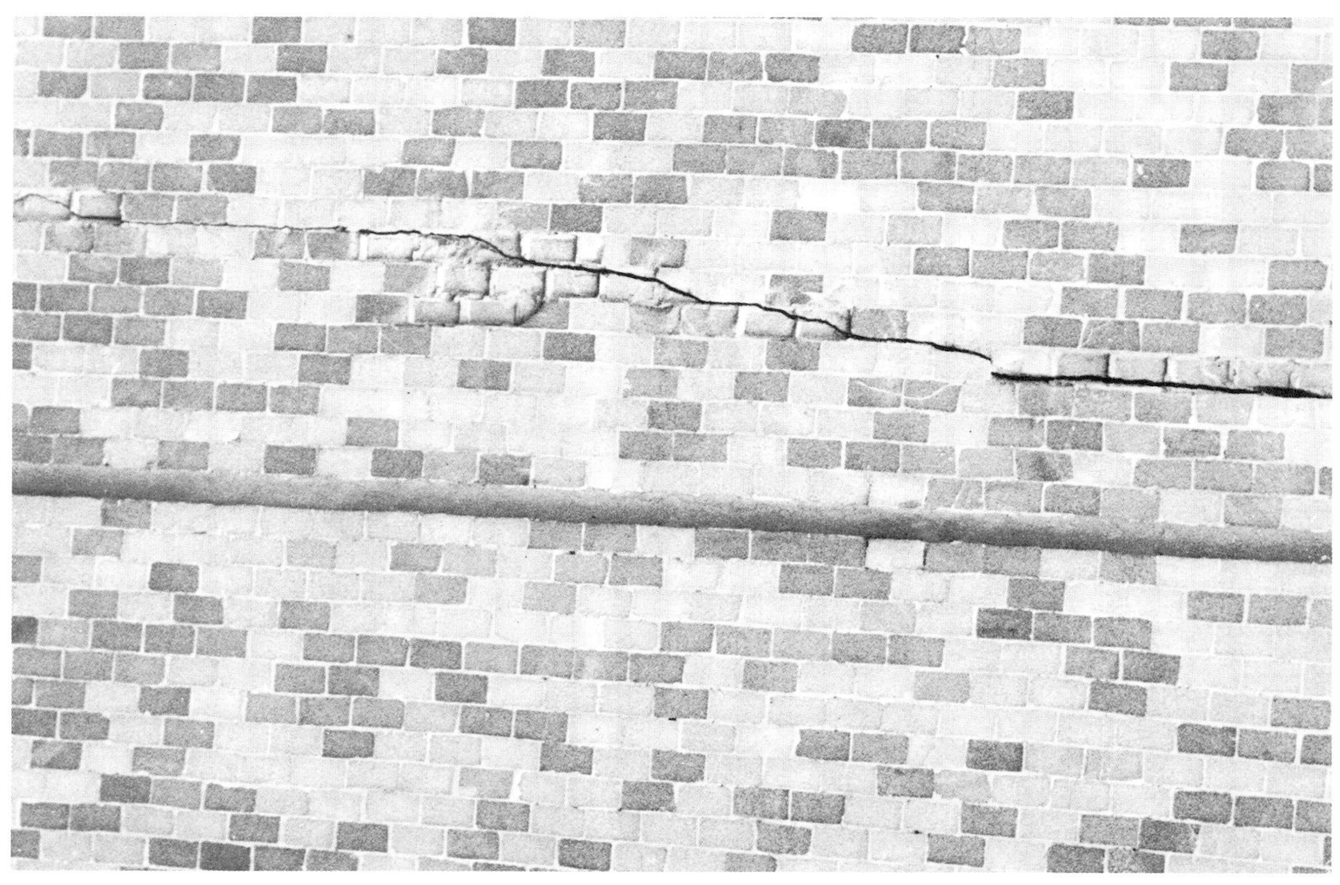

Photo 1 Mosaic cladding becoming detached at a crack.

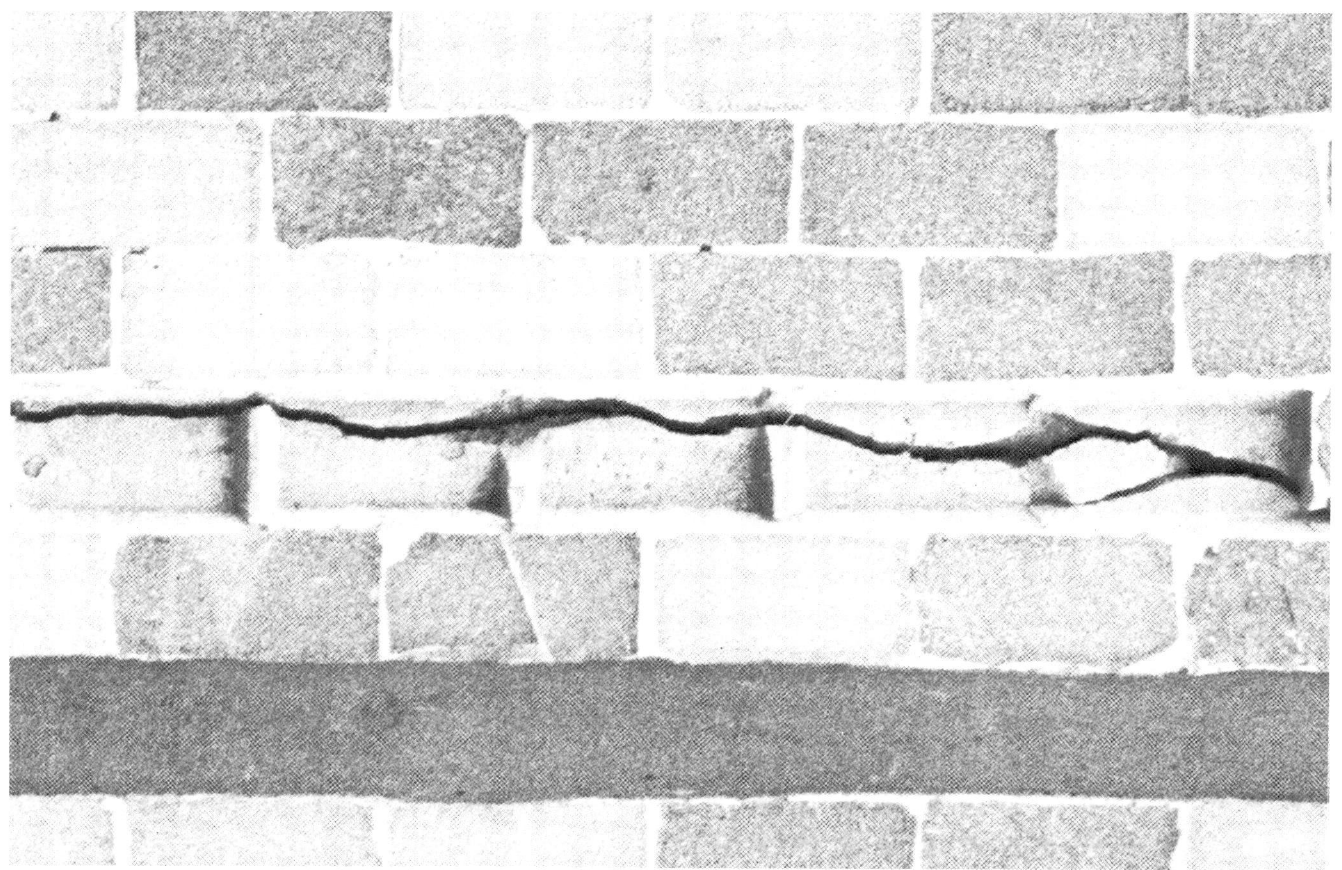

Photo 2 Close up.

5.3.18 EXTERNAL WALLS. TILE AND MOSAIC CLADDING. DETACHMENT

A very common defect which has contributed to the loss of popularity of this type of finish.

Symptoms

Areas of tile or mosaic finish crack, become loose and sound hollow or break off either as individual tiles or in large sheets.

Investigation

Examine the back of mosaic that has broken free to establish the place of failure of adhesion and note whether the break was clean or uneven. Check the bedding mortar for coadhesivity and thickness and if it is suspected that the mixture is particularly strong have it analysed for cement content. Note whether the mosaic has a backing fabric and check for evidence of a bonding agent having been used. Examine the background wall for cleanliness and note whether a keyed surface was provided. Where movement joints have been provided check their width, the distance between them, and ensure that they extend through the full depth of the tile and bedding.

Diagnosis and cure

Cracking will often occur where the mosaic is applied to a thick bedding which subsequently shrinks. Where the bedding has been applied to a concrete background that has not fully matured, the concrete may still shrink causing a break in the bond between the two surfaces. A bedding mixture containing a high proportion of cement will also shrink excessively and will crack the mosaic as it contracts. Even if it remains adhered to the background, there will be a danger of water ingress which will freeze in sub-zero temperatures and break the finish off as it thaws.

Cracking may also occur where there is a change of material in the background such as brick infil panels in a concrete framework. The finish will adhere more strongly to the keyed brickwork than the concrete, unless the concrete surface has been well broken-up with a bush hammer or similar. BS 5386 pt 2 advises that movement joints should be provided horizontally at each storey and vertically at approximately 3 m centres, and also at points where there is a change in background material. The joints must reach right back to the wall surface, be a minimum of 6 mm width and should be pointed with an elastomeric sealant.

Very occasionally, failures occur when unsuitably thin bed adhesives are used. Fixing mosaics is a highly skilled trade and should only be attempted by specialist operatives using approved materials.

Patchwork repair of failed areas can look very unsightly as it will be extremely difficult to match the existing mosaic for colour and the affects of natural weathering. It is also a slow laborious job which tends to make it costly. A more favoured approach is to disguise the defect by overcladding with a sheet material. However, this may present problems in obtaining planning permission.

REFERENCES

BS 5385: Part 2: 1978. *Wall and floor tiling* Part 2 *Code of practice for external ceramic wall tiling and mosaics*

Construction Feedback Digest 41: Winter 1982/83. Item 9 *Loose mosaic finish* B & M Publications (London) Ltd

5.3.19 RAPID FLAME SPREAD ON PAINTED SEMI-RIGID MINERAL WOOL SLAB LININGS

Although this defect deals with increased fire hazard, which has generally been excluded from the coverage of this book, it has been included here as it may become evident without a major fire.

Symptoms

Flames spread rapidly over the surface of a wall (or ceiling) where non-combustible semi-rigid mineral wool slabs have been painted (usually spray painted) with several coatings of emulsion paint.

Investigation

Establish the type of insulation slab and what surface spread of flame rating is required for the linings.

Find out the type of paint and how many coats were used on the slabs and whether the paint is classified as fire retardant.

Note the extent of the flame spread and the original source of ignition.

Diagnosis and cure

Normal coatings are combustible whether solvent or water based. However, on a solid surface, the fire only has access to the outer layer and its heat flows quickly into the substrate so that the coating makes very little contribution to the surface spread of flame. When a coating is used on a substrate that has a low thermal conductivity and a surface that the paint adheres to very poorly – such as rigid mineral wool slabs, then the combustibility of the coating becomes significant. The temperature of the coating builds up rapidly in a fire and it starts to flake away exposing more surface area to the flames. Even flame retardant coatings on mineral wool may have to be downgraded slightly from that of their rating on a solid substrate. Where slabs are already painted, a fire retardant coating over the surface will tend to consolidate the surface and will improve the surface spread of flame properties.

REFERENCES

PSA Feedback. *Painting mineral wool – fire hazard.* Building Technical File No.1

Typical time to next painting maintenance after initial application (see Notes column*)	General description of paint (see Notes column*)	Coating specification (see Notes column*)	Notes	Manufacturers
10 years plus	2-pack polyurethane paint aliphatic type (a) gloss, semi-gloss or eggshell	Depending on manufacturer's recommendations, may be 2 coats finishing coat, or may require a special primer, Preferably applied by brush or roller (b) to dry surfaces in good weather condition (c)	(a) Aliphatic type preferable for colour retention/non-yellowing properties, and gloss retention (b) Spray application not recommended on sites because of droplet toxicity hazard and possibility of contamination of distant surfaces in windy conditions. Brush/roller application is usually satisfactory but precautions are still needed to avoid toxic hazard to operatives and occupants (c) Dry weather, temperatures above 5°C needed for satisfactory drying/curing ie must not be applied when there is risk of rain within 6 hours	TKS Ltd Berger Paints Ltd International Paints Ltd ICI Ltd Paints Division Dufay Titanine Ltd Trimite Ltd Some of these may offer polyurethane paint to DTD 5580 Standard
8–10 years	1-pack moisture cured polyurethane paint, gloss, semi-gloss or eggshell	1 undercoat, 1 finishing coat or 2 finishing coats according to manufacturer's recommendations for condition of surface. Preferably applied by brush or roller (b). Tolerant of damp conditions at application, but weather should be dry as above for best results (d)	(b) See above (d) Whilst more tolerant of damp conditions, this property should not be exploited unreasonably and weather conditions should be preferably dry and settled	Crown Decorative Products Ltd
5–6 years	Silicone alkyd gloss or semi-gloss (e) single pack material	1 undercoat, 1 finishing coat or 2 finishing coats according to manufacturer's recommendations for condition of surface. Preferably applied by brush or roller. Dry surfaces necessary. Weather should be dry but risk of rain 2–3 hours after application is acceptable	(e) Available as high build coatings for greater film thickness ie about twice normal paint film thickness	Crown Decorative Products Ltd Berger Paints Ltd Goodlass Wall & Co Ltd
3–5 years	Drying oil alkyd conventional gloss paint and drying oil polyurethane full gloss (f)	1 undercoat and 1 finishing coat, or 2 finishing coats. Preferably applied by brush or roller. Dry surfaces necessary. Weather should be dry but risk of rain 2–3 hours hours after application is acceptable	(f) No advantages of drying oil polyurethane type in durability. No advantage in durability from non-drip types	Many manufacturers available through Building Paint Scheme (DEF STAN 80–59) – conventional type
3–6 years (g)	Emulsion paints – various exterior quality types (g)(h)	2 coats preferably applied by brush or roller. Dry surface necessary. Weather should be dry with no risk of rain for 6 hours and no risk of frost while film is wet	(g) Life dependent on types of emulsion coating and suitability in the particular circumstances (h) Ranging from standard exterior emulsion paints to specialist high performance coating.	Many manufacturers available through Building Paint Scheme (TS 473) Specialist material – Liquid Plastics Ltd Decadex or equivalent

*letters in brackets refer to Notes column

Table 1 Schedule of paint systems for overpainting weathered GRP cladding published in the Construction Feedback Digest No. 51 1985.

5.3.20 PEELING AND FLAKING OF PAINT ON EXTERNAL PLASTICS CLADDINGS ETC.

This defect can affect painted plastics windows, rainwater goods and trims as well as cladding.

Symptoms

Recently applied coatings start to flake and peel. There may be signs of dust or dirt on the inside surface of the coating as it peels away.

Investigation

Establish the type of paint used (if necessary by laboratory tests).

Identify the type of plastics. BRE have produced a kit of samples with explanatory notes to help with identification.

Check the surface of the plastics for dirt and grease – some of it may have been removed with the paint film.

Find out what preparations were done prior to painting.

Diagnosis and cure

If the paint and preparation is suitable, most plastics will provide a more stable base for painting than the majority of timbers or metals. However, plastics are usually more flexible than other substrates. It is important to note that the application of certain paints (eg conventional hard gloss) to plastics, can result in loss of impact strength and this is the case in particular, with rigid plastics such as uPVC, polycarbonate, PMMA and polystyrene. Flaking and peeling of paint may either be due to the incompatibility of the coating and substrate or to the inadequate preparation of the substrate. Compatibility should be checked with the manufacturers. Solvents in paint can damage certain plastics such as polycarbonate, polymethyl methacrylate (PMMA) or polystyrene (the solvents can cause stress crazing). Dirt, grime, traces of cleaning materials or harsh abrasion can all prevent the paint adhering properly to the substrate.

When this defect has occurred, it will be necessary to remove the remaining paint layers and start afresh. Paint cannot be burned from plastics. Fine abrasives or, on PVC and GRP special purpose trichloroethylene based strippers should be used. Conventional solvent based paint strippers (ie based on dichloromethane) are unsuitable for removing paint from plastics surfaces. Before repainting, the surfaces should be washed with detergent in warm water and thoroughly dried. Do not use dark coatings in sunny positions as this will raise the temperature of the plastics and cause additional thermal expansion.

REFERENCES

BRE information paper 11/79 '*Painting Plastics*'

BRE Monograph 1977 *Aids to the identification of plastics explanatory notes*

Construction Feedback Digest 51, Summer 1985 *Restoring appearance of weathered GRP cladding* B & M Publications (London) Ltd

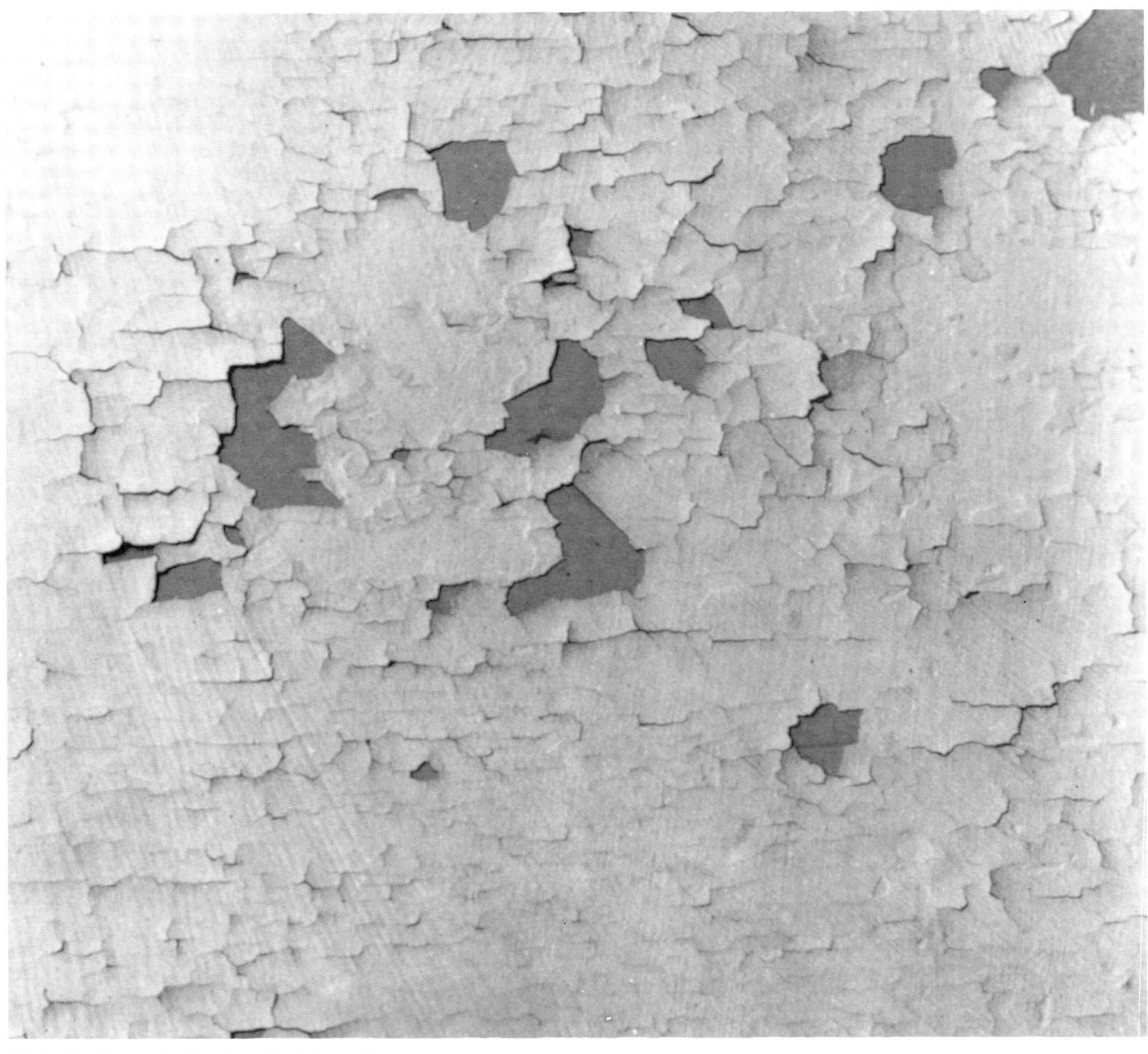

Photo 1 Loss of adhesion of paint coating.

Photo 2 The presence of moisture aggravates painting defects – similar defects can be seen on external as well as internal surfaces.

Photo 3 Drying out resealing and stabilising may be necessary if the coating has failed.

5.3.21 INTERNAL WALLS. BREAKDOWN OF PAINT COATINGS AND DISCOLORATION

The defects described here can affect various paint systems applied to gypsum and lime plasters or to cement render. They may also occur *externally* but the symptoms will be complicated by the influence of climate and exposure to sunlight.

Symptoms

Symptoms range from slight unevenness of the sheen on the surface (particularly noticeable when light falls on the wall obliquely), to serious crazing, flaking and detachment of the paint coating. There may be signs of mould or salt crystals and the plaster surface may crumble and 'rot'. It is always likely that dampness has been present at some stage in the development of the defect, although it may have dried before investigation starts.

Investigation

Establish the composition of the coating and substrate materials. If records are poor or deposits are present, it may be necessary to send samples to a laboratory for microscopic examination and analysis.

Investigate potential sources of dampness, ie damp penetration, leaks, condensation, construction moisture, rising damp etc. Use an electrical resistance meter to establish the extent of dampness in the substrate. Monitor the use of the rooms to discover whether and when humid conditions occur.

Record the extent of the defective coating and monitor development over several months. Photographs can be used to compare appearance at subsequent inspections.

Diagnosis and cure

The table describes the causes of breakdown of paint coatings on new plaster and cement. It also lists recommended cures in each case. The presence of water, or the too rapid drying out of a plaster substrate is the most common cause of failure and may be made more serious by the presence of salts and the alkalinity.

Breakdown of paint coatings can also occur where they are applied to previously painted walls. There will usually be some dampness present from condensation, washing down, renovation works etc. It may be that the new coating merely appears to be more susceptible to mechanical damage.

A common cause of coating breakdown is poor adhesion between new and existing coats. Washing down and sanding may not have been adequate with the result that the new coating does not adhere. Altneratively the new coating may have been incompatible with the previous coat, for example certain emulsion paints applied to previous oil bound gloss paint. A laboratory examination may be necessary to identify the coating that is breaking away. The total build up of paint layers over the years may have become brittle and more liable to damage. A few brightly pigmented paints can bleed through to differently coloured top coats.

The cure for defects affecting overpainting of existing coatings is to strip back sufficient coatings, either to the surface of the substrate, or to a firmly adhered previous coating. There are various ways to strip walls but large areas are very expensive to clear. A more robust textured overcoating such as replastering or dry lining is worth investigating because it could well be less expensive than stripping. Wood chip paper is sometimes used to disguise paint defects. Wood chip and lining paper are liable to mechanical damage but if they do have to be stripped off, they are likely to bring off much of the flaking paint with them, and so give a better start to subsequent coatings.

If damp walls have to be painted before they have completely dried out, use a porous coating such as emulsion, if necessary as a temporary finish.

REFERENCES

Construction Feedback Digest 35: *Painting new plaster and cement – defects and cures* B & M Publications (London) Ltd

Construction Feedback Digest 56: *The flaking of paintwork in corridors* B & M Publications (London) Ltd

BRE Digest 197: 1982 *Painting walls* Part 1 *Choice of paint*

BRE Digest 198: 1984 *Painting Walls* Part 2 *Failures and remedies*

BS 6150: 1982 *Code of practice for painting of buildings*

Defect	Cause	Prevention	Cure
All defects	Moisture contributes to all failures	Paint only when dry, or use porous temporary decorations until the wall has dried out	Allow to dry, remove loosely adhering and defective paint and repaint
Loss of adhesion Peeling, blistering and flaking. Powdering	Friable plaster, powdery or chalky surface, or water with alkali attack or efflorescence, or even water alone (eg condensation). Application of emulsion paint near or below freezing point.	Dry out before painting but avoid too rapid drying. Plaster that is powdering slightly may be bound back with alkali resisting primer. Do not paint below 5 °C (air or surface temperature), when frosty conditions can be soon expected, or on cold surfaces subject to condensation before paint dries	Remove loose paint, allow to dry and repaint. Slightly powdery surfaces should be primed with alkali resisting primer (thinned)
Alkali attack or 'saponification' In mild cases, peeling and blistering. In severe cases gives a soft sticky film, with water blisters and yellow oily runs. In all cases, some pigments are bleached or discoloured	Attack on oil paints by alkali, in the presence of moisture. All Portland cement products, lime plasters and lime-gauged plasters may cause this attack	Dry out before painting. If decoration must be done before drying is complete, use porous alkali-resistant flat emulsion paints, at least as first decoration. Non-porous paint should not be applied until the wall is fully dry. For oil paint systems, two coats of alkali resisting primer must first be applied	Strip, clean down, allow to dry, and repaint as in preceding column (Prevention)
Efflorescence A crystalline deposit, often white and fluffy, which may push off the decorations, or appear on top of them. Sometimes hard, firmly adherent, especially on cement based materials, and known as lime bloom	Salts from the structure are carried to the surface by water and deposited there as drying occurs	Wipe loose deposits away with a dry cloth or brush, and leave for a few days to see if any more appear. Repeat until no more efflouresence forms. Apply alkali resisting primer (thinned) before any other painting	If efflorescence is pushing off the decorations, strip, allow to dry, brush down. Repeat until no more appears after a few days. If efflorescence appears on top of the decorations, wipe off until no more comes. Thin films of lime bloom which cannot be removed are usually safe to paint over
Coloured spots or patches Often grey, green, black, purple, red, or pink	Moulds, mildew or fungi encouraged by damp conditions, especially where paper, paste and size are present	Remedy damp conditions. Use fungicidal wash at first signs of mould growth. Special paints containing fungicidal ingredients may be helpful but should not be used where they may contaminate foodstuffs improve ventilation	Strip off old decoration where affected. Treat walls with a fungicidal wash.* Dry. If the mildew continues to grow, give a second fungicidal wash. When dry, repaint. Special paints containing fungicidal ingredients may be useful, except where there are foodstuffs
Patchy colour or sheen	Uneven porosity in plaster, especially lightweight plaster	Apply a coat of alkali resisting primer, thinned, under any finish	Unless paint can be readily removed, apply (over existing paint) an oil paint system or alkali resisting primer plus emulsion paint
Brown stains or streaks	Soluble matter from the background, especially clay pots and clinker blocks — and from some kinds of emulsion paint		
Dry out In mild cases, plaster does not attain full strength and hardness. In severe cases plaster is powdery and friable on the surface, or throughout, and paint applied on it may peel	Too rapid removal of water from some gypsum plasters	Avoid rapid drying following plastering	In mild cases, if the plaster is allowed to dry and is *kept dry*, redecoration without replastering may be successful. Slightly powdery surfaces should be primed with alkali resisting primer (thinned). In severe cases, hack off and replaster and repaint
Delayed expansion Slight rippling and softening, or wholesale rotting, expansion and blistering of the plaster	If plaster suffering from dry out again becomes wet, delayed expansion occurs		

**Suitable toxic washes are listed in BRE Digest 139. The manufacturer's directions should be closely followed in making up and using proprietary materials. Domestic bleach solutions, used at the recommended dilutions, may also be used to wash down mildly affected surfaces but they may cause colour changes in the paint and redecoration may be needed.*

Table 1 Painting new plaster and cement-defects and cures.

Photo 1 Paint will rapidly peel off damp timber or undercoating.

Photos 3 & 4 Clear varnish is a very high maintenance finish on the exterior of a building. It is rarely successful.

Photo 2 Stains need regular maintenance generally at shorter intervals than paint coatings.

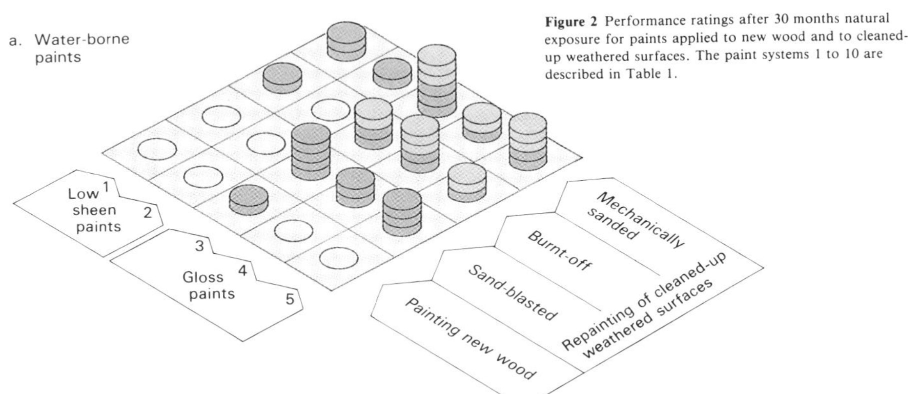

Sandblasted

Burnt-off

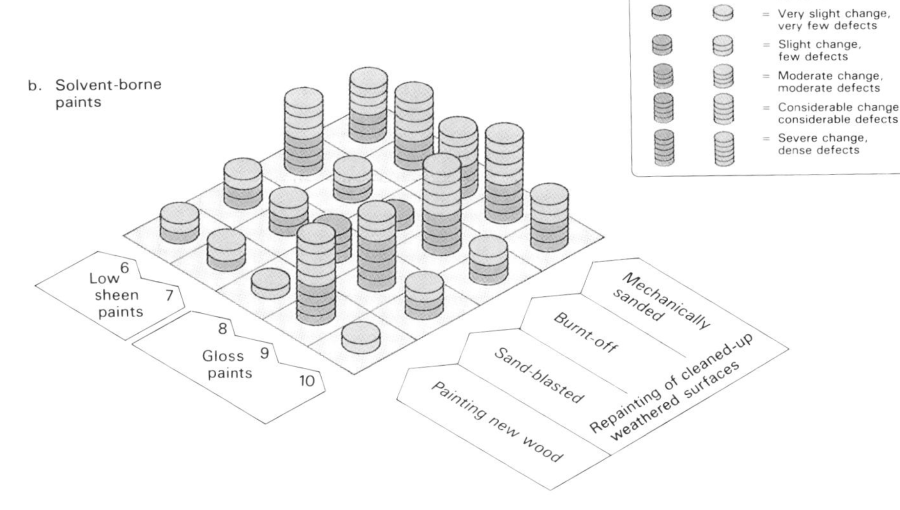

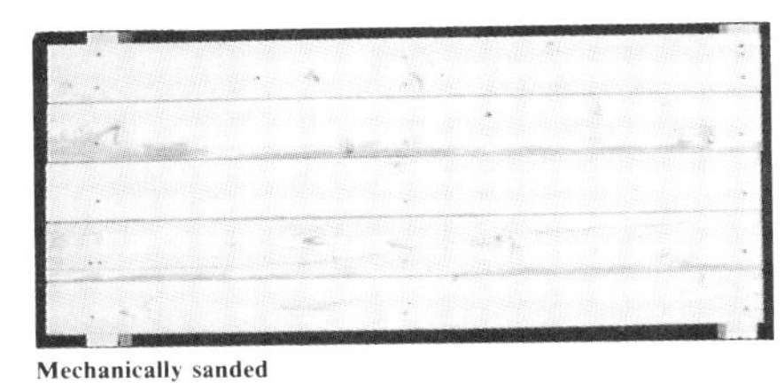

Mechanically sanded

Figure 1 Cleaned up weathered softwood cladding used in the trials summarised in Figure 2.

Figure 2 Performance ratings of paints on different surfaces in a BRE trial.

5.3.22 EXTERNAL TIMBER CLADDING. BREAKDOWN OF COATING

This defect appears first on exposed areas and will be more marked with clear varnishes and lightly pigmented coatings – see also 6.2.3 which discusses stains and varnishes on windows and doors.

Symptoms

Brushed, sprayed or dipped coatings on timber all have a limited life. For some clear or lightly pigmented coatings this can be between 1 and 4 years. Paint and high build stains are expected to last 3–5 years or even longer. Only when a paint stain or varnish fails to last as long as expected by a margin of a year or so can it be considered to be defective. Signs of failure are:

- discolouration of the timber showing through a clear finish.
- peeling, flaking, blistering and powdering.
- erosion of the coating showing the wood grain through the surface.

Investigation

Establish the nature and, if possible, the manufacturer of the coating. Find out the species of the timber and the conditions and location in which it was stored immediately before use. Find out as much as possible about the coating and the conditions under which it was supplied.

Ascertain the length of time the cladding had been in position and how long after that the first coat of paint was applied. What was the time interval between coats?

Examine the surface below the peeled and flaking coats to establish whether there is dirt or grease on the undercoat.

Record the moisture content of the timber.

Record the extent of the damage and the orientation of the affected claddings.

Send a sample of the coating and substrate to a laboratory for microscopic examination.

Diagnosis and cure

Dampness and exposure to UV light cause the deterioration (greying) of the external surface of uncoated timber. Coated timber is affected when the coating allows water to enter the timber. Small splits and cracks due to thermal movement are enough to let the wood become damp. Once the surface of the timber is affected, the bond between timber and the coating is weakened, and the coating can start to peel or flake away. Clear coatings are more vulnerable because they do not exclude UV light and because they can not be made more durable. Coatings perform much better when applied to dry, newly planed or sanded timber. It is vital to achieve a good bond between the primer and the timber surface. Even factory applied primers do not always achieve this. However, failure can also occur at other interfaces between coats in conditions where:

- preparation has not been adequate
- the types of paint or varnish are not compatible
- the surface is damp when the coat is applied

When a failure has occurred, the coating should be stripped down to the bare timber and a new coating applied. Stripping by sandblasting has been shown to give a good base for new paint coatings. Developments in paint technology are producing coatings that shed rainwater, but are vapour permeable so that timber can dry out more quickly if it does get wet. Stains and varnishes can be made opaque to ultra violet light. BRE recently tested water bound and solvent bound paints for repainting cladding and found that low sheen, water borne paints on sand blasted or burnt off surfaces give the best results. Light coatings generally perform better than dark or transparent coatings because they reduce heat build-up. However, careful consideration must be given to each case prior to the use of vapour permeable coatings. In situations where water can penetrate into timber, other than through the coated surface, such coatings can enhance the likelihood of decay. The testing undertaken by BRE and quoted here, also demonstrated that conventional hard gloss paints, complying with DEF STAN 80–59 gave an acceptable standard of performance over the test period.

REFERENCES

BRE Information Paper 16/87: *Maintaining paintwork on exterior timber*

BRE Information Paper 17/87: *Factory applied priming paints*

BRE Digest 286, June 84: *Natural finishes for exterior timber*

TRADA Wood Information sheet 1 Section 2/3: *Finishes for exterior timber*

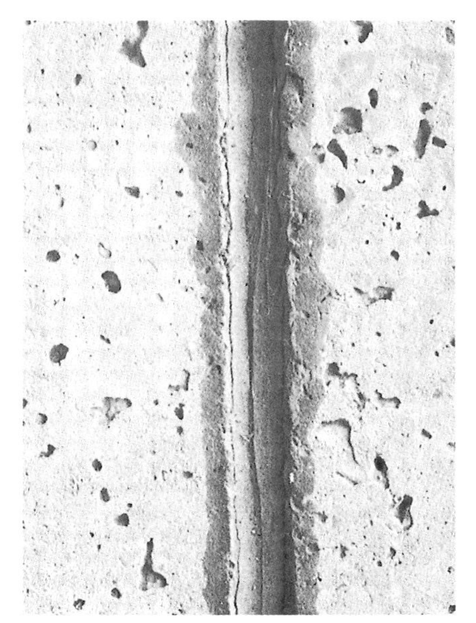

Photo 1 Splitting sealant between concrete panels.

Photo 4 Wrongly detailed sealant at edge of window frame see Figure 5

Photos 2 & 3 Sagging and extruding sealant.

5.3.23 EXTERNAL WALLS. FAILURE OF SEALANTS AROUND CLADDING PANELS AND WINDOWS

The defects commonly occur between large concrete panels or between masonry walls and window frames. Sealants used between glass and window frames are referred to in chapter 6.

Symptoms

The defect may merely affect the appearance of the joint if sealants sag, split or come loose, or it may first be noticed because the joint has leaked, and dampness shows on inside surfaces of walls. Sometimes the leak will only show under a particular combination of wind, rain and temperature.

Investigation

Establish the construction of the wall and type and section of the sealant that has been used. Record the shape and dimensions of the sealant and the temperature when the joint was inspected.

Where possible, find the name of the sealant manufacturer, the date and conditions under which the sealant was applied, and the conditions under which the defect first became apparent. A laboratory examination may help to identify the sealant.

Measure the cladding on either side of the sealant and try to establish the fixed points that would restrict the movement of the cladding. Examine the pattern of failure over the building if there are a number of faulty sealed joints.

Note whether the sealant is the one that was originally applied or whether it is a replacement. If it is a replacement, try to find out what it replaced and how the original sealant failed.

If the fault is occurring on an industrial building, attempt to discover whether substances used within the premises could have come in contact with the sealant.

Diagnosis and cure

Sealants have a limited life. BS 6213 divides them into three groups with expected lives of up to 10, 15 or 20 years. The components on either side of a sealant filled joint will generally last much longer than the sealant itself. Replacement of sealants is therefore to be expected during the life of a building, and may not be due to the defects described here.

It is necessary to make an assessment of the amount of movement that the joint is likely to sustain. A method of estimating thermal movement is given in BRE digests 227, 228 and 229. As an example, the joint between two 4 m wide concrete panels can have a movement range of 8 mm if the end of one panel is fixed. The temperature variation of external claddings is very wide. Dark coloured concrete can reach 65°C in summer sunshine.

Different types of sealant can accommodate different amounts of movement, and wide joints can accommodate more movement than narrow ones. A likely cause of failure is the inability of the sealant used to accommodate the movement that is taking place. This can result in splitting, folding or excessive extrusion.

The bond between the sealant and the edge profiles of the adjacent components can also be a cause of failure. If the edges are weak, loose, dirty or friable, or if the sealant is not forced into contact with the edge when it is applied, the sealant will break away from the edge when subjected to its maximum expansion limits. Resealed joints are vulnerable to this type of failure because previously applied sealants may not have been thoroughly cleaned off; or they may have been cleaned off with solvents; or may even have contained solvents that have contaminated the edge profiles. Manufacturers will recommend treatment and priming coats to be used before applying their sealants. Not all sealants are compatible with commonly used materials – such as preservative treated timber.

The shape of the sealant section greatly affects its performance. Different materials perform best when applied in different profiles. The Triangular profiles that are common around windows perform poorly in comparison with rectangular profiles, particularly when the exposed side of the triangle is concave. Where a rectangular profile is against solid material on three sides, a bond breaker tape is required on the intermediate side to ensure that the sealant is free to expand and contract across the backing material. The depth of a rectangular profile can be controlled by a back-up insert. The wrong profile can cause slumping, splitting, folding or adhesion failure.

Failures can also be caused by chemical attack in buildings where industrial processes are taking place such as breweries or dairies where chemicals may come into contact with sealants. A further possibility is

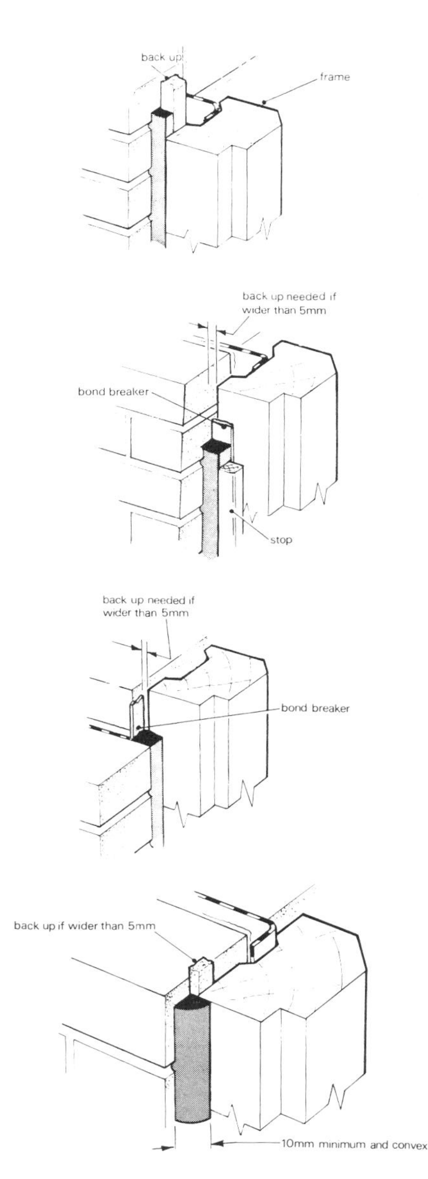

Figure 1 Alternative sealant details around windows. Versions of these details also apply to aluminium and plastic windows.

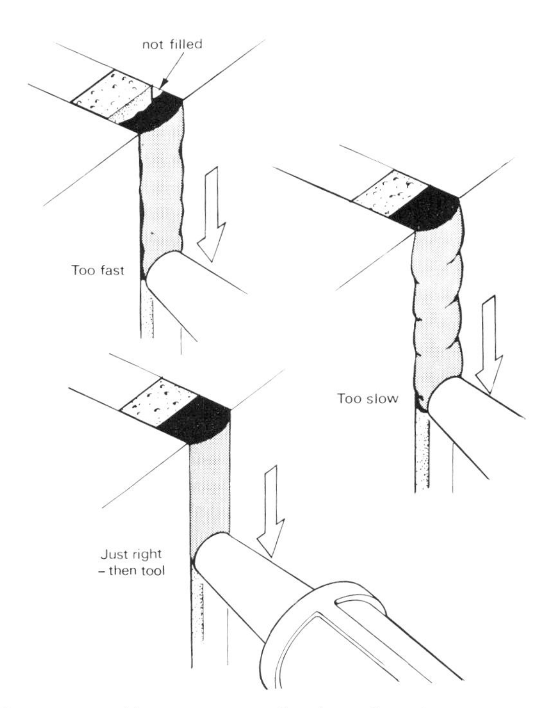

Figure 2 Correct, and incorrect, application of sealant.

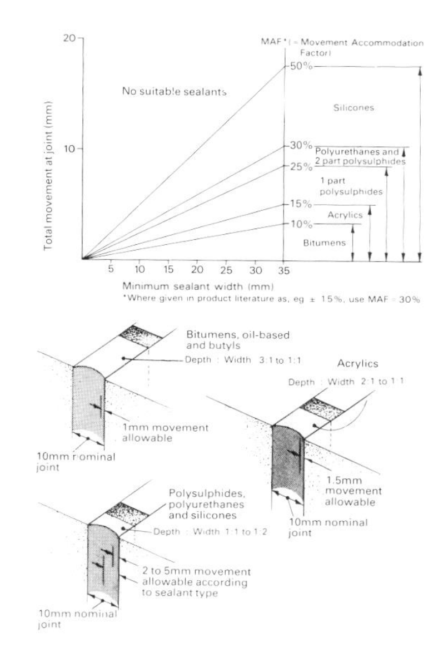

Figure 3 Dimensions of sealants.

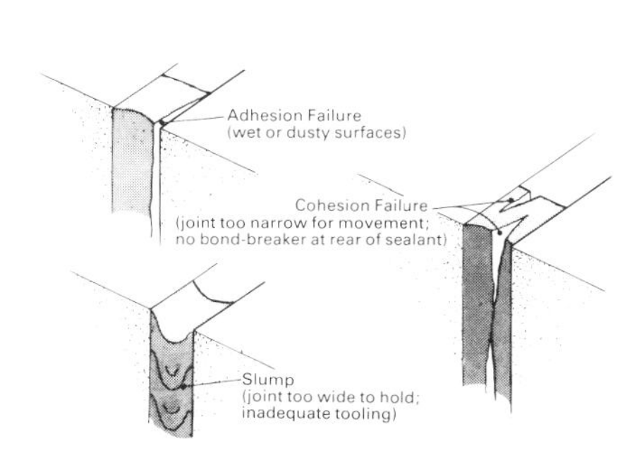

Figure 4 Modes of failure.

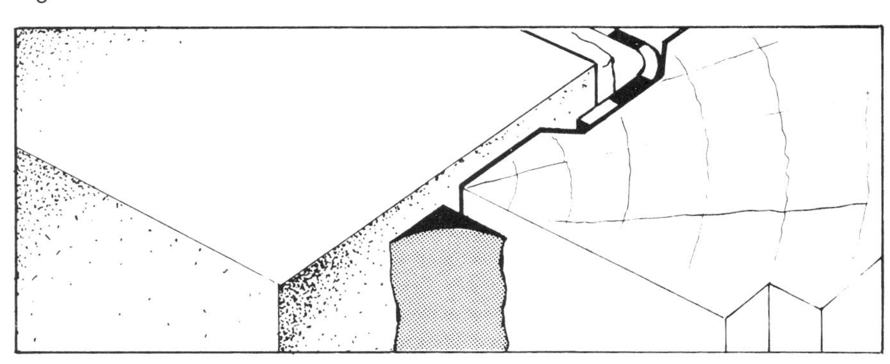

Figure 5 Wrong detail – perimeter sealants should not be concave and brought to a feather edge – See Fig 1 for correct application.

5.3.23 EXTERNAL WALLS. FAILURE OF SEALANTS AROUND CLADDING PANELS AND WINDOWS

continued

that the sealant has been displaced by vandals or when the building was being maintained.

Where a sealant has failed prematurely, the replacement may need to be in a different material, or have a different profile, or the edge profiles of the joint may need special treatment. An alternative to the triangular seal to the edge of a window is illustrated. Where possible, the replacement seal should be recessed to protect it from UV light. Where sealants cannot accommodate the movement that is likely to occur on a vertical joint, it may be possible to prevent water leaking into the building by the use of a cover strip.

If the joints are too narrow to fit a sealant that can cope with the movement that is occurring, a rebate can be made in the edge profiles, and the back face of the rebate can be isolated with a bond breaker to allow a wider rectangular sealant profile to be installed.

REFERENCES

BRE Defect Action Sheet 66: *Applying sealants in bad weather*

BRE Defect Action Sheet 68: Dec 85. *External walls: Joints with windows and doors – detailing for sealants*

BRE Defect Action Sheet 69: Dec 85. *External walls: Joints with windows and doors – application of sealants*

BRE Defect Action Sheet 97: March 87. *Large Concrete panel external walls: re-sealing butt joints*

Edwards, Marilyn J *Weatherproof joints in large panel systems: 1 Indentification and typical defects.* BRE Information Paper 8/86

Edwards, Marilyn J *Weatherproof joints in large panel systems: 2 Remedial measures* BRE Information Paper 9/86

Edwards, Marilyn J *Weatherproof joints in large panel systems: 3 Investigation and diagnosis of failures* BRE Information Paper 10/86

Beech, J C and Aubrey, D W *Joint primers and sealants: performance between porous claddings* BRE Information Paper 9/87

BRE Digests 227, 228 & 229: 1979 *Estimation of thermal and moisture movements and stresses.* Parts 1, 2 & 3

BS 6213: 1982 *Guide to selection of constructional sealants*

Guidance to Specifiers in the use of mastics and sealants on site Aluminium Window Association 1980

BBA Method of Assessment and Test No.14 *Directive for the assessment of building sealants* (Useful check list for performance of a sealant).

CIOB Technical Information Service No.13 *Sealants in remedial work* (Gives examples of solutions adopted in different situations)

Photo 1 Extruding dpc may be the result of heat and load but is more likely to be caused by structural movement.

5.3.24 EXTERNAL WALLS. EXTRUSION OF BITUMEN FELT DPC AT GROUND FLOOR LEVEL

This defect is dealt with as a wall defect but it may well be the first sign that there is a serious problem with the floor. The defects relate to slight movement of the wall at DPC level – see also 5.1.2.

Symptoms

Commonly, the bitumen felt is slightly more in evidence than normal, sometimes with beads of bitumen hanging from it. Any overlapping mortar may be displaced. There is often a slight outward movement of the wall immediately above the DPC. Very occasionally there is considerable extrusion of the bitumen felt up to as much as 40–50 mm.

Investigation

Note whether all the external walls of the building are affected or whether the effect is primarily seen on walls with a southerly aspect.

Examine the base of the wall for signs of movement and whether it is above or below the DPC. If the building has a solid ground floor this should be examined for signs of cracking or uplift.

If there is severe extrusion, the form of construction should be examined to ascertain whether there is undue freedom of movement of the wall on the DPC or whether the DPC is likely to be overloaded.

Diagnosis and cure

The movement of the bitumen felt may be due to small thermal movements of the walls, but is more often an indication of the expansion of a solid concrete ground floor – see 7.1.4. Expansion of the floor is often relieved by an uplift of the slab, but there may also be an outward expansion of the slab. It can first be seen on the outside wall at DPC level where it makes the DPC more visible.

Excessive extension may be due to an overloading of the DPC in walls that became heated by warm sun but as these cases are very rare, each one needs individual consideration. They appear to be very infrequent.

Unless the floor is found to be expanding and cracking, there is probably no need for remedial work, apart from throwing off very obtrusive DPC material. If the floor is suspect – for example if it is laid on burnt colliery shale, it should be inspected at 6 monthly intervals until the surveyor is confident that the defect will not prove to be progressive, (see 7.1.4). If the extension is due to overloading it is unlikely to effect resistance to moisture penetration.

REFERENCES

BRE Digest 77: 1985 *Damp proof courses*

CHAPTER 6 WINDOWS, DOORS & STAIRS

CHAPTER 6 Contents

CHAPTER 6 WINDOWS, DOORS AND STAIRS

This chapter deals with a wide range of manufactured components. Each has its own particular weaknesses which the manufacturer is quite likely to have encountered on previous occasions. When the source of the component is known, it is always worth seeking advice on how to tackle a particular defect. Some manufacturers can recommend specialists to deal with specific problems – such as the rejigging of steel windows.

The defects described here are of a general nature. They are related to the materials from which components are fabricated and to the surface finishes. Windows and doors often suffer from the failure of ironmongery items which cannot be covered here because the range of ironmongery used is so extensive. In most cases the best way to cure such failures is to replace the original ironmongery with something more robust and suitable.

When inspecting windows for defects it is essential to have access to the inside of the building. An external survey will not pick up problems with opening and closing the lights or missing items of ironmongery. Cracked and faulty glass is also easier to see from inside. However, provided there are some opening lights, it is normally possible to inspect the outside of a window from inside the building.

This chapter assumes that defective windows and doors will be repaired whenever this is technically feasible, but repair may not always be the most economical course of action. Increasingly windows and doors are being replaced in spite of their being repairable because the replacement components are easier to maintain and provide a better standard of performance. Sometimes replacement is chosen when repair would be more appropriate. The appearance of buildings can be changed for the worse by inappropriate replacement windows, and new problems of condensation or leakage can be introduced if the detailing of the replacement window is not well considered.

Protective coatings applied on site are particularly subject to the shortcomings of workmanship. In many cases it will be necessary to strip all existing coatings from a component to cure a defect. Redecorating the stripped component introduces additional risks. Any number of things can go wrong with the preparation of surfaces and the successive application of protective and decorative coatings. The process is vulnerable to the weather and to damage by other site operations. External painting must be specified in detail and thoroughly supervised to avoid initiating one defect while attempting to cure another.

Photo 1 Decay showing only as slight shrinkage of the underlying wood causing surface cupping.

Photo 2 Decay revealed as extensive when surface is broken away

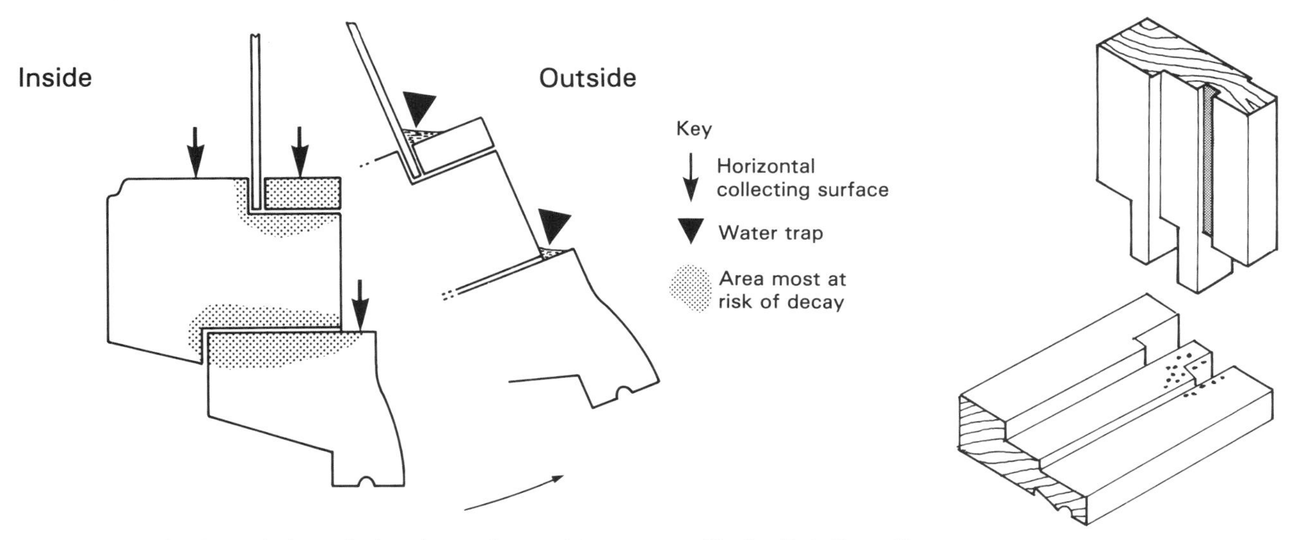

Fig 1 With a horizontal pivot window the surfaces of the bottom frame member may collect water when the window is open.

Fig 2 Detailing of frames can unintentionally promote dampness. Groove for weatherstrip feeds water into sill

Performance bands for watertightness. *Comparisons should not be attempted between similarly designated bands from each source*

	Maximum test pressure without 'leakage' *Pa*		
	DD4: 1971	AB-MOAT 1 1974	PSA/DHSS MOB 3rd Prog 1979
1	50*	50 to 150	100
2	150*	150 to 300	200
3	300*	300 to 500	300
4	–	over 500	–

* Linked to the design wind pressure

Tables 1 & 2 Different levels of performance are recommended for windows to suit exposure conditions

Performance bands for air permeability of opening lights. *Comparisons should not be attempted between similarly designated bands from each source*

Performance band designation	Source					
	DD1: 1974		AB MOAT 1 1974		PSA/DHSS MOB 3rd Prog 1979 (1)	
	Maximum air permeability					
	$m^3/h/m$ at 100 Pa	*Max test Pa*	*$m^3/h/m$ at 100 Pa*	*Max test Pa*	*$m^3/h/m$ at 100 Pa*	*Max test Pa*
1	12 (2)	100	12	150	12	150
2	12 (2)	150	6	300	9.5	200
3	12 (2)	200	2	600	7.5	300 (3)
4	–	–	–	–	2	600 (3)

(1) For fixed lights the air permeability should not exceed $1m^3/h/m$ at 300 Pa on PSA buildings

(2) $12m^3/h/m$ should not be exceeded at the maximum test pressure

(3) Only these grades acceptable for PSA buildings

6.1.1 TIMBER OPENING LIGHTS. DISTORTION AND LEAKING

This usually occurs in older windows, though it can happen at any time. It is more likely when joinery is poorly maintained, or not adequately treated prior to installation, or when there is a change in the conditions to which the joinery is subjected.

Symptoms

The opening light does not sit against the frame on all edges. This can lead to draughts and rain penetration, either of which may be related to certain wind conditions. In severe cases, the light or frame is so distorted that the two parts will not fit together. The light can then not be properly shut and secured.

Investigation

Look at the whole building. Distortion caused by structural movement will normally be evident in more than one element. If no structural cracks are obvious, test for the plumbness of walls with a plumb bob and the trueness of floors by rolling a marble.

Is the defect weather related? Note the conditions under which it occurs – wind direction and strength should be recorded. (BS 6375 Part 1:1983 gives guidance.) Could the defect be due to an air movement 'pull' from an internal heating appliance, chimney or lift shaft rather than a 'push' from outside?

Closely examine the paintwork: are there any runs or drips leading to a build-up of paint on what should be meeting surfaces? Is there cracking or unevenness in the paint film over the joints in the opening lights? Note that stained timber is especially vulnerable to water penetration at joints, because stain is not as thick as paint and therefore does not act as a 'filler'.

In new frames, examine other lights in similar situations to judge whether the components are of poor workmanship and/or not made to adequate tolerances (purpose made joinery is more likely to be affected by this problem).

Examine timber door leaves for bow and twist. Note whether the detail at the door sill and the bottom of the jambs of inward opening doors ensures that driving rain does not track back to the interior of the building.

Measure the diagonals of lights, doors leaves and frames for squareness. Flex the corner of hinged sashes to test rigidity.

Diagnosis and cure

If structural movement is suspected see chapter 5.

If the defect is only apparent under certain wind conditions, the weather-stripping may be defective, or non-existent. Alternatively, the detailing of the timber sections may be poor, i.e. inadequate rebates. Throatings may be filled with paint and therefore ineffective, and there may be insufficient slope on the sills. Check these points and consider renewing the weather-stripping, if possible in conjunction with a repainting programme.

To check if the defect is caused by a 'pull', seal the window temporarily with tape and note the effect. (An open fire will smoke and a boiler may fire more noisily.) An alternative, controllable, source of air should be provided if this is thought to be the cause. (See 9.4.1)

A common cause of the defect is moisture movement, often exacerbated by a decrease in the gap between the frame and the light caused by a build-up of paint. This diagnosis will follow the reported increase of the defect in wet weather, particularly if an examination of the paint films shows an inadequacy of paint coatings. The top and bottom of opening lights may be difficult for painters to reach and are often neglected, thus providing a surface by which moisture can enter the timber.

Timber external doors are vulnerable to moisture and thermal movement, particularly when conditions change in a building, eg new central heating or more moisture vapour released in the use of the building. The resultant bowing and twisting are difficult to cure. Weather-stripping and additional fastenings may help the problem, but it is more likely that the door will need to be replaced and a better protected door fitted in its place. Well maintained paint coatings (not the microporous type) give some protection. Light coloured coatings where exposed to strong sunlight will help to reduce movement and preserve the paint. Other measures that could reduce movement are the fitting of an external porch or the fitting of an inner door.

If the paint film is defective or there is a build up of paint, then the paint must be stripped back to the bare timber and recoated with a new paint system. Inadequate tolerance between frame and light requires them to be planed down of the lights until they fit, and then repainted.

Draught stripping can cope with gaps of between 1 mm

and 8 mm on windows, self adhesive silicone rubber sealant is effective on the tapering gaps occurring round distorted windows, but it can look untidy. If the lights are out of square but the timber sections used are of adequate size, glazing blocks may have been omitted and the glass pane may be too small and is therefore not acting as a stiffening panel. Check at top of glass for gaps or over-thick putty to cover gaps. Reglazing should cure this problem.

REFERENCES

BRE Digest 119 – *The assessment of wind loads*

BRE Digest 319 – *Domestic draughtproofing: materials, costs and benefits*

BRE Defect Action Sheet 14: *Wood windows: preventing decay*

BRE Defect Action Sheet 67: *Inward opening external doors: resistance to rain penetration*

BS 6262: 1982 *Code of practice for glazing of buildings*

BS 6375: *Performance of windows* Part 1:1983 *Classification of weathertightness (including guidance on selection and specification)*

Photo1 Steel opening lights can be distorted by attempts to close the window when its movement is obstructed by paint build up. (In this case a ventilator has been stopped up with paper)

Fig 1 Assessment of suitability – some draught strips for common types of doors and windows.

Type of draughtstrip *	Windows: Steel casement	Windows: Wood casement	Windows: Wood sliding	Windows: Wood pivot	External doors: Wood hinged	Remarks
Self adhesive foam strip (PVC)	◪	◪	□	□	□	Relatively stiff; if heavily compressed possible high force needed to close the component, hence match size of strip to sizes of gap.
Self adhesive V fin (Polyester)	◪	◪	□	◪	■	Durability may not be high, but is cheap and easily replaced on doors; somewhat noisy when opening and closing the component, can hum in high winds.
Self adhesive gap filler (Silicone rubber)	■	■	□	◪	□	Untidy appearance, but hidden on outward opening windows; high friction; avoid undercuts in sash frame sections; cannot follow appreciable seasonal changes in gap size.
Glued-on tubing (Silicone rubber)	■	■	□	◪	◪	Not very sightly; high friction; correct size of tubing relative to gap is important at points where sliding and compression occur together, otherwise possible high force needed to close the component.
Finned brush (Polypropylene) in housng (PVC, aluminium) (Nail/screw fixings)	□	■	■	□	◪	Brushes without a central plastics fin are less effective, but may be needed at some door thresholds; short brushes cannot follow substantial seasonal changes in gap size.

Suitability

■ General use on type of component

◪ Many specimens of type of component

□ Unsuitable, or of too limited usefulness

* The absence of other types of draughtstrip does not imply that they are unsuitable for any of these components, any more than the indications imply that all brands and models of a type are equally suitable for a component.

6.1.2 METAL WINDOWS. DISTORTION AND LEAKING

This usually occurs in older windows, although it can happen in windows of any age that are of poorly fabricated metalwork or where the metalwork is not adequately treated.

Symptoms

The opening light does not sit against the frame on all edges, which can lead to draughts and rain penetration, either of which may be related to certain wind conditions. In severe cases, the light or frame is so distorted that the two parts will not fit together. The light can not then be properly shut and secured.

Investigation

Look at the whole building. Distortion caused by structural movement will normally be evident in more than one element: test for the plumbness of walls and the trueness of floors if no structural cracks are obvious.

Is the defect in weather related? Note the conditions under which it occurs – wind direction and strength should be recorded (BS 6375 Part 1:1983 gives guidance). High temperatures can cause distortion (remember that dark colours in strong sunlight can induce higher temperatures in materials than light colours). Could the symptom be due to an air movement 'pull' from a heating appliance, chimney or lift shaft rather than a 'push' from outside?

Examine the paintwork to see whether there are runs or drips leading to a build up of paint on meeting surfaces (runs are especially common on the sharp edges of metal frames). Check for any bubbling under the paint with rust showing through.

Inspect the metal sections used and look at the method of forming corners. See that all the welds are good. Note that lights should be mitred on corners, but that the joint formed should not stand up through the paint. Measure the diagonals to check for squareness, and flex across the longest dimension to test for rigidity.

See whether metal sections are bowed or bent.

Diagnosis and cure

If the defect is apparent only under certain wind conditions, check the weather-stripping, if any is present (weather-stripping was not common on metal frames until the 1960's). Distortion under wind load is only likely to be the cause in aluminium windows, particularly large sliding aluminium windows. Decide whether the window design is suitable for the site, particularly sill rake and ironmongery. Consider weather-stripping as a remedy. Weather-stripping with a self-adhesive silicone sealant can be used to seal tapered gaps around lights provided the appearance is acceptable.

Thermal movement is more likely to be a problem in aluminium windows than in steel. (Note that some early aluminium windows were cast, and can be mistaken for steel – use a magnet to distinguish between them.) Where frame and light have been picked out in different colours, unequal movement can be expected in full sun.

Should the cause of the defect appear to be a 'pull' verify this by sealing the window temporarily with tape and noting the effect. An alternative, controllable, source of air should be provided if this is proved to be the cause.

Paint build up is a common defect and can lead to permanent distortion of the light through forcible closure. Strip off the paint, and if distortion remains, the light can be straightened in an in-situ jig after the removal of the glass. Rust must be removed by rubbing down until bright metal is reached. Prime and paint in accordance with a paint system designed for the bare metal.

If corner welds on either steel or aluminium windows are fractured the light must be removed and rewelded off site.

If the joints on aluminium windows are mechanically made, check that they are not overtightened as this can lead to distortion from the flat plane. Similarly it is possible to overtighten aluminium window fixings so that the members bow and leaks develop.

REFERENCES

BRE Digest 228: Part 2 *Estimation of thermal and moisture movements and stresses:*

BS 6262:1982 *Code of practice for glazing for buildings*

BS 6375 Performance of windows Part 1:1983 *Classification for weathertightness (including guidance on selection and specification)*

Photo1 Fusion welded corner joint in plastics window – A joint failure will allow water to enter the core and may possibly affect the core reinforcement.

Table 1

Typical properties of plastics used in building

Material	Density kg/m³	Linear expansion per °C Coefficient	mm/m	Max temperature recommended for continuous operation °C	Short-term tensile strength MN/m²	Behaviour in fire
Polythene *low density	910	20×10^{-5}	0·2	80	7–16	Melts and burns like paraffin wax
high density	945	14×10^{-5}	0·14	104	20–38	
Polypropylene	900	11×10^{-5}	0·11	120	34	Melts and burns like paraffin wax
Polymethyl methacrylate (acrylic)	1185	7×10^{-5}	0·07	80	70	Melts and burns readily
Rigid PVC (UPVC)	1395	5×10^{-5}	0·05	65	55	Melts but burns only with great difficulty
Post-chlorinated PVC (CPVC)	1300–1500	7×10^{-5}	0·07	100	55	Melts but burns only with great difficulty
Plasticised PVC	1200–1450	7×10^{-5}	0·07	40–65	10–24	Melts, may burn, depending on plasticiser used
Acetal resins	1410	8×10^{-5}	0·08	80	62	Softens and burns fairly readily
ABS	1060	7×10^{-5}	0·07	90	40	Melts and burns readily
Nylon	1120	8×10^{-5}	0·08	70–110	50–80	Melts, burns with difficulty
Polycarbonate	1200	7×10^{-5}	0·07	110	55–70	Melts, burns with difficulty
Phenolic laminates	1410	3×10^{-5}	0·03	110	80	Highly resistant to ignition
GRP laminates	1600	$2\text{–}4 \times 10^{-5}$	0·02	90–150	100	Usually inflammable. Relatively flame-retardant grades are available

Key: UPVC = unplasticised polyvinyl chloride GRP = glass-reinforced polyester PVC = polyvinyl chloride ABS = acrylonitrile/butadiene/styrene copolymer

* High density and low density polythene differ in their basic physical properties, the former being harder and more rigid than the latter. No distinction is drawn between them in terms of chemical properties or durability. The values shown are for typical materials but may vary considerably, depending on composition and method of manufacture.

6.1.3 uPVC OPENING LIGHTS. DISTORTION

Symptoms

The opening light does not sit tight against the frame on all edges, leading to draughts and rain penetration, either of which may be related to certain wind conditions. In severe cases, the light or frame is so distorted that the two parts will not fit together. The light can not then be properly shut and secured. Sliding lights will not move easily or not slide fully open.

Investigation

Look at the whole building. Distortion caused by structural movement will normally be evident in more than one element: if no structural cracks are obvious plumb the walls and see whether the floors are level.

Investigate the source and date of manufacture of the windows. Establish how the frame is fixed and whether it is free to expand if exposed to strong sunshine. Note the colour of the finish. Is the defect weather related? If so, what conditions cause leaks, draughts and difficulty in operating the window? Record the overall dimensions of the frame.

Examine the corner joints of lights for defects in the weld. (Note that there are 3 ways of forming the joints: most common are fusion welds, but solvent welds and mechanical fixings are also used.) Larger lights may have internal reinforcement or windows might be timber or metal cored. Any failure in the joints will lead to water build up inside the frame – test for this by drilling a small hole, or removing window furniture to examine the state of the core.

Measure the diagonals to check for squareness, and flex across the longest dimension for rigidity.

Diagnosis and cure

If structural movement is suspected see chapter 5.

If the defect is apparent only under certain wind conditions, weather-stripping (usually fitted as standard) may be defective and should be renewed; check with window manufacturer. If the defect is only evident in hot weather, thermal expansion (in dark colours especially) must be suspected. Consider over-painting with ligher colours, taking into account that this loses the low maintenance advantage of uPVC windows.

Should the cause of the defect seem to be a 'pull', verify by sealing the window with tape and noting the effect. An alternative controllable source of ventilation should be provided. If the frame has been pulled out of shape by thermal movement and an inadequate gap has been maintained around the perimeter, the fixings should be examined. Sleeved fixings are available for uPVC windows that allow the frame to expand and contract freely.

If the joints have failed, the light must be removed and returned to the manufacturer for repair. If the joint failure has enabled water to get inside the frame, unless the reinforcement is stainless steel or aluminium, it is recommended that the window be replaced. Note that it is the slimmer sections (which offer a smaller surface area to be welded) whose joints are most likely to have failed. Mechanically made joints (largely superseded by welding) have always been difficult to waterproof, so these frames must always be suspect. There is as yet an inadequate fund of experience of how reinforcement corrosion first manifests itself in failure.

If the light flexes excessively, check glazing material for suitability – some plastics sheets are particularly vulnerable. Reglazing with a more rigid material may be the answer.

REFERENCES

British Plastics Federation and Glass and Glazing Federation *A Trade Standard for uPVC Windows* (1986)

BRE Digest 69: *Durability and application of plastics*

BRE Digest 228: *Estimation of thermal and moisture movement and stresses* Part 2

Photo 1 Resin exudation from dark painted timber in strong sunlight

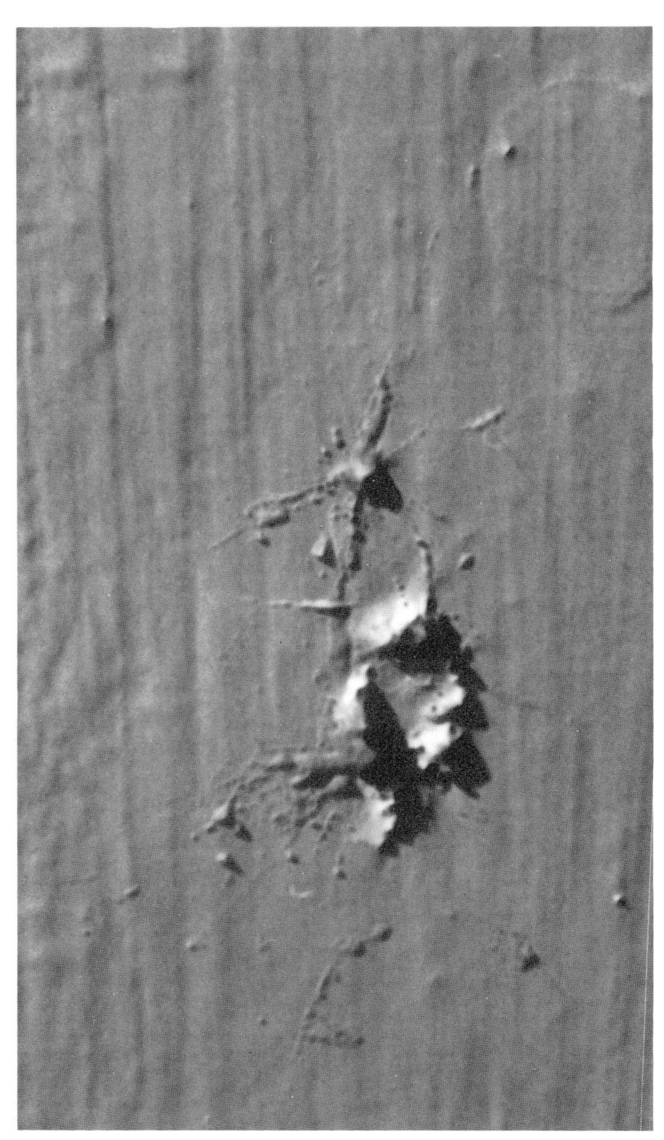

Photo 2 Typical damage to paint coating from resin exudation

Photo 3 Paintwork on South elevation deteriorates more rapidly – the North elevation was still in good condition.

Photo 4 Dampness at sill level causing paint failure

6.2.1 TIMBER WINDOWS AND DOORS. BLISTERING OF PAINT FINISH

Symptoms

The paint film may remain unbroken while lifting off the timber in small areas. Discolouration of light coloured coatings may occur. This is sometimes accompanied by softening of the paint surface and by fine blistering. The film can then be broken if the blister is compressed. Sometimes the blister bursts to exude a brown sticky liquid (resin) (see 6.2.2 for blistering cracking and flaking of the paint coating with associated deterioration of the wood below).

Investigation

Note the orientation of the joinery and whether it is exposed to full sunlight.

Establish the joinery, painting and preservative specification. Some types of timber, eg Douglas Fir, contain more resin than others and special precautions must be taken when they are painted. Solvent based preservatives can make timber more susceptible to this defect.

Note the colour of the paint, dark colours are more likely to be affected.

Use a moisture meter to establish the moisture content of the timber.

Diagnosis and cure

Two common causes of this defect are resin exudation from knots and resin pockets in the timber, and water vapour expansion from moisture in the timber. Both result from the timber being warmed, usually by strong sun. If the paint is dark, heating up by sunlight will be exaggerated.

If the problem is resin exudation, which will show as brown sticky resin or, when dried, as a whitish powder, then the cure is to strip affected areas and apply shellac knotting. An aluminium primer should be used before repainting, with a light colour. There can be no guarantee that the problem will not recur and require more than one treatment. If it is vital to avoid recurrence, the timber component should be replaced by one in a different species of timber, eg Scandinavian whitewood, that contains less resin.

If the problem is damp timber, the source of dampness should be identified and cured, the paint should be stripped or burnt off and the timber protected until it has dried out. In suitable conditions, where dampness is only likely to enter the timber from rain on the external painted surface, a microporous paint system can then be applied. It should allow the drying process to continue while preventing surface water entering the timber.

Occasionally, with larger frames it is possible to clean off paint by sand blasting, and in these instances an external wood stain can be used.

REFERENCES

BRE Digest 261: *Painting woodwork*

TRADA Wood information sheet section 2/3 sheet 1 *Finishes for exterior timber* revised May 1984

GLC Development and materials bulletin 114 Item 2 *Effect of dark colours on resin exudation from exterior joinery*

GLC Development and materials bulletin 118 Item 2 *Exterior painting of resinous timber*

Table 1

Some suitable joinery timbers and their properties

Timber		Sapwood distinct from heartwood	Natural durability of heartwood	Movement in service	Paint performance	Notes
Softwoods	European redwood *(Pinus sylvestris)*	Yes	Non-durable	Medium	Good	Sometimes resinous
	Whitewood *(Picea abies)*	No	Non-durable	Medium	Good	Not as resinous as European redwood
	Western hemlock *(Tsuga heterophylla)*	No	Non-durable	Small	Good	Non resinous
	Douglas fir *(Pseudotsuga menziesii)*	Yes	Moderately durable	Small	Very good	Sometimes resinous. A marked contrast between early- and late-wood can show through painted flat-sawn surfaces; advisable to use stainless or non-ferrous fastenings
Hardwoods*	Utile *(Entandrophragma utile)*	Yes	Durable	Medium	Very good**	Pronounced interlocked grain, best avoided in small sections
	American (including Brazilian) mahogany *(Swietenia macrophylla)*	Yes	Durable	Small	Very good**	–
	Red lauan, dark red meranti *(Shorea spp)*	Yes	Variable; mostly moderately durable, occasionally non-durable	Small	Very good**	–
	Teak *(Tectona grandis)*	Yes	Very durable	Small	Very good**	Non-ferrous fittings essential
	Iroko *(Chlorophora excelsa)*	Yes	Very durable	Small	Good**	Extractives may occasionally retard the drying of oil-based paints

* To avoid the risk of discolouration, it is recommended that stainless or non-ferrous fixings/fastening are used with hardwoods.
** Natural finishes are often preferred for these decorative timbers.

Table 2

Preservative and paint treatments

Exterior	Preservative	Primer *(to appropriate BS)*	Paint finish *(three additional coats: preferably one undercoat, two gloss)*
Window and doors Softwood eg *Baltic redwood*, *Hemlock*, *Whitewood (Norway spruce)*	ESSENTIAL vacuum/pressure, diffusion – WB; double vacuum, dip – OS, WRP	Solvent borne (oleo-resinous or alkyd)	Alkyd gloss (4–5 years)
Douglas fir	ESSENTIAL if sapwood present	Aluminium preferred	Alkyd gloss (4–5 years)
Hardwood eg *Teak*, *Afzelia*, *Afromosia*, *Utile*, *Oak*	ESSENTIAL if sapwood or timber of low durability present	Not usually painted unless used as cills: then use oil-based or aluminium primer (filler needed)	Alkyd gloss
Plywood		Aluminium preferred	Alkyd gloss
Cladding, including barge boards, soffits and fascias Softwood eg *Redwood*, *Whitewood*, *Hemlock*, *Lauan*	REQUIRED BY BUILDING REGULATIONS Types as for windows	Moisture regulating primer	Moisture regulating paint See note stains preferred
Western red cedar	Not required	Not usually painted, but can be painted as other softwoods	
Pitch pine	ESSENTIAL if sapwood present	Aluminium preferred	Alkyd gloss paint
Hardwood	REQUIRED BY BUILDING REGULATIONS for some hardwoods of low durability	Not usually painted	

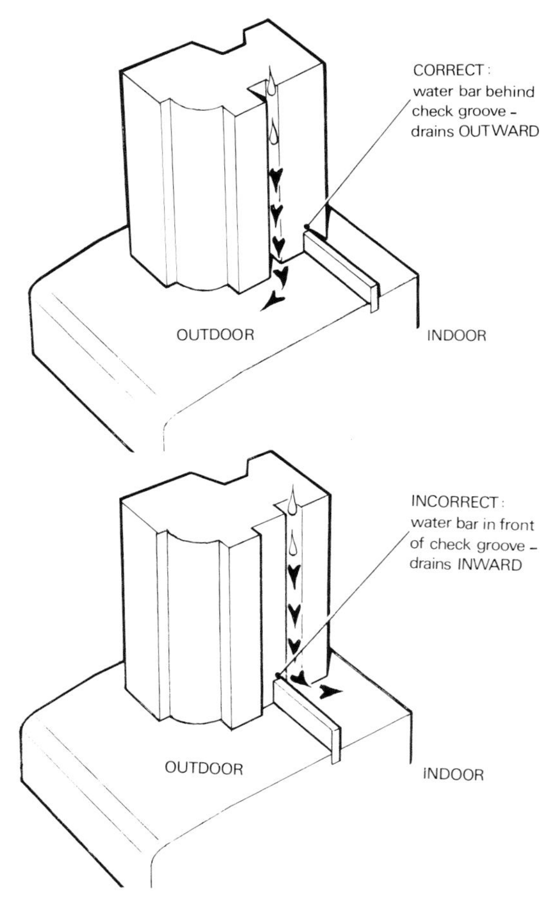

Figs 1 & 2 The water bar must be correctly positioned in relation to a groove in the jamb of an external door.

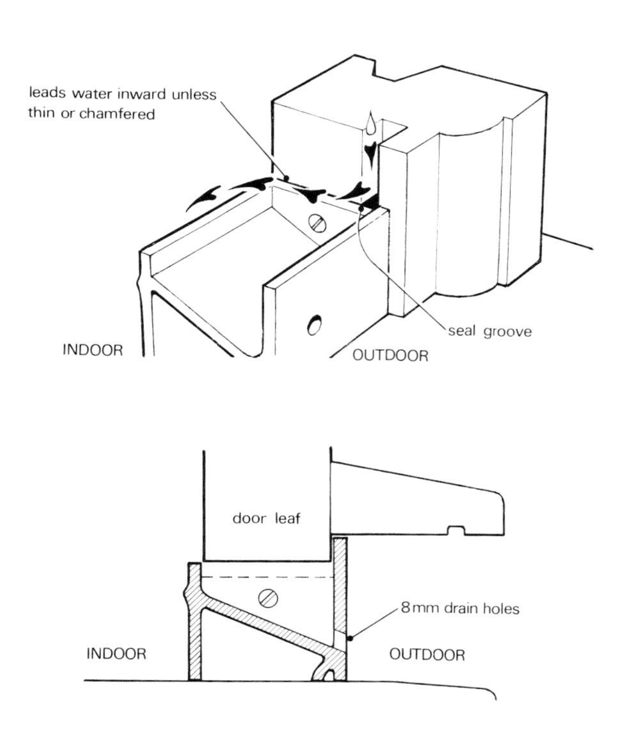

Figs 3 & 4 Even a purpose made self draining threshold can allow water to enter at the corner.

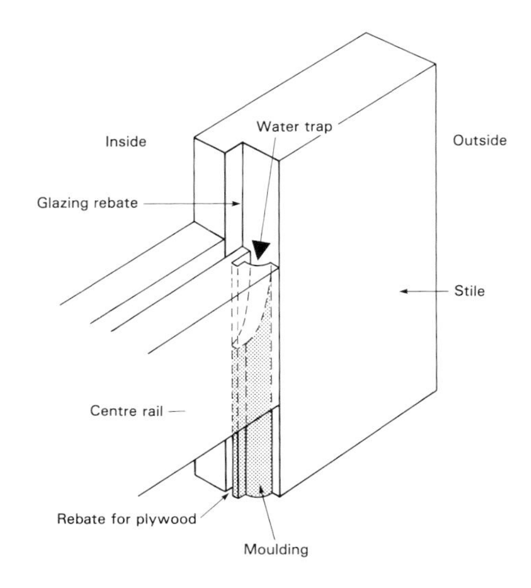

Fig 5 Water traps can be formed where machining leaves gaps

Photo 2 Weatherbar set too far forward

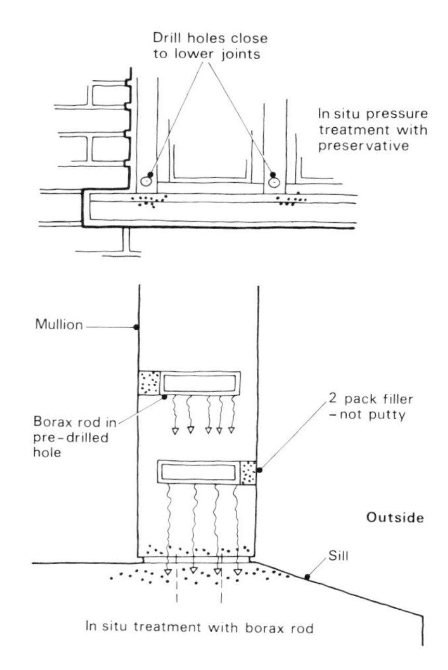

Figs 6 & 7 Treatment with boron rods reduces fungus attack if timber does become wet, as the dampness spreads the boron through the timber.

6.2.2 TIMBER WINDOWS AND DOORS. DETERIORATION OF PAINT FINISH AND OF THE TIMBER BENEATH

Symptoms

Blistering, cracking, peeling and flaking of the paint coating with cupping of the timber as the defect progresses. Eventually pieces of timber fall away and the inside of the timber members are seen to be soft and spongy or even hollow.

Investigation

Examine the condition of each component probing suspected areas with a pen-knife. Note where the timber is soft (horizontal members at the bottom of frames and the lower parts of vertical members are most likely to be affected).

Record the moisture content of the timber in the affected areas and adjacent parts of the component.

Find out whether the timber was preservative treated in the original specification. A laboratory may be able to establish the extent and type of treatment that was used.

Examine loose flakes of paint to establish the nature of the paint coating that was used. It may be possible to see whether the primer has been omitted. If there are doubts about the number or type of coating, pieces of paint can be sent for analysis.

Diagnosis and cure

This defect is caused by ineffective coatings, poor detailing of joinery and defective putties that allow water to enter untreated timber where rot can attack. The rot may be 'brown' or 'white' wet rot or occasionally dry rot. Dry rot is only likely to occur when there is an outbreak that extends into the adjacent windows or doors. Soft timber and sections of timber with a high moisture content are suspect and should be cut back or stripped of paint to establish the full extent of the rot. Rotten timber must be cut out and replaced with preservative treated timber, the new timber being spliced into the old or being fitted to replace complete sections. Adjacent timber can be injected with solvent based preservative or drilled and fitted with boron rods. When repairs are being undertaken, rebates and exposed end grain should be sealed with a primer coat.

Repair is only worthwhile on expensive components or components that show localised damage. It is often better to remove the component and replace it with a new preservative treated component. With such components it is essential that ends cut or trimmed on site are treated with preservative before being built into the construction.

Where condensation on the inner face of the glass and defective back putties have allowed water to enter the timber, a bead seal of clear silicone sealant to BS 5889 type B can be used to avoid a return of the problem once the window has been repaired.

REFERENCES

Sturrock, Doug *Repair preservation and maintenance of windows* Update on domestic windows 1986 Volume 1: Systems Design Publication.

BRE Digest 261: *Painting woodwork*

BRE Digest 304: *Preventing decay in external joinery*

BRE information paper 16/87 *Maintaining paintwork on exterior timber*

BRE Defect Action Sheet 11: *Wood windows and door frames: care on site during storage and installation*

BRE Defect Action Sheet 13: *Wood windows: arresting decay*

BRE Defect Action Sheet 14: *Wood windows: preventing decay*

Fig 8 Places where dampness is likely to be found (and where probing may reveal rot).

	DAYS 1	2	3	4	5	6	7
BURN OFF	*						
KNOTT & PRIME		*					
CUT OUT ROT & REPAIR			*				
INSERT BORON RODS			*				
HACK OUT CRACKED GLASS & PUTTY				*			
RUB DOWN & PAINT REBATE				*			
REGLAZE				*			
ALLOW PUTTY TO DRY – 48 HOURS MIN							
UNDERCOAT OVERALL						*	
FINISH OVER							*

TIMBER WINDOW – REPAIR, REGLAZE & REPAINT USING BORON RODS AS PRESERVATIVES & MICROPOROUS PAINT SYSTEMS.
FOR WATER BASED MICROPOROUS PAINT SYSTEMS ALLOW FOR ALKYD UNDERCOAT OVER FRESHLY RUN PUTTY TO PREVENT CISSING.

Photo 1 The sill mullion joint is vulnerable inside as well as outside.

	DAYS 1	2	3	4	5	6	7	8	9	10	11	12	13	14
BURN OFF	*													
KNOTT & PRIME		*												
CUT OUT ROT & REPAIR			*											
INJECT WITH PRESERVATIVE			*	ALLOW SOLVENT TO EVAPORATE										
HACK OUT CRACKED GLASS & PUTTY										*				
RUB DOWN & PAINT REBATE										*				
REGLAZE											*			
ALLOW PUTTY TO DRY														
UNDERCOAT OVERALL				ALLOW A FURTHER DAY FOR A COMPLETE BURN OFF OF EXISTING PAINT									*	
FINISH OVERALL														*

TIMBER WINDOW – REPAIR, REGLAZE & REPAINT USING ORGANIC SOLVENT PRESERVATIVE AND A CONVENTIONAL PAINT SYSTEM. WHERE AN ACRYLIC PRIMER IS USED TO PAINT REBATES, REGLAZING CAN PROCEED ON THE SAME DAY. (NOT APPLICABLE TO STEEL WINDOWS)

Photo 1 Deterioration of stain on timber window. Frame joints tend to open up when finish has gone due to wetting and drying of the timber.

Treatments suitable for exterior timbers (softwoods and hardwoods)

Application	Exterior wood stain			Clear varnish	Preservatives			Preservative requirements[1]
	Low-solids	High-solids	Madison formula		Creosote	CCA	Clear water-repellents	
Boarding Bargeboards Fascias Soffits	*	*	*		*	*	*[3]	Building Regulations (England and Wales) and Building Standards (Scotland) name timbers which require treatment when used as boarding and those which can be used untreated. For situations not covered by these Regulations, preservative treatment is advisable for timbers rated[2] as 'perishable' or 'non-durable' or if appreciable quantities of sapwood are present.
Plywood	*				*			
Joinery: Windows	*	*						Preservative treatment is necessary for redwood, hemlock and spruce (see PRL Tech Note 24). If sapwood is excluded, timbers rated as 'moderately durable' or better can be used without preservative treatment.
Doors		*		*				
Gates and fences	*				*	*	*[3]	Unless rated as 'durable' or 'very durable' and sapwood is excluded, all timber in contact with the ground should be preservative treated. For timbers out of ground contact, preservative treatment for 'perishable' and 'non-durable' timbers is advisable. Where systematic maintenance can be ensured, repeated application of the preservative finish, eg creosote or exterior wood stain may suffice.
Sheds	*	*	*		*	*	*[3]	
Benches	*						*[3]	
Handrails	*	*		*				

Notes: [1] Only recommendations of a very general nature can be given. More detailed recommendations will depend on life requirements and characteristics of the timber used.
[2] Durability ratings are described in PRL Technical Note 40.
[3] Only if it is desired to achieve weathered silver-grey effect.

6.2.3 TIMBER WINDOWS AND DOORS. DETERIORATION OF STAINS OR VARNISHES

Symptoms

The surface of stained or varnished coatings flakes and powders off, and the timber below becomes discoloured.

Investigation

Establish the type of coating used and when it was applied. Was the component left with the factory finish for some time before being given the final finish? Has the component been regularly maintained (4 years is the very maximum for a stained finish externally and 2 years for varnish)? Determine what species of timber was used, particularly if the component is of hardwood.

Diagnosis and cure

The life of varnish and of both high and low solid type stains varies with site exposure and pigment. Erosion of the finish is a natural weathering effect, (unless the process takes place over an unexpectedly short time) it is quite normal and should be expected. If it occurs within 1 year of installation, it may be caused by the poor adhesion of finishing coats when sunlight and dampness have degraded the surface of a component. This may have happened if it was inadequately protected for a long period on site before finishing coats were applied. Loose areas of coatings should be sanded down before retreating. Normal repeated maintenance will darken the timber and obscure the grain. This process can only be reversed by removing the finish and starting to build up the coatings from the cleaned timber.

REFERENCES

BRE Digest 286: *Natural finishes for exterior timber*

TRADA Research report 1979/3. *Exterior wood stain performance on hardwoods – natural weathering trials at 3 sites*

Photo 1 Condensate on glass can corrode the bottom of a steel frame – here a missing screw has not helped the situation

	DAYS 1	2	3	4	5	6
CHECK IRONMONGERY	*					
EASE OPENINGS		*				
HACK OUT CRACKED GLASS & PUTTY		*				
REMOVE RUST BACK TO BRIGHT METAL		*				
SPOT PRIME		*				
WASH DOWN OVERALL			*			
REGLAZE			*			
ALLOW PUTTY TO DRY						
UNDERCOAT OVERALL					*	
FINISH OVER						*

STEEL WINDOW – REPAIR, REGLAZE & REPAINT

Table 1 Time required for steel window – repair, reglaze and repaint

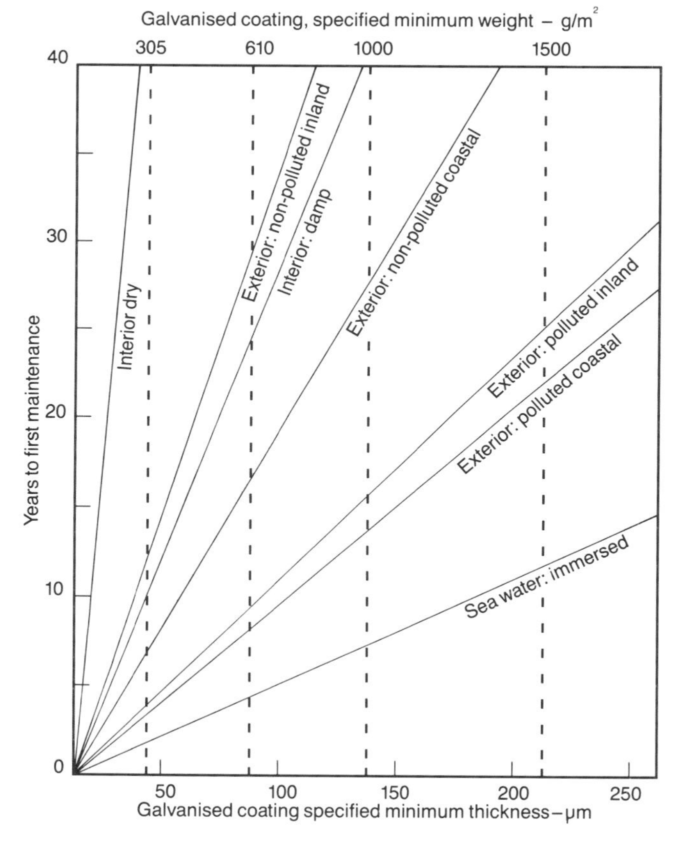

Fig 1 The thickness of the galvanised coating on steel must be related to the predicted environment.

6.2.4 METAL WINDOWS AND DOORS. DETERIORATION OF PAINT FINISH

Symptoms

Blistering and bubbling of the paint coating, particularly noticeable on light colours; pitting of the finish; rusting of steel where paint has flaked away.

Investigation

Establish the metal used to make the component. Use a magnet to identify ferrous metals. Note that opening lights in standard steel windows were sometimes made of cast aluminium (in the late 1930's).

Examine the metal where the paint has become detached for signs of corrosion.

Note the colour of the paint and whether the component is exposed to strong sunlight.

Note whether there are salts on the surfaces that are not regularly washed by rain (eg under soffits and beneath overhangs or porches) salts may originate from sea spray, de icing or adjacent materials.

Diagnosis and cure

Possible causes are: the rusting of steel components; the poor adhesion of the zinc phosphate priming coats to zinc coated steel components (occurs when they have not been previously treated with a two pack etch primer immediately before the zinc phosphate primer); the build up of heat in components exposed to strong sunlight. Salts on the surface of aluminium components can often explain the premature deterioration of factory finishes. These causes may act separately or in combination, though heat build up on its own is only likely to affect inferior finishes on metal.

If steel has rusted, it will be necessary to clean back the rust to bright metal. Zinc coated surfaces are likely to be damaged by abrasive cleaning of old paint coats. In both cases the exposed steel will need to be protected by a paint system that is suitable for uncoated steel. Where dark coatings are causing a build up in heat when exposed to the sun, and are suspected of contributing to the defect, then a lighter coating should be used as a remedy. Once the existing paint has been removed, repainting should be followed as soon as possible, but a minimum of 7 days should elapse before new metal glazing putty is overpainted.

Note that etch primers are not always effective as a pretreatment on zinc coated surfaces under oil based primers and undercoats, particularly in damp or exposed conditions. Manufacturers advice should be sought on special primers to suit particular situations.

REFERENCES

Sturrock, Doug *Repair preservation and maintenance of windows* Update on domestic windows 1986 Volume 1 System Design Publication

Construction Feedback Digest 56, Sept 86 *Corrosion of recently repaired shutter doors* B & M Publications (London) Ltd

BS 6150:1982 Code of practice for *painting of buildings*

GLC Development and Materials Bulletin 138 Item 4 *Overpainting of zinc and aluminium sprayed steel* Item 5 *Overpainting of sheradized components*

Photos 1 & 2   Maintenance of these coated aluminium windows was necessary within 4 years of installation because of insufficient heat curing of the organic coating.

Photo 3   Coatings on metal sills may be damaged during construction – subsequent remedial treatment is expensive.

Photo 4   Shows deterioration of mill finished aluminium block in anodised frame.

6.2.5 ALUMINIUM WINDOWS AND DOORS. DETERIORATION OF FINISH

Symptoms

Coated and anodised aluminium components are affected by erosion, pitting, fading and sometimes blistering of the finish. The damage may go unnoticed until it has spread across an area that is easily visible. Mill finished aluminium becomes coated with white rust or a combination of dirt and rust.

Investigation

Note the location of the damage and whether it is likely to have been caused by knocks, scratches or mortar droppings during construction.

Establish the finish of the aluminium. The most likely alternatives are:

- Mill finish, the natural surface of the metal
- Organic factory applied finishes of either the acrylic or polyester type. These are sometimes called powder coatings.
- Anodised finishes, which are an electrolytic thickening of the natural oxide coating. The anodised coating can be dyed to give it different metallic colours.

The thickness of the coating can be important and can be measured from a sample by a laboratory. If an anodised coating can be damaged by being scraped with the milled edge of a 10p coin, it is suspect and should be checked for thickness.

Diagnosis and cure

Deterioration of cills and exposed edges of components points to mechanical damage during construction. Damage which goes unnoticed until it has spread over wide areas of the surface is likely to be the result of pinholing of too thin or badly applied finishes. Flaking, with no corrosion present, suggests poor adhesion of the finish. Blistering, fading and loss of gloss suggest that the thickness of the coating should be measured. Bright patches on recently installed mill finished components are signs that mortar droppings have been removed from the surfaces. Thick, dirty, white rust on a mill finish comes from the failure to wash down the component regularly.

Organic coatings can be painted over areas that have been marked, provided that the original surface is thoroughly flatted down with wet and dry sandpaper to give a good base for the new coating. Though it should be recognised that the durability of paints applied to maintain or repair factory coated window frames, are likely to be of poorer durability than the original finish. When finishes fail to the extent that they have to be removed, or when a mill finish component is to be painted, the new paint system will have to include suitable preparation with etch primers. However, if weathered, an etch primer should not be used, an oil based, zinc chromate primer being quite adequate under an oil-based paint finishing system.

REFERENCES

BS 6150: 1982 *Code of practice for painting of buildings*

CP 3012: 1972 *Code for cleaning and preparation of metal surfaces*

Construction Feedback Digest 50 Spring 1985 *Coating failure on aluminium windows* B & M Publications (London) Ltd

Construction Feedback Digest 58 Feb 1987 *Damage to factory coatings on MOB Aluminium alloy window frames* B & M Publications (London) Ltd

Photo 1 uPVC windows are affected by dirtying and sometimes by staining particularly if beads are not of same material as frames.

Photo 2 Where plastic frames are set in painted timber subframes maintenance of the subframes is required. A colour match between the paint and the plastic cannot be expected to last.

6.2.6 uPVC WINDOWS AND DOORS. COLOUR CHANGE OF SURFACE

Symptoms

Obvious discolouration, yellowing and darkening of the surface, of part, or all of the component. White deposits on the surface.

Investigation

Note the extent of the problem. Establish the origin, history and age of the components.

Does the problem affect beads and frames to the same extent?

Record the orientation and exposure to direct sunlight of the affected components.

Diagnosis and cure

uPVC windows have generally been free from deterioration in appearance, however, chalking sometimes mars the appearance of coloured, particularly dark coloured, windows. It is caused by residual filler which remains on the surface as a white deposit and results from ultraviolet degradation leading to loss of polymer. It cannot always be removed by washing, though this should be tried. Acrylic faced sections are free from this defect.

Glazing beads are often made from a lower quality compound that is more prone to discolouration than the frame of the component. If this is the problem, the manufacturer may be able to supply replacement beads.

Discolouration adjacent to the glazing gasket is caused by the use of PVC gaskets in a uPVC unit. Manufacturers are now well aware of this problem and it is therefore only likely to affect older components. The solution is to replace the PVC gaskets with synthetic rubber gaskets.

Over-painting of plastics frames removes their principal advantage, low maintenance costs, because they must subsequently be regularly repainted. Repainting need not be avoided if it is essential to improve the appearance. However, paint coating on some plastics (eg uPVC window frames and rainwater goods) can lead to embrittlement and a reduction in impact strength.

REFERENCES

BRE Information Paper 11/79 *Painting plastics*

BRE Digest 69: 1977 *Durability and application of plastics*

British Plastics Federation and Glass and Glazing Federation – *A trade standard for uPVC windows* (1986)

GLC Development and Materials Bulletin 146 Item 4 *Plastics windows 3: Block Evaluation*

Photos 1 & 2 Advanced corrosion affecting stability on an external stair.

Photo 3 Fixings are not always protected from corrosion as effectively as the main members (photo of a gantry socket)

6.2.7 EXTERNAL FERROUS METAL STAIRS, RAILINGS AND DOORS. CORROSION

Symptoms

Rust usually shows on the surface. There may also be flaking and blistering of the paint finish.

Investigation

Establish the material used. Cast iron was commonly used for columns, railings and stair treads. Wrought iron was used for gates and railings. Both are less prone to corrosion than steel. If galvanising is present under the paint, send a sample to a laboratory for the thickness to be measured. Records of measurements of the galvanising taken for quality control purposes during construction may still be available. Examine corroded areas for signs that the galvanising is missing or has been damaged.

Examine past maintenance records if available. If the defect affects recently installed elements, check the specification for pretreatment and site work, and obtain laboratory confirmation of compliance with the specification. Pay particular attention to fixings, including nuts, bolts and washers as these may have been separately supplied to a less stringent specification.

Diagnosis and cure

Most ferrous metals corrode in damp conditions. Mild steel corrodes rapidly, particularly when dampness and certain chemicals combine to attack it (eg salt).

The protective metal coatings used to protect ferrous metals can be damaged by electrolytic action when other metals, such as copper, are adjacent to the steel (see chapter 1 & 2). The commonest cause of corrosion is an insufficient thickness of coating, that has not given adequate protection under the prevailing environmental conditions. Once corrosion starts, it leads to further deterioration of the finish. BS 5493 gives guidance on appropriate coatings.

An alternative cause of corrosion is the poor pretreatment of the metal base, such as the failure to remove all mill scale before priming, or the failure to make good any site installation damage to factory treatments before on-site finishing.

Corrosion can also be caused by over-zealous maintenance. Sand-blasting or other aggressive treatments used to clean off previous finishes can damage the galvanising. Unless the maintenance programme specified subsequent treatment as if for ungalvanised steel, the metal is likely to corrode. However, a more likely cause of failure is paint flaking due to the use on a clean galvanised surface of a paint system designed for ungalvanised steel.

Remedial measures consist of removing all corrosion and priming the bright metal without delay. A full paint specification, relevant to site conditions should then follow.

If a badly corroded element is structural, the advice of a structural engineer must be sought to establish that its strength will be adequate.

REFERENCES

BS 5493: 1977 *Code of practice for protective coatings of iron and steel structures against corrosion* (this standard is being revised at the time of going to press).

Construction Feedback Digest 56 September 1986 *Corrosion of recently repaired shutter doors*. B & M Publications (London) Ltd

Bickerdike, Allen, Rich and partners in association with Turlogh O'Brien *Design Failures in Buildings*, Second Series Sheet No. 23. *Paint on steel* George Godwin Ltd

Photo 1 Unpainted putty between metal and stone dries and cracks within a year or so.

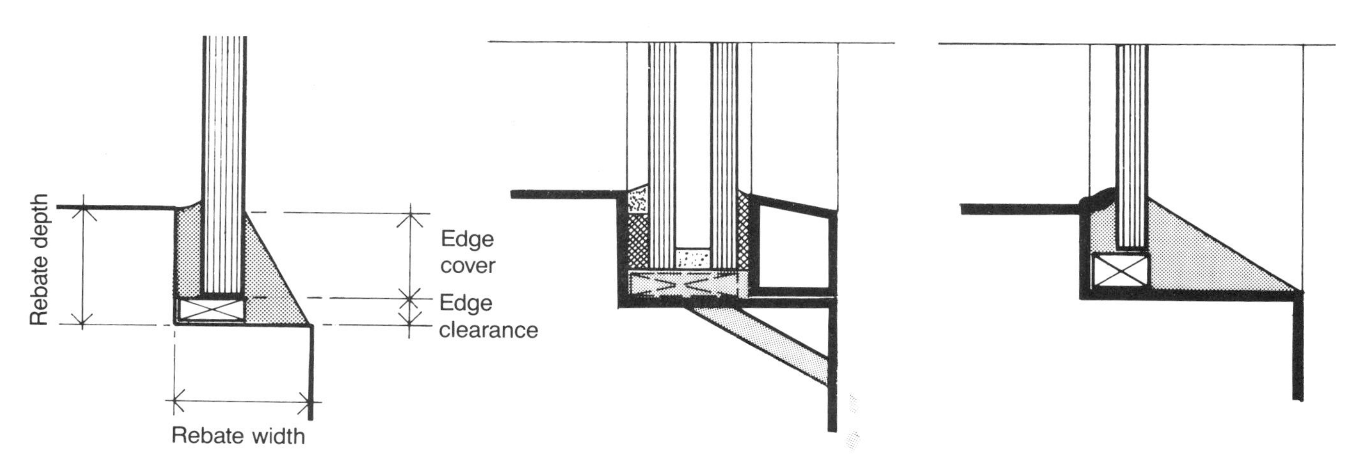

Figure 1 Rebate and putty dimensions
edge cover – varies with glass area and site exposure but for 6mm glass and under the minimum is 6 mm.
edge clearance – Approximately 3 mm all round but solar control glass needs more (consult the manufacturer) and greater tolerances make for easier handling of large panes.
rebate depth – minimum should be glass thickness x 2 for up to 6 mm thick. (Consult manufacturer for thicker glass).
rebate width – minimum should be glass thickness + 10 mm for back and front putties.

Figure 2a DRAINED DOUBLE GLAZING

Figure 2b BACK PUTTY PROFILE IMPROVED USING SILICONE BEAD — but consult manufacturer on priming requirements.

6.3.1 PUTTY FAILURE IN METAL AND TIMBER WINDOWS

Symptoms

Paint flakes or peels off the putty surface. There is a loss of adhesion between the putty and the rebate. The putty may crack and eventually fall out.

Investigation

If putty is cracked and is of recent installation, cut out a section and send to a laboratory for identification.

If the putty has poor adhesion to its surround, establish whether there is primer under the putty and check the moisture content of timber windows by using an electrical resistance meter (see chapter 4). Find out about weather conditions prior to glazing. Were the timber frames saturated on site, or after fixing, before being glazed? Where timber windows have been glazed using a putty system, check whether the putty has been painted after installation to protect it from drying out.

Examine the back putty. It should be flush, or above the bottom rail or transom, and free from gaps, in order to shed water from internal condensation.

Note the dimensions of the rebate and the dimensions of the putty section. A section that is small, or feathered out against the frame, will dry out and probably crack sooner than a more compact section.

Diagnosis and cure

Traditional linseed oil putty is not suitable for use on metal frames and quickly cracks and draws away from the frame. It should be replaced with metal glazing putty.

Linseed oil putty should not be applied to bare timber, but to a primed surface. Glazing should not be carried out in wet weather. Both of these practices may show up later as lack of adhesion between putty and frame, and can be diagnosed by looking at the surfaces of the rebate.

Stained timber frames should be glazed using internally fixed beads. If putty was specified, and has subsequently failed, it should be completely removed and replaced with a suitable sealant not requiring protection.

Back putty that collects condensation can be raised using a sealing compound (eg clear Silicone sealant to BS 5889 type B). However, if the defect is widespread, re-glazing will be necessary. Consideration should also be given to improving ventilation to reduce condensation on the inside of the glass.

Re-glazing should whenever possible take place as part of a maintenance programme that includes window repair and painting. Attention must be paid to manufacturer's instructions relating to intervals for drying out of putty prior to painting, eg BRE advise leaving the putty for about four weeks before painting with water-borne Acrylic paints (or overcoating the putty beads with solvent thinned paint after a shorter drying period). The painter should ensure coverage of putty edges by the paint film – ie by overlapping the top-coat (not primer or undercoat) onto the glass by a margin of 1–3 mm.

REFERENCES

Glass and Glazing Federation – *Glazing Manual*

BRE Digest 286: *Natural finishes for exterior timber*

BRE Information Paper 9/84 *Water-borne paints for exterior wood*

AJ Products in Practice Supplement May 1985 – *Glass and Glazing* The Architectural Press

BS 6262:1982 *Code of practice for glazing for buildings*

BS 6150:1982 *Code of practice for painting of buildings*

Photo 1 Staining is more likely to occur on sloping roof glazing

Photo 2 The alkaline run off from other parts of the building can cause damage to the surface of the glass

Photo 3 Glass can be damaged by storage that retains water on the surfaces. Close packed glass in damp conditions should be interleaved with absorbent non alkaline paper

6.3.2 GLASS. SURFACE DISCOLOURATION

Symptoms

The surface is seen to be discoloured when the glass is looked through at an angle.

The surface of the glass is obviously stained. The defect is more likely to affect rooflights and horizontal glass rather than vertical glass.

Investigation

Note how many panes are affected.

Is the glass washed naturally? If so, does the rainwater wash off any other material onto the affected glass?

The colour of the staining should be noted and whether or not it is removable with an abrasive cleaner.

Diagnosis and cure

If water is retained on glass, corrosion is induced by sodium hydroxide produced by the water dissolving sodium oxide from the surface of the glass. This causes staining, which is usually only visible when the glass is looked through obliquely. It may be possible to remove it using an abrasive cleaner or in more extreme cases by acid polishing.

Water run off from new concrete or other strong alkaline solutions can cause damage to the surface of the glass.

Sloping roof glazing is vulnerable to dirt build up. On a roof which gets little sun, algae may develop and spread, leading eventually to permanent staining. Regular maintenance cleaning is required.

REFERENCES

Construction Feedback Digest 41 Winter 1982/3 *Glass staining*
B & M Publications (London) Ltd

Glass and Glazing Federation – *Glazing Manual*

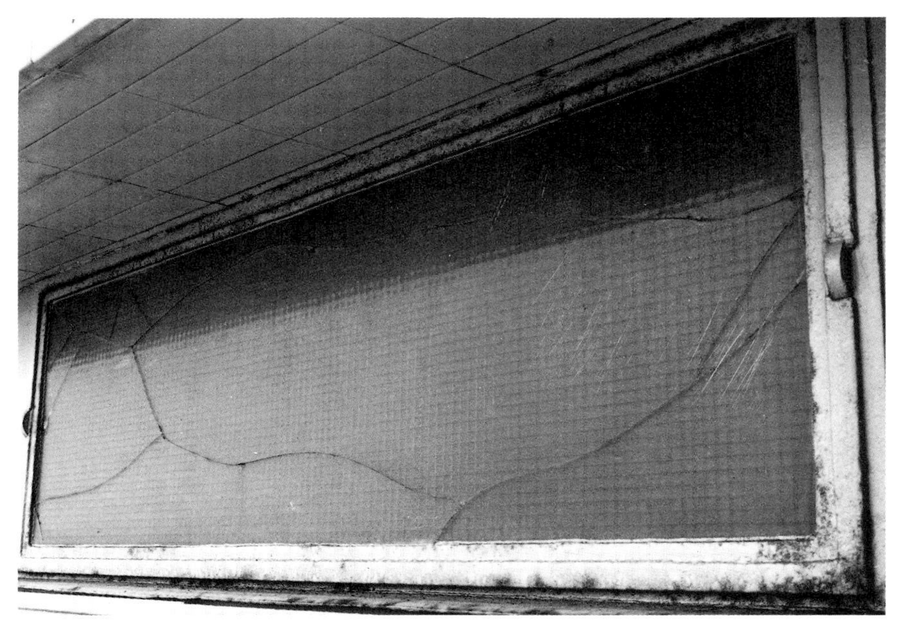

Photo 1 Rust in steel frame and thermal stress where glass is partially shaded can both cause cracking.

Photo 2 In this instance a sealed double glazed unit in a rooflight incorporated coated glass. There was insufficient allowance for expansion between the glass and a gantry fixing bracket.

Photo 3 Rooflights are susceptible to thermal stress cracking.

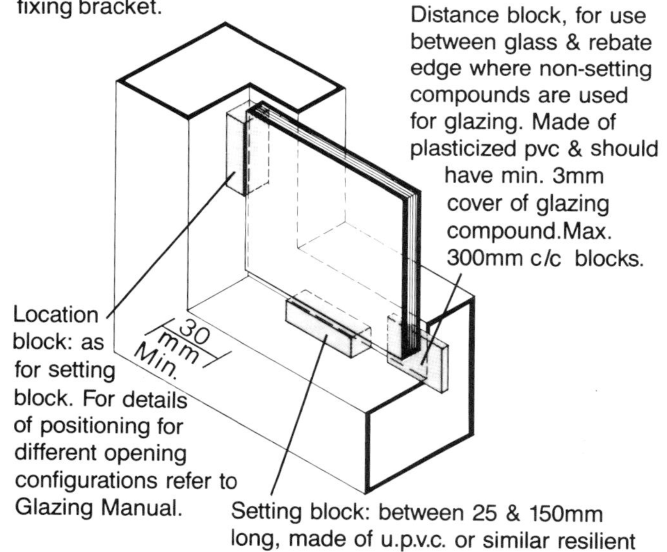

Fig 1 Blocks in glazing practice,

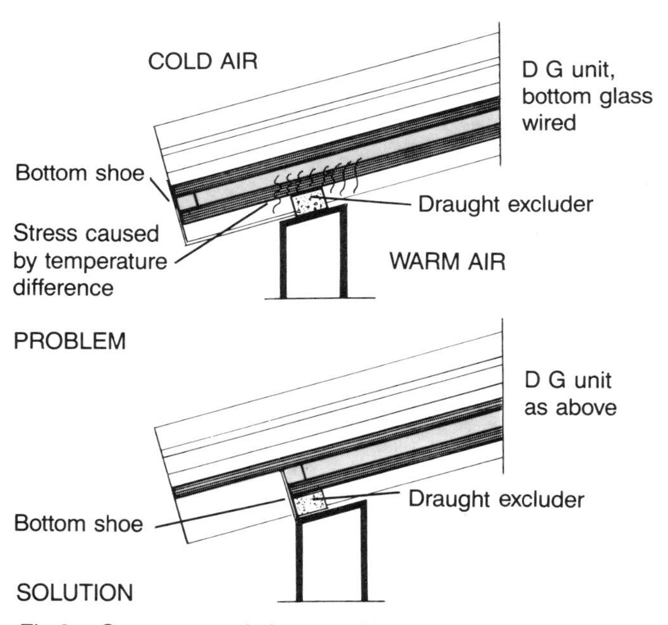

Fig 2 Stress caused glass cracking in double glazed rooflight.

6.3.3 GLASS. THERMAL STRESS CRACKING

Symptoms

Cracks develop in areas of glazing with no obvious signs of impacts on the glass.

Investigation

Examine the glass and the frame adjacent to the glass for signs of impacts that could be the cause of cracking. Note the installation technique, the frame material and the overall condition.

Check the type of glass. Is the glass tinted or coated? If the defect is in a double glazed unit, make a note of which layer is cracked. If both layers of glass are not affected, check the cut edge of the cracked pane.

See whether a strong shadow is thrown across the glass shielding part of it from sunlight.

Assess whether the external environment at the time the crack appeared was likely to have caused thermal stress.

Diagnosis and cure

The risk of glass breakages from thermal stress has become much greater with the advent of special heat absorbing and reflecting glasses, and the wider use of double glazing. Glass in any type of frame can crack if it has been fixed too tightly, or not sized to allow for thermal expansion within the glazing rebates. Once putty or beads have been removed for re-glazing, this defect will be obvious. Note that if reglazing with thicker glass, larger tolerances are required. Steel frames can cause glass to crack by rusting in the rebate behind the glass. The expansion of the rust presses against the glass and cracks it.

Glass is often more at risk from thermal stress at the coldest times of the year, for example when the sun is low in the sky on frosty mornings. The cold surround retards expansion of the glass perimeter which triggers cracking as the glass heats up.

Glass that is never cleaned allows the build up of solar heat under the dirt. This can cause sufficient thermal expansion to break the glass. Institute a regular cleaning programme after reglazing.

Wired glass is inherently not quite as strong as clear glass and is therefore susceptible to cracking: rust at the edges can cause expansion of the wires. If the detail of the rebate cannot be improved through draining or a better glazing technique, it may be possible to reglaze with a different type of glass, but fire protection and safety requirements will have to be satisfied. Wired glass is particularly at risk when used as the inner pane of double glazed units. Where it has failed it should be replaced by toughened glass.

Wired glass is often the inside sheet of a double glazed unit in a patent glazing roof. The thermal stress in the inside sheet where it is exposed to the external air at the overlap can cause the wired glass to crack. On reglazing, the detail should be amended and a stepped section unit provided for the bottom edge only.

Thermal stress can also occur where a strong shadow across the glass only allows part of the sheet to heat up in sunlight. Glass which has 'shelled' edges is inherently more likely to crack than well cut glass. Check that new material is free from shelling and from other defects before reglazing.

REFERENCES

BS 6262:1982 *Code of practice for glazing for buildings*

BS 5561:1977 *Code of practice for patent glazing*

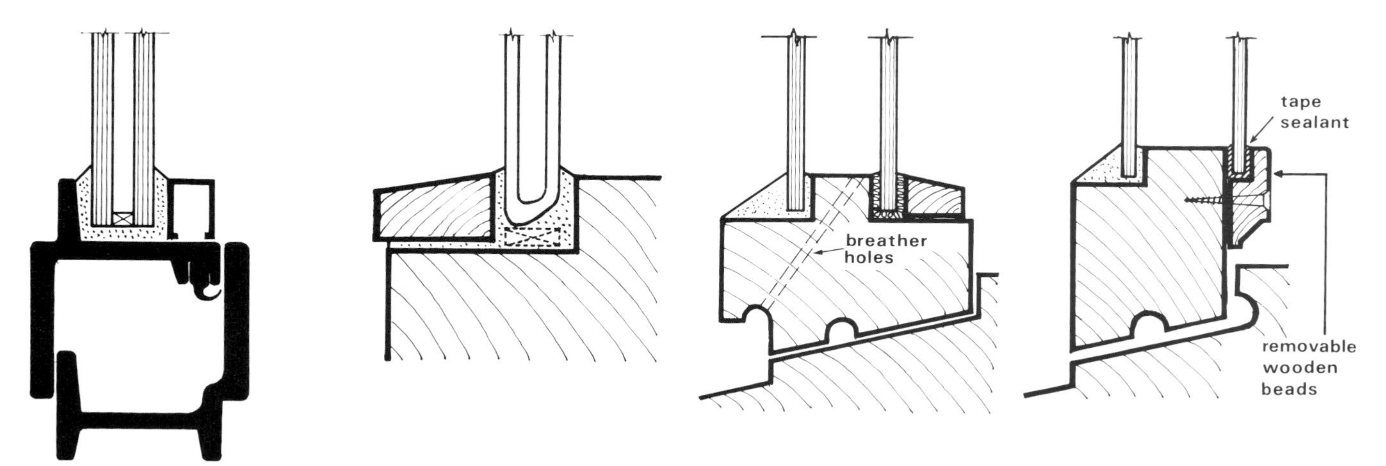

Fig 3 Typical factory sealed double glazing units

Fig 4 Typical glazed in-situ double glazing

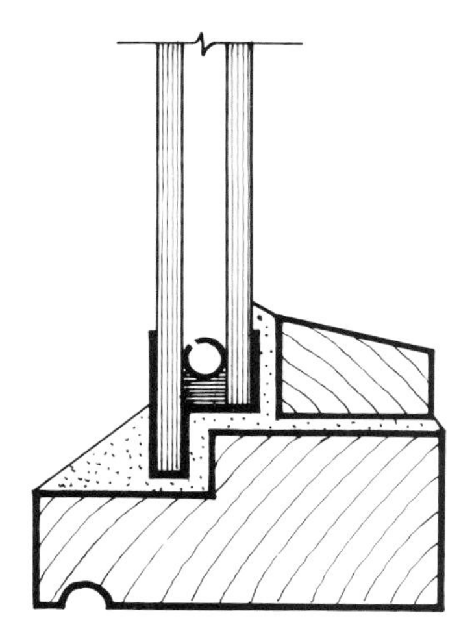

Fig 1 Typical stepped unit

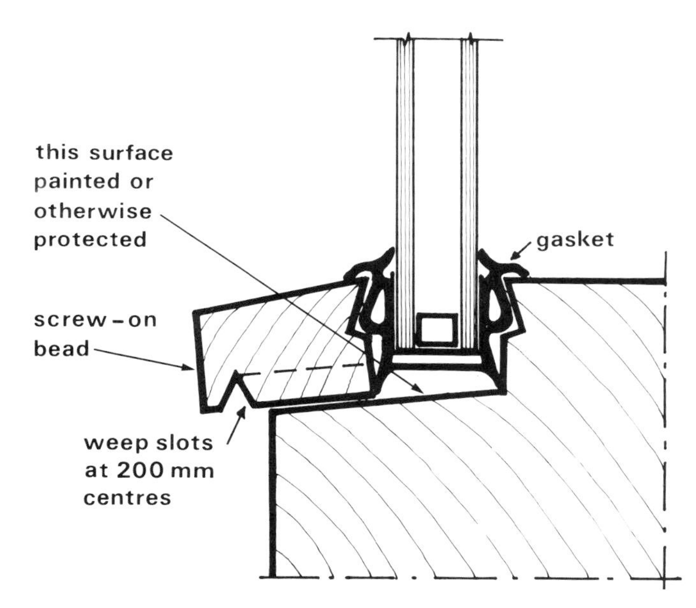

Fig 2 Drained glazing system

Photo 1 A breakdown of the edge seal of double glazed units causes misting and eventually staining within the unit.

6.3.4 DOUBLE GLAZED UNITS. MISTING AND DISTORTION

This defect covers some common failures in double glazed units. It does not apply to secondary glazing where a separate inner window is fitted.

Symptoms

Condensation mists some or all of the glass area inside the unit.

Glass is seen to be deflected. When looked at from the side, the pane may be seen to be bowed in across its surface, or it may show signs of stress by rainbow like refraction.

Investigation

Examine the edges of the unit for water ingress through cracked putty etc.

Note the temperature inside and outside when condensation appears and when it clears.

Measure the thickness of the glass and of the unit.

Diagnosis and cure

The most common double glazing failures are caused by water trapped against the foil back of the sealing block, which eventually deteriorates and allows water to break the seal. Once the seal is broken, moisture seeps between the sheets, and can condense within the unit, causing misting of the inner face of the external sheet of glass.

The cure is reglazing into a redesigned rebate with a condensation channel to draw off condensation from the bottom of the frame, and with putty or a glazing bead set to shed water.

If the glass can be seen to be deflected, and condensation occurs in the middle of the glass internally, either the incorrect pressure has been created within the unit during manufacture, or the glass is not thick enough to withstand the pull of the slight vacuum between the sheets (strongest when the air is coldest). The unit should be replaced with one that has been made to the correct pressure with the correct glass thickness.

Note that the organic seal used for the perimeter of double glazed units is liable to deteriorate in strong ultraviolet light. The exposed face is usually protected with foil but occasionally other faces are exposed by undersized beads etc. Where the seal is exposed, it may deteriorate and allow moisture to enter between the two glass sheets.

REFERENCES

BS 6262: 1982 *Code of practice for glazing of buildings*

Glass and Glazing Federation – *Maintenance recommendations for double glazing sealed units*

AJ Products in Practice Supplement May 1985 *Glass and glazing* Architectural Press

Construction Feedback Digest 58 February 1987 *Wet rot in windows – Replacement Recommendations* B & M Publications (London) Ltd

Photo 1 The degraded surface of glass reinforced plastic rooflights collects dirt that is hard to remove.

Photo 2 Damage caused by a cracked plastic rooflight (the light had become opaque – thermal stress or vandalism could have caused the cracking).

Photo 3 Damage by seagulls who habitually brought food onto these rooflights and either dropped it from a height or pecked at it on the surface.

6.3.5 PLASTICS ROOFLIGHTS. SOILING, HOLING, CONDENSATION AND LEAKS

Symptoms

On translucent GRP rooflights: there may be a loss of translucency and a dirt build up on the outer surface. Translucent PVC rooflights may be punctured. All types of plastics rooflights can be affected by condensation forming between double skins or by water ingress at fixings, or the perimeter.

Investigation

Establish the manufacturer and the material from which the rooflight is made.

Common materials are:

- Acrylic (Perspex); clear material cracks rather like glass and is easily scratched.
- Polycarbonate; hard and clear and thinner and tougher than acrylic.
- uPVC; easily scratched, softer and more easily bent than Acrylic.
- GRP; strong and flexible but not as clear as acrylic or polycarbonate.

Examine the surface and record the extent of the soiling or roughness. If the rooflight is cracked or holed, look on adjacent areas of the roof for objects that could have caused the damage.

Inspect the fixings to check they are well fitted, ie not too loose or too tight. Note whether they are corroded and whether the correct washers have been used in all cases. Note any damage or cracks close to the fixings.

Inspect the perimeter of the rooflight for defects.

Diagnosis and cure

Translucent GRP rooflights:
in an aggressive environment, the loss of surface by natural weathering can take place more quickly than originally predicted. The loss of surface, and the exposure of the glass fibres, leads to a build up of dirt that is not easy to remove. Manufacturers offer a polyester or PVF film to smoothen the surface of sheets not yet judged too dirty, but there is insufficient experience to predict their long term effectiveness.

Translucent uPVC rooflights:
this material is more brittle than GRP and is liable to impact damage from a severe hail storm. There is even a recorded case of seagulls bombing uPVC, leading to the gradual holing of an entire installation. Replacement with polycarbonate sheeting will avoid this problem.

With all plastics rooflights, fixings are suspect if water is getting in. Original drilling may have left uneven, torn, internal surfaces, that do not provide a proper sealing for the washer. Too much or too little torque applied to the fixing bolt can also lead to failure.

Thermal movement can lead to tearing of the rooflight around the fixing hole. Sheets that are affected should be replaced, and fixings with more tolerance should be specified.

Condensation that collects between the outer and inner layer of a rooflight often leads to problems. When it is not feasible to seal the void, permanent ventilation of the void should, if possible, be incorporated.

REFERENCES

BRE Digest 69: *Durability and application of plastics*

Construction Feedback Digest 44 Autumn 1983 *Rooflights under attack* B & M Publications (London) Ltd

Construction Feedback Digest 59 April 1987 *Translucent roof lights* B & M Publications (London) Ltd

CHAPTER 7 FLOORS & CEILINGS

CHAPTER 7 FLOORS AND CEILINGS Contents

CHAPTER 7 FLOORS AND CEILINGS Contents

continued

continued

INTRODUCTION

Floor defects can be broadly split between those occurring in the structure and those in the finish. Structural defects may only become apparent when the finish fails or an adjacent part of the structure behaves abnormally.

This chapter starts with structural defects that can be associated with either suspended floors or solid ground floors. The suspended floors may be of reinforced concrete or timber construction, defects in suspended floors are confined to occasional excessive deflection and to fungal and insect attack. Solid concrete ground floor defects are invariably related to one of three major causes:

- the ground fill
- the concrete mix from which the floor is constructed
- the incorrect detailing and application of the damp-proof membrane.

Defects in finishes are dealt with following the structural defects. A wide range of finishes is available to the specifier to suit most circumstances of use. Either the wrong specification or a change of use of the building can result in the failure of the floor finish. Water ingress from below ground slabs may adversely affect both screeds and finishes.

The third section of this chapter concerns ceiling defects. It only covers ceilings that form part of an intermediate floor. Those that form the soffit to a roof structure are dealt with in Chapter 8. The majority of defects are to do with finishes as the structural defects of ceilings tend to be more often related to the floor structure. A large number of proprietary suspended ceiling systems are available on the market and defects that occur in these systems often are peculiar to the particular system. In view of the variety of systems in use, they are not dealt with in this book.

The conversion of large dwellings into a number of smaller units has focussed attention on the last defect of this chapter concerning impact sound transmission through floors. It is a particular problem with suspended timber floor structures, but can also be a problem with solid concrete and beam and pot floors if they are not properly constructed.

General points

Although the first signs of a problem with a floor may be hidden by an applied finish, it is often the appearance of cracks or dampness in the finish that will draw attention to it. A thorough investigation of surface defects is essential to ascertain their root cause. Simple patching of a surface crack may allow a deeper rooted problem to worsen.

Defects in solid ground slabs fall broadly into two categories. Those associated with movement of the sub-strata and, more commonly, those associated with drying shrinkage of the concrete.

Sub-strata defects need to be investigated beyond the confines of the local problem. A detailed history of the site will need to be formulated. Old mineworkings, tree planting or felling, alteration of ground water levels are just a few of the things that can drastically effect the stability of the ground.

General movement of a ground slab may be accompanied by cracking in walls which indicates an overall ground problem. If, upon investigation, there are no other cracks appearing in the structure, the problem is more likely to be confined to the ground fill or hardcore on which the slab is built.

In spite of the well documented theory and practice of concrete construction, it is often forgotten that concrete shrinks as it dries out. This may cause it to crack and impose stresses on screeds and finishes. If ground water subsequently passes through the cracks due to an inadequate damp-proof membrane, the consequences and the problems of diagnosis can be further complicated.

The defects described here that apply exclusively to the floor or ceiling finish rather than the structure are of three types. In one the wrong material has been used. In the second, the layers of finish are not compatible with each other. In the third workmanship is at fault, mixing, preparation, material storage, or the time lapse between the application of the different layers having not been properly controlled. The investigation of defects in finishes will be helped in the diagnosis of failure if good records and first hand evidence can be obtained as to what materials were used, and how, and under what conditions they were used.

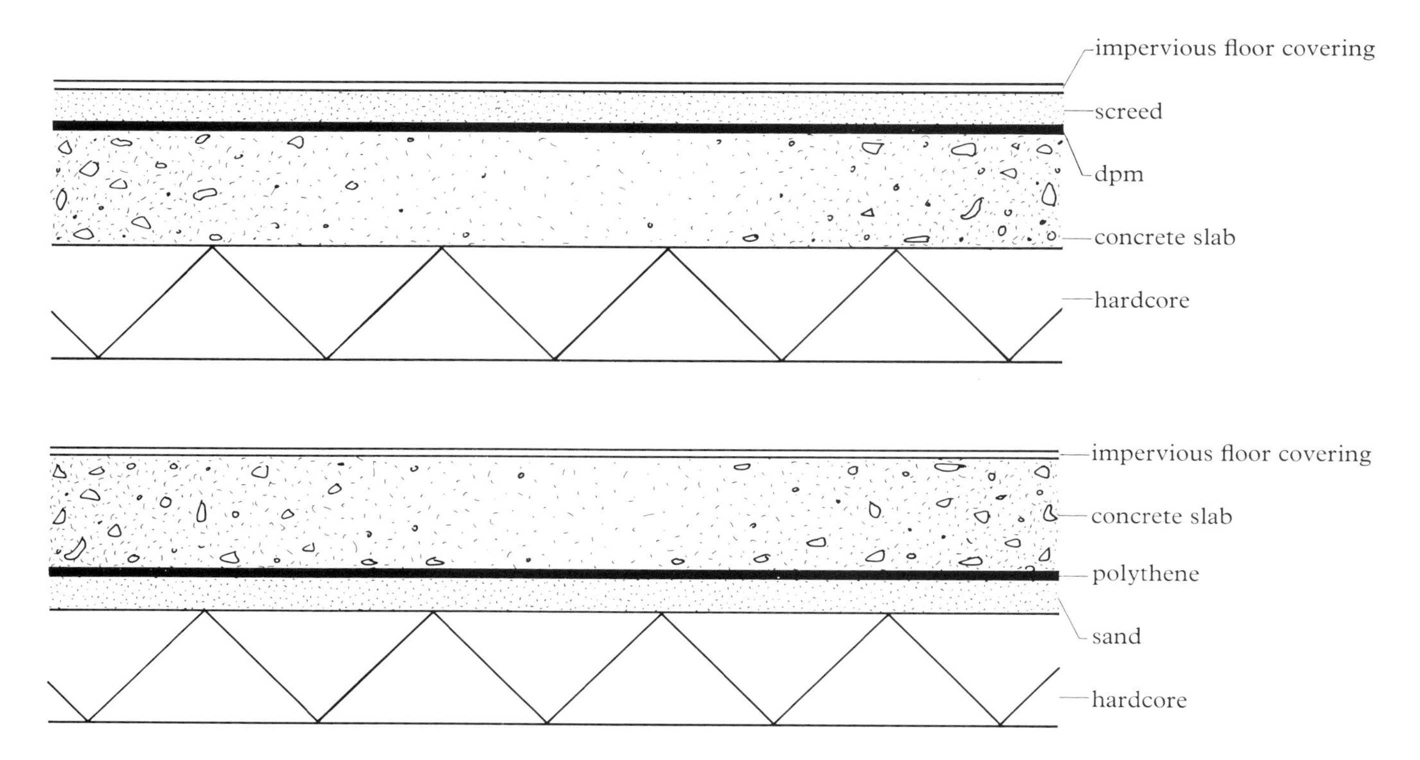

Figure 1 Damp proof membranes in solid floor construction.

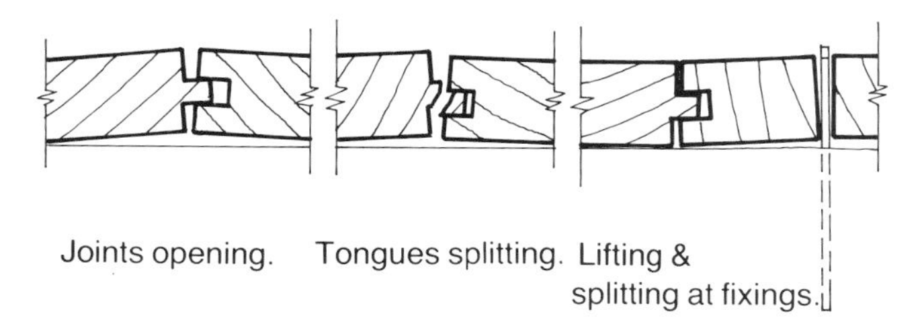

Figure 1 Distortion of boards due to moisture movement.

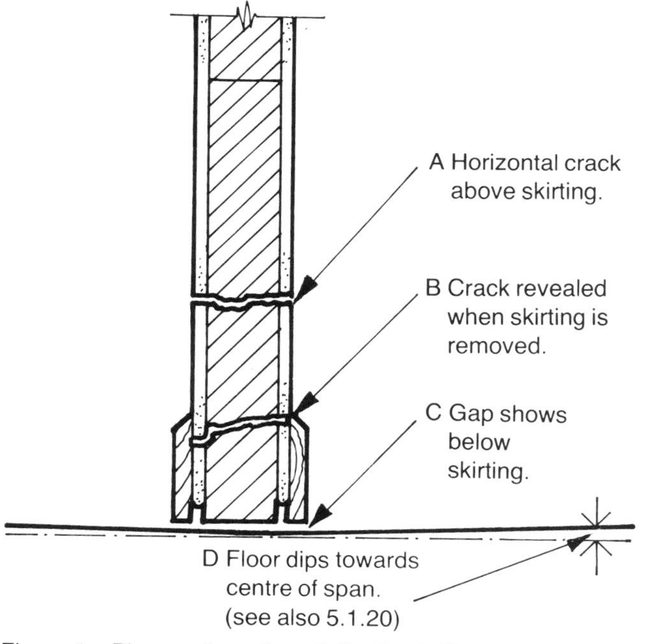

Figure 1 Places where floor deflection is likely to show.

7.1.1 EXCESSIVE DEFLECTION OF SUSPENDED CONCRETE FLOORS

In reinforced concrete floors it is quite normal for a limited deflection to occur which usually goes unnoticed. This defect describes a more conspicuous deflection that is most pronounced at mid span.

Symptoms

The deflection may be observed by eye but is more likely to be detected by the side-effects produced, of which the most typical is the failure to provide continuous support for partition walls. This often results in a horizontal crack in the partition at, or close to, floor level. It may be hidden by the skirting board, but assuming the board is straight, an obvious increase in space beneath the board towards the middle of the wall could be a telltale sign.

Investigation

Information should be obtained on the original design and construction of the floor. Check to see whether any partition walls have been added since the structural calculations were done.

Measure the actual deflection and compare it with the tolerances allowed for in BS 8110. Check to see what type of aggregate was used in the concrete mix, looking out for high shrinkage aggregates in particular. See whether any other structural movement has occurred elsewhere in the building.

Diagnosis and cure

Deflections in concrete floors may be caused by a single factor or several factors acting together. Inadequate structural design is a possibility, but should be unlikely in view of the design safety margins. Poor workmanship during construction may have imposed excessive loadings on the slab by the premature removal of temporary props, or the use of the slabs to store building materials and equipment. Eccentric movement of supporting walls or columns is a possible cause and should show its presence elsewhere in the structure.

High shrinkage aggregates, eg sedimentary rocks such as greywacke, shale and mudstone, will have a considerable influence on the drying shrinkage of the concrete which may be four times the shrinkage that occurs when using normal aggregates. The use of these aggregates is more common in Scotland than elsewhere in the British Isles.

If cracks have occurred in a partition wall, they should be cleared of any loose material that may impose compression forces in the wall by acting as a wedge.

If excessive deflection is even suspected, the advice of a structural engineer should be sought. If it is thought that there is a possibility of collapse, measures should be taken to reduce the loading on the floor and the risk of damage (or injury) in floors below.

REFERENCES

BS 8110 *Structural use of Concrete*. Part 1: 1985 *Code of practice for design and construction*. Part 2: 1985 *Code of practice for special circumstances*

BRE Digest 35: *Shrinkage of natural aggregates in concrete*

see also 5.1.20 *Cracks in internal walls*

Photo 1 Ceiling finishes damaged by water from ablution area above.

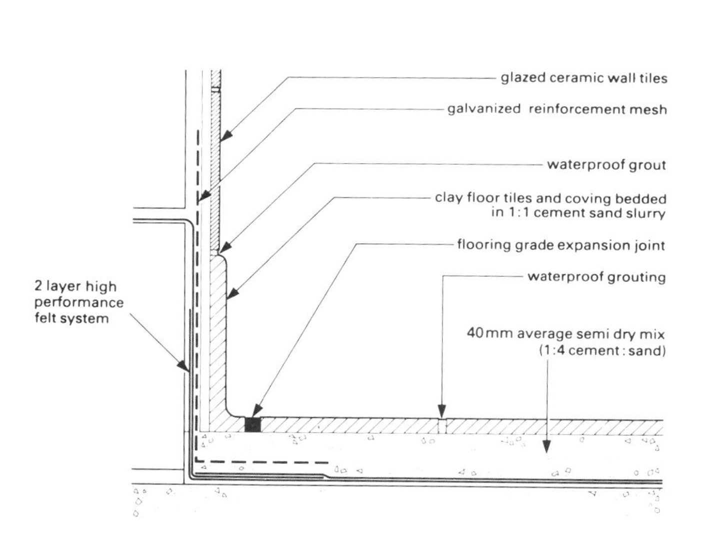

Figure 1 Alternative detail to fig 2 giving a tiled finish and allowing for some movement in the substrate.

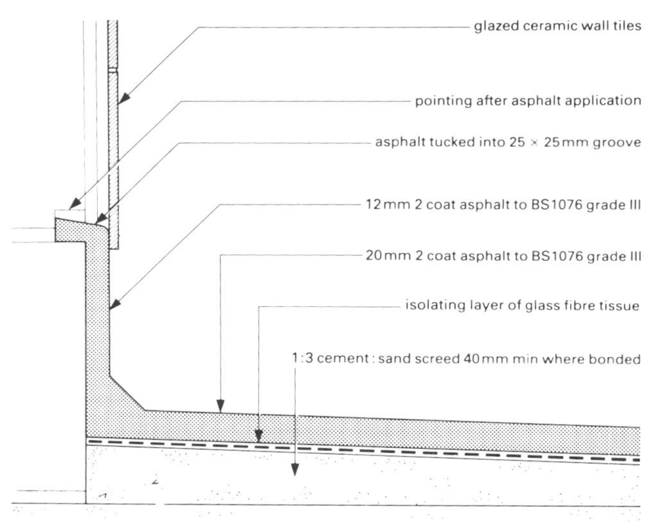

Figure 2 Asphalt detail used to replace failed lead sheet waterproofing to ablution area.

7.1.2 SUSPENDED CONCRETE FLOOR LEAKING WATER

Water seeps through the underside of a floor slab below an area that is frequently wetted, such as a washroom.

Symptoms

The paint finish to the ceiling starts to flake as it loses adhesion. Persistent damp patches appear on the ceiling which may be stained brown. If not attended to, cracks eventually start to form, running at right angles to supporting walls.

Investigation

Establish that the water ingress is from above and not the result of condensation from below. Record the time and extent of the leakage. Compare the records with the times that the floor gets wetted above.

Check whether the cleaning materials used contain harmful chemicals.

If brown rust staining is appearing on the ceiling, consult a structural engineer who may wish to take core samples to check the condition of steel reinforcement embedded in the concrete.

Diagnosis and cure

During the investigation procedure it may be necessary to strip back floor finishes to expose the structural reinforced slab. This will reveal the extent of any waterproof membrane in the floor. Chemical analysis of corrosion products will show if any reaction has been taking place between the concrete and the membrane. Lime in solution from wet fresh sand cement screed or mortar bedding can have a highly corrosive effect on lead. Water finding its way through joints in tiles will saturate any layers below. If the concrete is poorly compacted, any bridging of the waterproof membrane will allow the water to percolate through the slab. In time, the normal protective layer to the steel reinforcement is lost and the consequential corrosion will cause cracking in the slab allowing the water to pass right through. Although a certain amount of structural integrity will be lost by this action, significant weakening of the slab is unlikely to occur for some time after the cracks first appear.

The advice of a structural engineer should be sought for remedial action required concerning the structure. Treatment of a similar nature to that outlined in 5.1.23 is advisable. A new waterproof layer will need to be laid on the slab which will totally seal it from future wetting. Care should be taken to continue this layer up vertical surfaces to ensure its integrity. If tiles are relaid, waterproof grout must be used and enough space left at perimeter walls for expansion. This space should be filled with a good quality sealant (see BS 6213).

REFERENCES

BS 5385 *Wall and floor tiling* Part 4: 1986 *Code of practice for ceramic tiling and mosaic in specific conditions*

BS 6213: 1982 *Guide to selection of constructional sealants*

BRE Digests 263, 264 & 265: *The durability of steel in concrete*

Construction Feedback Digest 34 September '80 *Failure of sheet lead waterproofing in ablution floors* B & M Publications (London) Ltd

Construction Feedback Digest 54 Spring '86 *Defects in hollow pot floor construction* B & M Publications (London) Ltd

Classification of visible damage caused by ground floor slab settlement

The classification below attempts to quantify the assessment of floor slab settlement damage. It has not yet been used extensively to determine its applicability. It should be noted that the categorisation may be qualified by the possibility of progression to a higher category; this should arise only when examination has revealed the presence of voids or areas of loosely compacted fill (or degradable material) beneath the floor slab such that more settlement can be expected.

Category of damage	Degree of damage	Description of typical damage	Approximate (a) crack width (b) 'gap'[1] *mm*
0	Negligible	Hairline cracks between floor & skirtings	(a) NA (b) up to 1
1	Very slight	Settlement of the floor slab, either at a corner or along a short wall, or possibly uniformly, such that a gap opens up below skirting boards which can be masked by resetting skirting boards. No cracks in walls. No cracks in floor slab, although there may be negligible cracks in floor screed and finish. Slab reasonably level.	(a) NA (b) up to 6
2	Slight	Larger gaps below skirting boards, some obvious but limited local settlement leading to slight slope of floor slab; gaps can be masked by resetting skirting boards and some local rescreeding may be necessary. Fine cracks appear in internal partition walls which need some redecoration; slight distortion in door frames so some 'jamming' may occur necessitating adjustment of doors. No cracks in floor slab although there may be very slight cracks in floor screed and finish. Slab reasonably level.	(a) up to 1 (b) up to 13
3	Moderate	Significant gaps below skirting boards with areas of floor, especially at corners or ends, where local settlements may have caused slight cracking of floor slab. Sloping of floor in these areas is clearly visible. (Slope approximately 1 in 150) Some disruption to drain, plumbing or heating pipes may occur. Damage to internal walls is more widespread with some crack filling or replastering of partitions being necessary. Doors may have to be refitted. Inspection reveals some voids below slab with poor or loosely compacted fill.	(a) up to 5 (b) up to 19
4	Severe	Large, localised gaps below skirting boards: possibly some cracks in floor slab with sharp fall to edge of slab; (slope approximately 1 in 100 or more). Inspection reveals voids exceeding 50 mm below slab and/or poor or loose fill likely to settle further. Local breaking-out, part refilling and relaying of floor slab or grouting of fill may be necessary; damage to internal partitions may require replacement of some bricks or blocks or relining of stud partitions.	(a) 5 to 15 but may also depend on number of cracks (b) up to 25
5	Very severe	Either very large, overall floor settlement with large movement of walls and damage at junctions extending up into 1st floor area with, possible damage to exterior walls, or large differential settlements across floor slab. Voids exceeding 75 mm below slab and/or very poor or very loose fill likely to settle further. Risk of instability. Most or all of floor slab requires breaking out and relaying or grouting of fill; internal partitions need replacement.	(a) Usually greater than 15 but depends on number of cracks (b) greater than 25

Note 1 'Gap' refers to the space – usually between the skirting and finished floor – caused by settlement after making appropriate allowance for discrepancy in building, shrinkage, normal bedding down and the like.

Table 1 Classification of visible damage caused by ground floor slab settlement.

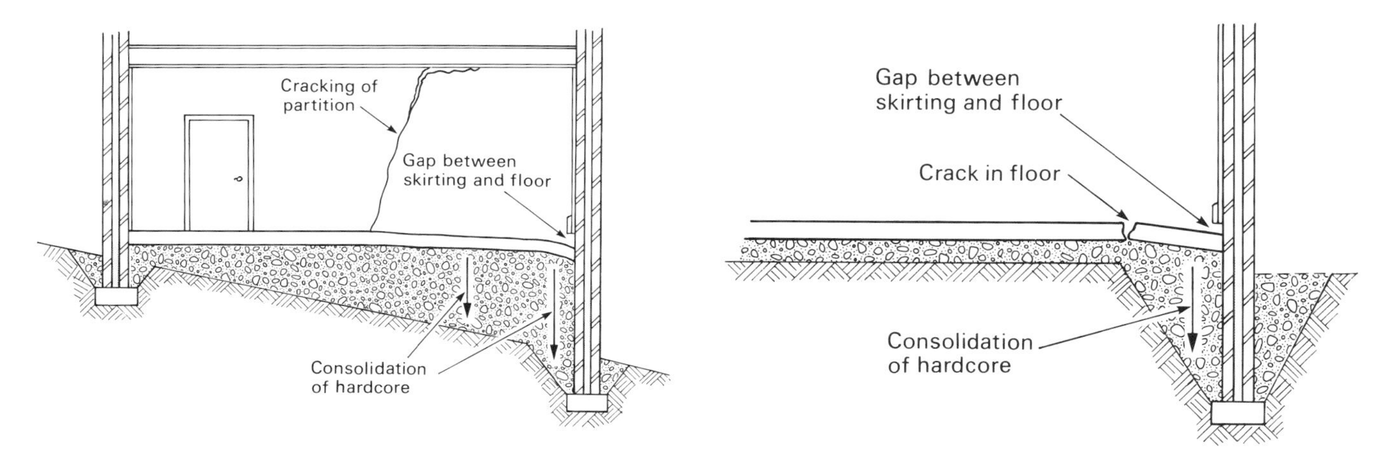

Figure 1 The consolidation of hardcore does not affect the foundations but may crack partitions.

Figure 2 Floors will show cracks and gaps below skirtings.

7.1.3 SOLID GROUND FLOOR SLAB. DISTORTION AND CRACKING

Consolidation of hardcore.

Symptoms

Gaps appear between the underside of skirting boards and the top of ground slabs. Slabs may crack adjacent to perimeter walls and in line with them. Partition walls built off the slab may crack where their support is disrupted. Loadbearing walls on foundations will not be affected.

Investigation

Remove a large enough area of the slab to enable a sample of hardcore to be excavated. Measure the depth of laid hardcore and make a visual inspection of its contents. Have a sample analysed for chemical content in a laboratory. Monitor further settlement by observing any change in the top level of the slab against a fixed point on a bordering loadbearing wall. If the safety of the slab is at all suspect, it is important to consult a structural engineer. Obtain a detailed history of the ground conditions in the area. Try to establish the extent of any excavation work carried out prior to construction.

Diagnosis and cure

The consolidation of hardcore should not be confused with swelling due to chemical change (7.1.4). Consolidation will result in a downward movement of the slab whereas chemical change will cause an upward movement. Widespread cracking of the slab can occur in both cases. Adequate compaction is required when placing hardcore, which is best achieved by applying it in layers. Each layer must be well compressed before laying the next. Past experience has resulted in the National House Building Council imposing a maximum depth of 600 mm for filling material at any one point beneath a ground slab. Beyond this depth they recommend that the slab should be designed as a suspended floor. Special attention is needed where there may be a sudden increase in depth of fill where trenches for wall footings have been excavated without vertical sides. Table 2 from BRE Digest 251, repeated here, categorises the degree of damage caused by ground floor slab settlement and suggests what remedial works will be needed.

REFERENCES

BRE Digest 251 – *Assessment of damage in low-rise buildings*

BRE Digest 276 – *Hardcore*

Photo 1 Distortion and cracking of thin floor slab caused by sulphate attack.

Requirements for concrete exposed to sulphate attack

Recommendations are for concrete in a near-neutral groundwater – for acid conditions refer to BRE Current Paper CP 23 77

Class	Concentrations of sulphates expressed as SO_3: In soil: Total SO_3 (%)	In soil: SO_3 in 2:1 water:soil extract *g/l*	In groundwater *g/l*	Type of cement		Requirements for dense fully compacted concrete made with aggregates meeting the requirements of BS 882 or 1047: Minimum cement content(1) *kg/m^3*	Maximum free water/cement(1) ratio
1	Less than 0.2	Less than 1.0	Less than 0.3	Ordinary Portland cement (OPC) or Rapid hardening Portland cement (RHPC) or combinations of either cement with slag(3) or pfa(4) Portland blastfurnace cement (PBFC)	Plain concrete(2)	275	0.65
					Reinforced concrete	300	0.60
2	0.2 to 0.5	1.0 to 1.9	0.3 to 1.2	OPC or RHPC or combinations of either cement with slag or pfa PBFC		330	0.50
				OPC or RHPC, combined with minimum 70% or maximum 90% slag(5) OPC or RHPC, combined with minimum 25% or maximum 40% pfa(6)		310	0.55
				Sulphate-resist Portland cement (SRPC)		280	0.55
3	0.5 to 1.0	1.9 to 3.1	1.2 to 2.5	OPC or RHPC, combined with minimum 70% or maximum 90% slag OPC or RHPC, combined with minimum 25% or maximum 40% pfa		380	0.45
				SRPC		330	0.50
4	1.0 to 2.0	3.1 to 5.6	2.5 to 5.0	SRPC		370	0.45
5	Over 2	Over 5.6	Over 5.0	SRPC + protective coating(7)		370	0.45

(1) Inclusive of content of pfa or slag. These cement contents relate to 20 mm nominal maximum size aggregate. In order to maintain the cement content of the mortar fraction at similar values, the minimum cement contents given should be increased by 50 kg/m^3 for 10 mm nominal maximum size aggregate and may be decreased by 40 kg/m^3 for 40 mm nominal maximum size aggregate.
(2) When using strip foundations and trench fill for low-rise buildings in Class 1 sulphate conditions further relaxation to 220 kg/m^3 is permissible in the cement content for C20 grade concrete.
(3) Ground granulated blastfurnace slag to BS 6699:1986.
(4) Pulverised-fuel ash to BS 3892: Part 1: 1982.
(5) Per cent by weight of slag/cement mixture
(6) Per cent by weight of pfa/cement mixture
(7) See BS CP 102: 1983: Protection of buildings against water from the ground.

Table 1 Requirements for concrete exposed to sulphate attack.

7.1.4 SOLID GROUND FLOOR SLAB. CRACKING AND LIFTING

Ground floor slabs constructed from concrete laid on hardcore are particularly vulnerable to attack by water-soluble sulphates in the hardcore. The effects of swelling due to other chemical constituent changes are similar but not as common.

Symptoms

The initial signs of trouble may take a number of years to develop after construction has been completed. Doors opening onto the floor will start to bind due to a slight uplift of the slab. Later, the arching and lifting becomes more pronounced and the top surface of the slab cracks in an irregular pattern. Partition walls built on the slabs become disturbed and in some cases the swelling forces may be great enough to cause an outward movement of the enclosing walls. The amount of uplift in the centre of the floor may be as much as 100 mm.

Investigation

A sample of the concrete should be analysed for sulphate content, and a sample of the hardcore examined to determine its contents. If possible, have the groundwater tested and analysed. Identify the aggregate used for making the concrete. If it is clinker, have it analysed for any unburnt coal content.

Diagnosis and cure

The signs of expansion are very obvious. Nevertheless, this defect is sometimes mistakingly diagnosed as being evidence of settlement or ground movement. There are three main causes:

1 The most common cause is the attack on concrete by water soluble sulphates. The likely source for the sulphates is from the hardcore, but in isolated cases it may be the groundwater. Whatever the source, the sulphates will chemically react with the Portland cement on the underside of an unprotected concrete slab in damp conditions.

  This causes a gradual breakdown of the concrete which makes it expand and lift. Colliery shale was a common culprit when used as hardcore, but nowadays a more likely source will be contaminated bricks or gypsum plaster. The latter can often be left adhered to broken brick or concrete from demolition works. The contents of hardcore from demolition must be carefully monitored for harmful elements.

2 Unstable aggregate may cause the concrete to expand. Some clinkers contain unburnt coal of a type that expands on oxidation. They may also contain sulphates which will react with the Portland cement. This is a much less common cause and the effects are usually not as severe as in cause 1.

3 The hardcore may expand due to chemical and volume changes in the material. All rely on the presence of water to support the reaction. The principal materials that have been identified are:

   a) unhydrated lime or magnesia in steel slags

   b) broken concrete or old, partially vitrified slags from old slag dumps

   c) colliery spoil with a high clay content

   d) oxidised pyrites which form soluble sulphates in the presence of air and moisture.

Whatever the cause, it will be necessary to remove the concrete floor if it has broken up badly. It may also prove necessary to remove the existing hardcore if tests prove it to have a high sulphates content. Any replacement hardcore material should be granular, of even composition and able to drain and consolidate easily. It should be overlaid with an impermeable moisture barrier, such as polythene sheet and should be at least 0.2 mm thick. It must be turned up the face of any perimeter wall and overlapped with the DPC. A new concrete slab can then be laid referring to Table 1 of BRE Digest 250 as a guide to the correct type of cement to use.

REFERENCES

BRE Digest 54: *Damp-proofing solid floors*

BRE Digest 250: *Concrete in sulphate-bearing soils and groundwaters*

BRE Digest 251: *Assessment of damage in low-rise buildings*

BRE Digest 276: *Hardcore*

Photo 1 Asphalt cracked over daywork/bay joints.

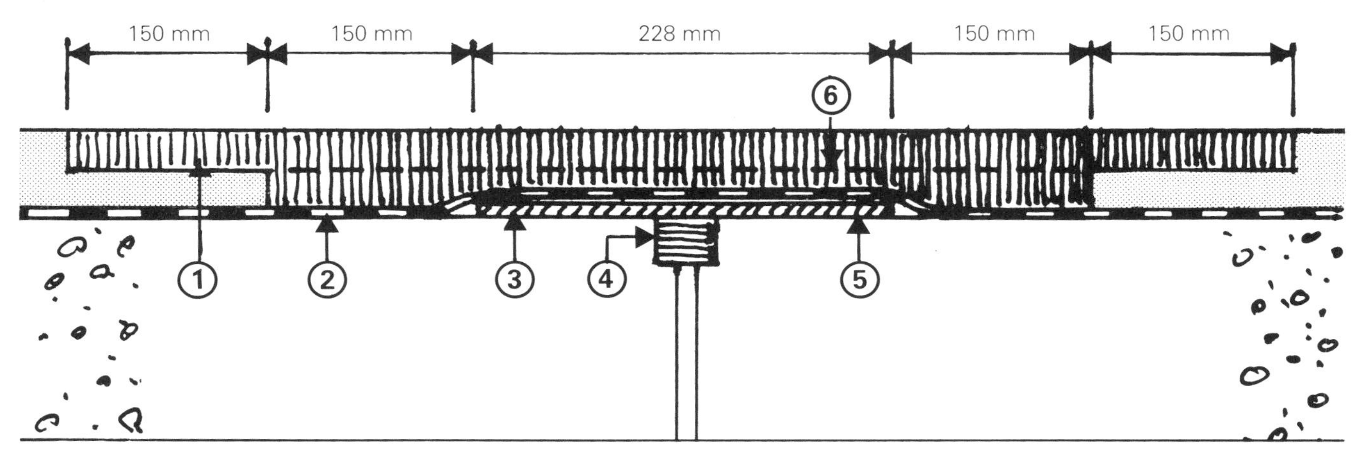

1 Existing asphalt cut back to dimensions shown

2 Original glass film tissue separating layer slit along bay line and laid back, and new section 528 mm wide inserted for full width and length of joint line (existing layer laid back-over prior to asphalting)

3 Any curling of concrete ground-off to produce flush adjacent edges and remove laitence

4 Concrete sawn-out sufficient to insert new 12 × 12 mm closed cell polyethylene foam 25 per cent compressed on insertion

5 Butyl rubber strip (228 mm wide) bonded to concrete slab both sides of bay joint with epoxy adhesive, strictly in accordance with manufacturer's instructions

6 New 2-coat asphalt to match existing

Figure 1 Method of cutting back and strapping daywork/bay joint to prevent cracking of asphalt.

7.1.5 SOLID GROUND FLOOR SLAB. CRACKING OF ASPHALT

Crack running the full length of the floor in a straight line, with occasional cracks running at right angles from it.

Symptoms

Cracks appear in asphalt flooring laid on concrete ground slabs a few weeks after laying. The cracks are uniform in width and may be curled at the edges. There may be a slight hollowness to the asphalt either side of the crack.

Investigation

Make a thorough check of the rest of the structure to establish whether there are other cracks in the walls, etc.

If possible, obtain details of the slab construction and the ground conditions beneath.

Check to see whether the floor is level and the same height either side of the crack.

Diagnosis and cure

The reason for this defect may be wrongly attributed to ground heave beneath the slab. In the absence of any cracking in the walls and if the slab is found to be level, the cause is almost certainly attributable to shrinkage cracking of the concrete. This will occur at its weakest point which in this case is the bay or daywork joints.

When asphalt is used as a floor finish, it will probably be also acting as the damp-proof membrane. Cracks in the asphalt should therefore not be ignored. It will be necessary to cut back the existing asphalt and 'strap' the slab joints with butyl rubber strips as shown in the diagram.

REFERENCES

Construction Feedback Digest 53 Winter '85/86 *Asphalt flooring defects*. B & M Publications (London) Ltd

BRE Digest 54: *Damp-proofing solid floors*

BS CP204: Part 2: 1970 *In-situ floor finishes*

BS 6925: 1988 *Specification for mastic asphalt for building and civil engineering (limestone aggregate)*

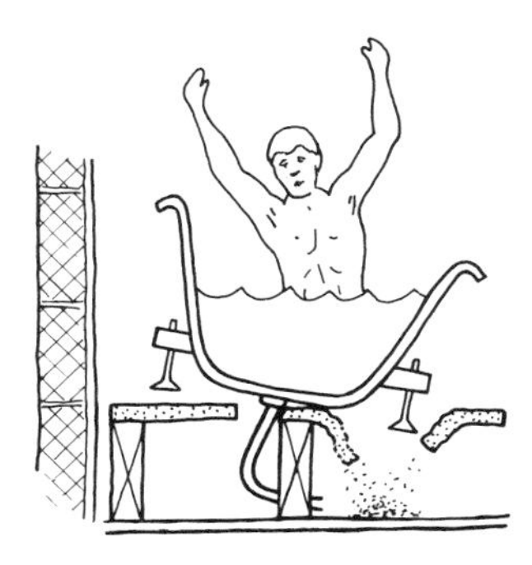

Tanking needed over whole floor, turned up at perimeter, and collared around services pipework.
Always provide bearers, but do not fix through the sheet

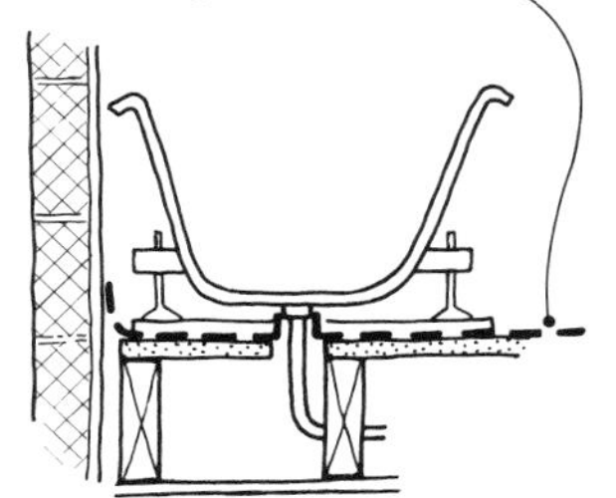

Figures 1 & 2 Support and waterproofing should be provided beneath the bath where a vulnerable type of deck has been used..

PSA do not recommend particle board in areas subject to wetting.

Photo 1 If particle board gets wet it can lose its strength, warp and delaminate.

Photo 2 Water seeping through the tile joints has caused the failure of this particle board.

Not less than
10 mm gap between
edge of flooring and wall

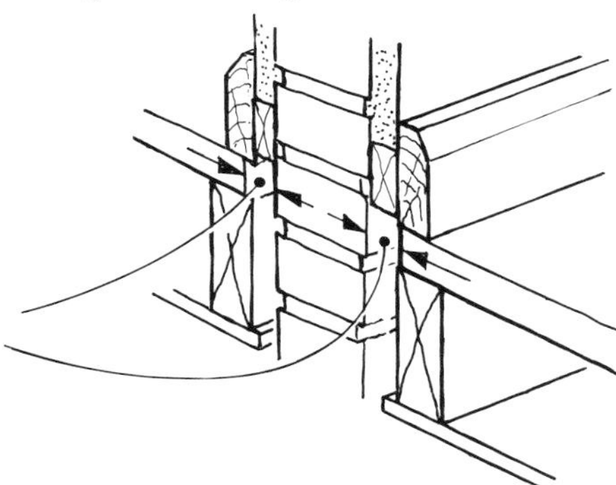

Figure 3 It is important to allow adequate tolerance around the edge of a particle board floor.

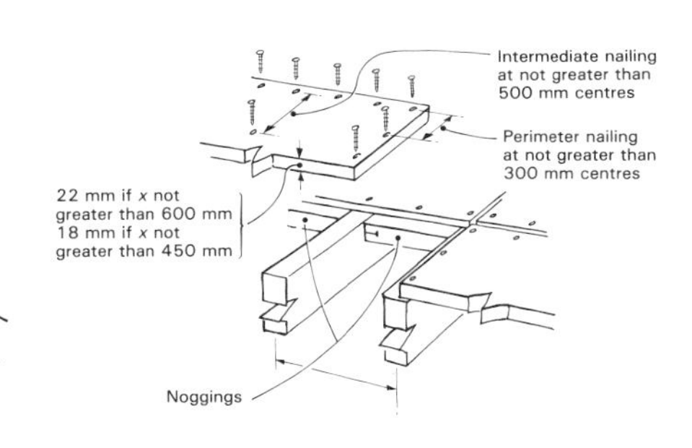

Figure 4 Screwing and nogging should be as shown unless type of board and manufacturers instructions indicate otherwise.

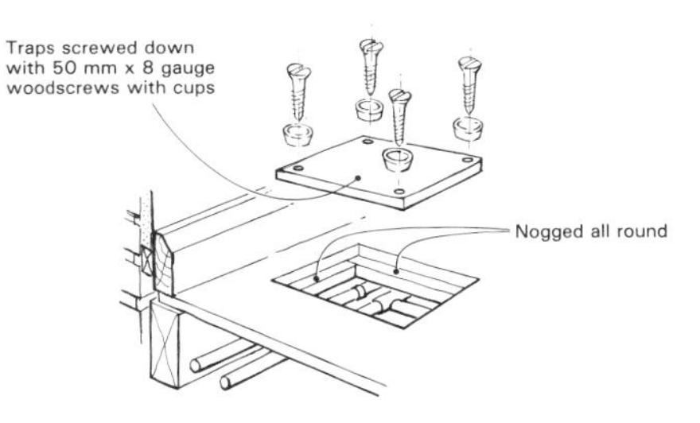

Figure 5 Service access holes should be provided to give access to pipe junctions.

7.1.6 SUSPENDED TIMBER FLOOR. LOSS OF STRENGTH AND STIFFNESS

This defect occurs where the incorrect grade of chipboard is used as flooring, or when the chipboard is installed incorrectly.

Symptoms

The floor sags and distorts, or at worst, punctures under point loads. Most likely to occur in areas that are subject to leakage or spillage of water such as kitchens and bathrooms.

Investigation

Remove any floor covering from the chipboard. Check to see whether the sheets have been fixed down correctly on all edges.

Check that the boards are the correct flooring grade and are marked BS 5669 type II or, where greater resistance to moisture is required, type II/III.

Check that the correct thickness of board has been used – 18 mm board for joist spacings up to 450 mm and 22 mm board for joist spacings up to 600 mm. If the boards are tongue and grooved, check that the long edge is at right angles to the direction of joist span.

Check that the boards are fully supported on all four edges.

Check all supply and drainage pipes for leaks, particularly areas under baths, basins, sinks etc.

If the boards are supported on battens on resilient strips (for sound insulation) check the condition of the resilient strips. See that they provide adequate support for the battens.

Diagnosis and cure

All chipboard that has got wet, must be replaced. Rapid deterioration will take place and will be accelerated if there is a covering of tiles or jointed sheet which traps the water. If the flooring is replaced with chipboard, type II/III boards must be specified and the correct method of laying, strictly observed.

An alternative is to protect the boards with an unjointed plastics sheet turned well up perimeter walls. Any service pipes that have to penetrate the floor must either pass through the walls or be fitted with a sealed waterproof sleeve. If the floor is at ground level, make sure that the underfloor is well ventilated to prevent a build-up of moisture from condensation.

If failure of resilient strips is found to be the cause of the defect, the floor will probably have to be replaced with a more durable sound insulation system – products that have British Board of Agreement certificates are more likely to be satisfactory.

REFERENCES

BRE Defect Action Sheet 31 – *Suspended timber floors: chipboard flooring – specification*

BRE Defect Action Sheet 32 – *Suspended timber floors: chipboard flooring – storage and installation*

BS CP 201:Part 1:1967 and Part 2:1972 *Wood flooring (board, strip, block and mosaic)*

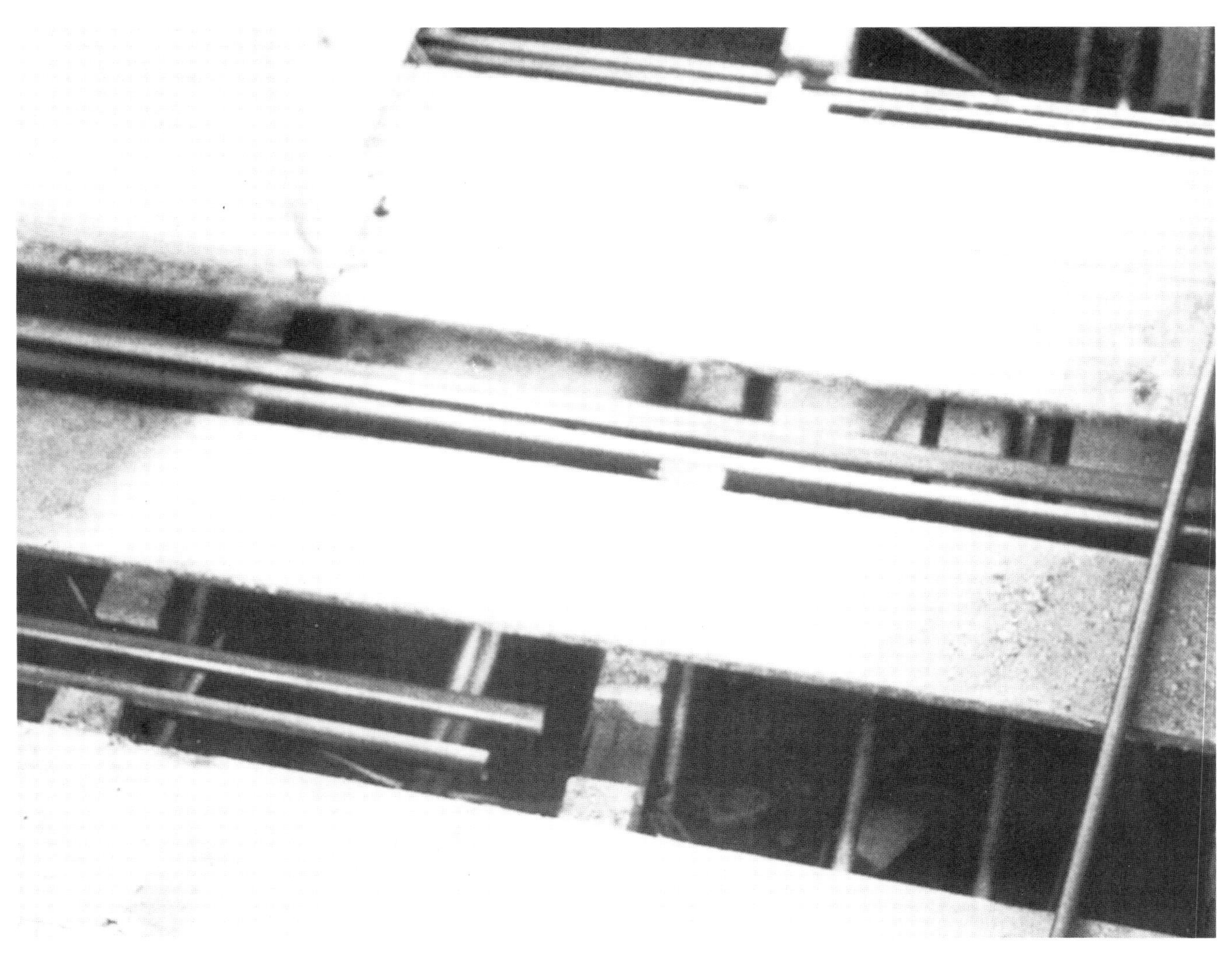

Photo 1 Notches must not be cut too close to each other or too deep or outside the permitted zones.

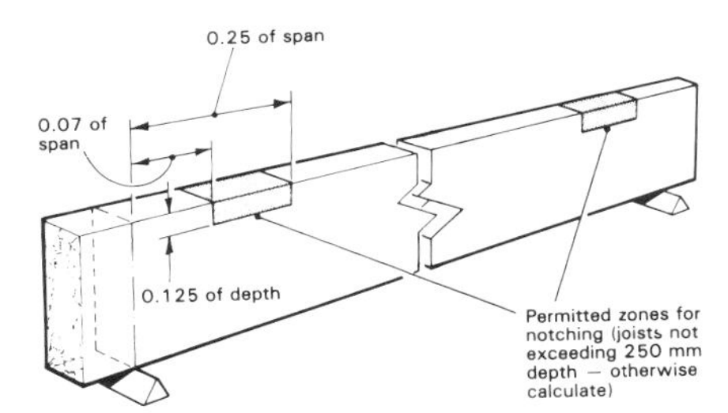

Figure 1 Permitted dimensions and positions for notches.

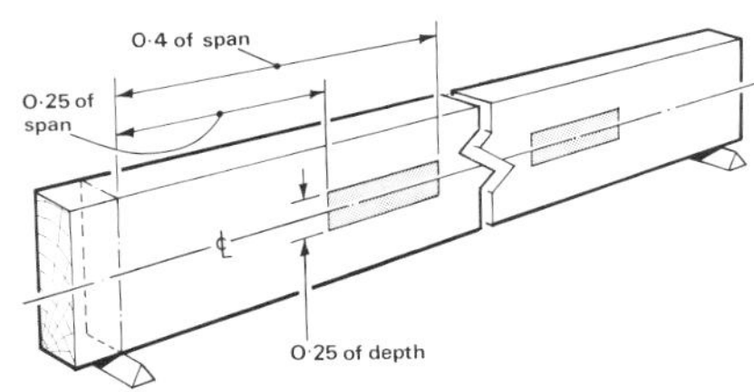

Figure 2 Permitted zones for holes.

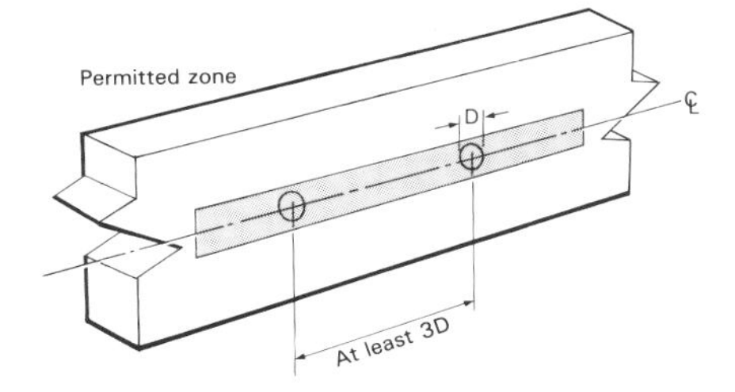

Figure 3 Distance apart for holes.

(Check first in existing buildings that joists are not undersized).

7.1.7 SUSPENDED TIMBER FLOORS. REDUCED DESIGN STRENGTH

The installation of central heating in an existing dwelling is probably the most common improvement carried out. With piped water systems, it is standard practice to hide the pipe runs beneath the flooring. This usually entails cutting notches and holes in the joists, often with little regard to the structural requirements of the floor. Such cutting may also have been undertaken for cables and conduits.

Symptoms

The floor bounces when walked on and deflects abnormally when heavy furniture is placed on it. Plaster ceilings crack and light fittings can be seen to move when a person walks on the floor above.

Investigation

Measure the actual deflection of the floor from a fixed point at the wall to the centre of the room. Remove flooring wherever there are services beneath and note the position of any holes or notches. Measure the depth of the joists and the sizes of all the holes and notches.

Diagnosis and cure

Both the Building Regulations and BS 5268 lays down strict limits on the size and position of holes and notches. The design strength of joists will be undermined if these limitations are exceeded and the floor will deflect. If this has happened, the cure is to replace or strengthen the joists.

REFERENCES

BRE Defect Action Sheet 99: *Suspended timber floors: notching and drilling of joists*

BS 5268: Part 2: 1984 *The structural use of timber*

BS 5449: Part 1: 1977 *Code of practice for central heating for domestic premises*

1985 Building Regualtions Approved Document for part A, *Structure* – gives tables of joist sizes.

Photo 1 Poorly supported or too widely spaced support battens are a likely cause of creakiness and uneven hardness of a gymnasium floor.

7.1.8 GYMNASIUM FLOOR. EXCESSIVE CREAKING AND VARIATION IN BOUNCE

Conflicts can arise over the use of gymnasium floors. The requirements for ball games are very different to those for gymnastics. The primary use of the hall will greatly influence the form of construction selected for the floor. Gymnastics requires a floor with spring in it to protect the gymnasts from excessive jarring. Ball game players require a solid floor which gives an even bounce response to the ball.

Symptoms

The floor creaks excessively during team activities. Balls bounce unevenly depending on which area of the floor they strike.

Investigation

Remove part of the flooring to expose the underfloor construction in an area where the problems are most pronounced.

Check the sub-floor for level and density of surface. If the density is at all suspect, have it tested with a BRE screed tester. Measure the spacing of support battens, and check the frequency of nailing.

Check the bounce response of the floor by dropping a ball on to it from a constant height and measuring the rebound height.

Diagnosis and cure

Hardwood is a natural material and will expand and contract due to seasonal variations in the weather. This will give rise to a limited amount of creaking which will occur whether or not the floor is in use.

The cause of more regular creaking will most probably be the uneven support of the floor battens. This can quite easily be cured by packing any spaces between the battens and the screed surface to provide an even level support. Another source of creaking is the incorrect spacing of the support battens. They should be 75 mm high by 50 mm wide spaced at 355 mm centres maximum. If battens are found to be at greater centres than this, the only remedy is to take up the whole floor and relay it correctly. Uneven bounce response may be the result of the above symptoms, but will occur to some degree on a suspended floor depending on whether the ball strikes over a batten or not. If the primary requirement of the floor is for ball games, the only cure is to remove the existing floor and relay it as a solid construction.

It may also be necessary to relay the screed if it is not found to be of adequate density. If this is done, ensure that a good quality DPM is provided to protect the timber from moisture.

REFERENCES

The Sports Council *Handbook of Sports and Recreational Building Design* Vol 2 *Indoor Sports* The Architectural Press Ltd

Penman Kenneth A *Planning Physical Education and Athletic Facilities in Schools*

National Playing Fields Association *Community Sports Halls*

BS CP 201: Part 2: 1972 *Wood flooring (board, strip, block and mosaic)*

Photo 1 Wet and dry rot can look very much the same. The top sample is dry rot and the bottom one wet rot.

Photo 2 The softwood floor; skirtings etc have been destroyed by dry rot which has had little effect on the hardwood floor finish.

Photo 3 Dry rot mycelium exposed by the removal of a timber panel.

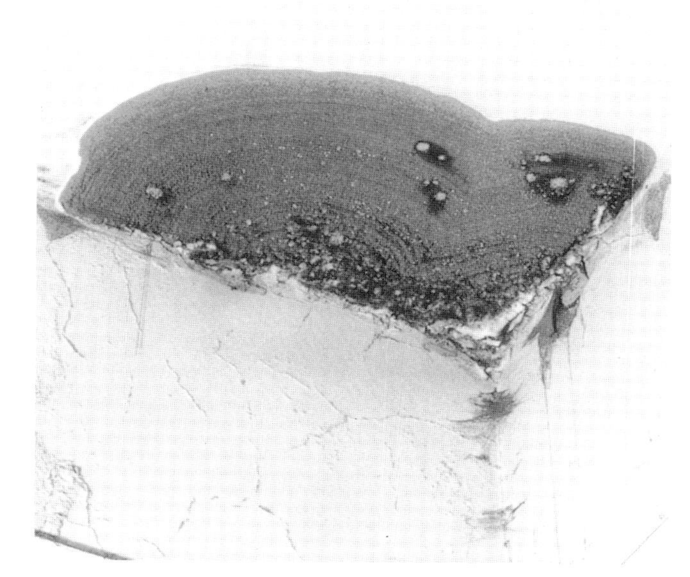

Photo 4 Dry rot fruit body – spores have collected on adjacent surfaces.

7.1.9 SUSPENDED TIMBER FLOOR. COLLAPSE

Collapse may be general or localised and occurs most frequently in suspended ground floors.

Symptoms

The wood is softened and may be breaking up. A change of colour occurs, usually to a dull brown and is accompanied by shrinkage and splitting into brick-shaped pieces. Larger structural members may sound hollow when struck with a hammer. An early symptom is a distinctive, musty, mouldy smell and insects which live on decaying timber may be present.

Investigation

Find out the age of the building and whether the floor is part of the original construction. Warn the client that the investigation may entail extensive removal of the floor.

It is necessary to find out why the timber became, or still is, wet or damp. The majority of cases of wood rot can be traced to an ingress of rainwater or rising damp. A comprehensive external survey will need to be done to establish the source of any leaks. Check for the presence of DPC's and make sure that they have not been bridged by a raising of the external ground level. Check that air bricks have not been blocked or buried below ground. Check all external renderings for soundness and note any associated signs of water ingress such as lichens and mosses. Check pipes for leaks. Make sure that no overflows are dripping.

Internally, remove sufficient flooring to establish the extent of rot. Take samples of the rotted wood and any associated mould growths, place in sealed containers and send them to a laboratory for analysis and identification. Check all plumbing pipes for leaks, paying particular attention to capillary and compression joints. Using a conductivity type moisture meter, measure the moisture content of all the wood and record any areas where it exceeds 20%. Where mould growth appears to be excessive, remove adjacent wall linings to establish the limits of invasion. The rot can run along the back of timbers embedded in masonry without showing on the exposed faces. Investigate the extent of the infection by exposing hidden faces of timber members.

Diagnosis and cure

Many different types of fungi will attack timber by growing on it, the particular type depends on the type of timber and its situation. The two common types of wood fungi found in buildings are wet rot and dry rot. Dry rot is more virulent than wet rot but the results of both can look very similar. Therefore the correct identification is important in order to ensure that the appropriate remedial measures are undertaken.

The dry rot fungus known as *Merulius Lacrymans* and also as *Serpula lacrymans* typically grows on timber remaining moist rather than very wet over long periods, eg joists in contact with moist brickwork. It prefers a moisture content in the region of 30–40 per cent. It cannot colonise saturated timber or timber with a moisture content below 20 per cent, but it can survive dryer conditions for up to 5 years becoming active as soon as there is enough moisture for growth. It has the ability to spread through damp bricks and mortar in the search for a food source, but cannot feed on these.

The humid conditions which give rise to dry rot are often caused by lack of ventilation. The fungus has the ability to moisten dry timber from an adjacent source of moisture.

The fungus grows from very fine air-borne spores, reddish brown in colour, which have alighted on damp timber or are on timber which becomes damp. It gradually spreads as delicate filaments that combine to form a silky white sheet or a felted skin, pearly grey in colour, sometimes with tinges of lilac, but more often patches of light lemon yellow. The fungus feeds on the timber, causing it to lose strength and reduces it to a dry state, hence the term 'dry rot'. The large fleshy fruiting bodies produce vast quantities of spores which are distributed by air currents and settle as a light rusty red dust which eventually darkens to reddish brown. It can only be differentiated from one of the wet rot fungi, *Coniophora cerebella*, by the fact that the timber is affected over a wider area, the cracking is deeper, the colour lighter, and the spores are redder in colour. There is a more frequent development of fruiting bodies and deposition of spores.

To treat the dry rot:

- establish the size and significance of the attack. In particular if structural timbers are affected, carry out or arrange for a full structural survey to determine whether structural repairs are necessary and, if they are, take appropriate steps to secure structural integrity.

continued

- locate and eliminate sources of moisture
- promote rapid drying of the structure
- determine the full extent of the outbreak
- remove all rotted wood, cutting away timber approximately 300–450 mm beyond the last indications of fungus
- prevent further spread of the fungus within brickwork and plaster by using preservatives
- use preservative-treated replacement timbers
- treat remaining sound timbers which are at risk with preservative (minimum two full brush coats)
- introduce support measures (such as ventilation pathways between sound timber and wet brickwork, or, where ventilation is not possible, providing a barrier such as a damp proof membrane or joist hangers between timber and wet brickwork)
- do not retain dry rot infected timber without seeking expert advice. There is always some risk in retaining infected wood which can be minimised by preservative treatment and subsequent reinspection.

Wet rot is the result of action by several species of fungi, the most common of which is *Coniophora puteana*, otherwise known as the cellar fungus. They all require persistently wet conditions in order to survive and reach their optimum growth in timber with a moisture content in the region of 50–60%. There are seldom any signs of a fruiting body or of the spores which are pale olive brown in colour. Any wood which is infected becomes very dark brown and rapidly loses strength. Splits occur mainly along the grain with occasional smaller cross cracks making a cuboidal pattern, which is similar to that caused by dry rot. This can cause great confusion when trying to identify the reason for decay. The exposed surface of affected timber may be free from attack as it is kept dry and, until collapse occurs, it may not be realised that the concealed timber in contact with adjacent damp surfaces has been weakened.

To treat the wet rot:

- establish the size and significance of the attack. In particular, if structural timbers are affected, carry out or arrange for a full structural survey to determine whether structural repairs are necessary, and if they are, take appropriate steps to secure structural integrity.
- locate and eliminate sources of moisture
- promote rapid drying of the structure
- remove all rotted wood
- use preservative-treated replacement timbers
- treat remaining sound timbers which are at risk with preservative (minimum two full brush coats)

REFERENCES

Building Research Establishment. *Recognising wood rot and insect damage in buildings*

BRE Digest 18: *Design of timber floors to prevent decay*

BRE Digest 299: *Dry rot: its recognition and control*

Oxley T A & Gobert E G *Dampness in Buildings*. Butterworths

BRE Princes Risborough Laboratory Technical Note 44. *Decay in buildings: recognition, prevention and cure*

Hollis M *Surveying Buildings*. Surveyors Publications, 1983

BRE Defect Action Sheet 73: *Suspended timber ground floor: remedying dampness due to inadequate ventilation*

BS 5268: Part 5: 1977 *Preservative treatments for constructional timber*

BS 5589: 1978 *Code of practice for preservation of timber*

BS 5707: *Solutions of wood preservatives in organic solvents*

Part 1: 1979 *Specification for solutions for general purpose applications, including timber that is to be painted*

Part 2: 1979 (1986) *Specification for pentachlorophenol wood preservative solution for use on timber that is not required to be painted*

Part 3: 1980 *Methods of treatment*

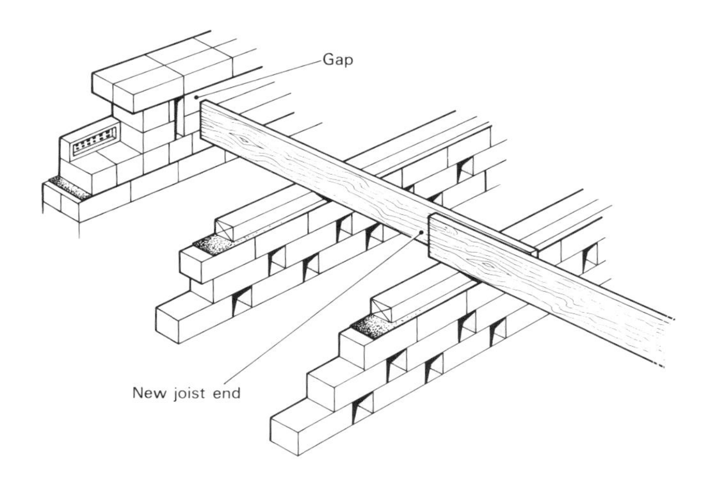

Figure 1 New joist ends should be supported clear of perimeter walls on RSJ's, hangers or new sleeper walls.

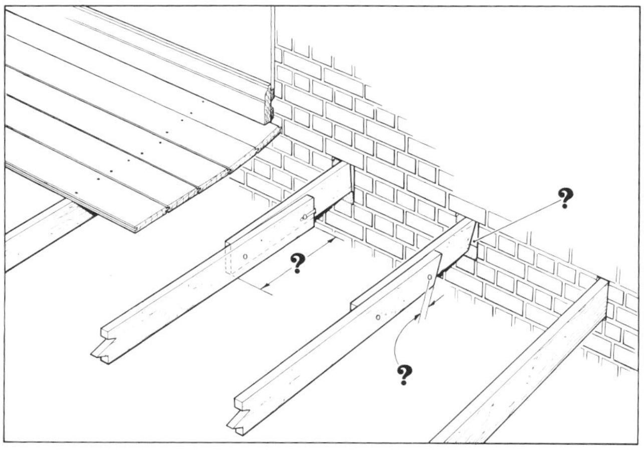

Figure 2 Common mistakes – insufficient overlap; ends loosely fitted in old cut outs; bolts too near ends of timbers.

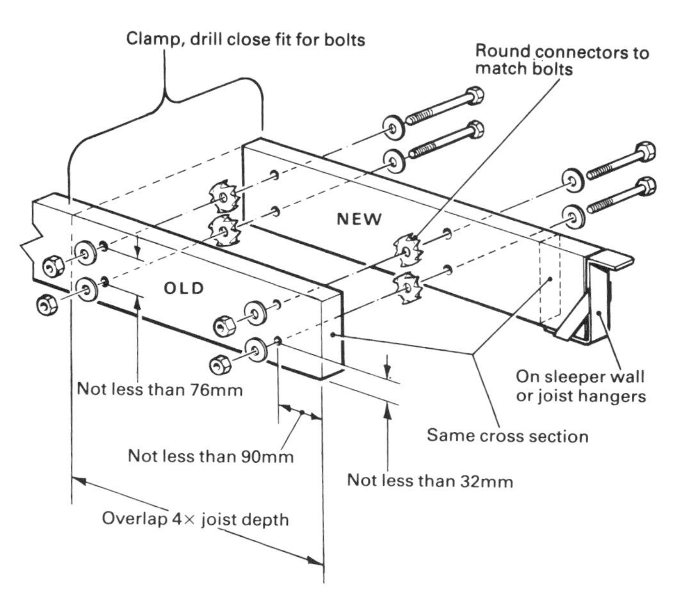

Figure 3 Spacing and number of bolts and connectors required to secure new joist ends.

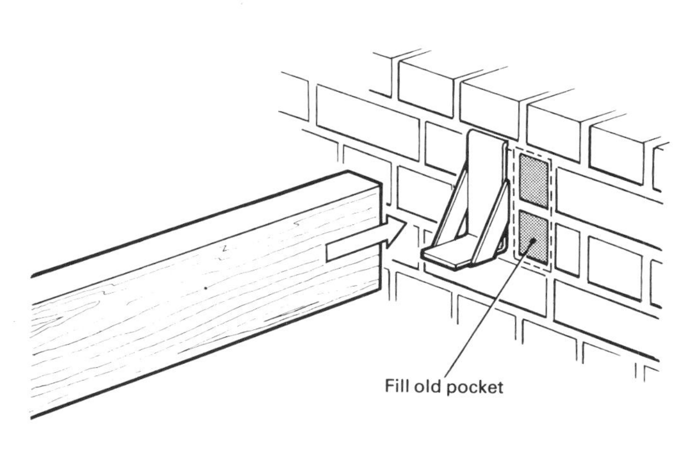

Figure 4 Hanger fitted alongside filled in pocket (check that adequate ventilation is provided).

7.1.10 SUSPENDED TIMBER GROUND FLOORS. UNEVEN OR LOCALLY UNSUPPORTED

Floor joists repaired after rot infestation fail to give adequate support.

Symptoms

The floor sags when walked on and being most pronounced adjacent to perimeter walls. Collapse may occur when heavy items of furniture are placed on it.

Investigation

Carry out an external survey in a similar style to that described in 7.1.9. Check whether any remedial works specification has been adhered to and that adequate ventilation is provided to the underside of the floor. Remove sufficient floor boards to expose all repaired joists. If joists have been lap jointed with new sections, check that the overlap is at least four times the depth of the joist. Check that the timber connectors comply with the British Standard and that they have been positioned correctly (figure 3). See that the cross section of the new joists is similar to that of the existing joists. Check that the ends are properly supported.

Diagnosis and cure

Rot can quickly recur in timber floors if adequate remedial measures are not taken. If rot is rediscovered it must first be eradicated using the methods outlined in 7.1.9. Consider increasing underfloor cross-ventilation by the installation of additional air bricks. The Building Research Establishment advise that for efficient cross-ventilation, there should be a clear depth of 150 mm between the underside of the floor joists and the site covering and sufficient airbricks to provide at least 3000 mm^2 open area per metre run of external wall.

Fill old pockets in the wall where joists have been removed. Keep new joists clear of external walls by either supporting them on joist hangers or new sleeper walls with a DPC under the bearers. When using new timber specify that it is pressure impregnated with preservative. Where ends of existing and new timber are exposed by cutting they must be treated with three full brush coats of an effective preservative. Allow sufficient drying time before relaying floor to keep the moisture content below 20%.

REFERENCES

BRE Digest 18. *Design of timber floors to prevent decay*

BRE Defect Action Sheet 74. *Susdpended timber ground floors: repairing rotted joists*

BRE Defect Action Sheet 103. *Wood floors: reducing risk of recurrent dry rot*

7.1.11 SUSPENDED TIMBER FLOORING. SHRINKAGE

This defect occurs with new timber flooring, especially when conditions are hot and dry.

Symptoms

Boards show signs of shrinkage, gaps open up between boards and tongues withdraw from grooves. Boards split at fixing points or fixings become overstressed.

Investigation

Find out the timber species and how it was stored before delivery. Check with the supplier whether the boards were air dried or kiln dried. Measure the moisture content of the boards as found and investigate what the moisture content was when they were laid. Record the relative humidity and temperature readings for the building interior over a representative sample number of days. Check the frequency of nailing of the boards and the spacing of the joists.

Diagnosis and cure

Timber will shrink as it dries out. The degree of shrinkage is dependant upon the species and its original moisture content. Most timber flooring is seasoned to a moisture content of 20–22% but this can increase if it has been stored in damp conditions or exposed to the weather in transit or on the building site. The British Standard Code of Practice allows moisture contents at the time of erection of up to 22% although specifications often call for less. However, the ultimate use of the building may produce conditions that are very different. In a well heated building with a dry atmosphere the moisture content of the timber may fall as low as 6–10%. In order to avoid excessive shrinkage, it is advisable to reduce the moisture content of any flooring timbers as close as possible to the conditions expected in the completed building. Correct fixing must also be carried out with each board being double nailed at each joist. It may be possible to rescue a lot of the flooring by relaying it when conditions have stabilised. Any replacement board will need to be well seasoned, preferably by kiln drying to a suitable specified moisture content.

REFERENCES

BS CP 112: Part 2: 1971, *The structural use of timber.* Part 2. *Metric units*

BS 1297: 1970 *Specification for grading and sizing of softwood flooring*

BRE Princes Risborough Laboratory. Technical Note No. 12. *Flooring and joinery in new buildings. How to minimize dimensional change*

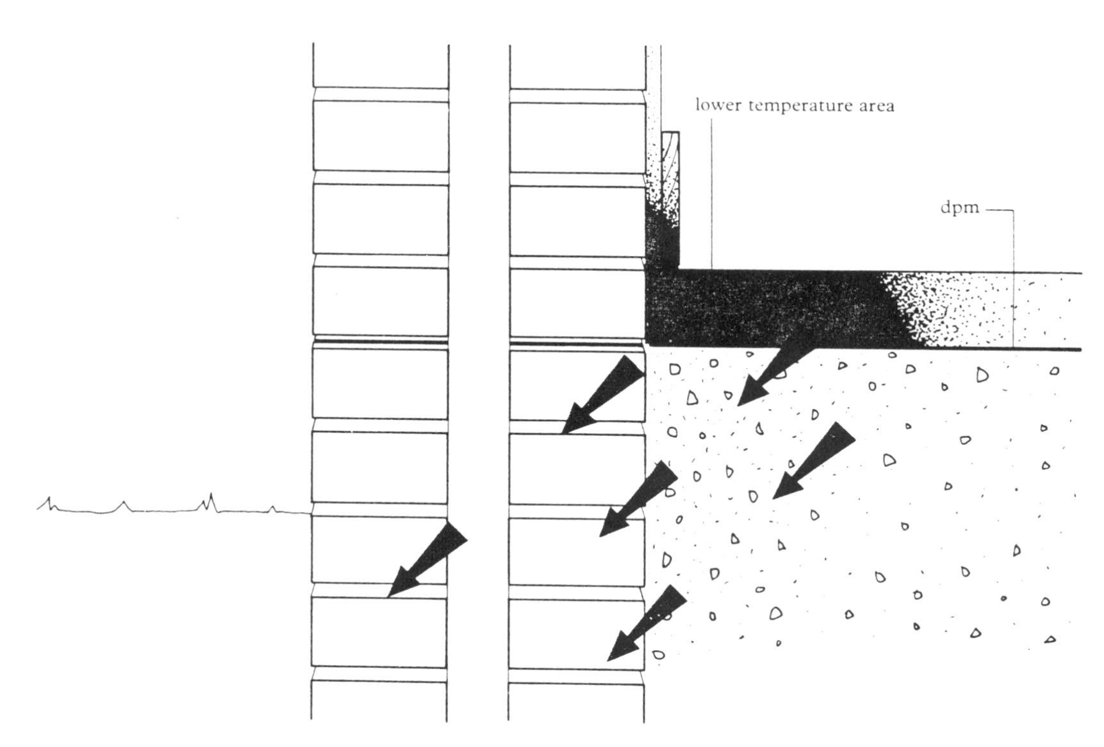

Figure 1 Cold bridge causing condensation at perimeter of floor.

7.1.12 SOLID GROUND FLOORS. OCCASIONAL DAMPNESS

Solid ground floors of intermittently heated buildings such as assembly halls and canteens are affected by sudden changes in use.

Symptoms

The surface of the floor finish becomes damp an hour or two after the room is occupied. It may then become either sticky to walk on or very slippery, depending upon the type of floor finish and the amount of dirt on the surface.

The dampness may be confined to the perimeter of the floor in contact with the outer walls.

Investigation

This defect is transient but recurring and this fact should be confirmed. The weather conditions at the times the defect occurs should be recorded, together with the form of heating, if any, and the amount of ventilation provided. The possibility of excess moisture vapour escaping into the area from an adjacent part of the building should be examined; eg servery hatches from a kitchen to a canteen.

Note the time of occupation of the building or affected parts.

If possible, examine the construction drawings to assess whether a cold bridge has been formed which will cause a loss of heat from the edge of the floor to the external walls and ground outside.

Diagnosis and cure

This defect is quite often wrongly diagnosed as rising dampness, despite the fact that the floor finish may be impervious to water. Rising dampness would be a relatively permanent defect whereas the one described here is temporary. The effects of the dampness generally show within a fairly short time of the building or room being occupied by a number of people. It is more common where flueless gas heaters are used for the intermittent heating and where there is insufficient ventilation to keep the level of water content in the air down to a reasonable level.

There are three possible causes:

1 In the type of building mentioned above, the floor is often the last part of the building to become heated. During cold weather periods it may remain cold enough for condensation to take place on the surface, especially if the humidity of the air rises quickly as a result of the number of people present. The condensed water sometimes becomes contaminated with dust or dirt from footwear, leading to tackiness. With some finishes, especially if highly polished, the floor surface then becomes dangerously slippery. Water vapour from flueless heaters, and from kitchens where the building concerned is a canteen and also only used at intervals add to this.

   To cure this defect, the aim must be to improve the heating of the floor surface and to reduce the humidity of the air. It may be necessary to change the form of heating, not only to decrease the amount of water vapour produced but also to concentrate or direct a greater proportion of the heating on to the floor. The provision of background heating can also provide a solution since it will tend to reduce the chance of the floor being the coldest part of the room.

   In the case of canteens with attached kitchens, extractor hoods with power ventilators should be used above any appliances or areas where steam is produced, so that moist air does not penetrate to other parts of the building.

2 If the floor is of magnesite the surface may become similarly damp, but in this case the incidence of dampness may be linked with damp weather conditions as well as with occupancy (see 7.2.21).

3 Perimeter dampness on parts of the floor in contact with the external walls may be due to condensation through a cold bridge effect lowering the temperature of the perimeter of the floor below the dew point of the air in the room. Alternatively it may be due to dampness rising up the walls and being transferred into the floor (see 7.1.13). It will not be easy to overcome this type of perimeter condensation, since the cure involves cutting a vertical groove between the edge of the floor and the wall, repairing or providing an effective DPC and inserting insulating material, which must be kept dry to be effective, and may have to be wrapped in polythene or other waterproof material.

REFERENCES

BRE Digest 110. *Condensation*

BRE Digest 145. *Heat losses through ground floors*

BS 5250: 1975 *Code of Basic data for the design of buildings: the control of condensation in dwellings*

Photo 1 Damp proof course bridged by plaster.

Photo 2 Lifting plastics tiles over defective damp proof membranes (see also 7.2.13).

7.1.13 SOLID GROUND FLOORS. PERSISTENT DAMPNESS

This defect concerns rising damp. Except when they are finished with asphalt, solid ground floors should always be provided with an adequate damp-proof membrane, even if the floor finish to be laid on it, is unaffected by moisture.

Symptoms

The surface of the floor finish is persistently damp except in very dry weather. The floor finish may become defective as described elsewhere (items 7.2.7, 7.2.11, 7.2.13 & 7.2.14 refer).

The underside of an impervious sheet or tile flooring material may be found to be quite damp, or even wet, although it had been laid on a floor which had previously always been dry. Other finishes, such as linoleum or carpets similarly laid on an apparently dry floor, are found to be covered with mould on the underside and may have rotted.

In rooms with a low level of ventilation there will be a musty damp smell, and condensation develops very easily when the room is heated.

Investigation

Ascertain the construction of the base on which the floor finish has been laid, particularly noting whether a DPM is present and the materials from which it is made.

Note whether the dampness is confined to any particular part of the floor. Note also whether the floor finish is the original finish or a later one laid on the original.

Check for plumbing leaks – embedded heating pipes are a common cause of dampness in screeds (see 7.1.14).

Diagnosis and cure

There are two common causes for this defect:

1 Fairly persistent dampness is generally the result of the absence of a DPM, although sometimes one may be present but ineffective. Dampness rises from the ground, passing through the base concrete and any pervious finish sometimes causing them to become defective. (For details of these defects see items outlined in symptoms.)

Some floor finishes such as natural stone, tiles and bricks without a DPM beneath, may allow the rising dampness to evaporate sufficiently fast to maintain the surface of the floor in a dry condition. When another finish is applied on top, eg carpets or linoleum, the rate of evaporation is reduced and the new floor finish becomes damp, initially only on the underside but later causing the finish to become defective.

If the base concrete has been damaged it will be advisable to remove it, insert a DPM and lay a new concrete base. If it is undamaged it may be sufficient to remove only the floor finish and replace it with either a new finish laid on a DPM or one that is not affected by moisture such as asphalt. Adequate time should be allowed for a new concrete base to dry out before laying the finish on it. The DPM must be connected to the DPC in the walls and if not at the same level, the connection should be made with a vertical DPC.

2 If the dampness is confined to the perimeter of the floor, it is possible that it is the result of rising dampness by-passing the DPM in the floor structure because this has not been linked to the DPC in the walls. Occasional dampness in the same position is more likely to be due to condensation (see item 7.1.12).

If there is perimeter dampness due to an ineffective or interrupted DPC/DPM, a groove must be cut around the perimeter of the floor deep enough to prevent the transfer from the wall into the floor. An impervious material or a vertical DPC should be inserted and the remainder of the groove filled with mortar or sheet material cut to size.

REFERENCES

BRE Digest 54. *Damp-proofing solid floors*

BS CP102:1973 *Code of practice for protection of buildings against water from the ground*

Photo 2 Condensation and leaks are often associated with internal rainwater downpipes.

Photo 1 Dampness on ceiling close to service duct.

Photo 3 Condensation forming on internal downpipe.

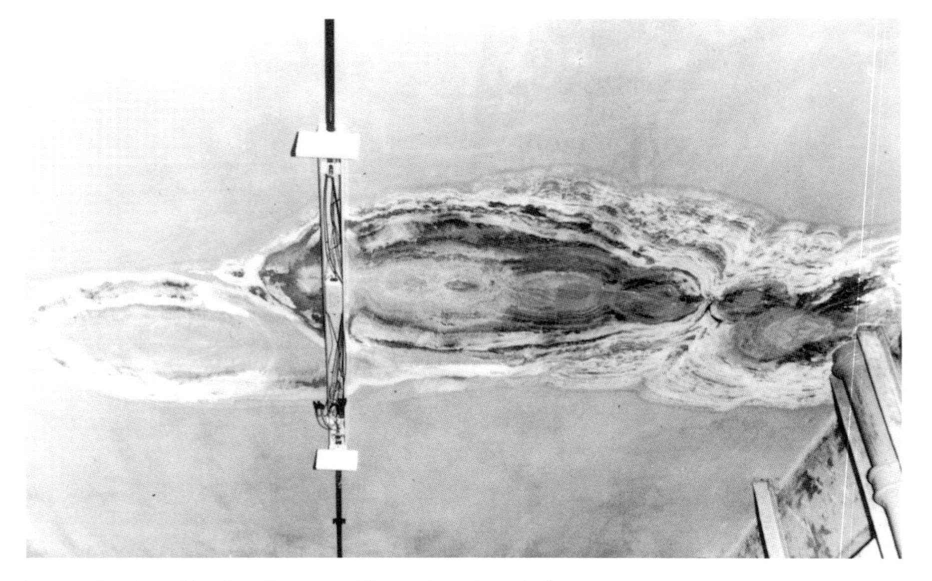

Photo 4 Localised stains on ceiling related to drainage.

7.1.14 FLOORS AND CEILINGS. DAMPNESS NOT ASSOCIATED WITH RAINFALL

Inadequate forethought to future maintenance requirements at the design stage can cause problems of access to service pipes.

Symptoms

Damp patches appear on walls and ceilings, often some distance from the external walls of the building. The dampness may be spasmodic in occurrence.

Investigation

It will be a considerable help when attempting to trace the source of dampness if copies of the original drawings can be obtained. In forms of construction where this defect occurs there may be passage of water for some distance before it makes its presence known on the surface. Examine all parts of the building where there is a source of water, particularly service ducts where long pipe runs can aid the passage of leaking water through several floor levels.

If the services are not contained in ducts, the service runs should be traced. Check whether any pipes were buried in the screed if the building is of reinforced concrete construction. Where damp patches appear on ceilings, check the floor above, particularly if it is a toilet or washroom area.

When the appearance of patches is spasmodic, log the use of water using appliances and other movements of water.

In domestic buidlings where central heating pipes have been installed, or other improvements carried out, it will be necessary to lift floor boards to expose any joints in the pipe runs.

Similarly, where the building has been sub-divided in to smaller units, check all new pipework installations.

Diagnosis and cure

In many examples of this defect the water appears some distance from its point of origin. Diagnosis of the fault can be very time-consuming, particularly when it is intermittent. The sources of water will be varied and each will need consideration. Apart from leakage from easily accessible pipes and appliances, it is necessary to consider leaks from pipes that have been hidden from view, particularly since these are not always easily inspected for regular maintenance.

Leaks from joints or defective pipes that run in ducts or under floorboards will allow water to migrate down the pipes until it meets a joint in the pipework or the pipes come in contact with the enclosing structure.

There should be no problem dealing with leaks from pipes and appliances not associated with ducts, but the repair of pipes and joints within a duct or under the floor may present problems. This will largely rely upon the position of the joints in relation to the access points. When access is severely restricted it may be advantageous to reroute the pipes to make future maintenance more convenient.

Where the leaks are occurring from toilet or workroom areas above, the fault may be very easy to diagnose but difficult and expensive to remedy (see item 7.1.2).

REFERENCES

BRE Digest 81. *Hospital sanitary services: some design and maintenance problems*

BS CP413: July: 1973 *Code of practice for ducts for building services*

Greater London Council. *Detailing for building construction: fittings and services* The Architectural Press Ltd

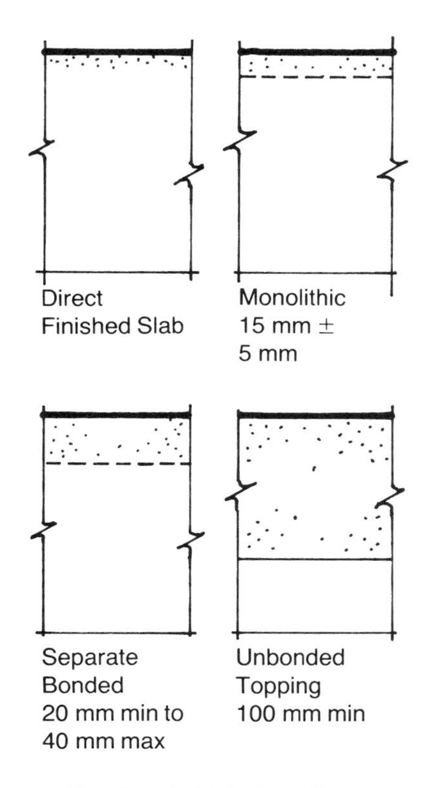

Toppings in high strength concrete (BS 8204 Part 2: 1987)

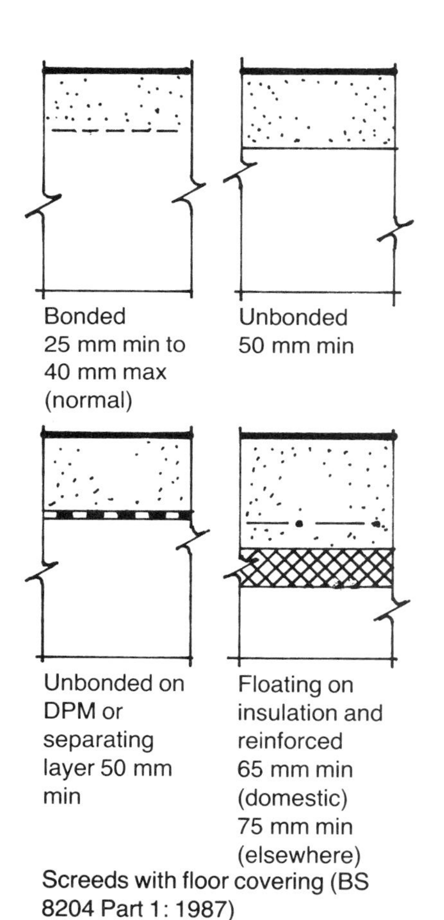

Screeds with floor covering (BS 8204 Part 1: 1987)

Figure 1 Thicknesses for screeds & toppings

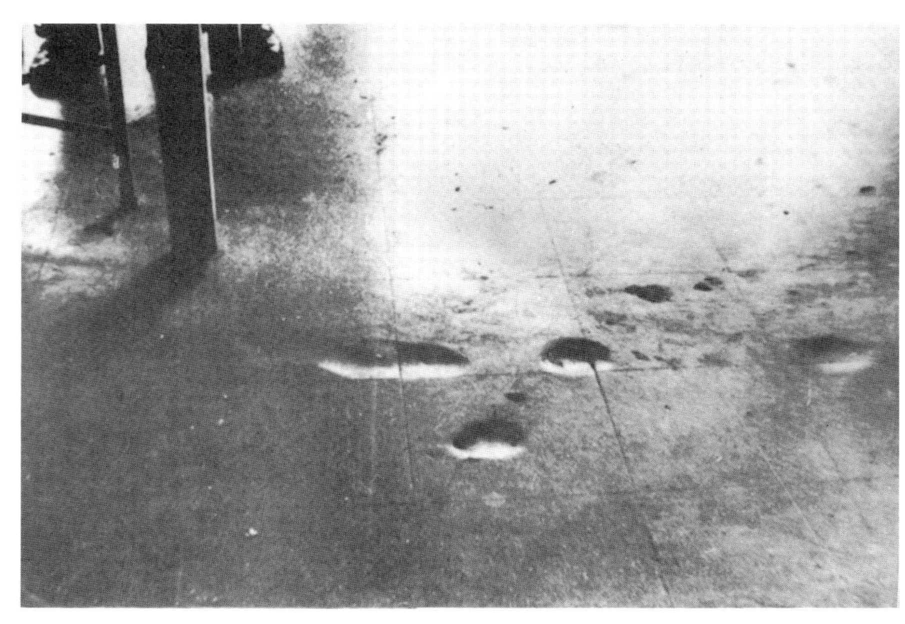

Photo 1 Indented floor finish over weak screed.

Photo 2 Section of screed that has not been properly compacted.

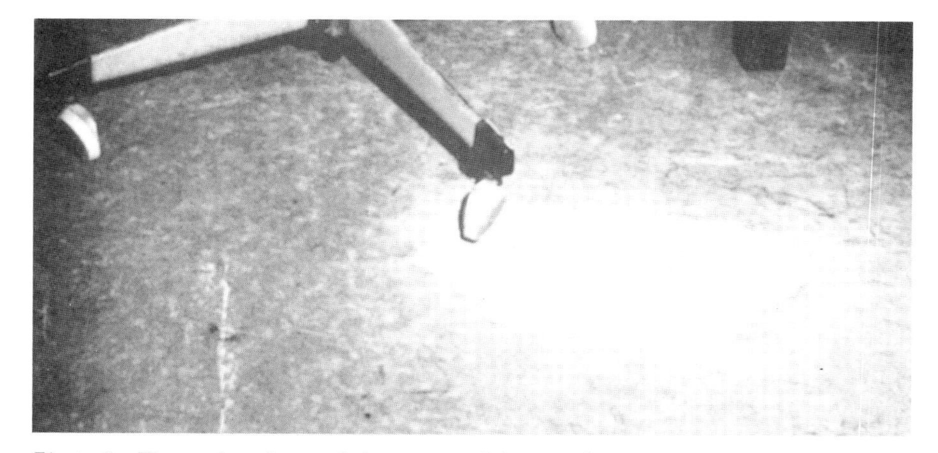

Photo 3 The surface layer of strong material can collapse under point loads with normal use.

7.2.1 FLOOR SCREEDS. HOLLOWNESS, LIFTING, CURLING

Faults occur in sand cement screeds for a variety of reasons. A lot of them are directly attributable to poor workmanship and supervision.

Symptoms

The observed effects depend somewhat upon the nature of the applied finish. In-situ finishes may crack along the line of bay joints in the screed and may be accompanied by cracking at the edges. Tiled finishes and sheet coverings may also crack and split at the same point, or they may be indented at random points, particularly in heavily trafficked areas. The floor may sound hollow when tapped or walked upon: however, this also occurs when screeds are laid on membranes or insulation materials.

Investigation

Remove some of the floor finish at the point of failure to expose the screed surface. Check whether any hollowness is associated with the screed becoming detached from the base slab or whether a separating membrane is installed.

Remove a sample of the defective screed and send it to a laboratory in a sealed container for chemical analysis of constituents and tests for mix and void content.

Examine the condition of the surface of the base slab and note the presence of any extraneous material such as plaster droppings. This is important if any large areas of hollowness exist, but will not apply in the case of floating screeds or where there is a separating membrane.

Check the screed for soundness using a BRE screed tester.

Measure the depth of the screed and compare it with the specification.

Where no damp-proof membrane has been used in the system, consider the possibility of sulphate attack (see 7.1.4).

Diagnosis and cure

There are many different factors that can cause screeds to fail and they may act singly or in combination. This often makes it difficult to identify the actual cause, especially since any one effect may have been produced by different causes and the information required may not be available.

The factors include poor preparation of the concrete base; incorrect design, including unsuitable mix specification; poor workmanship and supervision, either by having the mix too wet or too dry; too rapid drying of the screed.

Debris left on the base can be identified and incorrect mixes can be detected by analysis, but poor workmanship and wrong drying rates are more difficult to diagnose.

Where cracking and curling is not excessive and where soundness testing with the BRE Screed Tester suggests that the screed is otherwise sound, it may be sufficient just to cut out and replace the worst affected areas. Grind down all edges and make good with a proprietory levelling compound, following the manufacturers instructions.

Localised indentation may also be dealt with in this way, but more general indentation would indicate that the whole screed is faulty and the only possible cure is to replace it.

When replacing screeds, a modified cement:sand screed should be considered, as it will require less water to obtain a workable mix and will dry more quickly than ordinary dense screeds. Experience suggests that defects are less likely to occur with modified sand:cement screeds, largely because they are laid by approved specialist contractors.

REFERENCES

BS 8204 *In-situ floorings* Part 1: 1987 *Code of practice for concrete bases and screeds to receive in-situ floorings*. Part 2: 1987 *Code of practice for concrete wearing surfaces*

BS CP 204: Part 2: 1970 *Code of practice for in-situ floor finishes*

BRE Digest 104: 1973. *Floor screeds*

Pye, P W CChem. MRSC. *BRE Screed tester classification of screeds, sampling and acceptance limits* BRE information paper IF 11/84 July 1984

BRE Defect Action Sheet 51: May 1984 *Floors: cement-based screeds – specification*

Photos 1, 2 & 3 Cracks in straight lines and at right angles indicating relationship with construction operations, board materials or services.

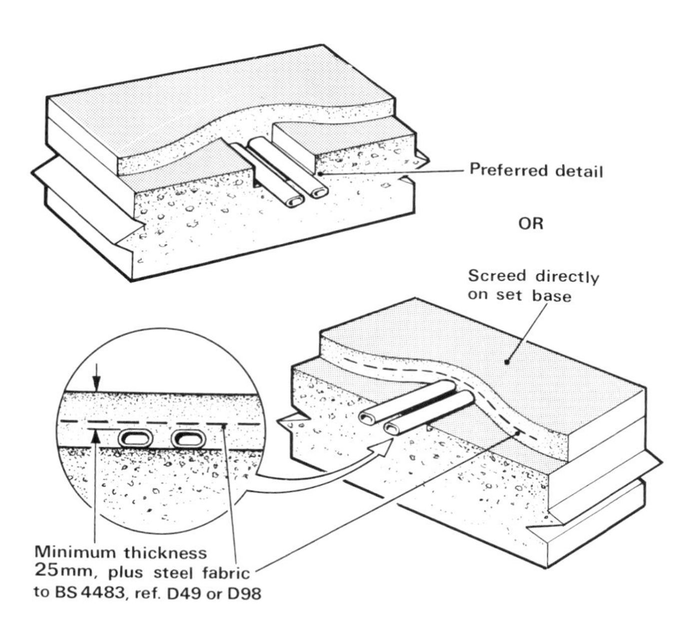

Figure 1 Avoidance of weakness over embedded services.

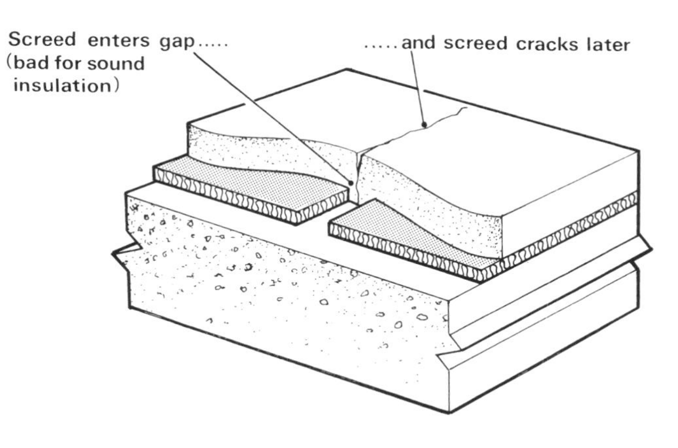

Figure 2 Construction defect creates a weakness.

7.2.2 FLOOR SCREEDS. CRACKING AND BREAKING UP

In this defect screeds crack in well defined lines.

Symptoms

Sheet floor finishes laid on the screeds are cracked or split or there is unevenness along lines, the finish on one side of the line being slightly proud of that on the other. There may be a slight difference in the levels of tile and block flooring, also along a line that is generally well defined. If the tiles are brittle they may have cracked along the line. The cracking may occasionally turn through ninety degrees.

Investigation

Remove any floor covering along the affected area to reveal the screed beneath. If any of the screed is breaking up, remove a sample to be analysed in a laboratory and have the screed tested for soundness using a BRE Screed Tester. Examine any cracks in the screed and note their position, width and depth.

Establish whether any service pipes or conduits are placed either in the screed or immediately below it.

Check whether any insulation has been laid below the screed, and if it has, what form the insulation takes.

If the screed has been laid on a solid ground slab, check whether any movement or daywork joints were incorporated in the slab and also whether these joints were repeated in the screed.

Diagnosis and cure

Most screeds provide a satisfactory base for the laying of a floor finish even if there is slight loss of adhesion between the base concrete and the screed. However, apart from the disturbance of the screed due to defects in the base concrete (see 7.1.4 & 7.2.4), the screed may be defective for one or more of the several reasons outlined in the previous item.

When bay or daywork joints are included in the screed, cracking will almost certainly take place along the same line. If the shrinkage is different in adjoining bays this will form a small but definable step. If any of the bays are subjected to drying out too quickly they may curl up at the edge.

If the screed has broken-up extensively, there will be something far more radically wrong which should be reasonably easy to determine by studying the results of laboratory analysis tests.

Where floating screeds have been formed by laying sheets of insulation material on the base slab and there are gaps between these sheets, the screed will crack in line with those gaps. The sound insulation value of the slab will also be lost.

It will be necessary to carry out remedial work only if the floor finish has been badly affected, or the defect has brought about a difference in levels, which is likely to be a hazard to foot or wheeled traffic.

The extent of the work will depend upon the state of the screed; if it is weak, it will be best to take it all up and relay; if it is a floating screed and the sound insulation has been seriously undermined, again the best course will be to relay; if cracked, but lifted slightly, it may be sufficient to grind one side down to the level of the other. Between these extremes it may be possible to cut and patch, though it may not be possible to get a good-looking final result.

REFERENCES

BRE Digest 104: 1973 *Floor screeds*

Pye P W and Warlow W J BRE current paper CP 72/78 *A method of assessing the soundness of some dense floor screeds*

BRE Defect Action Sheet 51 *Floors: cement-based screeds – specification*

BRE Defect Action Sheet 52: *Floors: cement-based screeds – mixing and laying*

BS 8204: *In-situ floorings* Part 1: 1987 *Code of practice for concrete bases and screeds to receive in-situ floorings*. Part 2: 1987 *Code of practice for concrete wearing surfaces*

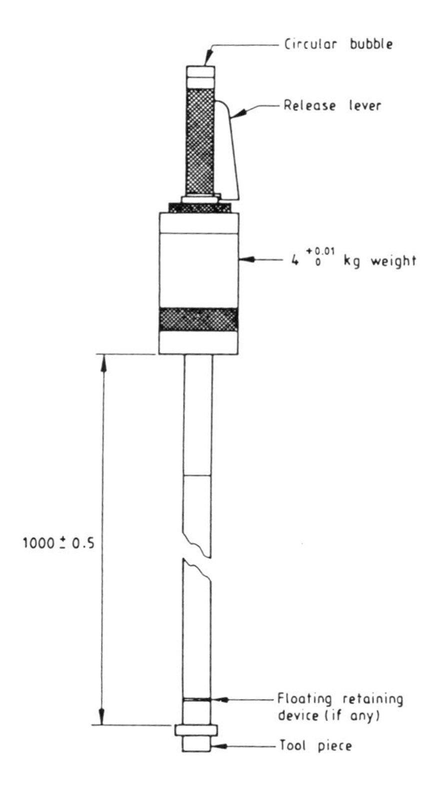

Figure 1 BRE screed tester. The weight is released down the stem until it hits the stop and the tool piece is driven into the surface of the screed.

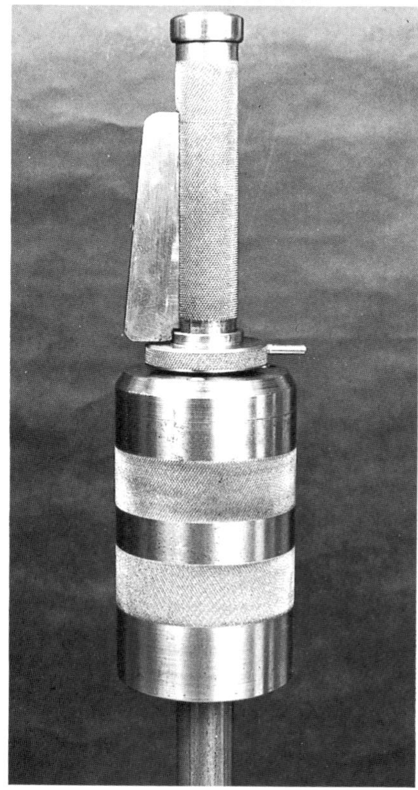

Photo 2 Release trigger for BRE screed tester.

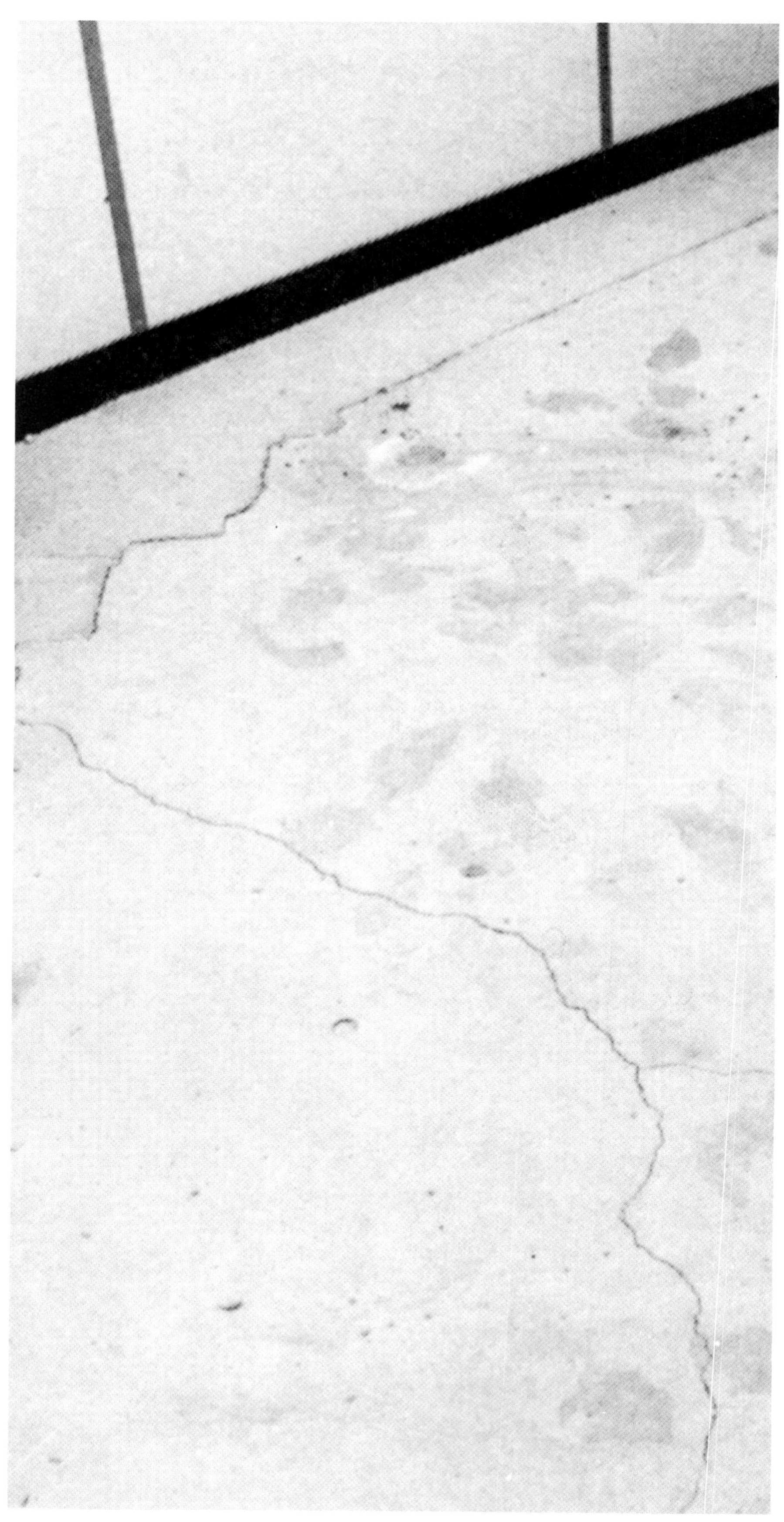

Photo 1 Cracking of new cement sand screed in an old building where the thickness, mix and compaction of the screed were variable, and the existing floor could not be adequately wetted because of the risk of damage to lower floors – In this case the remedy was to overlay the floor with a floating screed.

7.2.3 FLOOR SCREED. CRACKING

This defect occurs when sand cement screeds are laid on existing suspended concrete floors.

Symptoms

Soon after laying, random cracking occurs in the screed. The cracks are uniform in width and there is some curling of the screed at bay edges.

Investigation

Establish the construction of the floor and examine the underside for any signs of cracking. If any doubts exist about the structural integrity of the floor, consult a structural engineer for advice.

Arrange for core samples to be taken of the screed at random positions on the floor. Send the samples for contents analysis by an independent laboratory. Measure the thickness of the screed and compare with the original specifications.

Check the screed for hollowness by tapping with a hammer and record where the hollowness occurs. Examine the top surface of the structural slab and note whether the screed is adhering to the surface.

Have the screed tested for soundness using a BRE screed tester.

Diagnosis and cure

The most likely causes for the cracking are incorrect preparation of the slab and inadequate casting procedures.

The concrete slab needs to be thoroughly wetted prior to the laying of the screeds. If this is not done the concrete will draw the water out of the screed like a huge sponge. There may be a reluctance to soak an existing floor on the basis that any water passing right through the slab will damage areas below.

Curing of the screed requires care to prevent it drying out too quickly and before it gains enough strength to resist shrinkage cracking. This is best achieved by covering it with a fully waterproof sheet immediately after laying and allowing at least a week before removal to allow drying. If the floor is to be covered with flooring materials, any form of curing compounds must be avoided as they will prevent adhesion.

In both these cases the necessity to maintain an even thickness of screed is important; the thinner the screed is, the more liable it will be to shrinkage.

If patch repairs are not possible and if levels permit, the best cure for the condition would be to overlay all the existing screed with a 25 mm thick synthetic anhydrite floating screed supplied and laid by a specialist. This type of screed has a very low drying shrinkage as almost all the water is used up in the chemical reaction as it hardens. This means that large areas can be laid at one time without joints and with no risk of curling.

REFERENCES

BRE Digest 104: 1973. *Floor screeds*

Pye P W and Warlow W J BRE current paper CP 72/78 *A method of assessing the soundness of some dense floor screeds*

BRE Defect Action Sheet 51: *Floors: cement-based screeds – specification*

BRE Defect Action Sheet 52: *Floors: cement-based screeds – mixing and laying*

BS 8204 *In-situ floorings* Part 1: 1987 *Code of practice for concrete bases and screeds to receive in-situ floorings*. Part 2: 1987 *Code of practice for concrete wearing surfaces*

Photo 1 Breakdown of terrazzo finish at a buffet checkout caused by a combination of wear and spills of milk and other food stuffs.

7.2.4 CONCRETE SLABS AND FINISHINGS. DISINTEGRATION

This defect occurs in industrial premises where the floor is subject to spilled water and/or chemicals.

Symptoms

Concrete slabs with or without either granolithic or permeable finishings gradually disintegrate. Finishings with joints that allow liquids to pass through fail because the concrete breaks up at these joints.

The defect is mainly confined to floors in industrial buildings in which spillage of liquids used in manufacturing processes is liable to occur.

Investigation

If the surface of the concrete/granolithic finish has merely dusted, refer to item 7.2.6. In the areas of attack, identify any liquids that may get spilled on to the floor.

Check on all maintenance procedures and note the type of cleaning materials used.

Diagnosis and cure

Concrete slabs constructed using normal Portland cement will absorb liquids if they are not protected with an impervious finish. Granolithic finish will be attacked for the same reason but may offer better short term protection due to its smoother finish. It is not normally possible to leach out or neutralise the material responsible for the damage.

Surface grease stains can be removed using a solution of sodium metasilicate, caustic soda or proprietory product but care must be taken to remove all the cleaning solution as it may in itself cause trouble if allowed to remain in contact for long.

Portland cement is attacked by all acids, therefore milk products used in the manufacture of foods can be troublesome because lactic acid is formed by the action of bacilli on stale milk.

Weak alkali solutions have little effect, but strong aggressive alkaline cleaners should be avoided. Long term exposure to salts, organic materials such as oils, fats, greases and solvents will damage concrete and is likely to damage any protective dressing.

Cure can only be effected by cutting out any defective areas and replacing with a suitably resistant material. Where possible, it would be helpful to install better drainage and organise more frequent washing-down of areas where spillage is inevitable.

REFERENCES

BS Code of Practice CP 204, Part 2: 1970 *In-situ floor finishes* Part 2. Metric units

BS 8204 *In-situ floorings* Part 1: 1987 *Code of practice for concrete bases and screeds to receive in-situ floorings.* Part 2: 1987 *Code of practice for concrete wearing surfaces*

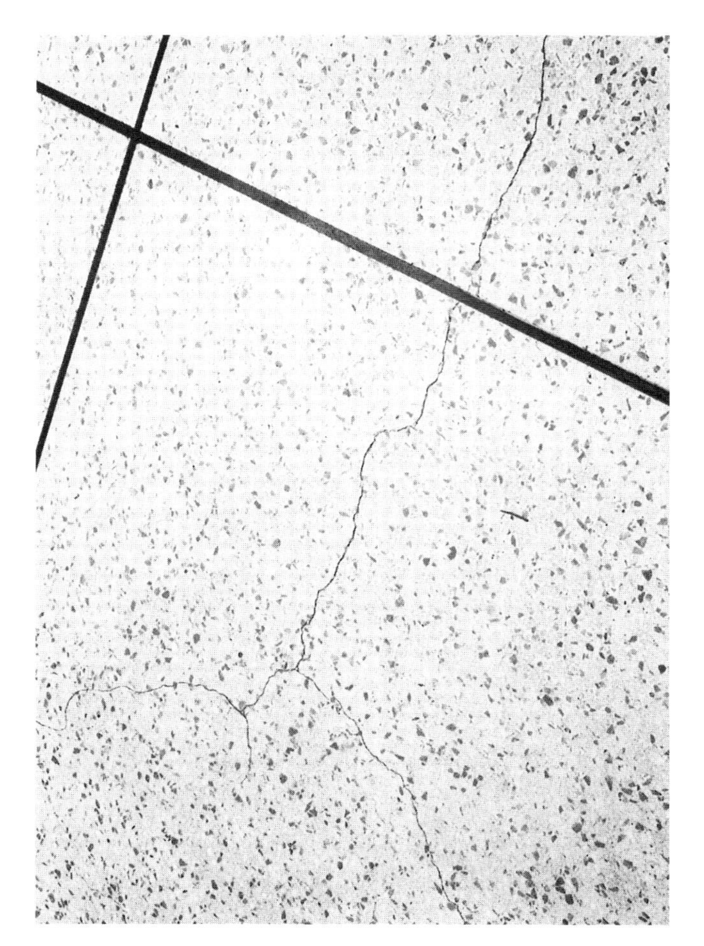

Photos 1, 2 & 3 Cracking of dense, in situ, floor finishes.

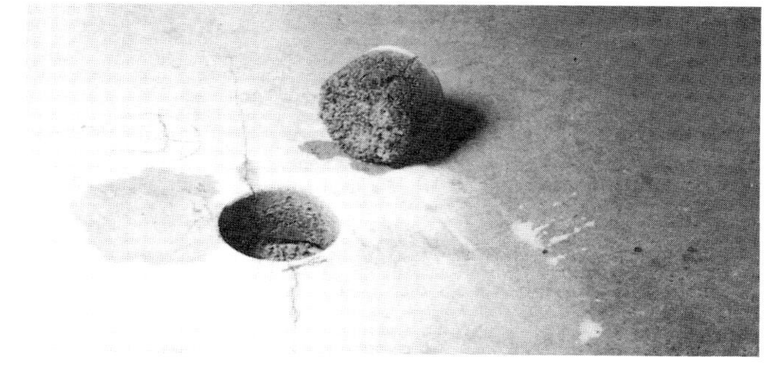

Photos 4 & 5 Cores drilled from a granolithic floor topping showing poor compaction and a high percentage of voids.

Photo 6 Good core sample from an area in sound condition.

7.2.5 DENSE CONCRETE FLOOR FINISH. FAILURE

This item refers to failure of granolithic and terrazzo floor finishes due to hollowness, lifting, cracking and curling.

Symptoms

The floor sounds hollow when tapped or walked upon with hard soled shoes, more especially at the joints of the bays in which it was laid. The amount of cracking may be appreciable, with a tendency to curling or uplifting of the edges of the cracks. Any cracking should not be confused with surface crazing which is commonplace and very difficult to avoid with these types of finish.

Investigation

If the cracking is severe, the concrete slab should be checked for cracks. In the case of ground slabs, have the floor checked for level in case there is any heave or settlement taking place (see 7.1.3 & 7.1.4). With suspended slabs any severe cracking will need to be examined by a structural engineer. Have core samples taken for analysis.

Diagnosis and cure

There should be little difficulty in distinguishing this defect from the lifting of concrete floors attacked by sulphates originating in the hardcore, since that is more pronounced and accompanied by excessive cracking of the concrete.

Failures of this nature are nearly always attributable to poor workmanship. The base may not have been cleared of debris and dust. The mix proportions and the quality of the aggregates may differ from those in the original specification. The mix may have been poorly compacted leaving a high proportion of voids. The cracking may have occurred in line with cracking of the base slab. Any of these faults will be revealed by investigation and by the core samples.

Surface crazing of terrazzo may be due to using too rich a mix, to inadequate grading of the aggregate or to too rapid drying by not following the correct curing procedures. Crazing or minor surface cracking may often have to be tolerated because of the difficulty and high cost of removal and replacement.

An alternative to total replacement is to pressure grout the underside of the topping using a thixotropic epoxy resin injection system. This will need to be done by a specialist sub-contractor. This system cannot be guaranteed to fully cure the problem but should be adequate in areas subjected to light traffic.

Any uneven edges can be ground down to produce a level surface.

REFERENCES

BS Code of Practice CP 204, Part 2: 1970 *in-situ floor finishes* Part 2. Metric units

BS 8204 *In-situ floorings* Part 1: 1987 *Code of practice for concrete bases and screeds to receive in-situ floorings*. Part 2: 1987 *Code of practice for concrete wearing surfaces*

Photo 1 Dusting of granolithic floors becomes evident early in the life of the building and is due to the mix being too wet when laid.

7.2.6 GRANOLITHIC FLOOR FINISH. DUSTING

This condition may occur in bays or larger areas and cannot be cured by simply sweeping.

Symptoms

The surface is covered with fine dust and wears rapidly. It may also be cracked, sometimes deeply.

Investigation

Little investigation is really needed unless there is a dispute to resolve, in which case core samples will need to be taken for analysis. A copy of the original specification will be useful along with any information that can be obtained about the laying and curing.

Diagnosis and cure

The defect is self-evident and unlikely to be confused with defects resulting from chemical attack (7.2.4). These would be much more penetrating, causing the disintegration of the floor. The dusting will be most evident in areas subject to heavy traffic, but all areas will produce clouds of dust when swept with a stiff broom. Analysis of the dust will show it to consist mainly of cement and very fine aggregate.

The whole problem arises because wet concrete mixes are easier to lay than dry ones. This leads to the formation of excessive laitance and a relatively weaker concrete with a higher drying shrinkage, hence the possibility of cracking and rapid wear rate.

If the surface is very weak and friable, or the floor is defective in any other way, it will be best to take the finish up and re-lay.

If the dusting is only very slight, or is discovered at a very early stage, it is possible to improve the wearing quality by the surface application of sodium silicate solution or other proprietary treatment. If this is done, it is imporatnt to remember that the treatment may only have a limited wear life, and retreatment will be needed at regular intervals. An alternative would be to overlay the whole area with a proprietary modified cement screed laid by a specialist.

REFERENCES

BS Code of Practice CP 204 part 2: 1970 *In-situ floor finishes* Part 2. Metric units

C & CA Advisory note 58. 040 *High-strength concrete toppings for floors, including granolithic* British Cement Association

7.2.7 MAGNESIUM OXYCHLORIDE (MAGNESITE) FLOOR FINISH. DISINTEGRATION

This defect relates to the susceptibility of magnesium oxychloride floor finishes to moisture. The finish is commonly referred to as 'magnesite' or 'composition' flooring.

Symptoms

The surface of the floor finish starts to disintegrate and if measures are not taken quickly to arrest it, the disintegration becomes general.

The disintegration may start from the underside, but this will not become apparent until the top surface deteriorates.

Investigation

If the floor finish has been applied to a concrete ground slab the effectiveness of the damp-proof membrane may be suspect. Attempt to obtain details of the original construction, but if these are not available, the floor will need to be opened up to establish what has been done to combat rising damp.

If the surface alone is breaking up, investigate the possibility of spilt fluids being allowed to stand on the finish for long periods.

Check on the methods of routine maintenance and on the frequency and the types of chemicals used.

Diagnosis and cure

Magnesite flooring finish is particularly susceptible to moisture. Disintegration starting on the surface is usually due to inefficient maintenance of the floor finish so that spilt fluids are not cleaned up quickly or cleaning water becomes absorbed. If the disintegration starts below the surface, it is most likely that the source of water will be rising damp. However, when this does occur on ground floors, it should not be assumed that this is the only source of moisture. It is generally better to assume that there is both rising damp and poor maintenance, even if there is no damp-proof membrane.

On upper floors it is safe to assume that the sole cause for the defect is poor maintenance. If the floor finish has disintegrated it will be necessary to renew it completely. Surface treatments will be satisfactory only if they are applied as soon as disintegration starts to show. When renewing the finish, consideration should be given to an alternative material with greater durability.

For ground floor slabs affected by rising damp, a choice must be made between providing a floor finish that is not susceptible to moisture such as asphalt, or providing a damp-proof membrane before applying a new magnesite floor finish.

Whatever curative measures are taken, it will be necessary to establish regular and effective maintenance procedures.

REFERENCES

BS CP102: 1973 *Code of practice for protection of buildings against water from the ground*

BS CP 204: Part 2: 1970 *In-situ floor finishes* Part 2. Metric units

BS 776: Part 2: 1972 *Specification for materials for magnesium oxychloride (Magnesite) flooring*

7.2.8 EPOXY FLOOR PAINT. LIFTING

A new two-pack epoxy paint applied to a prepared concrete ground slab fails after a few days.

Symptoms

The new paint coating starts to break up and lift from the substrate. The defect occurs mainly on floors that have been previously coated.

Investigation

Obtain a copy of the decorating specification and establish precisely which types of material were used for the surface preparation. Check which paint finish was used and establish its suitability for use against the manufacturers recommendations. Lift a suitably large area of the paint using a flat paint scraper and check for any odour of solvent.

Examine the concrete surface for any residue of previous painting or moisture.

Have a sample of the failed paint checked for coating thickness and possible delamination.

Diagnosis and cure

The successful coating in the past would indicate that rising damp is not a problem. Any presence of moisture on the concrete surface could be the result of using a water soluble paint stripper before recoating. The temperature of the ground slab would probably have been much lower than the ambient air temperature. This would have made drying-out of the concrete difficult and would also have made it particularly vulnerable to condensation. The use of solvents in the preparation would have reduced the surface temperature as they evaporated.

Microscopic examination of the paint flakes will reveal any detectable interface between the first and second coats. Difficulty in detecting this will indicate slow curing of the first coat with some resoftening by the second. The slow curing rate would indicate that the floor was cold at the time of painting.

The presence of any paint remover residue would also prevent satisfactory adhesion.

The remedial work first requires the complete removal of existing coatings. This is best achieved by dry grit blasting which will both remove all the coatings and clean the concrete surface at the same time. It has the added advantage of not introducing any moisture to the slab. All loose particles should be removed from the prepared surface by careful sweeping, or preferably vacuum cleaning. Check the surface with a moisture meter to make sure that it is dry enough for recoating. Apply a good quality floor paint suitable for the location carefully following the manufacturers instructions. Allow at least 24 hours between coats and keep the area well ventilated and dry.

After painting, keep the floor out of use for as long as specified by the coating manufacturer to allow the finish to harden.

REFERENCES

Concrete Society Current Practice Sheet 41a/c *Painting of concrete*

Photo 1 Blisters that appeared some six weeks after the floor was laid.

7.2.9 EPOXY RESIN FLOORING ON OLDER CONCRETE SLABS. BLISTERING

Epoxy resin flooring is widely used and normally provides a reliable and tough finish. This defect has occurred on several occasions although it is still comparatively unusual.

Symptoms

Some weeks after laying, blisters start to appear in new epoxy resin flooring. Fresh blisters continue to erupt on the surface over the months that follow. The blisters are not confined to any particular area but they can be closer together in some areas. They tend to occur on older slabs and not at all on new concrete. Generally the blisters are about 30 mm in diameter and 3 mm high although some may be smaller and only discernible against back lighting.

If a hole is drilled in the blister, a dark straw coloured liquid oozes out under pressure.

Investigation

Puncture several of the blisters; collect samples of the fluid from within and send away for analysis of contents.

Establish whether there is any likelihood of damp penetration from beneath the slab.

Check on the past maintenance procedures for the floors taking note of any chemicals used.

If chemicals have been used for any production processes in the affected area, have them identified.

It may be useful to take a core sample of the slab.

Diagnosis and cure

The defect (which also happens occasionally with other thin sheet floorings, such as flexible PVC and rubber) is probably caused by osmosis. This cannot occur without the presence of the following three pre-conditions:

1 The formation of a salt concentration.

2 The formation of a semi-permeable membrane or layer.

3 A source of water.

Osmosis is the passage of water molecules through a membrane which is more permeable to water molecules than to dissolved solute molecules. The process continues until the concentration of dissolved substances on both sides of the membrane are equalised, or sufficient pressure is generated to prevent further movement (of water molecules) across the membrane.

It is only possible to speculate why blistering does not occur in areas of relatively new concrete. The most likely explanations are that either the pore sizes in the concrete are too large to act as a semi-permeable membrane, or a soluble salt concentration has not been formed at the surface.

Though a semi-permeable membrane and the formation of a salt concentration are necessary for the occurrence of the osmotic blistering, they are not sufficient to produce it. A water source is also essential. If there is no damp proof membrane in the floor, water can move upwards from the ground or hardcore below the concrete.

Experience at the Building Research Advisory Service indicates that osmotic activity does not usually extend beyond two years, and that the formation and increase in the size of blisters will slow down and finally cease at the end of that period.

Epoxy resin flooring is very tough and abrasion-resistant, and it is likely that some years will pass before blisters will break down in service.

Further osmotic action seems extremely unlikely, but as an insurance against the small risk that osmotic action could continue to some extent, the replacement of the floor with a trowelled epoxy resin and sand mix might be considered. Such mixes have not been known to suffer blistering under similar conditions. This is because there is sufficient permeability between the aggregate grains for pressure to be released without harming the flooring.

Floors should be tested for dryness, before laying the finish, using the test described in the BS Code of Practice. Floors in contact with the ground should incorporate a damp-proof membrane.

REFERENCES

BS 8203:1987 *Code of practice for installation of sheet and tile flooring*

BRE Digest 33: *Sheet and tile flooring made from thermoplastic binders*

Photo 1 Dampness and an inadequate underlay cause this ridging of the sheet flooring (see also 7.1.11).

7.2.10 SHEET FLOORING. PARALLEL RIDGING

Ridging occurs in sheet flooring laid on timber boards.

Symptoms

Regular parallel ridges form on the surface of the sheet flooring which worsen over a period of a few weeks. The problem usually occurs soon after the floor is constructed but can occur if the heating of a building is improved.

Investigation

Lift the sheet flooring to reveal the timber board floor below. Check that the boards are adequately fixed. Measure the moisture content of the boards on both the top and bottom faces. Check whether the boards were laid in a wet condition and note whether the humidity of the air in the building is likely to be unusually low at any period.

Diagnosis and cure

When the boards are revealed they will be seen to have curled up at the edges. Moisture meter readings will probably show the boards to have a greater moisture content on the underside than the topside, especially if the floor is a suspended ground floor. The boards will have been laid whilst still too wet. The top sides will naturally dry more quickly than the bottom causing the boards to curl. The movements of the floorboards will affect any finishes laid on them, unless a thick underlay has been used.

The boards should be planed or sanded to an overall flat surface or to provide a surface flat enough to take a hardboard underlay, which must be properly fixed down. This may be done without removing them, or alternatively take the boards up, plane them flat in a machine and re-fix them.

All fixings should be checked for tightness and replaced as necessary.

REFERENCES

BS Code of Practice CP209: Part 1. *Wooden flooring*

BS 8201 Code of practice for *Flooring of timber, timber products and wood based panel products*

Photo 1 The lifting and arching of the blocks is caused by construction water, use of the building, leaks and flooding or an inadequate damp proof membrane.

7.2.11 WOOD BLOCK FLOORING. LIFTING OR ARCHING

This defect occurs mainly on concrete ground slabs but can occur on upper floors in certain humidity conditions.

Symptoms

The wood blocks lift up over a large area or along two or three rows. The defect may occur at any time but although the point of failure of the block and the base may vary (ie the block may come away from the adhesive or the block and adhesive may come away from the screed) the cause will be the same. It is a very common defect, its incidence being greater than the loosening of individual blocks due to poor bonding of the adhesive to the block.

Investigation

Check the moisture content of the loosened blocks. This must be done without delay or the blocks must be placed in an airtight container until they can be tested. The moisture content of some of the blocks still in position should also be measured.

If the moisture content of the blocks is above 18% an examination should be made of the base concrete to find out if it is passing ground water into the floor finish.

If the loosened blocks are mainly around the perimeter of the floor the possibility of moisture transfer from the walls should be investigated.

The blocks should be examined for signs of fungal attack.

Diagnosis and cure

The primary cause is an increase in the moisture content of the blocks which causes them to swell and sets up compressive forces. These are initially contained by compression of the blocks but if continued for any time they force the blocks up from their bedding.

The increase in moisture may arise from construction water still present; high humidity conditions in the building; flooding; lack of a damp-proof membrane between the ground and the blocks; or reliance upon the adhesive used for fixing the blocks to provide a barrier.

The trouble also arises when blocks are laid in a very dry condition in a building where subsequently the atmospheric humidity is high.

If the defective area is examined soon after the defect was noticed it will often be found that because of their increase in size the blocks cannot be set back into their original position, though this will be possible when the whole area, as well as the loosened blocks, has dried out.

If it is known that the building will have relatively high humidity conditions internally, the floor blocks should be laid with a moisture content higher than normal, so that any further increase in moisture content will have only a small effect. It should be appreciated that gaps may occur between the blocks if conditions subsequently become much drier.

If the damp conditions are only temporary, the blocks can be relaid when they have dried out, but when this is not the case some preventive measures must be taken to reduce the dampness before relaying the blocks. If the normal humidity conditions in the building are high, it will be advisable to condition the blocks to a higher moisture content than normal, so that they approximate to the moisture content of the air.

The provision of a compression joint in the floor finish around the perimeter of the floor is recommended.

REFERENCES

BS 1187: 1959 *Wood blocks for floors*

BS 8201: 1987 Code of practice for *Flooring of timber products and wood based panel products*

Wood Information sheet 28 Section 1. December 1985. *Timber and wood based sheet materials for floors* Timber Research and Development Association

Photo 1 Irreversible expansion of clay tiles, thermal movement or shrinking screeds may all work together to produce this effect.

Photo 2 Arching of clay tiles.

Photo 3 Long term expansion of the tiles can cause failure.

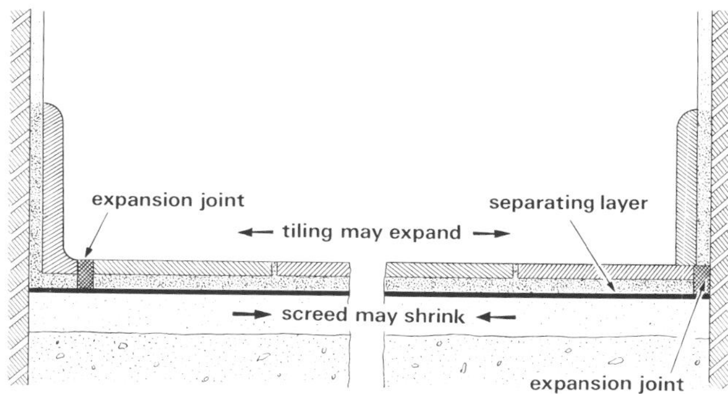

Figure 1 Expansion joints can be provided at the skirting edge or wall (intermediate joints should be provided every 4.5 m).

7.2.12 CLAY TILE FLOORING. LIFTING OR ARCHING

Symptoms

The tiles lift over a large area by arching or along two or three rows by 'tenting up'. Before this, they may show signs of hollowness when tapped, the final failure often occurring suddenly and with an appreciable noise as the tiles part company from the bedding. The break from the bedding is usually clean.

Investigation

Obtain information about the make and origin of the tiles. Find out how they were stored before laying and check on their tendency to moisture expansion.

Establish the floor construction and whether or not a damp-proof membrane was provided.

Check on the frequency and maintenance methods used for the floor.

Find out if there are great fluctuations in the humidity of the room.

Diagnosis and cure

There are three causes which singly or together, can lead to the tiles becoming loose and lifting up. The majority of cases will be due to more than one cause.

Clay tiles, like other ceramic products, may be liable to an irreversible expansion as a consequence of the absorption of moisture. This may come from the air, from cleaning operations, or from constructional defects which allow water to pass into the screed on which the tiles were bedded or from construction water initially in the floor. The extent of the expansion depends upon a number of factors. Most of it takes place in the early life of the tiles but it may be some years before compression forces have been built up sufficiently for the stress to be relieved by an uplifting of the tiles.

Thermal movement may also be enough to break the bond, particularly if the floor becomes cold, the screed will then contract more than the tiles.

Freshly-laid screeds shrink as they dry. The shrinkage will be greater than the contraction of the tiles and the bond between the tiles and the screed may be broken. Usually this action occurs within a year of the tiles being laid.

If a large area of tiles has become loose it is advisable to take up all the tiles and relay them using the methods described in the British Standard Code of Practice. If only a small area has been affected it may be possible to relay those that have loosened, but check the whole of the tiled area for signs of hollowness.

REFERENCES

BS CP 202: December 1972 Code of practice for *Tile flooring and slab flooring*

BRE Digest 79. February 1976 *Clay tile flooring*

Photo 1 Strongly alkaline moisture from the concrete floor and cement screed below can cause the failure of the tile adhesive.

7.2.13 THERMOPLASTIC AND VINYL TILE FLOORING. LIFTING; EDGE DETERIORATION

Tiles become loose particularly on ground slabs.

Symptoms

The edge of individual tiles lift or the whole tile becomes loose. There may be white salts present on the edges of the tiles or adhesive may have squeezed out from between tiles. The tiles may be delaminating and have a powdery appearance.

Investigation

Evidence should be sought of the presence of a damp-proof membrane in the slab if at ground level. Check for the presence of moisture in the slab by probing with a moisture meter.

Check the slab for cracking if the defect is localised.

If the defect is only occurring at the perimeter walls, check for continuity between the slab damp-proof membrane and the wall damp-proof course. Check also that any wall DPC has not been bridged.

Find out what maintenance procedures are in use and whether there is any extensive spillage of water in the area.

Diagnosis and cure

The primary cause of the defects is water which has been made alkaline by passage through the concrete base on which the tiles are laid. Strongly alkaline water may attack the adhesive or effect its adhesion to the concrete, causing the tile to become loosened so that it is liable to be damaged or displaced by foot traffic or by normal cleaning operations.

Sodium carbonate from the alkaline solution may appear at the edges of the tiles as a result of the evaporation of water through the joints and may spread slightly across the surface. In extreme cases the edges of the tiles disintegrate.

Excessive use of water for cleaning the floor surface or gross spillage of water onto the floor may also be to blame. This would certainly be most likely on upper floors free from the effects of rising damp.

Whatever the source of the water, there is usually no point in having any salts analysed.

It is essential to find the source of the water reaching the tiles and to eliminate its presence in the future. Where there is no DPM and the cause is rising damp, it can be remedied by applying a surface DPM of asphalt to the base, though this could also be used as the finish instead of tiles.

Excessive quantities of water on the floor surface usually leads in time to a failure of adhesion. The swilling of floors for cleaning must be avoided and where possible cleaning should be restricted to wet mopping.

If it is difficult to keep the concrete dry and prevent water from reaching the finish, it may be preferable to relay with a different finish. The new finish must be one that is less susceptible to water than the original. Because of the difficulty of cleaning off the adhesive from the original tiles, replacement will generally require new tiles. Salts can be cleaned off the surface of the tiles, but if the edges have been affected it will be better to replace.

Occasionally the tiles may not have been stuck down properly in the first place and as a result single tiles become displaced.

REFERENCES

BS 8203: 1987 Code of practice for *Installation of sheet and tile flooring*

BRE Digest 33: *Sheet and tile flooring made from thermoplastic binders*

7.2.14 SHEET FLOORING. GENERAL LOSS OF ADHESION

This defect occurs most commonly on concrete ground slabs.

Symptoms

The sheet flooring loses adhesion with the slab. This may be localised causing surface blistering or rippling or more generalised failure over large areas.

Investigation

If the slab is a concrete ground slab, establish the presence of a damp-proof membrane. Lift the failed sheeting and examine the underside for moisture beads on the adhesive. If the adhesive is still tacky it should be smelt to detect the presence of solvent. If the adhesive is dry and hard and perhaps brittle, take note of the area where the bond has failed. Obtain as much information as possible on the type of adhesive used, the conditions under which it was used, and in the case of a screeded slab, the time interval allowed between the casting of the slab and the laying of the finish.

Take moisture content readings of the slab.

Diagnosis and cure

There are two possible causes for this type of defect; failure of the adhesive or damp penetration.

Where an adhesive is used for fixing the sheeting, loss of adhesion may be caused by entrapped solvent contained in the adhesive. Moisture may be affecting the adhesive or the adhesive may be unsuitable for the particular application. The manufacturers of both the flooring and the adhesive should be consulted for advice on suitability and method of use before laying any replacement material.

The most probable cause for failure is moisture movement. In the case of a ground slab it is quite likely to be due to rising damp. Unless an effective damp-proof membrane is installed, sheet flooring will not provide a satisfactory finish as it is dimensionally unstable in the presence of water.

Another source of moisture is residual construction water from the casting of concrete slabs and screeds. To enable better workability, there is always an excess of water used in the mix beyond that required to hydrate the cement. The only way that this excess water can be eliminated is by evaporation. There are no hard and fast rules to calculate the time required to complete evaporation. It can vary considerably, depending on the weather, heating and ventilation and the thickness of the slab.

Occasionally the moisture movement is caused by osmosis. This phenomenon is dealt with in section 7.2.9.

Any of these causes may be aggravated by heavy trucking over the floor when equipment is being installed. Excessive use of cleaning water which is allowed to stand on the surface will also increase the risk of damage.

REFERENCES

BRE Digest 33: *Sheet and tile flooring made from thermoplastic binders*

BRE Digest 104: *Floor screeds*

BS CP 102: 1973 Code of practice for *Protection of buildings against water from the ground*

BS CP 8203: 1987 Code of practice for *Installation of sheet and tile flooring*

7.2.15 FLOOR TILING. LIFTING AND SLIPPING

Occurs particularly in non-domestic kitchens and processing plants that require frequent washdowns.

Symptoms

The tiled surface becomes very slippery when wet. The grout between the tiles erodes and individual tiles loosen and lift. Areas of tiles crack even though there is no arching of the tiles.

Investigation

Record the extent of loose tiles. Check the floor for sufficient 'fall' and adequate drainage. Investigate the type of tiles used and note whether they have any moulding of the surface. Establish the construction of the slab and the position of the damp-proof membrane, if present. Take moisture content readings of the screed and have a sample analysed for content and mix.

Check on the maintenance procedures adopted, paying particular attention to washing-down methods employed.

Diagnosis and cure

Some finishes are intrinsically slippery to certain types of footwear and some are only slippery when wet. Tiled floors in 'clean' areas are very likely to be wet a lot of the time because of routine washing-down. Synthetic rubber or similar footwear, whilst providing good water resistance, may slip easily on wet floor tiles. Even so-called non-slip tiles can be slippery unless provided with a dimple or other type of raised pattern finish.

If the only problem with the floor is the slipperiness, the best effective cure will be to replace the tiles with a more suitable type.

If the joints between the tiles have eroded it will most likely be caused by incorrect specification of the bedding grout. Analysis of the grout will establish the type used. If water has penetrated to the underside, it will be necessary to lift the tiles to allow the base to dry out. If the screed is also to the incorrect specification, it will also need to be replaced.

The use of harsh chemicals in the washing down water will exaggerate any weakness in the grout. When relaying the tiles, establish with the manufacturer the suitability of the grout to be used.

REFERENCES

BS CP 202: December 1972, Code of practice for *Tile flooring and slab flooring*

BRE Digest 79: February 1976 *Clay tile flooring*

Photo 1 The cure is to change the finish and possibly also the furniture.

7.2.16 SHEET AND TILE FLOORING. INDENTATION

This fault can occur in any location but is most common in areas such as canteens.

Symptoms

These are self-evident as depressions in the surface under the legs of furniture where they are kept in the same place for some time. They are most noticeable where chairs are set at tables with a limited amount of variation of position.

Indentation may also be seen on parts of the floor subject to foot traffic. Those caused by narrow shoe heels will be easily distinguished from indentations caused by furniture. For other foot traffic see item 7.2.18.

Investigation

Record the position of the indentation and confirm any point loads that the floor has to sustain. Establish the composition of the flooring and the base upon which it is laid. If the indentation is excessive relative to the loading, evidence should be sought of possible localised heating of the floor or of the surface being softened by the spillage of liquids or the use of unsuitable cleaning and polishing materials.

Diagnosis and cure

Indentation occurs when concentrated loads are applied to the finish for an appreciable length of time; frequently the case with furniture. Short term very high concentrated loads such as those from stiletto heels or people tilting back in dining chairs can be equally damaging. If the correct finish has not been chosen to cope with these loads, indentation will occur and will be further magnified if the sub layer is not of sufficient density to resist the applied loads itself. Some indentation can be tolerated, but if it is excessive it may eventually lead to the surface of the flooring being torn.

It is not generally possible or advisable to attempt to fill the depressions in the surface. Further indentation by furniture may be prevented by placing spreader cups under the legs, but these are generally thought to be unsightly. The only real cure is to replace the flooring with a material more suitable for the application. Most manufacturers are willing to advise on the suitability of their products for a given use.

REFERENCES

BS 8203: 1987 Code of practice for *Installation of sheet and tile flooring*

Photo 1 Adhesive combing showing on the finished surface-probably indicating that the adhesive had partially set before the tiles went down.

7.2.17 SHEET AND TILE FLOORING. DISFIGURATION, PATTERN

This defect only occurs with floor finishes which have been stuck down on a base, generally of concrete.

Symptoms

The disfiguration shows the pattern formed when combing the adhesive prior to laying the sheet or tile floor finish. It is often very resistant to cleaning.

Investigation

Close examination of the surface will reveal the presence of scratches filled with dirt, the scratches being on those parts of the finish which are in contact with the adhesive used for fixing. This can be checked by taking up part of the finish, eg a tile, and making a comparison of the top and under surfaces.

Areas of sheet or individual tiles may be loose through inadequate bond between the adhesive and the finish. Check the whole area for adhesion.

Use a straight edge to highlight the extent of any variation in the level of the finish.

Diagnosis and cure

The pattern is that of the adhesive prior to laying the floor finish on it. All non-impact adhesives are combed to assist spreading and to ensure the correct quantity of adhesive is applied. The pattern may be an indication that the adhesive became hardened before the finish was laid. The floor finish then rests on the tops of the ridges produced by the combing of the adhesive instead of flattening them under pressure. The high spots, which are generally in the form of curved parallel lines, are only marginally higher than the remainder of the floor finish, but they become preferentially worn and liable to be scratched. Dirt collects in the scratches and produces the pattern which is very difficult to remove.

The performance of the flooring finish will only be marginally affected. A change in the method and frequency of cleaning the floor may result in some improvement in the appearance. However, if the result is aesthetically unacceptable the only cure will be to replace the finish.

REFERENCES

BS 8203: 1987 Code of practice for *Installation of sheet and tile flooring*

Photo 1 Wear marks may merely show where traffic is heaviest but they may also indicate that the finishing materials are defective.

7.2.18 FLOOR COVERINGS. UNEVEN WEAR

Occurs in areas subjected to heavy traffic and where concentration of the traffic occurs such as in corridors.

Symptoms

A discernible wear pattern is generated on the floor finish where people are concentrated in a particular line of travel. It is particularly noticeable in passageways and on corners or entrances to rooms. It is often accompanied by a change of colour of the worn area.

Investigation

If the wear has been very rapid, remove a sample and have it checked for constituents. Arrange a traffic count to establish the required performance specification for the flooring.

The techniques and frequency of maintenance should be checked.

Diagnosis and cure

Traffic is often guided along defined paths, by doors, narrow corridors, etc. Under these circumstances the selected floor finish may not have sufficient resistance to wear or the workmanship in laying may not have produced the expected wear characteristics. The information should enable a decision to be made, though the evidence may not always be clear-cut.

The cheapest remedial action will be to replace only the worn areas with new flooring. This is a comparatively easy exercise with tiles and it should provide a good match with the surrounding areas. If the flooring is in sheet form, it will be difficult to effect the same remedy without it being readily apparent. It may be preferable to wait for the floor to wear to a point where the whole area can be economically replaced.

Occasionally a new finish can be laid on top of the existing one, but care must be taken to ensure that the worn areas provide a satisfactory base for the new finish, or that they are levelled up. Flexible sheet and tile finishes must be laid on a level surface and where adhesives are used, the surface must be smooth and free of any loose particles.

Whenever the floor finish is replaced, it is advisable to contact the manufacturer to ascertain the suitability of the products for a high traffic situation. Consider using a hard wearing finish such as synthetic studded rubber or terrazzo.

REFERENCES

BRE Digest 33: *Sheet and tile flooring made from thermoplastic binders*

Photo 1 Shrinkage widens the gaps between tiles.

Photo 2 Shrinkage of sheet flooring opening a gap of approximately 20 mm between sheets allowing dirt to accumulate in the gap.

7.2.19 PVC FLOORING. SHRINKAGE

This defect is most frequently seen in hospitals, schools and restaurants in areas subjected to greatest soiling and therefore frequent wet cleaning. It can occur in any location where PVC flooring is used.

Symptoms

Over a period of time, gaps appear at the joints of tiles or sheet PVC flooring. In some cases the tiles may move under traffic causing very wide gaps to appear.

Investigation

If the floor is subject to only occasional washing, try to establish the type of adhesive used to stick the flooring to the base. Lift a tile or part of the sheet and note whether the adhesive has set completely.

Where it is known that the flooring is subject to frequent washdown maintenance, check on all cleaning agents used and the frequency of the cleaning.

Diagnosis and cure

The shrinkage of PVC flooring is caused by the loss of plasticisers from the flooring. This can be from two distinct causes.

1 The choice of the wrong adhesive for bonding the flooring to the base. Some adhesives are known to cause the movement of plasticiser from the flooring into the adhesive. The plasticiser softens the adhesive, which makes tiling subject to movement by foot traffic with large gaps opening between tiles. This fault can be easily identified by examination of the underside of the flooring as the adhesive will be sticky and unset.

2 More recently, it has been observed that the chemicals used in cleaning agents remove plasticisers from PVC flooring. A Building Research Establishment study into this phenomenon found that all the cleaning agents investigated, including the neutral detergents normally recommended for PVC flooring, were found to cause shrinkage. They found that in addition to the fluids tested, it was necessary to pay particular attention to gel cleaners which were associated with some of the more serious shrinkages of both sheet and tile PVC flooring. They concluded that their use on such floorings could not be recommended and indeed should be discouraged unless a protective coating is employed.

The only cure for either of these failures will be to remove the existing flooring and replace it with new. Any residual adhesive will have to be thoroughly cleaned off before the new flooring is laid. If PVC flooring is relaid in the same location, it should be protected with a coating, such as emulsion polish, which must be renewed on a regular basis.

REFERENCES

Warlow W J & Pye P W *The effects of some cleaning materials on PVC flooring* BRE Current Paper CP9/79

BRE Digest 33: *Sheet and tile flooring made from thermoplastic binders*

BS 6263: Part 2: 1982. Code of practice for *Care and maintenance of floor surfaces Part 2 sheet and tile flooring*

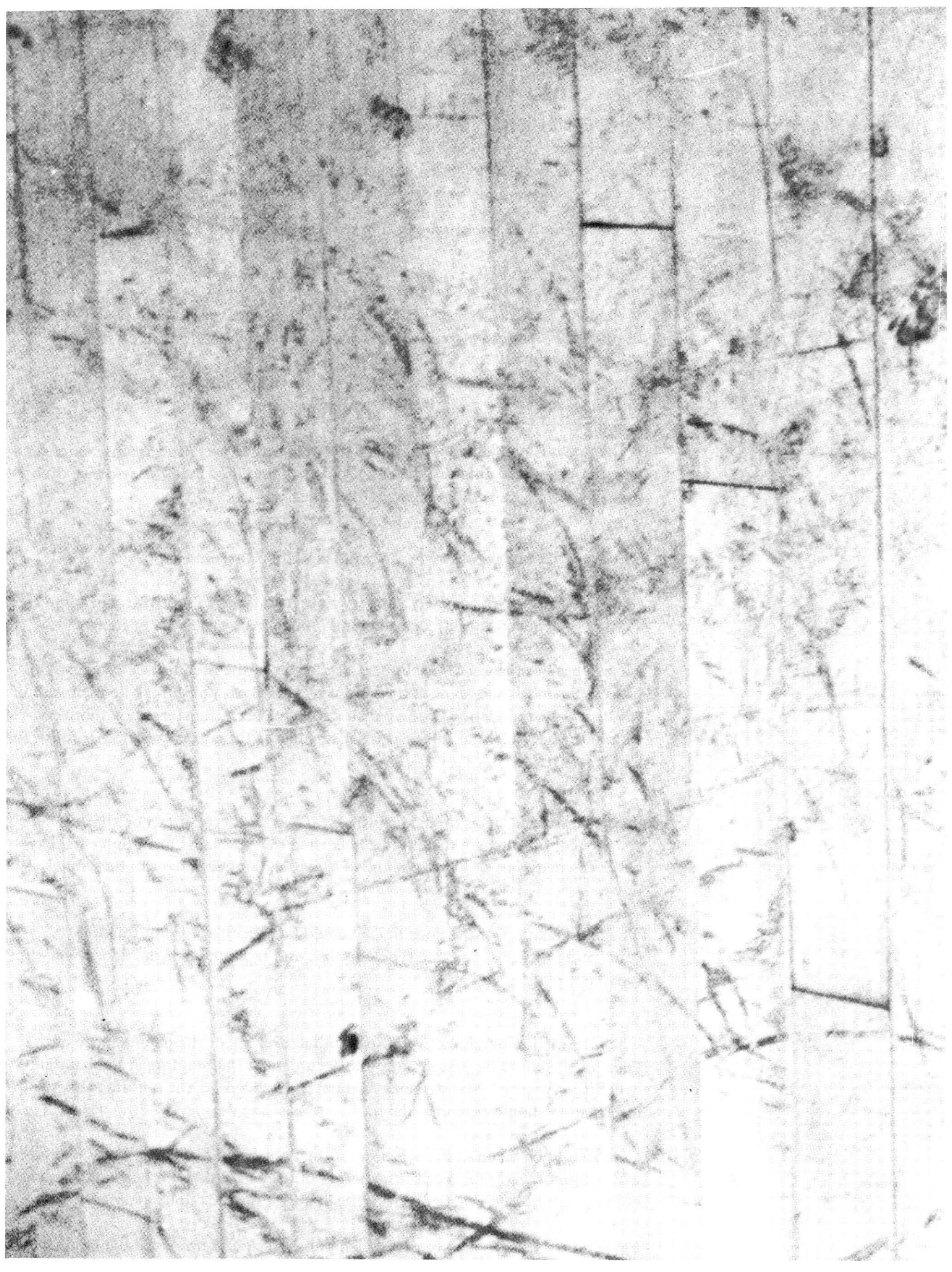

Photo 1 Rubber "burns" – regularly sealing the floor with an emulsion polish should at least make such marks easier to remove.

7.2.20 SHEET AND TILE FLOORING. DISFIGURING MARKS

Refers to random marking as opposed to distinctive patterns (see 7.2.17).

Symptoms

Many floor finishes become marked, particularly by rubber footwear, and their appearance suffers accordingly. Most sheets and tiles will be marked by cigarette burns, this defect does not deal with the control of such marking.

Investigation

Check whether any protective finishes are, or have been, applied to the flooring.

Note the type of footwear most regularly worn in the areas under investigation.

Diagnosis and cure

Footwear, especially when shod with rubber, has a tendency to leave black marks ('rubber burns') on many hard floor finishes due to anti-oxidant staining.

The degree of marking will vary as some floor finishes are more susceptible than others. Thermoplastic sheet or tiles are usually the worst affected.

'Rubber burns' may be impossible to remove as they tend to weld themselves to the surface. Other markings can be removed with scouring powder and steel wool, rubbing gently to avoid damage to the floor finish itself. It may be possible to remove the marks with detergent or by wiping with a cloth which has been dampened with a solvent such as white spirit, but this should never be done to PVC tiles or sheet.

If there are any doubts at all as to the correct treatment procedure, the manufacturers should be consulted.

After cleaning, the floor should be sealed with an emulsion polish and this should be reapplied on a regular basis to help prevent future marking.

REFERENCES

BRE Digest 33: Sheet and tile flooring made from thermoplastic binders

BS 6263: Part 2: 1982. Code of practice for *Care and maintenance of floor surfaces* Part 2. *Sheet and tile flooring*

7.2.21 MAGNESIUM OXYCHLORIDE (MAGNESITE) FLOOR FINISH. SWEATING

This defect is peculiar to magnesite flooring and is often confused with condensation.

Symptoms

Beads of moisture appear on the surface of the floor during damp weather which may, if the air within the building is damp, combine to form a film. Usually the whole area of the floor will be affected.

Investigation

Note the areas affected and the conditions prevailing when sweating occurs. Use a hygrometer to measure the relative humidity, a thermometer to record the temperature and a commercially available 'Damp Check' meter to indicate whether condensation is occurring.

If required, send a sample of the beads of moisture for laboratory analysis to check for the presence of magnesium chloride.

Check how the floor surface is maintained.

Diagnosis and cure

The magnesium chloride used in the preparation of this floor finish has the property of absorbing moisture from the air, giving the impression of sweating. The likelihood of this occurring may be increased if excessive water was used in the mixing process to aid laying. It will also be increased if this surface is not adequately maintained using suitable floor polishes.

This defect can very easily be confused with condensation which can in fact occur at the same time. The presence of magnesium chloride salts in the moisture will identify the sweating. Careful consideration of the incidence of the defect relative to atmospheric conditions which are conducive to condensation should also enable a distinction to be made. Sweating will even take place in warm weather if the relative humidity is high, whereas condensation will only occur if the floor surface is cold.

If the sweating is only slight, the application of a wax polish or a drying oil dressing may be enough to prevent it recurring except under very damp conditions. This treatment should be reapplied on a regular basis.

Keeping the room well ventilated and heated will also reduce the problem in most circumstances. In extreme cases, where sweating is severe and frequent, the floor surface may start to disintegrate (see item 7.2.7) in which case the only remedy will be to replace the finish.

REFERENCES

BS 776 *Specification for materials for magnesium oxychloride (magnesite) flooring*

BS CP 204 Part 2 1970 *In-situ floor finishes* Part 2. Metric units

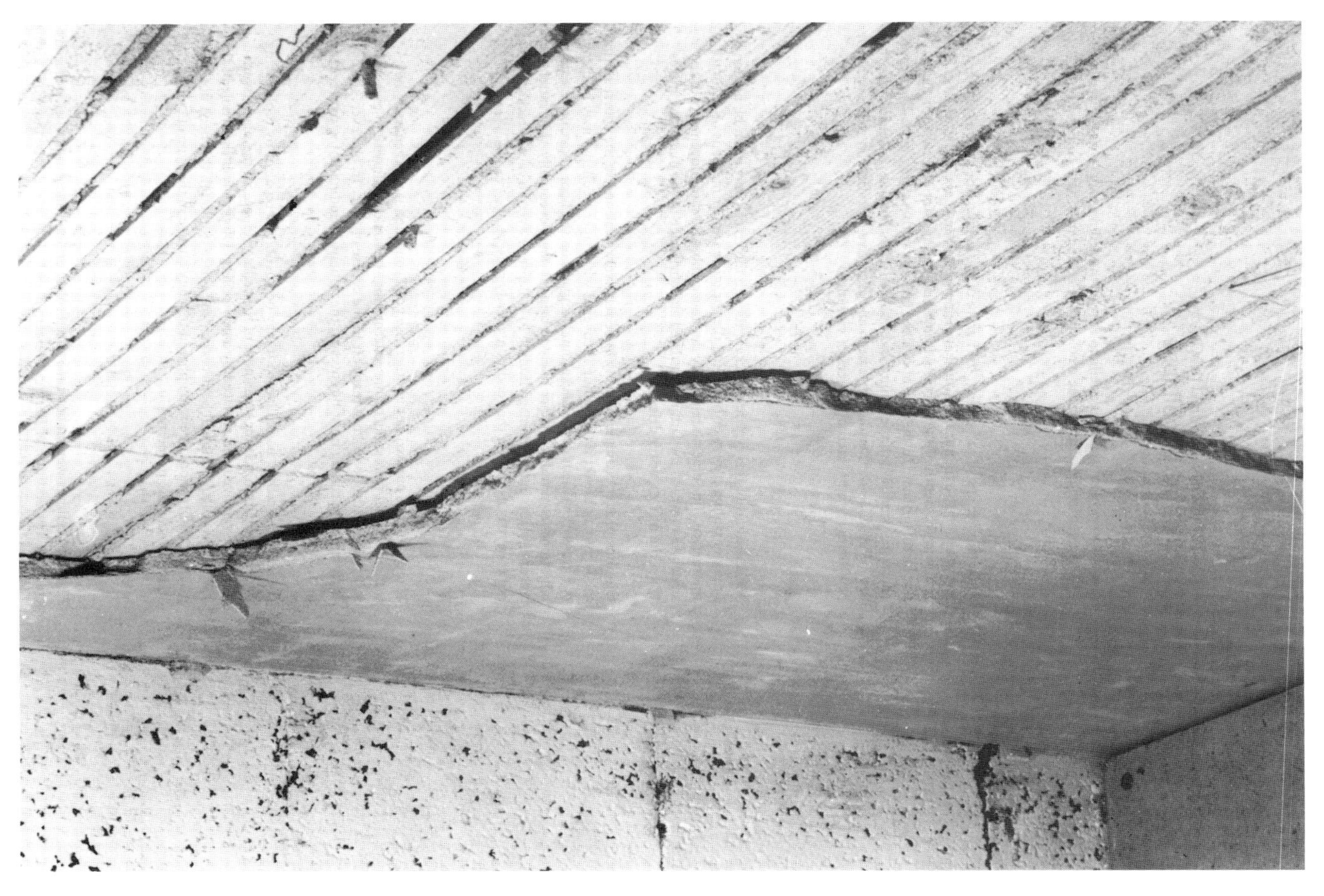

Photo 1 Typical collapsed plasterwork on ceiling with laths left intact.

7.3.1 PLASTERWORK CEILING ON LATHS. CRACKS AND/OR COLLAPSE

This defect occurs most commonly in domestic buildings with suspended timber floors constructed before the use of plasterboard linings became widespread.

Symptoms

Cracks appear running across the ceiling in a random pattern. Part of the ceiling may be at a lower level than the remainder with a discernible step at the change of level. Part or the whole of the ceiling may have collapsed.

Investigation

Try to determine the age of the plasterwork, the type of plaster used and the background on which it was applied.

If the ceiling is of some age it is quite likely that the background key is timber lathing. If these laths have fractured, attempt to establish whether the fractures are recent by noting whether there is any dirt in the depth of the fracture or if the break is clean.

Where the ceiling is only cracked or partially collapsed, carefully survey the remaining plasterwork by tapping it gently with a hammer or similar instrument over the entire area. Note the position of any places that sound hollow.

Establish whether the ceiling is subject to excessive vibration either from the floor above or from any external sources.

Check whether the supporting joists have been notched or drilled for the installation of pipework or wiring associated with conversion or modernisation. Check also whether any excess loads have been placed on the floor above which would cause increased deflection.

Diagnosis and cure

It is quite common for older plaster ceilings to crack. Occasionally it is attributable to the materials used but the most likely causes are vibration and/or structural deflection of the floor.

Plastered ceilings were at one time based solely on lime and sand which generally cracked. The addition of Portland cement did little to avoid this, though it gave an improvement in the early strength, enabling the plasterwork to better withstand any vibration from other work being done in the building, especially any on the floor immediately above.

Old timber lathing dries out with age and becomes quite brittle, which renders it unable to absorb even small movements. Increased heating standards with subsequent lower humidity levels are likely to exacerbate this problem, as the whole of the internal structure will become much dryer.

Generally it will be found that the plaster key has been broken between the laths but sometimes the finishing layer of plaster will have become detached from the base coat but unless the area of detachment is small (less than 1 m^2) replastering on an old base coat will not be a practical proposition.

If the ceiling is cracked but otherwise sound, it may be enough just to fill the cracks before redecorating. Alternatively the ceiling may be covered with lining paper or tiles. It must be remembered that any repair works will be ineffective if the primary cause for the initial failure is not rectified first.

Where there has been partial collapse of the plaster, it will usually be more economical to replace the whole ceiling then attempt a patch repair, as it is difficult to clean or replace the existing lathing. Any hollowness noted when tapping the ceiling will indicate that the ceiling plaster is detaching itself from its backing and therefore potentially liable to collapse. Obviously if the whole of the ceiling has collapsed it will need to be replaced, though this need not necessarily be by plasterwork.

REFERENCES

BS 5422: 1977 *Code of practice for Internal Plastering*

BRE Digest 213: *Choosing specifications for plastering*

British Gypsum Ltd. The White Book. *Technical manual of building products*

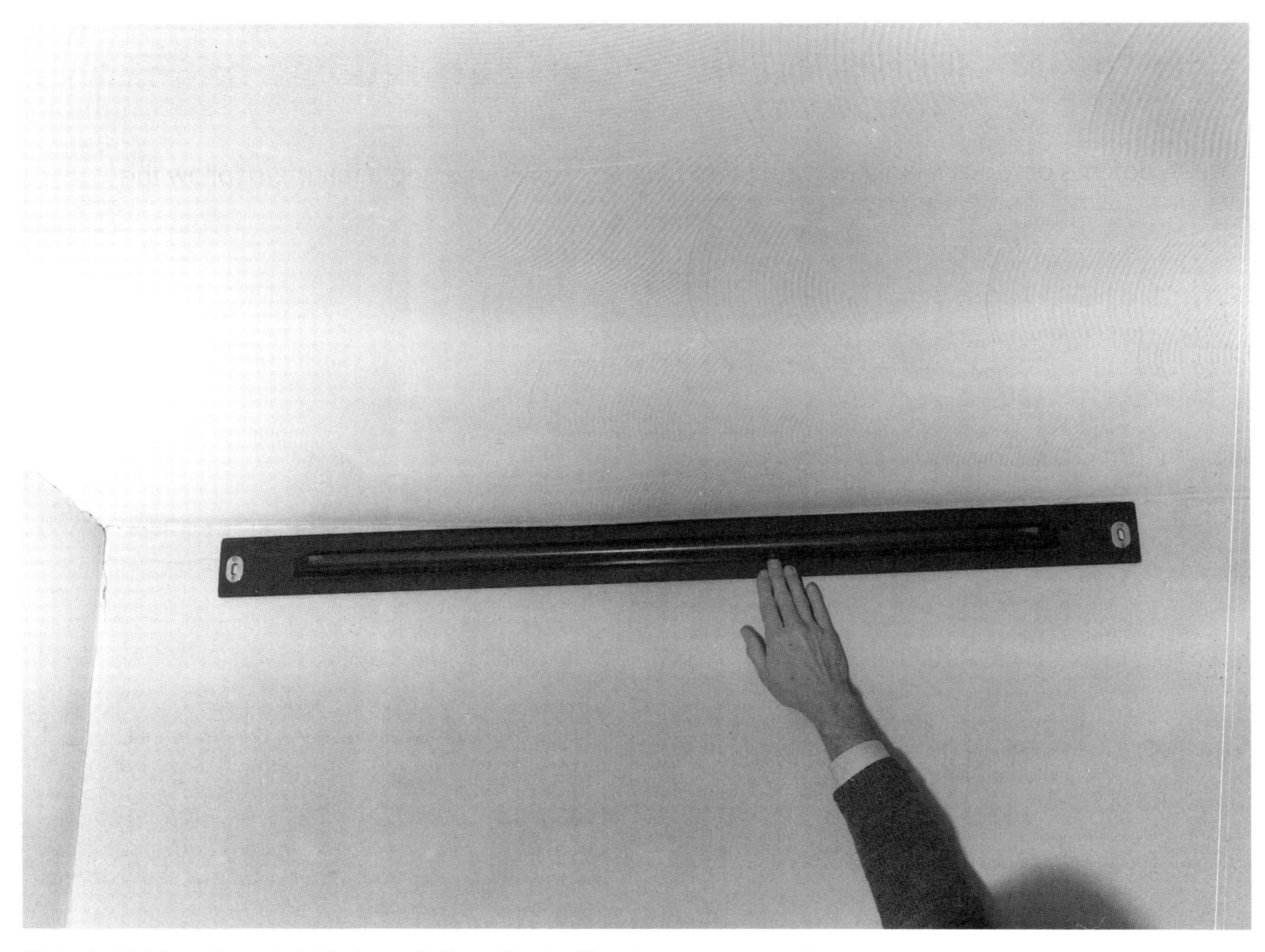

Photo 1 Holding a long straightedge up to the ceiling/wall junction reveals lack of nogging to the perimeter.

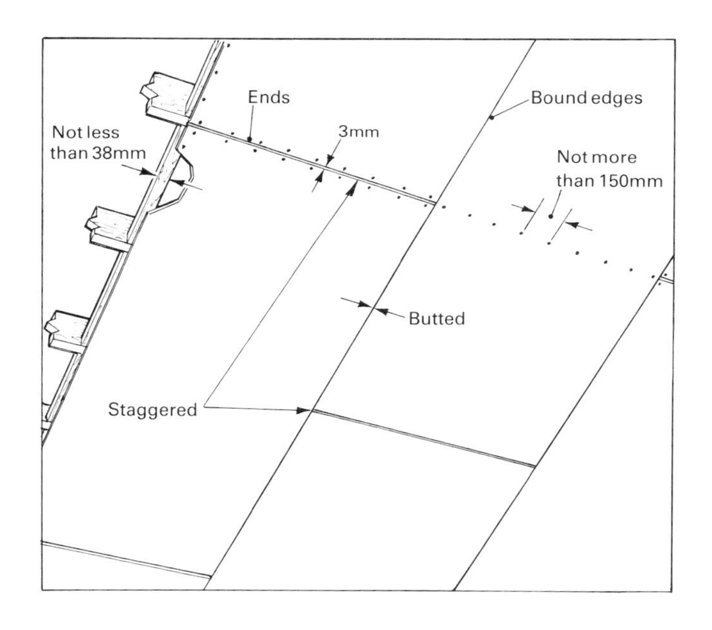

Figure 1 Adjacent boards must be staggered on their long edge.

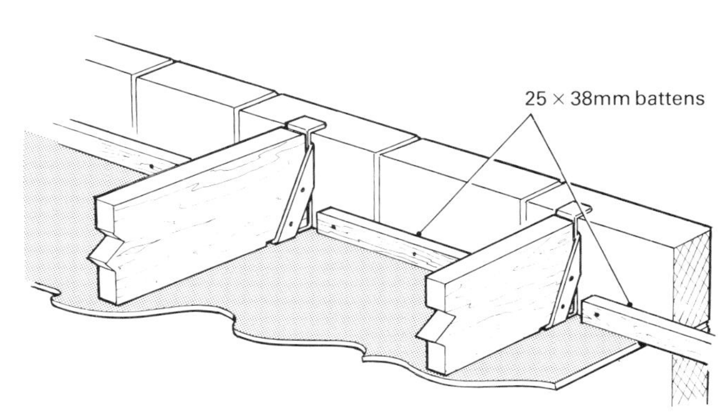

Figure 2 Perimeter noggings should be formed from 25 × 38 mm battens and fixed flush with the underside of the joists or main noggings.

7.3.2 PLASTERWORK ON PLASTERBOARD CEILING. CRACKS, LOSS OF ADHESION, SAGGING

These defects occur primarily as a result of poor workmanship and a failure to follow the manufacturers recommended procedures.

Symptoms

Cracking occurs which may take the form of fine hair cracks but is more often well-defined, running in straight lines and at right angles to each other.

If there is loss of adhesion, it may be restricted to one part of the ceiling, the plaster generally coming away cleanly from the boards.

Boards may sag appreciably, particularly at perimeter walls.

Investigation

If the cracks are straight, note their position and measure them in relation to the size of the boards. If it is suspected that the cracks are occurring between adjacent boards, check whether there is any jointing tape by passing a thin bladed knife through the cracks at various points. If there is a step between the adjacent surfaces at the crack, press the lower surface firmly upwards to establish whether any board fixings have worked loose or are missing.

Check the boards for sag by placing a long straight edge on the underside. Where boards are found to be sagging at the ceiling perimeter, check for the presence of a fixing batten.

If the plaster has fallen off, examine the surface of the boards and note the cleanliness of the face of the plaster that has been in contact with the boards. With single-coat work it may be necessary to investigate whether any lime was added to the plaster mix, while with two-coat work the strength of the undercoat and the strength of the finish coat should be compared.

Note whether one or two layers of plasterboard have been used. Joints should be staggered between the layers. It may be possible to see whether joints in the upper surface, visible in the floor void, coincide with straight cracks in the plaster below. Locating the fixings for the lower layer is more difficult than it is for the first layer, and longer nails are required.

Diagnosis and cure

Straight line cracking will be found in most cases to follow the joints between boards and can usually be attributed to incorrect jointing techniques. Where boards are butt jointed without either a scrim tape cover or a cover strip, there will always be a danger that cracks will appear if there is any structural movement or heavy vibration on the floor above.

Where boards have worked loose, check that the timber joists or battens on which they are fixed have not split or otherwise deteriorated. Refix the boards using zinc coated nails of the correct length (40 mm for 12.7 mm board, 30 mm for 9.5 mm board) fixed at not more than 150 mm centres as specified by the board manufacturer, for the particular grade of plasterboard.

Where there are straight cracks at joints between boards, hack off any remaining plaster back to board level for sufficient width to allow the joint to be overlaid with 90 mm width jute scrim, before making good with fresh plaster of the correct grade. It should be remembered that even this method is not a guarantee against future failure unless the original cause of the cracking is eliminated.

Slight hairline cracking is likely to occur only in two-coat plasterwork and may be due to applying the finish coat before the undercoat has set or to too much lime in the finish coat. Covering the plaster with lining paper is the most effective way of disguising this defect.

Where plaster has lost adhesion with the board it may be because the board surface was not clean before the plaster was applied. Any signs of dust or dirt on the detached plaster base surface should reveal this. Before replastering, remove any remaining plaster, thoroughly clean the board surface and apply a bonding coat of a suitable PVAC adhesive, in a manner recommended by the manufacturer. With single-coat work, the more likely causes are the use of the wrong plaster for the type of board forming the ceiling or the addition of lime to the plaster, either of which may contribute to poor adhesion. With two-coat work the most likely cause is that the finish coat is too strong for the undercoat. With both methods the boards will need to be cleaned of all existing plaster, sealed and replastered.

Boards may sag where they have not been correctly supported at all joints and in particular at the perimeter. Where this has happened it may be possible to cure it

Photo 2 Where the ceiling board edges are unsupported they sag causing cracking.

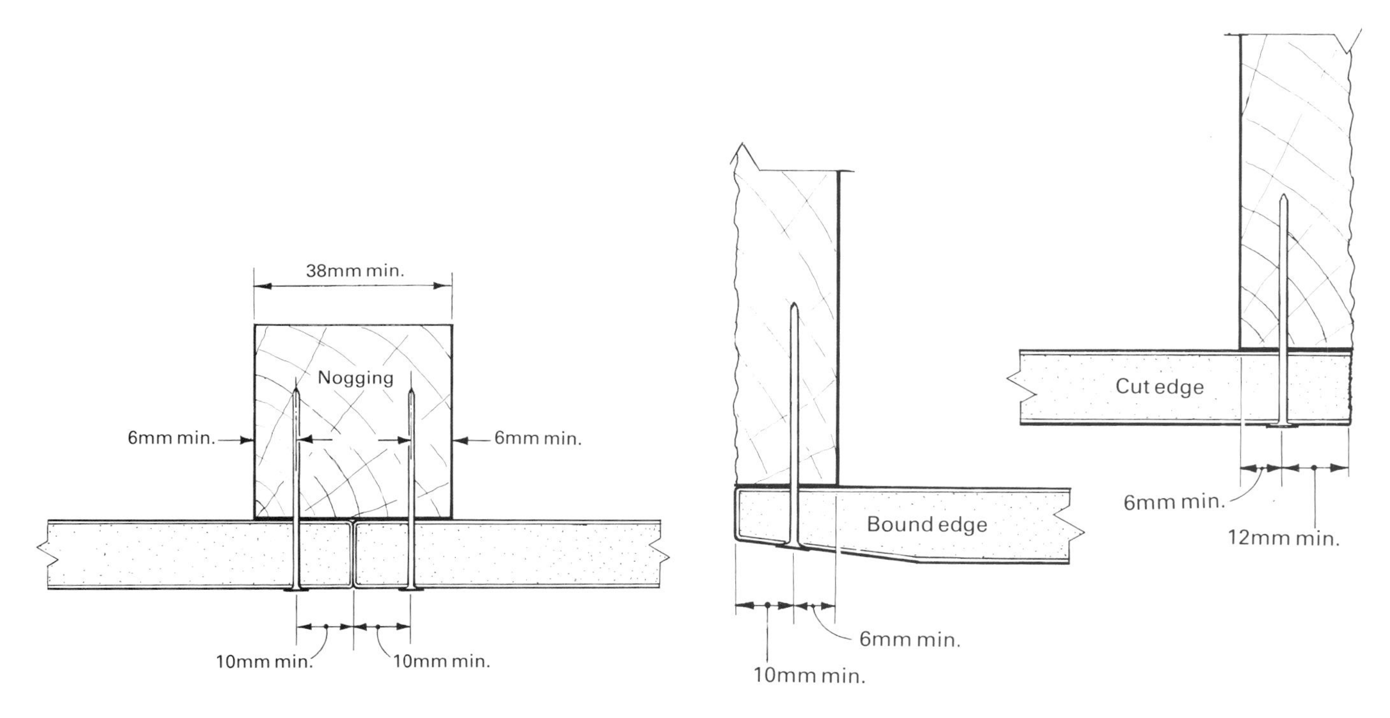

Figure 3 Minimum dimensions required for a secure fixing to a main nogging, joist or truss tie.

Figure 4 Bound edges should be nailed a minimum of 10 mm from the edge and cut edges 12 mm.

7.3.3 PLASTER CEILING ON CONCRETE SOFFIT. CRACKS AND/OR LOSS OF ADHESION

Plasterwork on concrete soffits can fail for a variety of reasons.

Symptoms

Any cracking of the plasterwork will be of two distinct types. Fine hairline cracking with a random pattern probably spread over the whole ceiling area, or more prominent cracks more clearly defined and extending across the ceiling in a fairly straight line.

Loss of adhesion may result in the whole of the plasterwork falling off or it may be confined to the finish coat shelling off the undercoat.

Investigation

If the plasterwork has cracked but not broken free, note the position and pattern of the cracks. If the cracks are wide and long, measure their depth and check the floor for excessive deflection. If there is any doubt about the structural integrity of the floor, consult a structural engineer immediately for advice.

If part of the plasterwork has fallen note the plane of failure. If this happens to be the surface of the concrete, examine it for the presence of any deleterious matter and for smoothness. The remaining plasterwork should be tapped all over and a note made of any areas that sound hollow.

Obtain details of the floor construction and if possible ascertain the direction of span.

Investigate the types of plaster used and the methods of application.

Diagnosis and cure

Cracking

Hairline cracking will usually be confined to the finish coat and any investigation should confirm this. The base coat will normally be cement or lime-based and the cracking is due to shrinkage caused by the application of the finishing coat before the undercoat has dried and completely shrunk. It is extremely difficult to treat these fine cracks and the best remedy will be to cover the ceiling with lining paper or tiles.

Larger cracks that take a definite line are generally caused by background movement and may be due to shrinkage of the in-situ concrete or to structural movement. If this is caused by excessive deflection of the slab, it may be necessary to replace the whole slab or if investigation proves the slab to be seriously overloaded, the removal of the excess load should return the deflection to an acceptable level. The cracks may be cut back and filled, but it is quite likely that further cracking will occur.

Loss of adhesion

Where the finish coat has shelled off the base coat, it is likely to have been caused by the application of a very strong finishing coat to a relatively weak undercoat. Alternatively the finishing coat may have been applied to the base coat before it had set properly or there may have been a lack of a satisfactory mechanical key.

If the undercoat has fallen off the concrete, it is possible that the wrong plaster or mix was used, especially if, as is usual, the concrete surface was smooth. The difference in the thermal expansion of gypsum plaster mixes and concrete, especially concrete made with a limestone aggregate, may be such that any marked changes in temperature will lead to a failure of the adhesion of the plaster to the concrete.

If the finish coat alone has failed, it should be stripped from the base coat and replaced with fresh plaster. The undercoat should be thoroughly roughened and brushed clean with a damp brush. It may be advantageous to treat the surface with a PVAC bonding agent to ensure good adhesion of the finishing coat.

If the undercoat is weak, it must be completely removed from the background. After cleaning the surface of the concrete it should be wetted with water before being replastered with a plaster which should be of the bonding type.

REFERENCES

BS 5422:1977 *Code of practice for internal plastering*

BRE Digest 213 *Choosing specificationf for plastering*

British Gypsum Ltd. The White Book *Technical manual of building products*

Photo 1 Paint applied to concrete that has not fully dried-out or ceiling that is damp, causes very quick failure.

Photo 2 Failure of new matt white emulsion paintwork due to inadequate preparation of original sound (but discoloured) gloss paint surface.

7.3.4 PAINTED CEILING FINISHES. PEELING AND FLAKING

Paint failure on ceilings usually occurs fairly soon after decoration but the reasons for it are not always obvious.

Symptoms

The decorative material may fall away in small flakes or in large sheets. The defect sometimes occurs to the decoration of new ceilings but is mainly restricted to re-decoration applied to ceilings of buildings that have been in use for some years.

Investigation

Where the ceiling is new and has not been previously painted examine the surface and the back of the peeled paint film for any signs of efflorescence, using a magnifying glass if necessary. Attempt to obtain as much information as possible about the weather conditions at the time of decoration. If the ceiling surface is concrete or plastered, establish the time lapse allowed for drying-out before it was painted.

If the ceiling is one that has been previously decorated, obtain as much information as possible on the type of paint last used and any previous coats or treatments. Find out which, if any, type of preparatory treatment was employed before the latest repainting. If there is any dispute over responsibility for the failure or if there is any doubt about compatibility, have some of the paint flakes analysed in a laboratory for content and thickness of coating layers. Examine the back of the paint flakes for any signs of dirt etc. and have the paintwork remaining on the ceiling adjacent to the defect tested for any signs of greasy material.

Diagnosis and cure

New Ceilings

The most common causes for failure are efflorescence and dampness, particularly where the ceiling is the underside of a concrete slab that has not been allowed enough time to dry out before decorations were applied.

Efflorescence is caused by the migration of salts to the surface of the concrete or plaster during the drying out period, and may take the form of a bulky fluffy layer which is easily removed, or of a hard thin film which adheres tenaciously to the surface and is extremely difficult to remove.

Plaster and concrete can take a long time to dry out, even when weather conditions are favourable. BS 6150 suggests, as a rough estimate, allow one week of good drying conditions for each 5 mm thickness of wet construction. During adverse conditions the drying process can be accelerated with the aid of heaters and dehumidifiers, but care must be exercised to ensure that the heaters do not produce their own combustion moisture. It is important that the area is kept well ventilated and that excessive heat, which could cause cracking of plasterwork, is avoided. A moisture meter may be used to establish that the surface is completely dry. Any remaining efflorescence should be removed with a dry cloth or a stiff brush. The surface should then be wiped with a frequently rinsed damp cloth, allowed to dry and then redecorated. The adhesion of paint on new plaster will be greatly assisted by the application of a sealer coat first.

Old Ceilings

There are two main causes of failure and several secondary ones. The most common one is insufficient preparation before re-painting. The next most common is incompatibility of the new material with that already on the ceiling.

As new films of paint dry they shrink thereby exerting considerable force on existing paintwork. Any weakness of adhesion between previous layers will eventually fail under the pressure of this force. Even if the failure appears to be localised, it should be assumed that the problem affects the whole area and all the paint should be stripped, despite the fact that it may appear to be firmly adhered. The reason for this is that any application of fresh paint will accentuate existing weaknesses.

There have been many reported cases of paint flaking off a background of distemper. This may involve several layers applied over a number of years until the weight and stress exerted becomes too great. In these cases it is necessary to strip the paint back to the distemper which should then be removed by thorough washing.

Very few paints are actually incompatible with others, the main problem being the adhesive bond between coats. Gloss surfaces should be lightly abraded to provide a key. Powdery surfaces should be washed down and treated with a penetrating sealer. If there are doubts about the nature of the existing paint and the new finish is likely to be of a different composition, an expensive failure could be avoided by the painting of a trial patch.

REFERENCES

BS 6150: 1982 *Code of practice for painting of buildings*

BRE Digest 163: *Drying out buildings*

BRE Digest 198: *Painting walls* Part 2: *Failures and remedies*

Fulcher A, Rhodes B, Stewart W, Tickle D, Windsor J, *Painting and decorating. An information manual* 2nd edition 1981. Collins

Photo 1 An example of the type of construction with exposed edge concrete floors that is prone to this defect.

Photo 2 Close-up of slab face with vertical crack clearly visible.

7.3.5 DAMP CEILINGS UNDER SUSPENDED CONCRETE FLOORS

This defect concerns isolated patches of dampness and should not be confused with condensation or other forms of water ingress. (See 7.1.14.)

Symptoms

Dampness associated with rainfall appears on the underside of concrete floors along lines which run predominantly at right angles to the external wall and up to distances of 1 m from it.

Externally, there are vertical cracks at irregular intervals on the exposed edge of the concrete floor slabs which have been exposed on the outside face as a band or string course.

Investigation

Examine the soffit of the slab for the presence of cracks, which may be very fine and difficult to detect with the naked eye.

The external exposed edge of the floor slab should be similarly examined, though the cracks will be more evident because dirt will tend to accumulate in them, nevertheless some of the cracks may again be very fine so a mist of water sprayed on to the surface of the concrete may help to make them stand out.

Obtain as much information as possible about the construction of the building, particularly the type of aggregate used in the concrete mix for the slabs.

Diagnosis and cure

Concrete shrinks as it dries and hardens, the degree of shrinkage being directly attributable to the type of aggregates used. Most of the aggregates used in England and Wales are of the low shrinkage type and are not expected to produce a shrinkage of the concrete greater than 0.045%. However, in Scotland, where many of the aggregates used are shrinkable, the shrinkage of the concrete may be as much as four times higher.

Any shrinkage is likely to result in the formation of cracks which are widest at the edge of the slab and diminish in width away from the edge. Rain running down the face of the building is liable to enter these cracks and to pass into the building along them, moving downwards by gravity so that it appears on the underside of the slab. The likelihood of water running down the face of the building is greatest when the external wall has little absorption and there are no projections to shed the water. It may therefore appear immediately underneath windows taken down to floor level or areas of light cladding taken to the same point. For this reason it is often mistakenly assumed that it is the windows or cladding that is leaking.

The remedial treatment will depend upon how wide the cracks are and the rate of water ingress. If the cracks are only narrow and comparatively unsightly it may be best to do nothing as the water will probably dry quickly. Wider cracks may be filled with one of a number of proprietary fillers but the rate of success may be low as some are not suitable for vertical surfaces.

Where the rain penetration is considerable or frequent, it will be best to either cover the slab edges with an impervious material making sure to provide continuity of any damp-proof courses, or have a concrete repair specialist cut out the cracks and fill them with an epoxy grout or similar repair material.

REFERENCES

BRE Digest 35: *Shrinkage of natural aggregates in concrete*

BRE Digest 75: *Cracking in buildings*

BRE Digest 237: *Materials for concrete*

Figure 1 Concrete base slab with floating screed on resilient layer. Screed must be totally isolated from floor slab and perimeter walls.

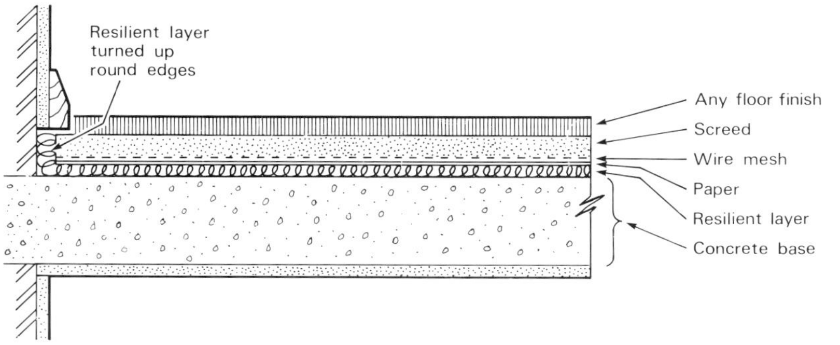

Figure 2 Concrete base slab with floating timber raft. The air space created by the support battens must be maintained.

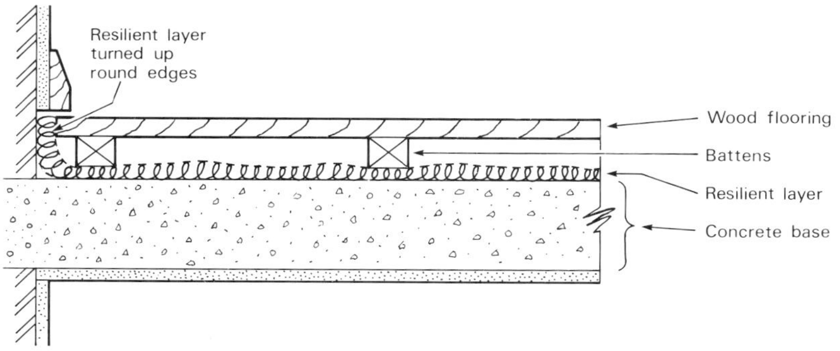

Figure 3 Timber joist floor upgraded with two layer raft on battens. The two layer plasterboard ceiling is desirable but not essential if access to the lower dwelling is restricted.

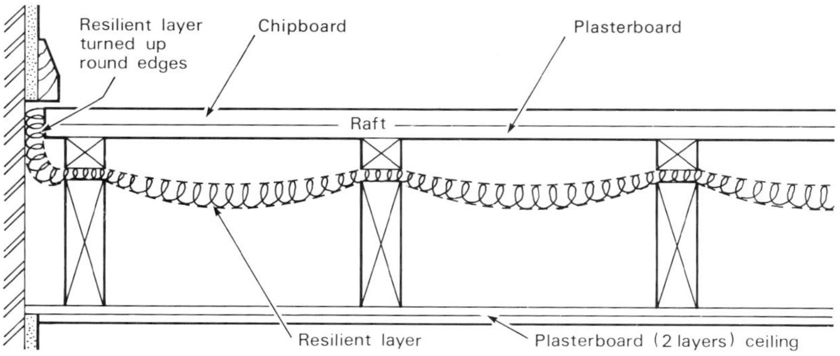

Figure 4 Existing timber joist floor upgraded. The resilient layer reduces the impact noises and the fibrous pugging the airbourne noise. This method increases the floor height the least.

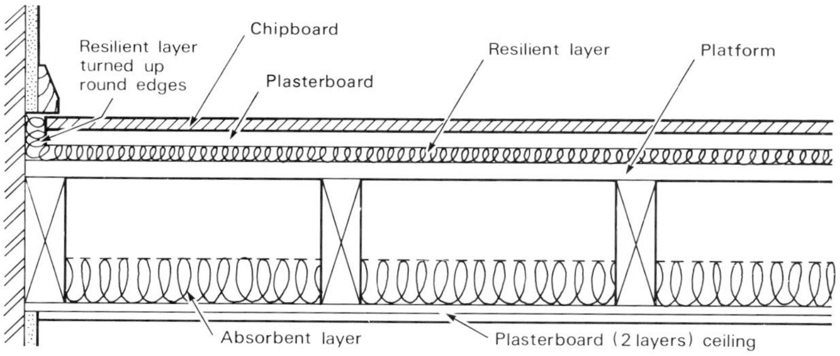

Figure 5 The best way of achieving the maximum sound insulation in a timber joist floor. Only possible when the structure can cope with the extra load from the sand.

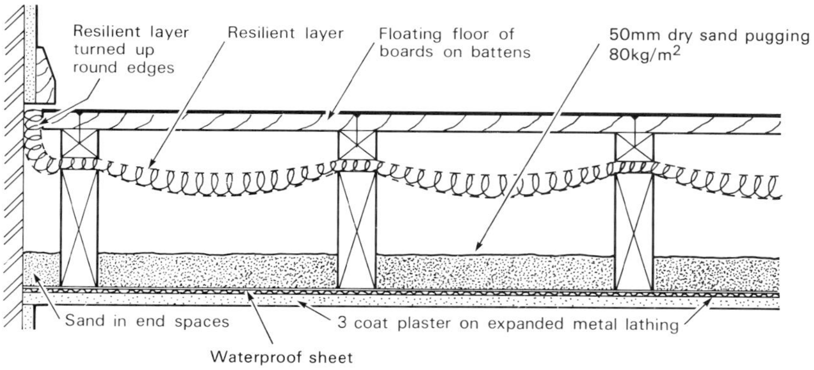

7.4.1 PARTY FLOORS. SOUND TRANSMISSION

As more existing houses are converted into smaller dwellings this defect is becoming commonplace. The use of lighter materials in new built housing since the early 1970s has also highlighted the problem in multi-occupation buildings.

Symptoms

Sounds are heard on or through the floor separating two vertically adjacent dwellings. The sounds may be in the form of impact sounds from above such as footsteps, articles being dropped, furniture being moved etc. or airborne sounds from above or below such as conversation, music playing etc. These sounds are frequently the source of extreme annoyance between neighbours.

Investigation

Establish the construction of the floor. If it is concrete very little further investigation will be needed beyond establishing the precise build up of layers in the construction. If it is a suspended timber floor, several points will need to be checked. Is the deck in the form of sheets or boards? If it is floorboarded, are they plain edged or tongue and grooved? Are the boards nailed directly to the joists and is the ceiling below similarly fixed? What type of joist hangers have been used and is the floor fitted tight up against the flanking wall? Check whether any sound insulating materials have been used. Are there free air spaces surrounding service pipes where they pass through the floor? Is there a coving or cornice on the ceiling perimeter below?

There are standard tests for sound transmission between spaces in a building (BS 2750). In cases where liability for the defect is being investigated it is important to note that not all forms of construction that comply with building regulations stand up to the in-situ tests.

Diagnosis and cure

Concrete floors

Floors constructed from concrete, no matter whether they are precast, hollow or solid, offer a reasonable barrier against airborne sound transmission because of their density. They should also provide a construction form that is free from air gaps and provided there are no unsealed gaps around service pipes there should be no direct sound transmission paths. However the hard surface means that they can be particularly vulnerable to impact sound transmission. This can be reduced in three ways.

1 If there is sufficient headroom in the rooms below, it may be possible to construct an independent ceiling under the existing soffit. The minimum space required to do this is 175 mm. It may therefore be difficult to resolve details at window heads. The construction should consist of new joists supported on edge bearers covered on top with 25 mm minimum quilt and finished below with two layers of 12.7 mm plasterboard. It is most important that no contact is made between the old and the new ceilings.

2 Alternatively a floating floor may be fitted in the room above provided that the existing floor can cope with the extra load. A 25 mm layer of mineral fibre of a suitable density (see manufacturers literature) is laid on top of the existing floor followed either by a layer of screed or by a timber raft, neither of which should touch the perimeter walls. Care must be taken that the resilient layer is not bridged. If using a concrete screed, make sure that there are no gaps in the quilt where the screed can make contact with the base. With a timber floating floor, remove all nails used as temporary fixings during construction and use only the minimum length of nails possible for permanent fixing. Proprietary battens fitted with resilient pads are available as an alternative to the overall mineral fibre quilt under battens. As this method will raise the floor level, detail problems at doorways, tops of stairways, services, sanitary fittings etc. will need to be sorted out.

3 The use of soft floor coverings will dramatically reduce most of the common impact sounds.

Timber floors

It is more difficult to insulate timber floors because they do not have the same mass as concrete. A much quoted method of increasing the density of a timber joist floor is to use sand pugging. In practice this poses several problems. Dry sand is extremely difficult to obtain but is an essential requirement for pugging otherwise rot may be started (7.10.9). For the same reasons the sand must always be kept dry when in place. Any services concealed in the ceiling void will become difficult to access. The existing floor joists will be most unlikely to have the strength to support the considerable extra weight of the sand. It will almost certainly be necessary to replace the existing ceiling with a stronger material.

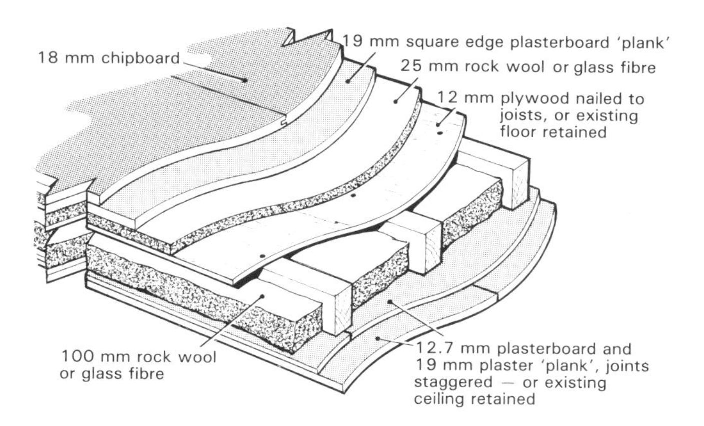

Figure 6 'Exploded' view of soundproofed timber floor. Details at doorways must be carefully considered to cope with the extra 62 mm top covering.

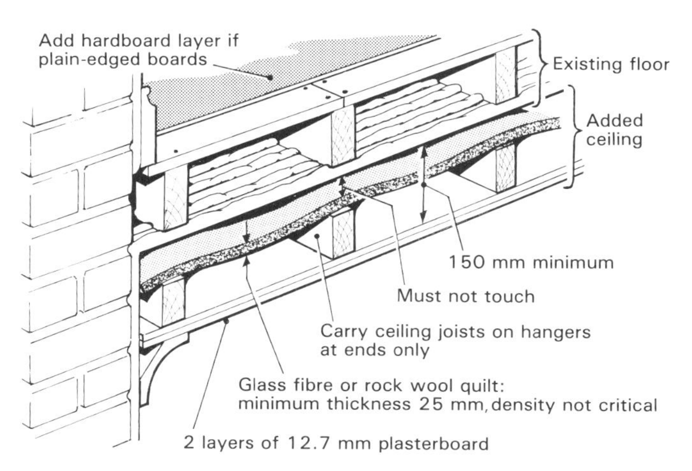

Figure 7 Where headroom permits it may be possible to construct a totally independent extra ceiling structure. Care must be taken to seal all air gaps, particularly at the perimeter.

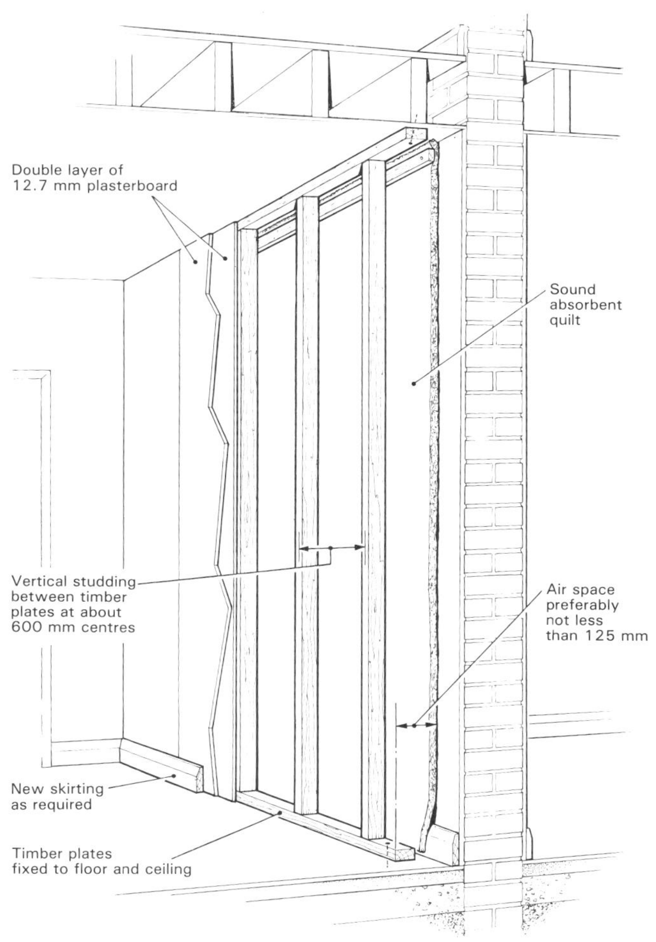

Figure 8 It may be desirable to also insulate against sounds passing through the separating walls by the installation of an independent leaf.

continued

Methods 2 & 3 suggested for concrete floors will work equally well with timber floors but with minor differences for the floating floor solution. Screeds are not suitable for use with timber floors therefore a suggested method is to use a combination of 18 mm chipboard laid on 19 mm square edge plasterboard plank over 25 mm mineral fibre. The spaces between the joists should be filled with 100 mm thick mineral fibre with a density not more than 36 kg/m^3. The plasterboard and chipboard must have a perimeter gap of not less than 10 mm. A skirting should be used to cover the gap and should be fixed just clear of the floor surface. Fill any gaps with flexible sealant both at the perimeter and at any other point where there is a free air sound path.

An alternative method where lightness of construction is important is to remove the existing flooring, lay 25 mm thick mineral wool over the joists, place battens over to run on the same line as the joists but not fixed to them. The original flooring may be refixed to the battens taking care not to penetrate the resilient layer. This method can be improved by using a combination of plasterboard and chipboard for the flooring, using two layers of plasterboard on the ceiling and/or including 100 mm mineral fibre between the joists. There will be an increase in the floor level by the thickness of the battens and the resilient layer.

REFERENCES

BRE Defect Action Sheet 45: *Intermediate timber floors in converted dwellings – sound insulation*

BRE Digest 143: *Sound insulation: basic principles*

BRE Digest 293: *Improving the sound insulation of separating walls and floors*

BRE digest 334: *Sound insulation of separating walls and floors* Part 2: *floors*

BS 2750 *Methods of measurement of sound insulation in buildings and of building elements.* Part 3: 1980 *Laboratory measurements of airborne sound insulation of building elements.* Part 6: 1980 *Laboratory measurements of impact sound insulation of floors.* Part 7: 1980 *Field measurements of impact sound insulation of floors.* Part 8: 1980 *Laboratory measurements of the reduction of transmitted impact noise by floor coverings on a standard floor*

BS 5821 *Methods for rating the sound insulation in buildings and of interior building elements.* Part 1: 1984 *Method for rating the airborne sound insulation in buildings and of interior building elements.* Part 2: 1984 *Method for rating the impact sound insulation*

BS 8233: 1987 *Code of practice for sound insulation and noise reduction for buildings* (formerly CP3: chapter 111)

Architects Journal Technical File AJ June 1986 pp 51–57 *Sound through floors in housing conversions*

Greater London Council GLC Bulletin 146 Item 8 July 1985 *Insulation of floors in converted dwellings against sound transmission*

CHAPTER 8 ROOFS

CHAPTER 8 ROOFS Contents

continued

Flat roofs

continued

INTRODUCTION

The roof is the most vulnerable part of a building. It is exposed to the elements, particularly to extremes of temperature, solar radiation, snow loadings and wind action. It can also be susceptible to movement from the structure below, and to chemical and biological agents. Roofs can also be subject to special risks due to the design of the building such as services passing through the roof covering, and internal gutters. Damage may also be caused by inappropriate methods of access over a roofing membrane whilst a roof, chimney or surrounding structure is being repaired. The consequences of a small area becoming defective are often far greater than the effects of a similar defect in a wall or other element.

In this chapter, as far as possible, roof defects are divided into structural defects and defects in the waterproofing material.

Most roofs are structurally stable, and roof structures do not provide many maintenance problems. The majority of those that do occur are confined to the results of deflections in the structure which manifest themselves as distortion of either the roof or of the walls at roof level. Most structural defects can be identified from a visual inspection.

All roof finishings must initially provide a waterproof covering to a building and they need to be durable under a wide range of exposure conditions. This durability is not always achieved for a variety of reasons, most of which are peculiar to the individual material.

Most finishings for pitched roofs fall into one of two categories: those in which the waterproofing material consists of overlapping small individual units such as slates or tiles, and those in which it is in the form of sheets, which may be either rigid or non-rigid. The most common roof covering outside these categories is thatch which, because of its comparative rarity, has not been dealt with in this book.

Finishings for flat roofs are either in the form of flexible sheets or in the form of materials applied in-situ in plastic form, which subsequently harden, eg asphalt.

Roof coverings can become defective because the ancillary materials used for fixings, flashings, etc. may also become defective long before any deterioration of the roof finishing itself. The remedial work required for such defects may be considerable.

Although defects in the covering material may be identifiable by a visual inspection, roofs are seldom regularly inspected. Pitched roofs are rarely examined, even where their condition can be readily checked with a pair of binoculars from ground level. In many cases, particularly with flat roofs, defects can be more difficult to identify. The first manifestation is likely to be the presence of dampness in a ceiling, by which time the roof structure may also have suffered damage. Minor defects may not cause immediately identifiable problems, but in due course most deteriorate and permit the ingress of water to the roof structure with consequential damage to other components.

INTRODUCTION

continued

A careful and systematic investigation of any roofing defect is essential. This should include, where possible, obtaining copies of drawings and specification prior to examination, together with details of any changes that have occurred since erection (such as the introduction of cavity barriers). When examining a defective roof, try to get the original roofing contractor to participate – if opening up of a roof is necessary a builder's presence will be required. Discuss with the occupants of the building the nature of any leaks, when the first leaks were noticed, the position of any drips, and whether they think they are weather related. It may be necessary to obtain meteorological data such as intensity and duration of wind and rainfall related to periods of water ingress. Remember that many moisture sensitive lightweight flat roof decks, such as chipboard and woodwool slabs, lose strength when wet, so exercise caution when walking on them if saturation and degradation is suspected. This can often be ascertained by examination of the underside from inside the building.

Internal leaks invariably manifest themselves at some distance from the point of entry. Water may travel along the surface of vapour checks, structural members, metal deck troughs, or along services and trunking.

Roof coverings are sometimes disturbed or even blown off by high winds. In most cases this will be readily apparent, but metal fixings should always be examined for signs of corrosion which could be a contributory factor. In areas where the risk of high winds is greater than normal, check that the specification, the selection of materials and the method of fixing are appropriate.

1 Gable wall that has been displaced because the lack of bracing has caused the roof to move sideways.

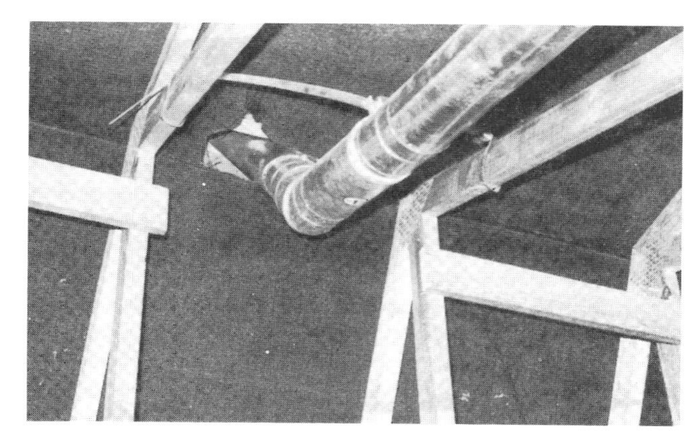

2 Service outlets such as this flue may interrupt bracing and/or binding. The location of flues and water tanks should always be included in the roof truss design and it is essential that service installers are prevented from cutting members.

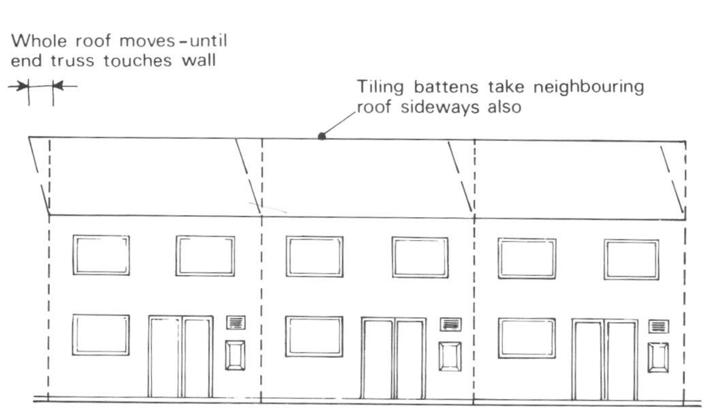

3 Roofs with inadequate diagonal bracing can move sideways until the roof truss meets the wall. Displacements of 100mm or more may occur. In terraces the tiling battens may take the neighbouring roof sideways.

8.1.1 ROOF LEANING AND DISTORTED

Typically gables and/or separating walls are out of vertical. The defect described here relates to timber roof structures.

Symptoms

The roof is seen to be leaning sideways, often with a displacement of 100 mm or more at gables and separating walls. The roof covering may also be displaced.

The problem may occur when tanks, flues and other items are present in the roofspace and are situated where binding and bracing should be. For spreading of the roof see 5.1.14 and 8.1.2.

Investigation

Record when the defect was first noticed and any changes that occur in the condition.

Check that no alterations have been made to the structural timbers.

Plumb the gable and separating walls at roof level.

In trussed rafter roofs – check the verticality of the roof trusses. Check also for the presence of bracing and binders as set out in BS 5268: Part 3 1985.

Check that any bracing and binders have been adequately fixed, being twice nailed to each truss member it passes.

Check that when bracing and binders are not long enough, that they are in two overlapping lengths and are lapped and nailed over at least two rafters.

Check that 3.35 × 75 mm galvanised nails have been used.

In purlin (and truss) roofs check the verticality of any trusses and check whether joints in purlins or trusses have opened.

Check that the end of the purlin is tied in to the gable.

Diagnosis and cure

This defect is caused by structural movement due to insufficient bracing and binding such that the roof structure functions as a single three dimensional structural unit.

Remedial work will depend on the findings of the inspection. If inadequate bracing and binders have been installed or wrongly positioned they should be corrected. New members fitted should be to BS 5268 Part 3.

Inadequate fixings should be replaced.

If any truss or wall is displaced 20 mm or more seek structural engineer's advice.

REFERENCES

BS 5268: Part 3: 1985 *Structural use of timber: Code of practice for trussed rafter roofs*

BS CP 112 *Code of practice for structural use of timber*: Part 3: 1972 *Trussed rafters for roofs of dwellings*

BRE Defect Action Sheet 83 *Dual-pitched roofs: trussed rafters; bracing and binders – specification*

BRE Defect Action Sheet 84 *Dual-pitched roofs: trussed rafters; bracing and binders – installation*

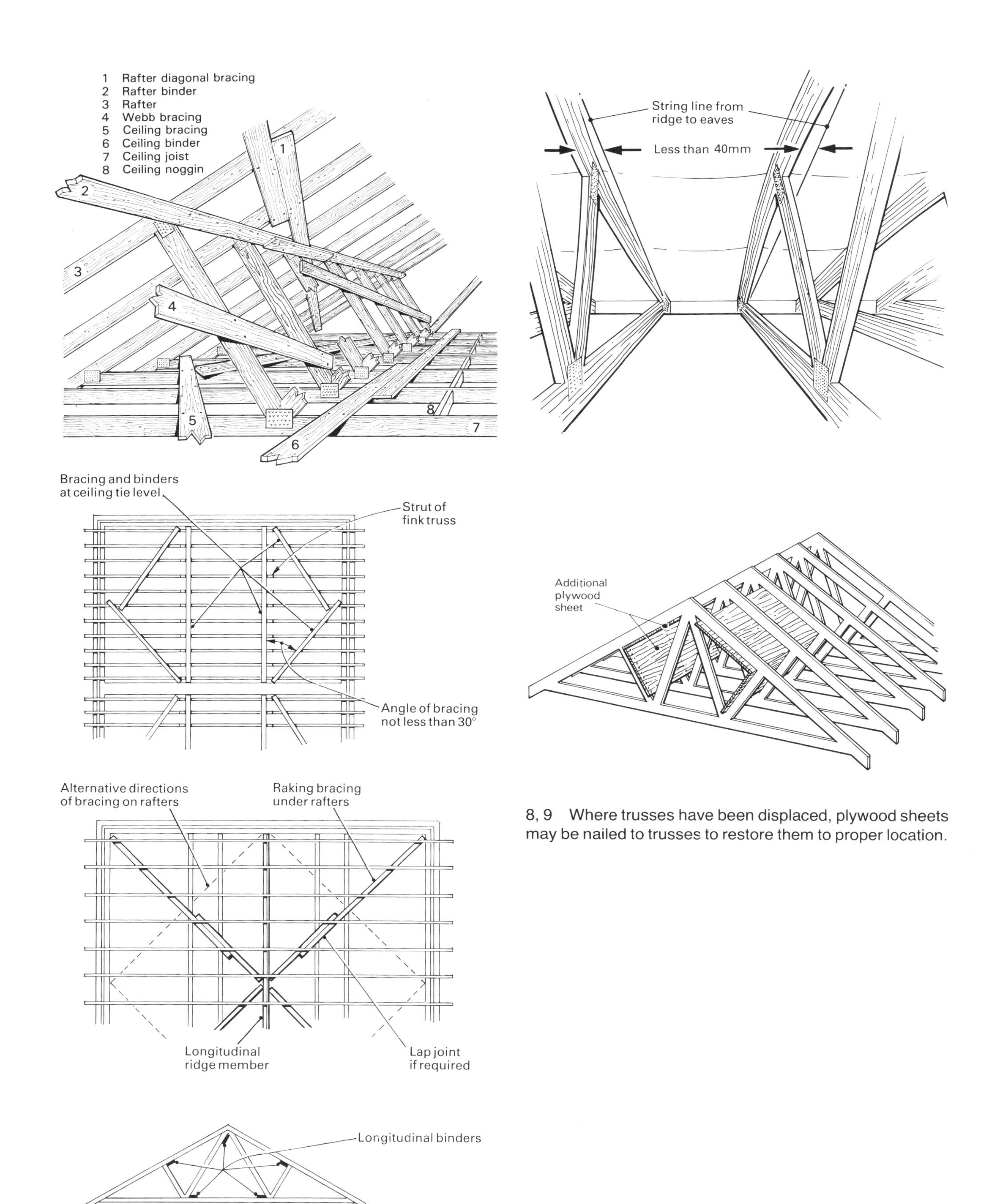

4–7 The bracing and binders required for fink truss roofs.

8, 9 Where trusses have been displaced, plywood sheets may be nailed to trusses to restore them to proper location.

1 Deterioration of the tie beam is causing this roof to spread at the eaves.

8.1.2 PITCHED ROOF SPREADING AT EAVES

This defect is more common in older buildings as modern buildings should have been designed with roof members so well connected that they are incapable of movement.

Symptoms

Horizontal cracks in the external walls of buildings near eaves level. These cracks are generally first noticeable on the inside face of the wall, and there is an outward movement of the portion of wall above.

Investigation

Record when the defect was first noticed and any changes of condition that occur.

Ascertain whether there has been any alteration made to the roof structure or change in the type of covering to the roof.

Plumb the walls at roof level to check the degree of outward movement that has taken place. Similar cracks may appear due to shrinkage of the wall plate.

Examine the walls for cracks on both faces.

Examine the roof structure for any signs of movement and weakness, especially at the joints of various members, and look for signs of beetle or fungal attacks.

Diagnosis and cure

The nature and extent of the cracking will give indications of the nature of the defect.

A single crack along the line of the timber wall plate is generally an indication of shrinkage of the wall plate, rather than roof spread.

Multiple cracks, particularly if there is also outward movement of the wall, are more likely to indicate roof spread.

Remedial work will depend on the findings of the inspection. Structural members with timber decay should be removed and renewed with fresh, preservative treated timber. The rest of the roof timbers should also be preservative treated.

If the wall is displaced 20 mm or more seek structural engineer's advice.

REFERENCES

BS 5268 *Code of practice for structural use of timber*: Part 2 1984 *Code of practice for permissible stress design, materials and workmanship*

BS 5950 *The structural use of steelwork in building*: Part 1 1985 *Code of practice for design in simple and continuous construction; hot rolled sections*

1 Concrete roof tiles on a sagging structure. The ridge on this roof is affected as well as the rafters.

2 Typical bowing rafters.

8.1.3 PITCHED ROOF SAGGING

This defect is more commonly found in older pitched and tiled or slated roofs.

Symptoms

The slating or tiling has a dished appearance, and the ridge may have also sagged.

Investigation

Record when the defect was first noticed and any changes of condition that occur.

Ascertain whether there has been any change in the covering material of the roof. Check that any new covering has not imposed a heavier loading on the roof than the previous one.

Examine the roof structure. Ascertain that no alterations have been made and look for any signs of movement and weakness, especially at the ends and joints of various members, and look for signs of beetle or fungal attack.

Check that the nails and other fixings used to join the structural timbers together are still sound.

Diagnosis and cure

There may be several causes for the sagging, acting independently or together.

Roof timbers may become bowed as a result of long term loading, and because of the culminated effects of dimensional changes (eg thermal expansion and contraction) to which a roof is particularly susceptible over time.

Nails and other fixings joining structural timbers may have corroded.

Deterioration of the ends of structural members has the effect of shortening the effective length of the timbers and causing displacement of other structural timbers attached to the affected member. This happens, for example, when timbers have become affected by wet rot, because they have been in contact with masonry which had become wet.

Timbers infested by beetle or fungal decay may have been weakened to the degree that they are no longer able to support the required loading.

Where the original roof covering has been replaced, the new covering may be so much heavier than the original that the design loading of the timbers has been exceeded.

Remedial work will depend on the extent of the defect, and the findings of the inspection. Unless the roof leaks, or there is a likelihood of the roof covering becoming detached or displaced, there may be no need for remedial work until the roof is approaching an unacceptable state of structural instability. The roof should, however, be inspected at regular intervals and measurements taken to ascertain whether the sagging is increasing, and at what rate.

If structural fixings have perished, they should be replaced, or new nails driven in to restore the connection.

If damp masonry is present, it should be dealt with (see chapter 5).

If fungal decay or beetle attack are present, they should be treated without delay. Any decayed sections of timber in structural members should be removed. In the case of wet rot attack, damaged ends may be replaced by splicing in new preservative treated timber pieces, and the rest of the roof timbers should be preservative treated. However, if dry rot has occurred, much of the timber may need to be replaced. Expert advice should be sought.

Once bowed, timbers often take on a permanent set. However, it may be possible to fix struts in suitable positions to support the roof and prevent further sagging, though they are unlikely to be useful in restoring the roof to its original condition.

If the roof covering has been replaced by one which has exceeded the original design loading of the timbers, it may need to be removed and a lighter covering substituted.

REFERENCES

BS 5268: *Code of practrice for structural use of timber*: Part 2 1984 *Code of practice for permissible stress design, materials and workmanship*

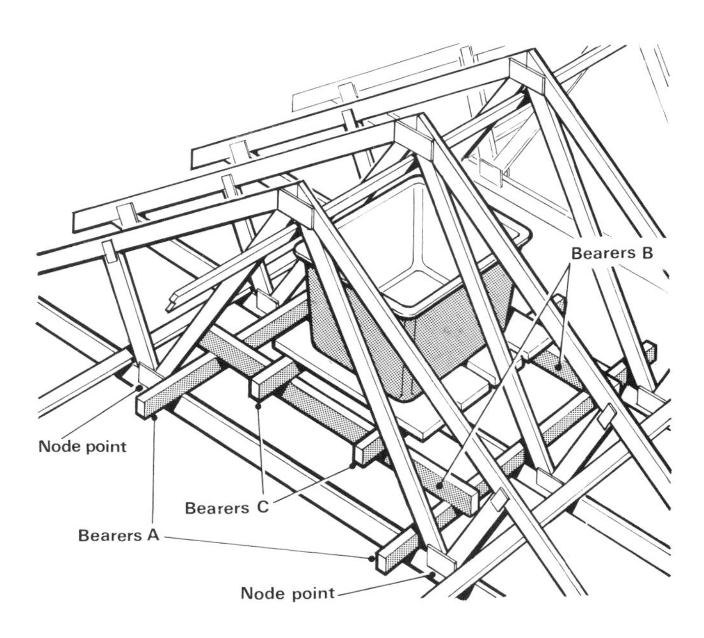

1 How a 230 litre water tank should be supported over three trusses.

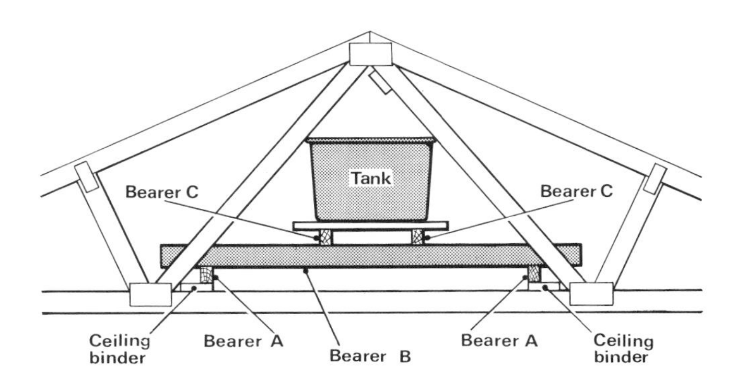

2 Section showing the locations of the bearers.

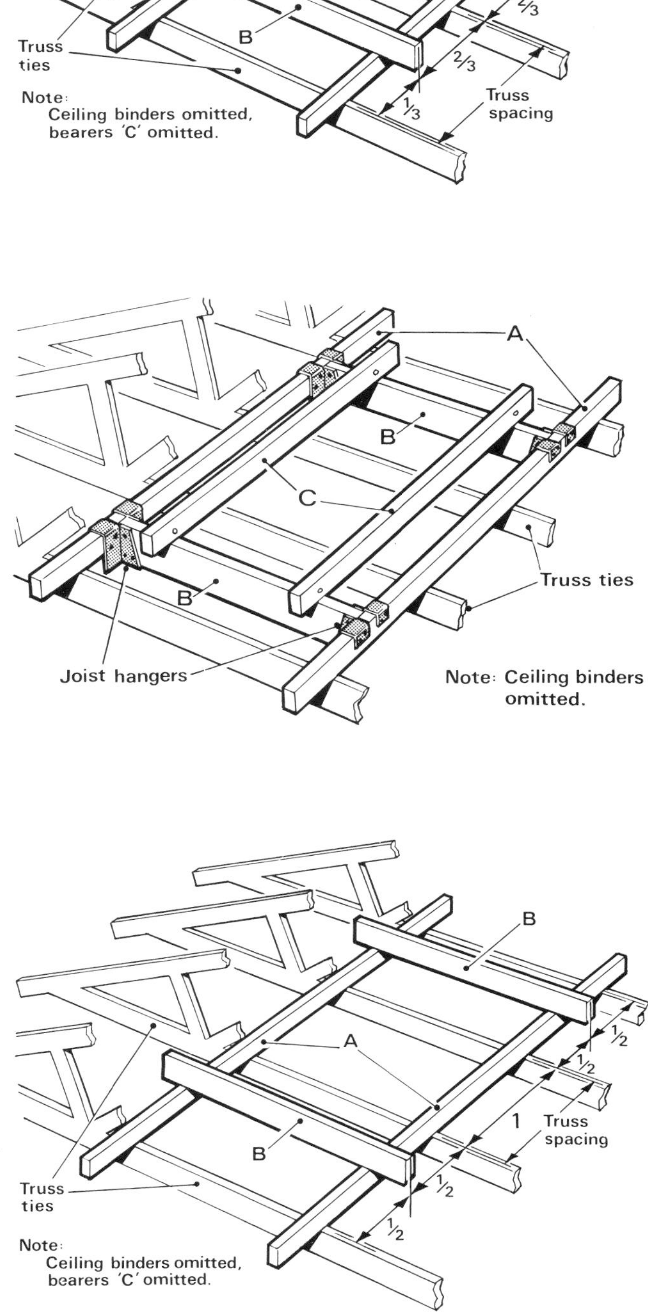

3 4 & 5 Alternative bearer layouts showing how they are supported and relate to trusses.
Bearers A sit on the trusses, Bearers B support Bearers C on which the tank sits.

8.1.4 TANK BEARERS IN TRUSSED RAFTER ROOF. INCORRECTLY SIZED AND POSITIONED

Symptoms

The trusses have distorted and may be damaged. The tank bearers may also have deflected and caused plumbing connections to leak.

Investigation

Check that water tanks are being carried on a sufficient numbers of rafters. 230 litre tanks should be supported by a minimum of three rafters, and 330 litre tanks by a minimum of four rafters.

Check that the tank bearers are the appropriate size for the tank, have been laid on edge and are positioned close to node points of trusses.

Check the type of material that the tank platform is constructed from. Chipboard should not be used as it may become wetted by condensation, plumbing leaks or rainwater ingress and lose its strength.

Diagnosis and cure

This defect is caused by a lack of adequate support for the water tank. The tank size and location should have been included in the structural design of the trusses.

If the bearers are of inadequate size or wrongly positioned, they should be corrected.

Specify that bearers and cross bearers are skew nailed to each other to prevent rotation.

REFERENCES

BS 5268 *Structural use of timber*: Part 3: 1985 *Code of Practice for trussed rafter roofs*

BRE Defect Action Sheets 43 *Trussed rafter roofs: tank supports – specification*

BRE Defect Action Sheets 44 *Trussed rafter roofs: tank supports – installation*

International Truss Plate Association Ltd. Technical Bulletin no 4 *'Tank supports, details and limiting spans'*

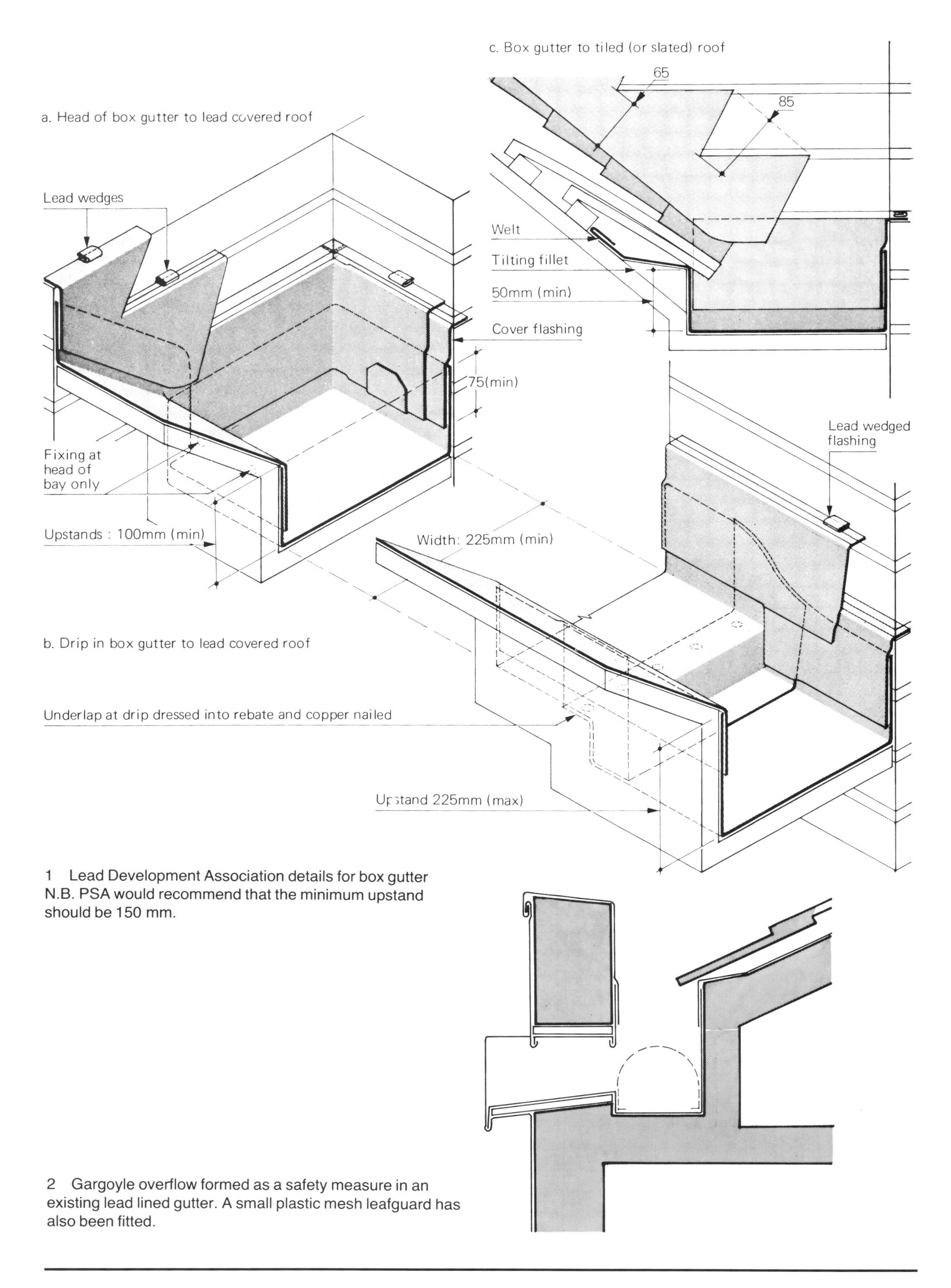

1 Lead Development Association details for box gutter N.B. PSA would recommend that the minimum upstand should be 150 mm.

2 Gargoyle overflow formed as a safety measure in an existing lead lined gutter. A small plastic mesh leafguard has also been fitted.

8.1.5 LEAKING LEAD-LINED PARAPET GUTTERS

Parapet gutters tend to be vulnerable to problems, particularly when they become blocked with debris or snow. It is essential that when such features are designed, flow calculations are undertaken to ensure that not only the gutter(s) are of sufficient size but also that an adequate number of outlets are provided. Where possible gutters should incorporate self cleansing falls.

Symptoms

Rain penetration where rainwater is being allowed to build up in the gutter(s) and seep through the joints.

Investigation

Record when the defect was first noticed and any changes that occur in the conditions.

Identify the reason for the gutters becoming blocked. If debris has accumulated in the gutters, eg leaves from overhanging trees, can measures be taken to avoid this happening in the future?

Check that the design capacity of the gutter is adequate, and that sufficient outlets have been provided.

Check that joints in the leadwork have been properly formed.

Diagnosis and cure

This defect usually occurs because gutters have been inadequately designed. The problem can be exacerbated by the presence of debris and inappropriate falls to the gutters. Excessive falls can lead to problems when outlets become blocked, while lack of sufficient fall can cause water to lie in the gutter.

Seal vulnerable joints in leadwork around cess boxes and in gutters by leadburning, or by the application of a suitable sealant.

If a gutter is vulnerable to blockage by the presence of fallen leaves, the provision of removable small mesh perforated plastic guards can ensure that the flow is maintained.

Consider providing gargoyle outflows through the parapet wall at each rainwater outflow position to provide an overflow relief system.
See also 8.2.3 for problems of condensation under leaf roofs.

REFERENCES

PSA Feedback Digest: *Parapet Gutters – Rain Penetraton.* Building Technical File Number 1 April 1988

Lead sheet in building – a guide to good practice Lead Development Association.

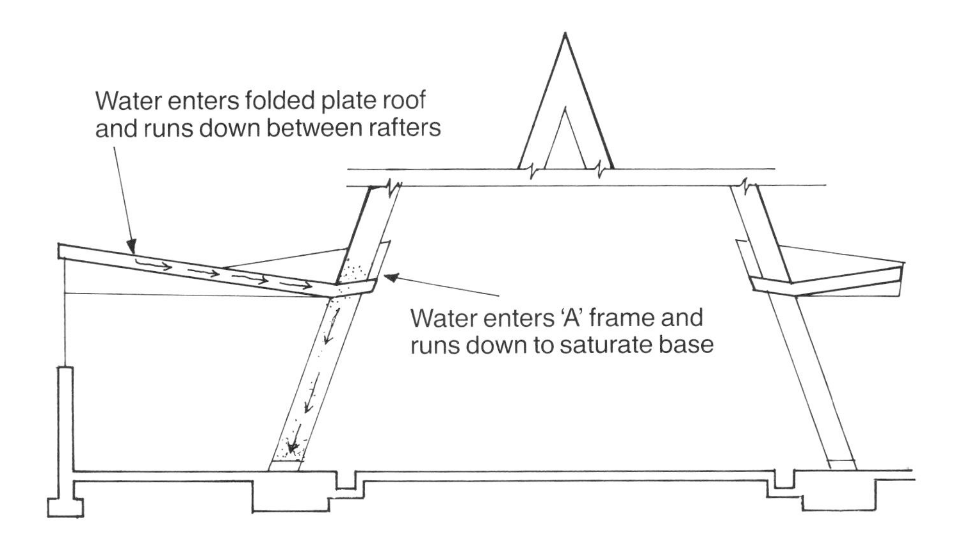

1 Diagrammatic section showing the way the folded plate lower roof drained towards the 'A' Frames.

2 Base of an 'A' Frame before being opened up. Note the screwdriver has penetrated the whole depth of the blade.

3 The rot behind when the facing timber was removed.

8.1.6 ROTTING STRUCTURAL 'A' FRAMES

Water leaking from a defective roof or roof junction, can be responsible for damage to other building components at some distance from the leak.

Symptoms

Serious rot occurring in structural timber 'A' frames beneath folded plate roof.

Investigation

Ascertain the construction of the building.

Record any visible signs of dampness and use an electrical resistance meter to establish the moisture content of structural members.

Locate the source of the dampness. Check whether it is caused by ingress of water from the roof or rainwater pipes from the roof. If rainwater pipes are encased, all the casings should be opened up.

Investigate the nature and extent of the rot in the structural members. For details of types of rot and treatment see 7.1.9.

Examine the timbers carefully for beetle attack.

Diagnosis and cure

Rot and other biological attack on timber occurs in the presence of dampness. Leakage from any roof over the structure and rising damp are the two most likely causes. Leakage from overhead tends to occur more frequently when the roof drains to internal outlets, or to valley gutters rather than to the perimeter. The rot is most likely to occur where dampness collects, ie at the lower end of the member which may be at some distance from the point of entry higher up. Where rainwater pipes are encased, and a leakage develops, damage may not be apparent until the casings are opened.

Where two roof planes meet and drain internally, investigate the feasibility of changing the fall and the rainwater outlets, such that rainwater drains away from the structure.

Seek specialist advice on treatment for rot or beetle attack.

If there has been degradation in the structural members seek advice of structural engineer.

REFERENCES

Construction Feedback Digest 54 *Repairing Rot in Structural Timber Frames*. B & M Publications (London) Ltd.

See also 7.1.4 for Diagnosis and treatment of biological attack on timber. See also 5.2.10 for dampness in Frame members of external walls

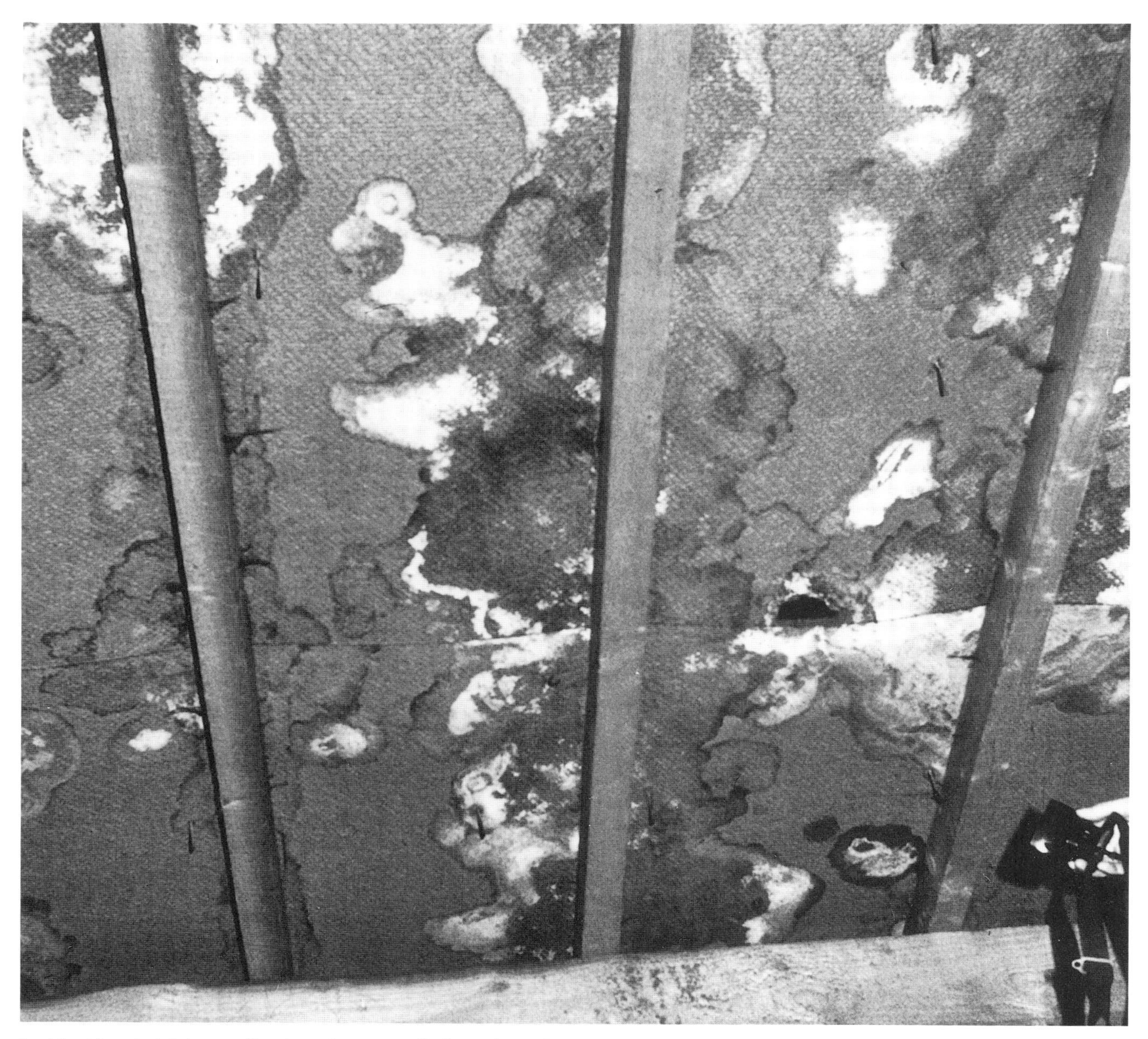

1 Mould and staining on fibreboard as a result of condensation.

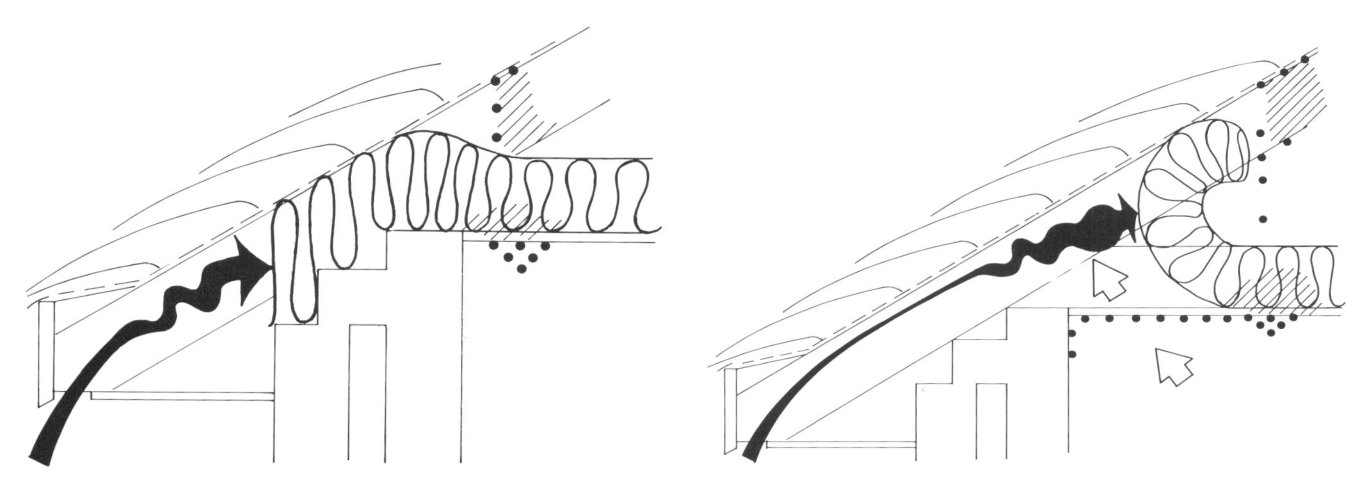

2, 3 Insulation which blocks off ventilation at the eaves may cause a build up of condensation in the roofspace, sufficient for the condensate to drip off the underside of the roof and to wet the ceiling below.

8.1.7 HIGH MOISTURE CONTENT OF THE AIR IN A ROOFSPACE

Typically this appears as dampness at ceiling level, often concentrated at the junction with the external wall. In some instances there may be decay in timber structural members.

Symptoms

Dampness at ceiling level, particularly where there are gaps in the vapour check or around roof hatch openings and service penetrations of the ceiling membrane.

Mould growth and other infestations of timber members within the roofspace.

Investigation

Record when the defect was first noticed and any changes in the condition that have occurred. Installation of insulation in the roof space may have aggravated the situation.

Check the amount of ventilation that there is in the roof space. Investigate whether the installation of insulation has blocked ventilation at the eaves.

Establish whether there has been any rain penetration into the roofspace.

Establish the form of wall construction and investigate whether the brickwork cavities have been sealed at the top. If there are insulated tanks in the roofspace, check that the lids are properly fitted.

Diagnosis and cure

The defect is caused either by condensation within the roofspace aggravated by insufficient ventilation or by rain penetration.

If there is timber decay in structural members, remove the decayed timber, treat the rest of the existing roof timbers with a preservative treatment, and renew the decayed members with preservative treated timber.

Seal the top of the brickwork cavity.

Provide ventilation at the eaves by means of openings along two opposite sides equivalent to a continuous opening of 10 mm for roofs above 15° pitch and 25 mm for roofs of 15° pitch and less.

If insulation is blocking ventilation at the eaves, install spacers to maintain a ventilation channel.

Provide tightly fitted covers to water tanks.

Install a vapour check in the ceiling and ensure all gaps are properly sealed, particularly where items penetrate the ceiling membrane, by the application of a sealant (a tight fit is not sufficient).

If there is a trap door into the roofspace it should be weatherstripped to prevent the transfer of warm moisture laden air from below.

REFERENCES

BS 5250: 1975 *Code of basic data for the design of buildings: The control of condensation in dwellings*

BRE Defect Action Sheet 3 *Restricting the entry of water vapour*

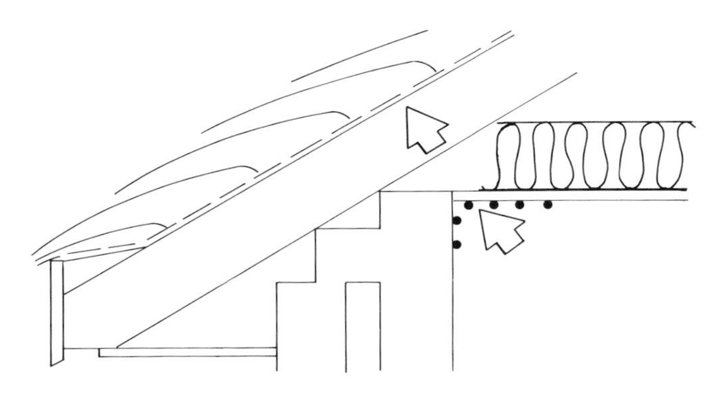

4 Insulation that finishes short of the wall structure may create a cold bridge with condensation forming at the junction of the ceiling and the wall.

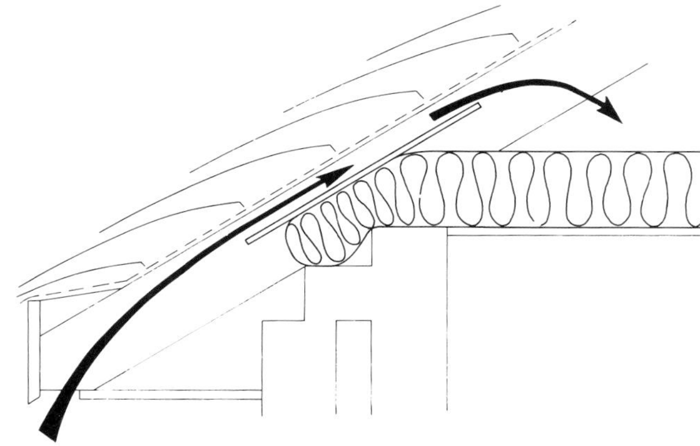

5 The correct way to terminate the insulation at the eaves, using a proprietary spacer tray that maintains the ventilation to the roofspace.

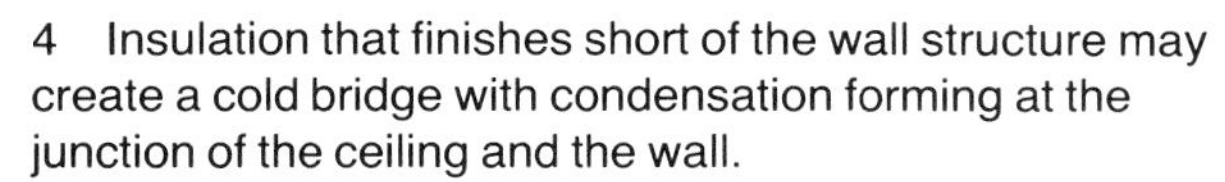

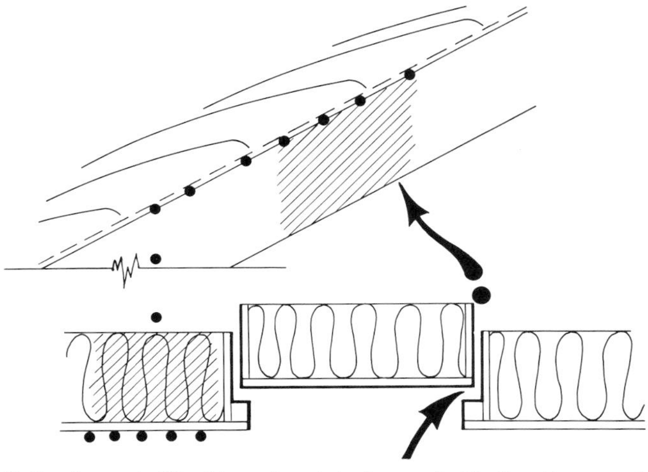

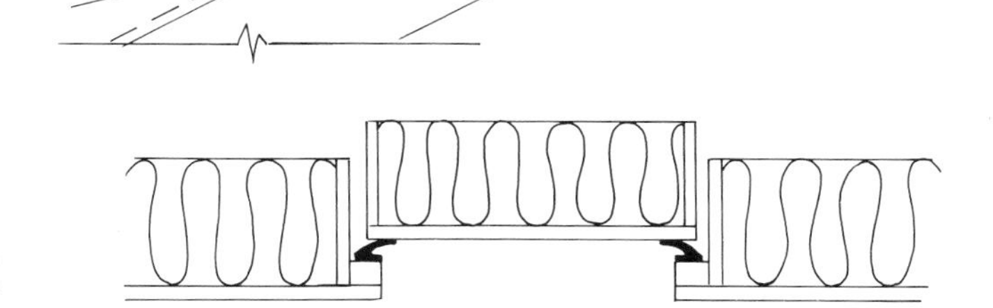

6, 7 A poorly fitted trap door into the roof will allow the penetration of warm moist air into the roof space that will condense when it meets colder temperatures. Weatherstripping the trap door and providing latches which keep the door tight to the frame should prevent this.

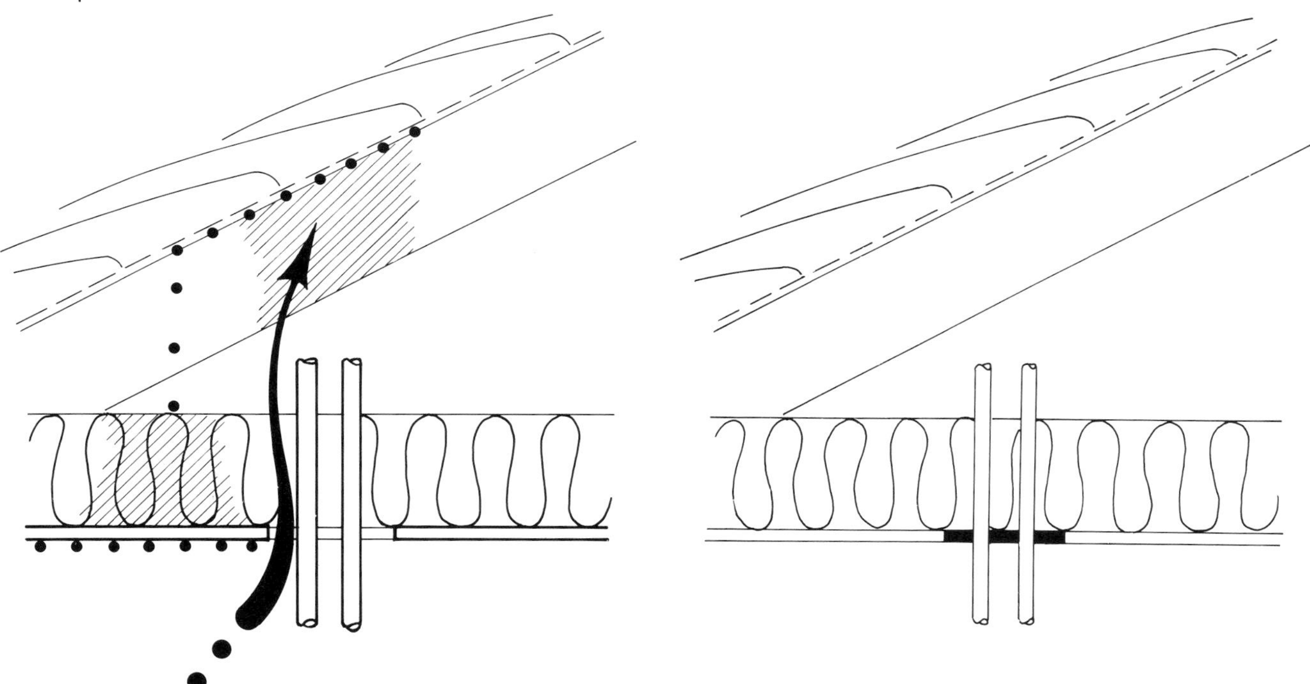

8, 9 When pipes penetrate the ceiling membrane a sealant should be applied to prevent any seepage of warm moist house air into the roofspace.

1 When sarking felt is not supported at eaves, water will collect and find its way into the eaves construction.

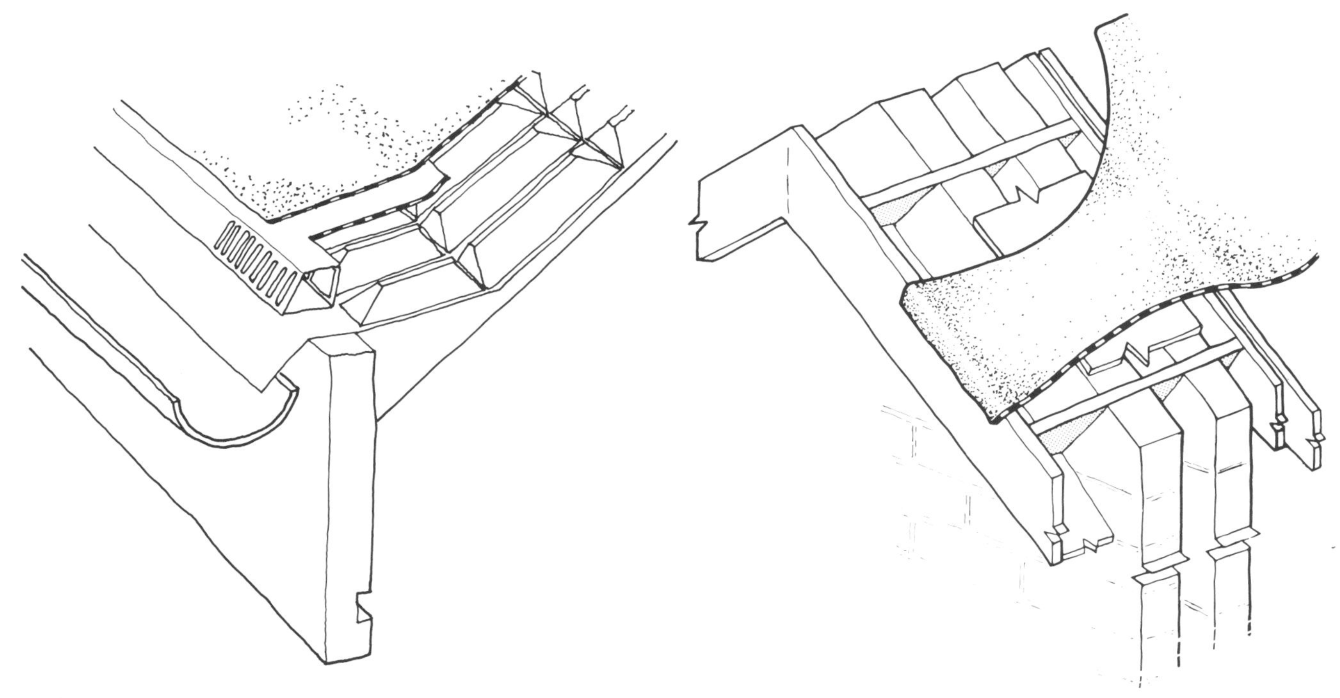

2 Sarking felt carried over eaves ventilating tray and fascia grill unit. Continuous support should be specified for the felt at the eaves.

3 Sarking felt should be carried over the top edge of bargeboards.

8.1.8 WATER PONDING ON SARKING FELT

Symptoms

Dampness and staining of ceilings.

Early and repeated paint failure on bargeboards.

Water penetration into eaves construction and sometimes into the internal walls of the building at the eaves.

Degradation of the facia board and other timber members at the eaves.

Investigation

Record when the defect was first noticed and any changes that occur in the condition.

Examine eaves and barge boards for presence of dampness using an electrical resistivity meter.

Remove bottom courses of slates or tiles to see if the sarking felt is properly supported behind the eaves facia board and around any penetrating soil vent pipes. Look for the absence of tilting pieces and for the presence of water ponding on felt behind the facia or bargeboard.

Check that laps in sarking felt are 100 mm minimum.

Diagnosis and cure

Faulty installation of the sarking felt and the omission of tilting pieces to support the felt over the bargeboard are the most likely causes of this defect.

If ponding is present install eaves sprockets or a continuous tilting fillet at the eaves under the sarking felt to support it.

Ensure the sarking felt is carried over the barge board.

If there is significant degradation of the felt at the eaves, renew it with a strip of more durable material, eg high performance felt or bituminous DPC material.

If sarking felt is damaged around soil vent pipes, install a further piece overlapping the original, with a cross cut for the pipe to pass through, and with a sealant bandage wrapped around the tongues.

REFERENCES

BRE Defect Action Sheet 9 *Pitched Roofs: Sarking felt underlay – drainage from roof*

BRE Defect Action Sheet 10 *Pitched Roofs: Sarking felt underlay – watertightness*

BS 5534 *Code of Practice for slating and tiling* Part 1: 1978 (1985) *Design*

1 Decay of nibs of roofing tiles due to magnesium sulphate efflorescence.

2 The displacement of the tiles adjacent to the hips in this photograph is due to the roof construction. The tiles have been laid on feather edge boarding with hip and ridge tiles solidly bedded. When the tiles were laid (during the 1930s) the timber was relatively green and over the years the boarding has dried out and shrunk. The solidly bedded hip and ridge have acted as an arch and have lifted the tiles.

8.1.9 SLIPPING AND DISPLACED ROOF TILES

This can either be the secondary effect of a structural roof failure (see 8.1.1–8.1.5) or can be caused by the failure of the fixing.

Symptoms

Tiles are seen to be slipped and displaced.

Investigation

Record when the defect was first noticed and any changes that have occurred.

Note the location of the tiles involved, and the nature and age of the fixing. Check that the fixing is appropriate to the pitch and exposure of the roof.

Examine samples of tiles that have slipped to check the condition of the nibs. Look for the presence of white crystals.

Examine the tiling battens to ascertain whether they have been subject to fungal attack.

Diagnosis and cure

If the failure is the secondary effect of a structural defect, this should be dealt with and then the affected tiles re-fixed.

If inappropriate or deteriorated nails or other fixings are present, they should be changed.

Ridge and hip tiles can move because nails or other mechanical fixings have not been used and they have been attached with mortar that has deteriorated. If this has happened the mortar should be replaced, and mechanical fixings installed.

If the nibs of tiles are found to have disintegrated, it is usually the result of crystallisation of salts that have been transferred by water from the exposed part of the tile to the upper part which is protected. This defect usually occurs when there is an excessive amount of soluble salts present in the tiles, which is likely in under fired tiles. Only some tiles in a batch may be affected. Examination of the roof will indicate whether the defect is general or confined to a few tiles. If only a few tiles are affected, they alone can be replaced, but if the problem is extensive, it may be preferable to replace all of them.

If tiling battens have been damaged by fungal or beetle attack, the reason for the attack should be ascertained, and damaged battens replaced with preservative treated ones.

REFERENCES

BS 5534 *Code of practice for slating and tiling* Part 1: 1978 (1985) *Design*

BS 402: 1979 *Specification for clay plain roofing tiles and fittings*

BS 473 and 550: 1971 (1980) *Specification for concrete roofing tiles and fittings*

1 The rear building shows a typical presentation of slipped and displaced slates.

2 Example of slipped stone slates caused by deterioration of the nail heads.

8.1.10 SLIPPED AND DISPLACED SLATES

Symptoms

Individual slates have slipped from their original position.

Investigation

Examine the condition of both the slipped slates and the others on the roof.

Check whether or not the nail holes are broken in the slipped slates.

Examine the fixings and note the material from which they are made and whether or not there are signs of corrosion.

Inspect the slating battens; look for signs of fungal attack.

Diagnosis and cure

When the nail holes in the slates are intact, the nails will usually have corroded and therefore no longer be able to hold the slates in position. If the nails over the whole roof are similar, it is likely that all the slates will be affected in time, and eventually the whole roof will have to be stripped and refitted using corrosion resistant nails. If the problem is found at an early stage and the roof has not been felted, consideration could be given to proprietory remedies that stick slates together from the underside.

Where the nail holes are broken, it is likely that the slates are deteriorating and unable to resist the chaffing action of the nails when disturbed. The condition may be aggravated by corrosion of the nails. The examination will show the extent of the problem. It is likely that the roof will need to be stripped eventually, and replaced with new slates conforming to BS 680.

If the battens have been weakened by fungal attack or woodworm, damaged battens will need to be removed and burnt. Specialist advice should be sought on treating the roof, and all new battens should be preservative treated.

REFERENCES

BS 680 Part 2: 1971 *Specification for roofing slates*

BS 5534 *Code of practice for slating and tiling* Part 1: 1978 (1985) *Design*

1 Typical example of a roof with spalling roof tiles.

2, 3. Example of pitted and spalling tiles. (Note the large particle of aggregate in the base of the pit.)

8.1.11 CLAY TILES LAMINATING AND SPALLING

Symptoms

Spalling of the top surface of the tile, and or lamination of the body of the tile.

Investigation

Ascertain the nature of the tiles, their age, and the weather conditions before the defect was observed.

Examine samples of spalled tiles and look for pieces of aggregate at the bottom of the 'pit' where the surface has been broken away.

Diagnosis and cure

This defect usually manifests itself following frost action and is most common with tiles that have a laminated structure, because of the method of manufacture. When a few tiles fail from this cause, it is probable that the majority will similarly fail in time.

Alternatively, tiles may fail because of defective manufacture. If tiles are underfired, there can be an excessive amount of soluble salts present in the tile which can lower their frost resistance. Only some tiles in a batch may be affected by underfiring, so replacement of the damaged tiles with sound ones may be a viable solution.

When tiles contain large pieces of limestone or silica aggregate caused by inadequate screening of the aggregate during manufacture, it can lead to spalling of the surface of the tile. This defect can be identified by the presence of the aggregate at the base of the spall pit. It is unlikely that existing sound tiles will suffer in the future as the changes which produced the spalling are probably complete. However, a very severe winter might cause additional deterioration.

All replacement tiles should be scrutinised before fixing to ensure they are sound and free from latent defects.

REFERENCES

BS 402: 1979 *Specification for clay roof tiles and fittings*

Construction Feedback Digest 34 *Defective clay roof tiles.* B & M Publications (London) Ltd.

Construction Feedback Digest 51 *Defective clay roofing tiles.* B & M Publications (London) Ltd.

1 Typical delaminating slates.

8.1.12 SLATES DELAMINATING

Symptoms

Pieces of the surface of the slate have fallen off.

Investigation

Whenever possible ascertain the country of origin of the tiles.

Note the likely presence of polluted air in the neighbourhood.

Carry out the BS 680 tests for water absorption, wetting and dryness, and for sulphuric acid immersion, on samples of slates removed from the roof.

Diagnosis and cure

The principle cause of this defect is attack by polluted air. Most British slates are very resistant to such attack, but some slates of European origin contain calcium carbonate (chalk) as an impurity. Acids in the air, attack the calcium carbonate leading to a breakdown in the structure of the slate.

Poor quality slates may be damaged by frost action, but such occurrences are rare.

The BS test will help identify whether all the slates in the roof are likely to be affected.

Damaged slates should be replaced with slates conforming to BS 680.

REFERENCES

BS 680 Part 2: 1971 *Specification for roofing slates.*

8.1.13 DETERIORATING CEDAR SHINGLES

This defect only occurs with some species of Western Red Cedar.

Symptoms

Instead of becoming silvery grey, the shingles become progressively darker. The shingles are weak and porous and leakage through the roof may be occurring.

Investigation

Examine the general condition of the shingles.

Diagnosis and cure

Some species of Western Red Cedar used for roofing shingles and shakes, have been found to be subject to fungal attack, particularly when specified in areas with high rainfall. This causes a breakdown of the wood so that it no longer prevents the entry of rain into the building.

The defect takes some years to reach a stage where action is needed. However, the only cure is to strip the roof and replace with new shingles that have been preservative treated. All shingles intended for use in the UK should now be supplied already preservative treated.

REFERENCES

Timber Research and Development Association PRL Technical Note No 3 *The preservation of Western Red Cedar shingles.*

1 Damage to copper due to chimney fumes.

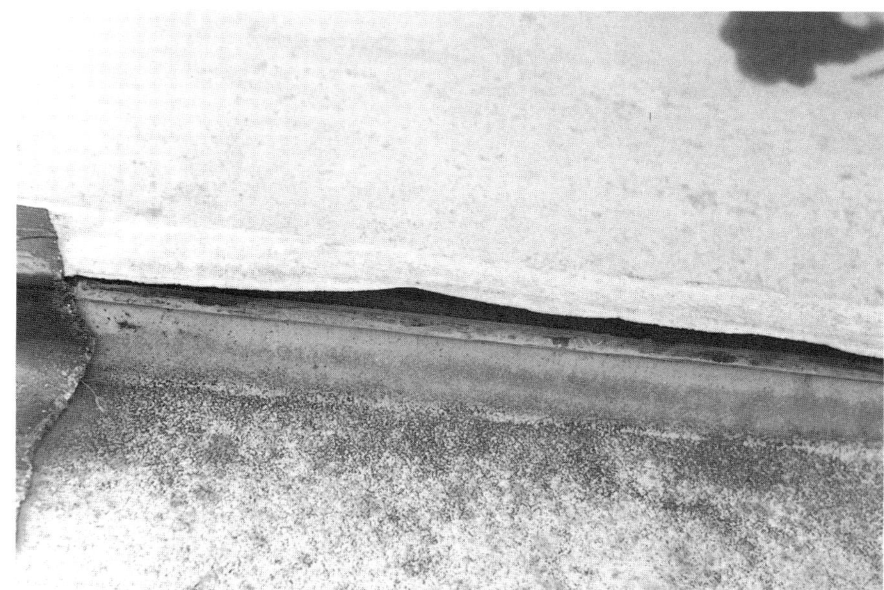

2 Detached copper sheet due to inadequate fixings combined with wind action.

3 Continual contamination from seagull droppings and inadequate patch repairs has resulted in the deterioration of this copper clad roof.

8.2.1 FAILURE OF COPPER ROOFING MEMBRANE

Deterioration and breakdown of the copper membrane of a roof will often be found to be aggravated by the installation being in an exposed location.

Symptoms

The copper membrane has corroded and thinned or split such that rain penetration has occurred. Apart from the splits, discolouration, surface pitting and copper roll joints perforating at welted edges are usually readily observed.

Investigation

Establish the construction of the roof and check that fixings appropriate to the exposure of the site have been installed.

Note the direction of the prevailing wind, and whether it is getting under the lap joints.

Check whether the roof has been subject to any pollutants. Particular attention should be shown to areas around any flues where acidic discharges may occur, and observe whether other chemical or biological agents are present. Seagull droppings on copper roofs in coastal locations have been known to cause severe deterioration of the metal.

Check whether a bitumen finish (either roofing sheets or paint) has been applied above the copper component, eg a flat bitumen roof on the top of a dormer where the sides have been copper clad. Bitumen reacts with the atmosphere and forms aggressive acids which attack the metal beneaath.

Have Red Cedar roof shingles been used where run off water might contaminate a copper component? Copper guttering under Red Cedar roofs has been found to be corroded within two years.

Diagnosis and cure

This defect is caused by; mechanical damage due to inadequate fixings for the exposure of the site, which have allowed wind under the membrane, or to chemical attack causing deterioration of the copper membrane, or a combination of the two. In either case the location and spread of the damage should give clues as to the cause.

If wind damage is established as being the prime cause of the defect, all cracked and damaged copper sheets should be removed and replaced.

Sheets adjacent to the gables should also be replaced even if they show no apparent damage, as they are the most vulnerable to wind uplift.

Replacement sheets should have batten roll joints at 400 mm maximum centres, and a minimum of two cross welts with central fixing clips. Double sided butyl or similar tape should be applied between welts to avoid the possibility of water ingress by capillary action.

If further defects start to develop in other sheets, small copper plates should be fixed centrally down all runs.

If the copper is damaged to the degree that replacement of the whole membrane is needed, consideration should be given as to whether an alternative covering should be specified, such as high performance felt or aluminium. If a copper finish is required, a fully bonded copper foil faced high performance felt may be specified. It should be fixed with two rows of copper nails at every 50 mm lap and at the head, and be laid on a fully bonded underlayer of high performance felt on a plywood substrate with the laps being made to protect the joint from the prevailing wind.

If the damage has occurred because of chemical attack, consideration should be given to whether the source of the contamination can be removed. For example in the case of a roof damaged by the discharge of acidic deposits from an oil burning appliance chimney, conversion to a gas burning appliance would dramatically reduce the deposits.

If the contaminant can not be prevented, for example seagull droppings, then the replacement of the copper with a more resistant membrane will be the only long term solution.

REFERENCES

Technical Note 25 *Jointing of copper and copper alloys.* Copper Development Association.

Technical Note 32 *Copper in roofing design and installation.* Copper Development Association

BS CP 143 *Code for sheet roof and wall coverings*: Part 12: 1970: copper metric units

1 Damaged flashing and membrane alongside box gutter.

2 Copper membrane removed from bed of gutter to show rotted support underneath. Although this example is of a copper lined gutter similar damage can occur under lead lined gutters.

8.2.2 UNSUPPORTED BOX GUTTER

This defect often occurs in conjunction with deterioration of the roof membrane. See also 8.1.5.

Symptoms

Distortion of the gutter due to lack of support from failed timber structure.

Investigation

Establish the construction of the roof, and examine for possible water penetration through the membrane and surrounding structure.

Remove sections of the roof covering and gutter in the vicinity of the defect to establish the state of the underlying structure.

Establish the measures that have been taken to ventilate the roof void.

Diagnosis and cure

Leaks in the roof membrane and through the surrounding structure such that the timber supports of the gutter are permanently wet, are likely to lead to rotting of the timber, particularly when accompanied by lack of ventilation in the roofspace.

Make sure the design of the gutter is adequate for the location and that sufficient outlets have been provided.

All decayed timber should be removed and burnt, and every part of the void sprayed with a fungicide to destroy any possible spores prior to reinstatement of the gutter-bed with preservative treated timber.

If the ventilation to the roof void is found to be inadequate, air movement should be created beneath the gutter deck.

REFERENCES

BS CP 143 *Code for sheet roof and wall coverings*: Part 12: 1970: *copper* metric units.

Construction Feedback Digest 57 *Copper Roof Failure: MOD Church. B & M Publications (London) Ltd.*

Technical Note 32 Copper in roofing design and installation. Copper Development Association

1 Nail left in bottom of roll after dressing to the roll, has held the lead too tight and lead has split at nail and cracked at bottom of the roll letting water to the timber structure.

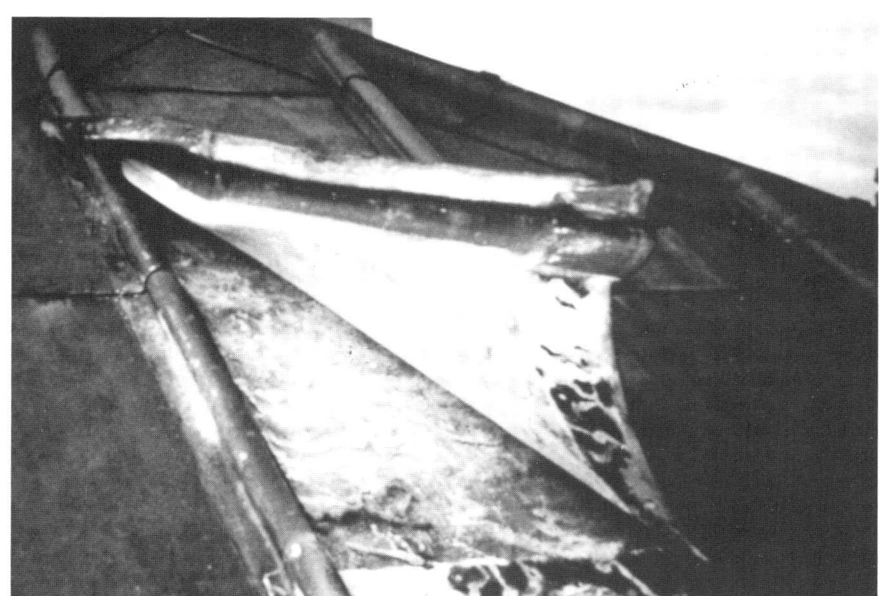

2 Lead carbonate on the underside of a lead sheet roof caused by interstitial condensation. The roof deck is also degrading.

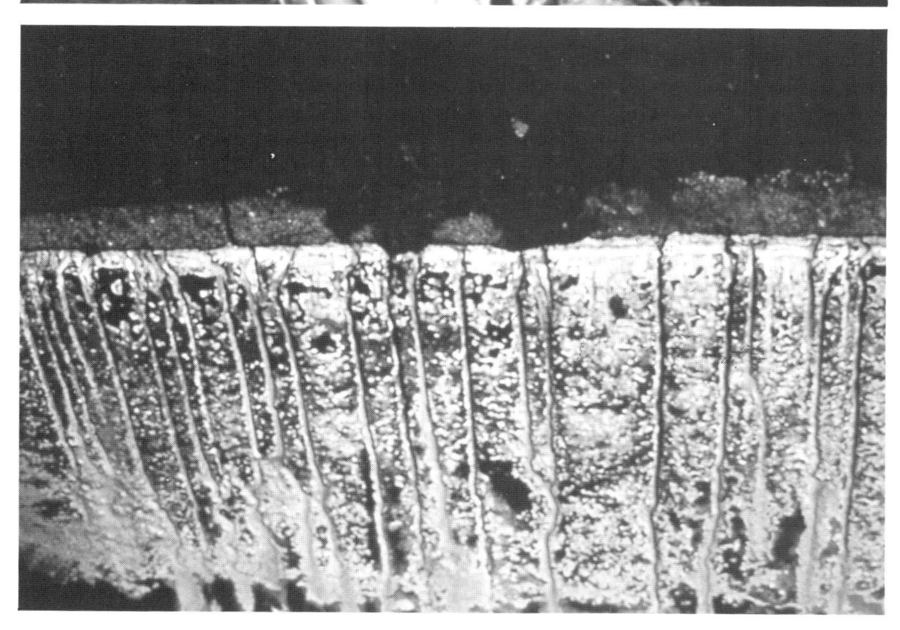

3 Lead flashing corroded by lichen run off.

8.2.3 FAILURE OF LEAD ROOFING

These failures detailed have all occurred within ten years of the roofing being installed.

Symptoms

Cracking of the lead.

Leakage occurring during heavy rain and high wind.

Blistering on the external surface of the lead with white salts that readily fall off, appearing on the underside.

Narrow clean cut grooves or round spots at water drip off points.

Investigation

Establish the construction of the roof.

Ascertain the code of lead used (ie the thickness) and measure the sizes of the sheets.

Note the details that have been used at joints, (pitches under 10° should incorporate 50 mm high drip joints, those of 10° or more, lap joints); and the location and form of the fixings.

If traditional wood cored rolls have been used, ascertain whether they have been installed correctly and whether it is possible for water to build up against the rolls under the splash lap.

Establish the nature of the material that has been used as an underlay.

Investigate the provision of heating in the building and the insulation and vapour barriers that are present. Has any ventilation been provided in the roof construction?

If the lead has corroded, arrange for a laboratory analysis of a sample to ascertain which agent has caused the failure. If the sheets have been sized and fixed correctly and failure has still occurred, arrange to have the lead analysed for the percentage of copper content.

Check that rainwater outlets are free from blockages.

Diagnosis and cure

Lead is normally an extremely durable material but inappropriate detailing and specification may cause premature failure.

All lead detailing is weathertight, rather than watertight, therefore if standing water is allowed to build up in gutters or at abutments, leakage is inevitable.

The various reasons for failure are outlined in turn.

Fatigue caused by insufficient allowance for the movement of the lead.

These are the most frequent form of defect in lead roofing. Lead has a coefficient of expansion of 0.0000293 per °C. The two common causes of such failure occur because either the sheet used is too large a size for its thickness; (see BS table of maximum sizes for sheets according to lead Code No) or because too many fixings have been used. The thinner the lead the smaller the sheet that can be used. When old lead roofs are being replaced with modern milled lead, a thinner code is sometimes used. The detailing will need to be changed to take account of smaller sizes of sheet possible with the thinner code (this is sometimes overlooked). When fatigue joints occur in gutters because lead has been used in lengths too long for its code, it may be necessary to redesign the substrate to accommodate more drips. If this is not possible, a heavier code lead should be specified.

Nail fixings in lead should be restricted to the top third of the sheet. It is important that the fixing pattern at the top of the sheet is of two rows of nails 25 mm apart, at staggered centres of 75 mm. The top row should be 25 mm from the edge of the sheet. If the fixings are further apart, or the top row of nails is too close to the edge of the sheet, premature failure is very likely. Nails should be copper or stainless steel clout nails with jagged or annular ring shanks, of not less than 10 gauge and not less than 25 mm in length. Fixings made of other metals are likely to corrode too quickly and not be as durable as the lead.

Nail fixing of undercloaks should be restricted to the top third of the bay. Check that nails inserted to hold the lead while the lower ends of rolls are dressed have been removed.

Fatigue cracking caused by restricting the movement of the lead can be repaired with a welded lead patch. However, if no action is taken to remedy the underlying cause(s), ie oversizing and over fixing, the defect will recur.

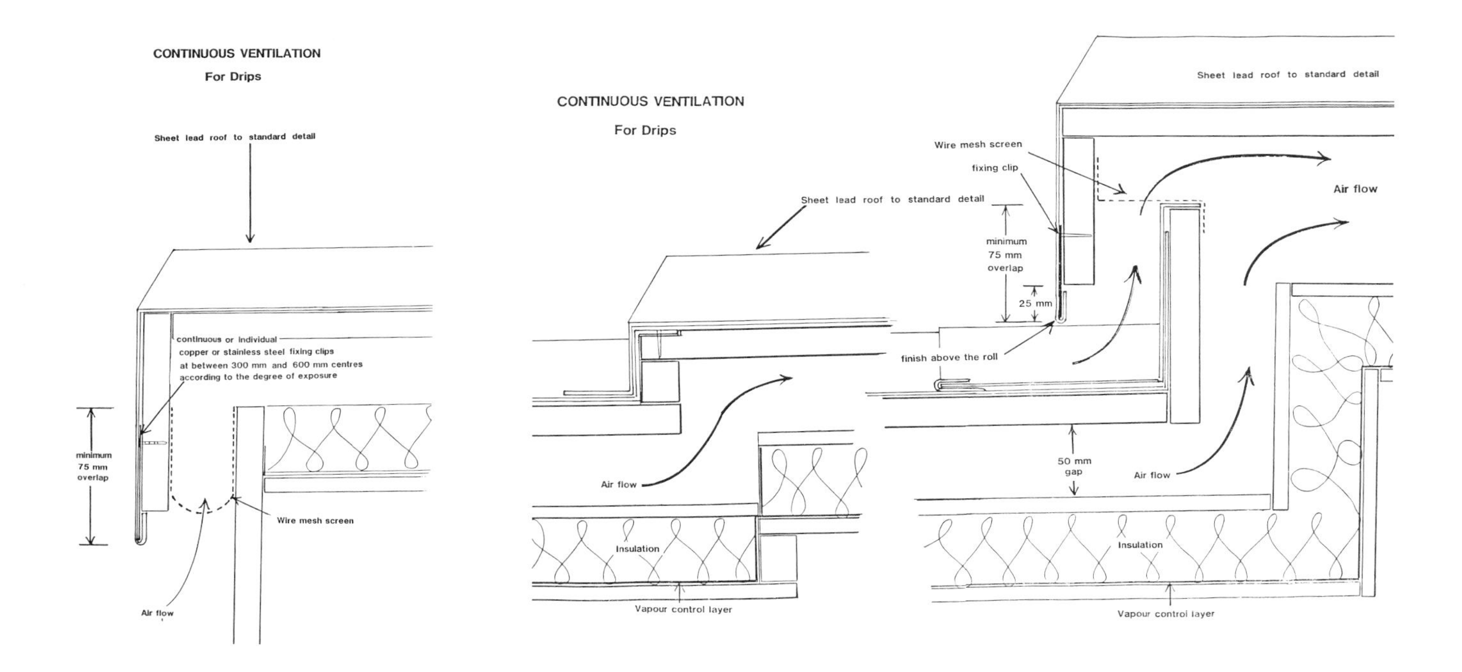

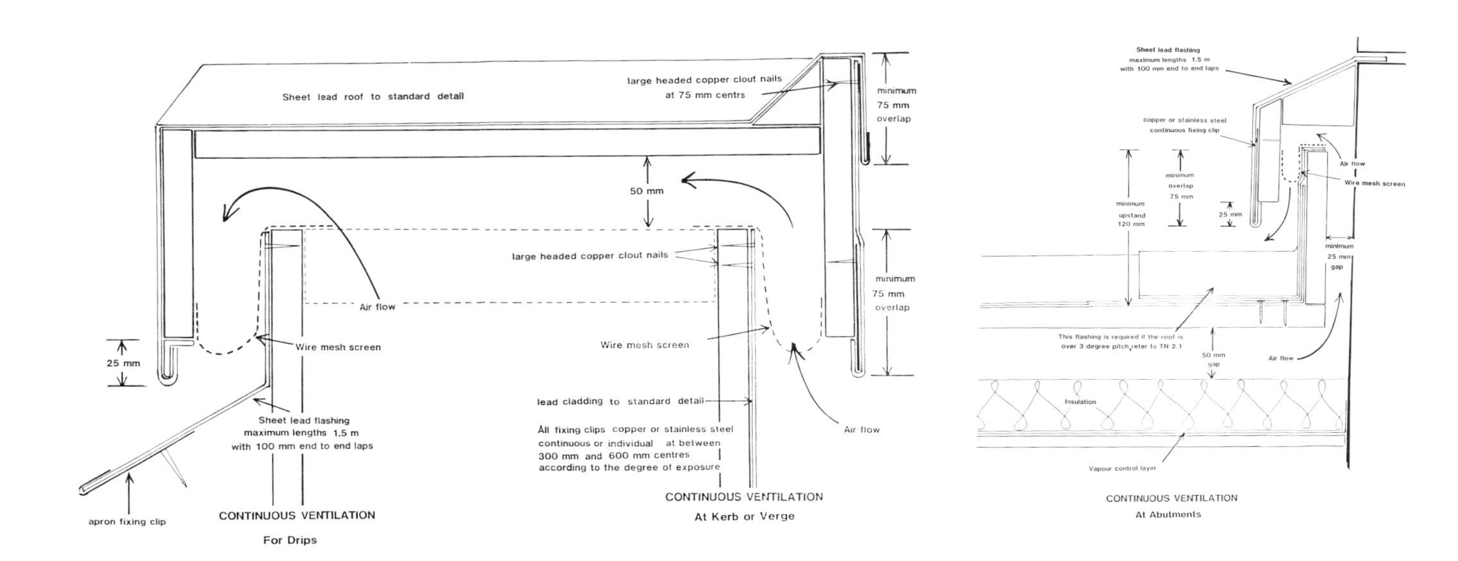

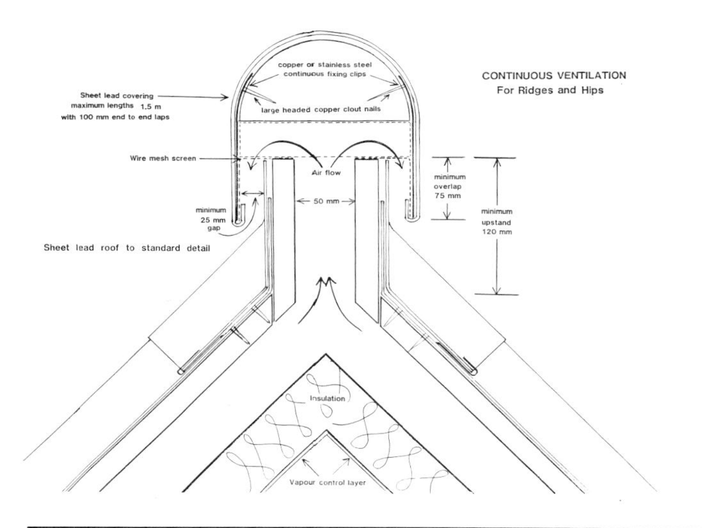

4–8 Schematic detail showing how to provide ventilation over insulation for sheet lead roofs.

continued

A further contribution to cracking due to restricted movement can be traditional felt underlay. In recent years these have been found to be increasingly unsatisfactory, for two main reasons. First, the organic fibres in the felt may rot due to condensation (see below). Second, because of the higher temperatures that lead sustains resulting from better insulation, the resin binding in the felt can be leached out, causing adhesion between the metal and the substrate. This can lead to failure. If this has occurred, the lead will need to be stripped and replaced, using either a BS 1521 Class A building paper or a needled non woven polyester textile (of a weight of not less than 210 g/m^2). The polyester textile (sometimes referred to as a geotextile) is considered to have better moisture permeability, and is also thought to aid the drying out of condensation that might form on the underside of the lead.

Poor detailing and workmanship.
When leaks are weather related, and not due to fatigue cracking, it is probable that the construction of the roof is faulty.

If the timber rolls have been installed such that they finish short of the roof edge, it is probable that rain is penetrating at the point where the rolls are terminated. This should be corrected by stripping the roof and installing rolls of the correct length.

If water can build up against wood cored rolls because for example, the timber substrate has been laid with incorrect falls, a combination of wind velocity and capillary attraction can force it over the roll and create leakage. This can be remedied by cutting off the splash lap, and welting the trimmed edge of the overcloak to stiffen the free edge and engage the sheet lead retaining clips. The overcloak should finish well above the likely water level.

A third problem that occasionally occurs with wood cored rolls is that the lead is overworked. When forming the detail, the lead becomes stretched as it is tucked in at the bottom of a roll. The stretched area can reduce the thickness of a Code 7 sheet locally to say Code 2 or 3 which then fails. A welded patch of the appropriate Code for the roof will provide a satisfactory repair.

Leakage at drips often occurs because the drip has been formed with less than 50 mm height.

Leakage occasionally occurs through fixing holes. This is most likely when fixings are exposed or when they are located under flashings, but within the reach of capillary movement of rainwater. This can be prevented by small lead patches being welded over the fixings.

Condensation

Where blistering of the lead and the formation of salts on the underside is occurring, it is most likely that entrapped water or condensation is the prime cause. The latter often happens following the installation of a new or improved heating system. Two processes may then occur. The action of the condensate (which is distilled water) can slowly convert the lead to lead carbonate which, in time, can eat its way right through the lead sheet. Alternatively, the condensation may attack a timber roof structure. In the case of hardwood structures, the condensate can leach acids out of the wood (eg tannic acid from oak, which may be in the form of fumes when the timber is new), whereas with softwood soluble sugars are leached out. Both will cause corrosion of the lead. This problem should be remedied by providing a vapour barrier with correctly sealed joints on the warm side of the construction, along with sufficient thermal insulation to prevent the formation of condensation. It is important however, that when specifying a warm deck roof under lead, that an additional ventilated air space is provided between the insulation and the boarding on which the lead is fixed (see illustration). The reason for this is that when a hot metal roof covering is suddenly cooled by a summer shower, air present beneath the lead is also cooled and a sub-atmospheric pressure created by the warm roof construction may develop, which can result in rainwater being pulled through the expansion joints – particularly lap joints. The additional ventilation space prevents this sub-atmospheric pressure occurring, and will also allow for the dispersal of any condensate that does arise. The air flow through this ventilation space should be continuous from eaves to ridge (see details). Vents located solely at the top of the roof should never be specified, as they can create negative pressure zones that can cause warm and moist air from underneath to be 'pulled' into the roofspace.

continued

Corrosion

Apart from corrosion occurring because of condensation or from contact with a timber substrate detailed above, the only prevalent source of corrosion to lead roofs is attack by the acid run off water from lichen. The problem occurs primarily in gutter linings or on flashings, and takes the form of either grooving or spotting. The acidic water dissolves the protective patina. Repeated dissolving and reformation of the patina eventually results in failure, although it may be many years before failure occurs. The ideal solution is to get rid of the lichen – see BRE Digest 139 for suitable toxicants. An alternative method of removing lichen is by the insertion of copper strips at every ten courses of tiles or slates. The copper salts present in the rainwater run off prevent the formation of lichens and mosses. If for aesthetic reasons it is considered desirable to retain lichen or moss growths, a sacrificial flashing should be provided at the drip off points to protect the weatherproofing lead sheet. Such flashings, which are simply strips of sheet lead inserted over the gutter lining, will often last for thirty plus years, and can be renewed at a fraction the cost of the weatherproofing sheet underneath.

Where cracking has occurred in the lead, and fatigue cracking as described above is not the cause, analyse a small sample to check the copper content. If the copper content is less than 0.03%, the lead may be too pure for use in roofing. It can become prematurely work hardened which results in fatigue cracking caused by thermal movement. BS 1178 was amended in 1982 to take account of this factor, so this type of failure should not occur with post 1982 roofs.

REFERENCES

Construction Feedback Digest 35 *Lead roofing problems.* B & M Publications (London) Ltd.

BMCIS Design Performance Data. Building Owners' Reports: 2. *Flat roofs.* No 15 – *Defective construction of lead rolls.* No 18 – *Incorrectly lapped lead sheet roof.* Building Maintenance Cost Information Service

Lead sheet in Buildings. 1978 Lead Development Association

Leadwork No 5 January 1988

BRE Digest 139: *Control of lichens moulds and similar growths.* HMSO

BS 1178: 1982: *Specification for milled lead sheeting for building purposes*

BSCP 143: *Code of practice for sheet roof and wall coverings* Part 11: 1970. *Lead.*

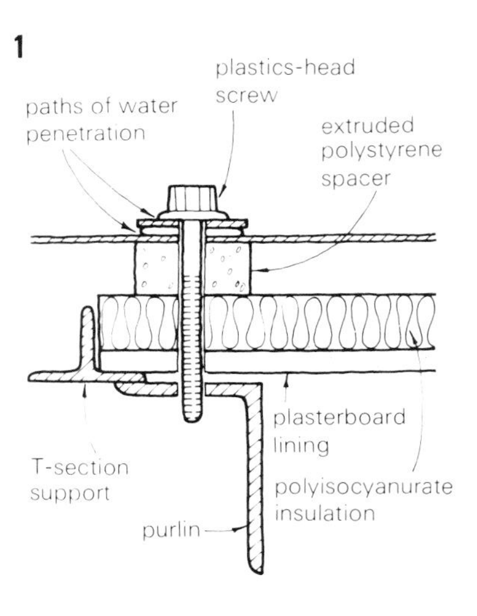

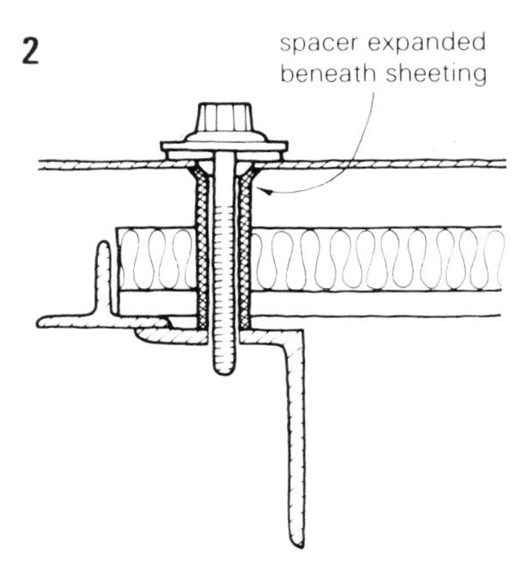

1 Detail of how water penetrates inadequate fixing.

2 The use of 'spacer' screws prevents overtightening of the fixing.

8.2.4 WATER INGRESS THROUGH PROFILED METAL ROOF AT FIXING POINTS

Profiled metal roofs are most vulnerable to water leaks at fixing points and at lap joints.

Symptoms

Rain penetration through fixings.

Investigation

Establish the construction of the roof, particularly the nature of the fixings and any spacers used.

Check that fixings have been correctly tightened.

Check that the pitch of the roof is appropriate for the material and the exposure rating of the site.

Note whether air pollution is present as it may cause corrosion of either the sheeting or the fixings.

Diagnosis and cure

Leakage through fixings usually occurs because the fixing has loosened. This can be caused by:

- corrosion of the fixings
- corrosion of the metal sheet at the location of the fixings
- the use of inappropriate fixings
- the use of inadequate spacers (such as expanded polystyrene)

Examination of the roof and fixing should determine the most likely cause. Even if only one sheet in a roof appears to have loosened, all the fixings on the roof should be checked, since it may be only a matter of time before more are affected.

If bimetallic corrosion has occurred, it may be necessary to use more suitable washers, or to use sleeves to prevent contact between the metals.

It would be wise to consider whether the washers are of adequate size for the particular conditions.

If the roof covering has become corroded to the extent that the fixing holes are enlarged, a new roof covering will usually be required. Careful consideration should be given to the suitability of the replacement cover in the light of the conditions that caused the original defect.

REFERENCES

BS CP 143 *Sheet roof and wall coverings.* Part 1: 1958: *Aluminium corrugated and troughed.* Part 15: 1973 (1986) *Aluminium* metric units

Construction Feedback Digest 41 *Aluminium profiled sheet roofing failure.* B & M Publications (London) Ltd

Construction Feedback Digest 43 *Roof leaks via fixings.* B & M Publications (London) Ltd

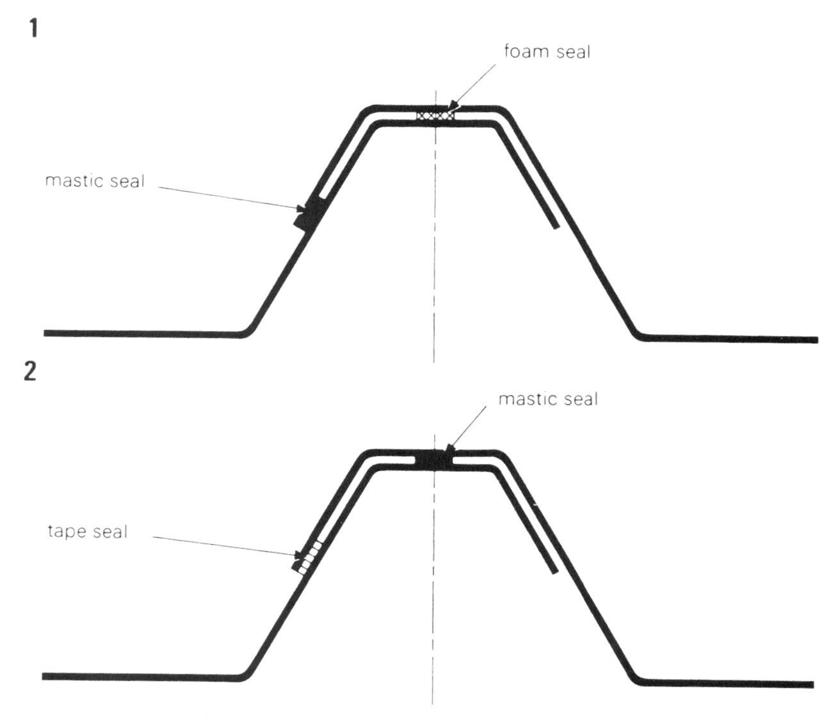

1 Incorrect location of mastic sealant in lap joint.

2 Correct location for mastic sealant in lap joint.

8.2.5 WATER INGRESS THROUGH PROFILED METAL ROOF AT LAP JOINTS

Profiled metal roofs are most vulnerable to water leaks at fixing points and at lap joints.

Symptoms

Rain penetration through lap joints.

Investigation

Establish the construction of the roof, particularly the extent of overlap of sheets and the measures that have been taken to seal the lap joints.

Check that the pitch of the roof is appropriate for the material and the exposure rating of the site.

Note whether air pollution is present as it may cause corrosion of either the sheeting or the fixings.

Diagnosis and cure

Leaks through the lap joints of sheets are commonly caused by inadequate overlaps, or sealing of overlaps, particularly when the pitch is shallow and the roof exposed to above average weather conditions. Alternatively, if the roof becomes subject to loading not envisaged at the design stage, such as foot traffic, this can cause laps to gape. Whenever leakage occurs through laps, it will be necessary to lift the upper sheet and reapply sealant between the sheets to obtain a satisfactory repair. Applying beads of sealant to the edge of lap joints will only effect a temporary repair, and therefore is not considered to be good practice.

REFERENCES

BS CP 143 *Sheet roof and wall coverings.* Part 1: 1958: *Aluminium corrugated and troughed.* Part 15: 1973 (1986) *Aluminium* metric units

Construction Feedback Digest 41 *Aluminium profiled sheet roofing failure.* B & M Publications (London) Ltd

Construction Feedback Digest 56 *Lap joints in low pitched profiled roofs.* B & M Publications (London) Ltd

1 Top of steel roofing sheet showing holes from rusting.

2 The reverse of the same sheet showing major rusting.

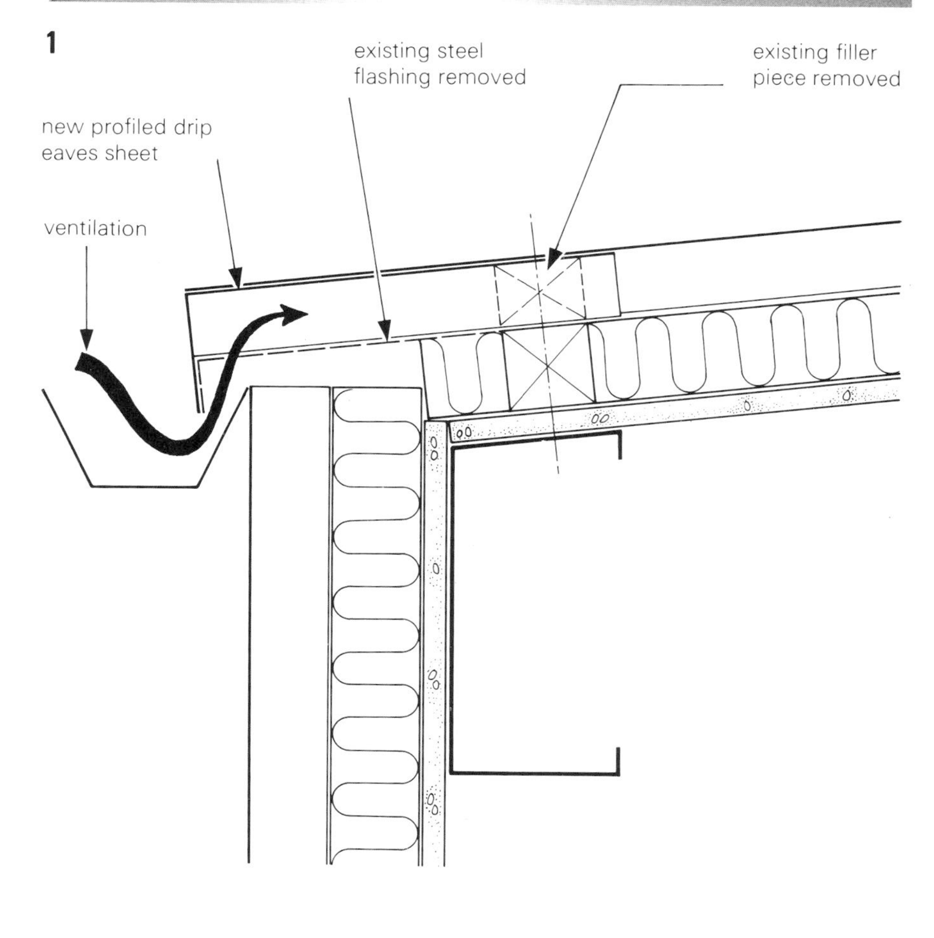

3 Detail showing provision of eaves ventilation for profiled metal roofs. This was remedial work to counter condensation within the roof void, similar ventilation provision was made at the ridge.

8.2.6 CONDENSATION ON UNDERSIDE OF PROFILED METAL ROOFING

Symptoms

Dampness and staining of ceilings.

If the underside of the metal sheeting is not suitably protected, the condensate can cause corrosion and deterioration of the sheet, which may not show before failure has occurred.

Investigation

Establish the construction of the roof and where present the ceiling below. Note whether a vapour barrier has been provided and whether it is fully continuous.

Examine the provisions that have been made to ventilate the underside of the sheeting.

Check whether sufficient ventilation has been provided to dissipate any water trapped in the roof space, either during construction or by previous leaks.

Investigate whether there is any water vapour generating activity within the building, and the measures that are in use, if any, to disperse the vapour.

Diagnosis and cure

Generally, condensation in this position can be caused either; by moisture vapour which has arisen within the building (such as that produced by showers penetrating the ceiling membrane); or by the moisture present in external air. During clear cold nights the metal sheeting can be at a significantly lower temperature than that of the external air, leading to the formation of condensation on the surface. This will be aggravated when an insulation layer is separate from the sheeting, and positioned beneath it.

It is likely that this problem is the result of both forms of condensation described above.

If moisture generated within the building is considered to be a major factor, install local ventilation to reduce the problem.

Increase the ventilation of the roof void by installing improved ventilating filler pieces in the eaves, and where appropriate, in the ridges.

If the maximum ventilating filler pieces are already installed and the problem still persists, mechanical ventilation of the roof void should be installed.

REFERENCES

Construction Feedback Digest 41 *Condensation on profiled metal roofing.* B & M Publications (London) Ltd

Construction Feedback Digest 59 *Improved specification for profiled steel and cladding* B & M Publications (London) Ltd

BS 5250: 1975 *Code of basic data for the design of buildings: The control of condensation in dwellings.*

1 Deteriorated coating on profiled steel roof with some rusting occurring.

8.2.7 CORROSION OF STEEL SHEET ROOFING AND CLADDING

This is primarily a problem that occurs with steel sheet roofs when the protective coating breaks down. It is more common with shallow pitched roofs.

Symptoms

When the protective coating on the upper (external) surface of a steel sheet roof breaks down, rusting will show. It will cause the failure of newly applied coating systems if the rust is not adequately removed before the coating is applied.

However, some proprietary coated steel sheets have different coatings to the external and internal faces. These sheets are more susceptible to damage on the underside caused by either rain penetration at side and end laps, or condensation (see also 8.2.5). The damage is often not apparent until the corrosion has eaten right through the metal.

Investigation

Establish the construction of the roof and the type of protective coatings that have been applied to both sides of the metal.

If the roof surface has had remedial protective coatings applied, ascertain the specification, and establish the method of preparation of the surface undertaken prior to applying the coating.

If condensation is suspected as a cause of the corrosion, see 8.2.5.

Diagnosis and cure

The cure will be determined by the extent of the corrosion. If it is very extensive, the steel sheeting will have to be replaced. Replacement sheeting should have a full protective coating to both sides of the sheet.

Where rusting is occurring, but the material is still basically sound, a proprietory remedial protective coating should be applied. Great care should be taken to ensure that all rust is totally removed by hammering, chipping and rotary wire brushing prior to the application of the coating. The application should be supervised and the coating manufacturer's instructions properly followed.

REFERENCES

Construction Feedback Digest 53 *Paint failure on steel plate hangar roof.* B & M Publications (London) Ltd

Construction Feedback Digest 59 *Improved specification for profiled steel and cladding.* B & M Publications (London) Ltd

BS CP 143 *Sheet roof and wall coverings.* Part 10: 1973 *Galvanised corrugated steel*

8.2.8 INSULATED SANDWICH PANELS ATTACKED BY BIRDS

Symptoms

The exposed insulation core of the sandwich panels at the eaves show a general random pattern of deep clean cut holes.

Investigation

Establish whether measures have been taken to seal the damaged edges of the sandwich panels.

Investigate whether birds have been noticed attacking the damaged edge.

Diagnosis and cure

Apart from aesthetic reasons, there is usually no functional reason to permanently seal the ends of sandwich panels, as attack by rain, sun and frost will not cause premature deterioration. Manufacturers do apply a silicone sealer to provide temporary protection during carriage, storage and erection, but this is expected to fall off within two years.

Where damage has occurred, the remaining temporary silicone sealant should be removed and all holes filled with a waterproof filler. When the filler has dried the ends of the panels should be coated with one coat of hot applied bitumen. Light coloured chippings could be applied immediately after the bitumen coating to achieve a more acceptable appearance.

The problem can be avoided by ensuring that when such panels are specified, the specification includes for sealing exposed ends. Manufacturers advice should be sought, as many make snap on capping strips.

REFERENCES

Construction Feedback Digest 63 *Birds attack roof insulation.* B & M Publications (London) Ltd

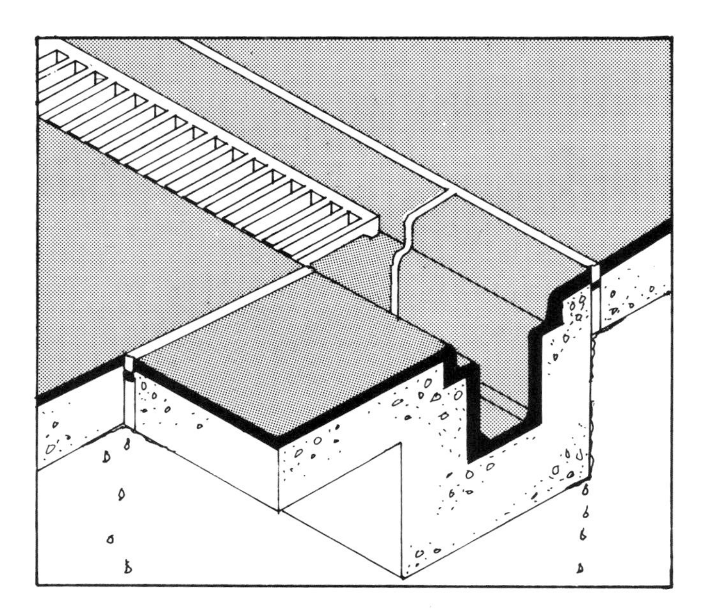

1 Badly designed roof with defective movement joints, one of which crosses gutter.

8.3.1 FAILURE OF MOVEMENT JOINT IN CONCRETE ROOF DECK

Symptoms

Leaks within the building coinciding with polysulphide filled movement joints in the slab and drainage gutter.

Investigation

Establish the construction of the roof and also that of the movement joint.

Note the material used for sealing the movement joint and investigate its compatibility with the waterproofing material.

Diagnosis and cure

The defect described here is commonly caused by the use of a mastic such as polysulphide that is not compatible with bituminous materials. Such failures can be exacerbated by the location of the movement joint. Joints located at lower positions on the roof are most vulnerable, as they are more likely to be under standing water. A movement joint should never be carried across a drainage channel. Wherever possible, such joints should be located clear of any standing water.

Defective movement joints should all be raked out and the waterproofing removed from the edges of the joints to allow for the insertion of a neoprene sheet which should be bonded to the waterproofing. Proprietary systems that provide two layers of neoprene are preferable, for if the top layer fills with debris and hardens, the lower layer should remain sound.

Protection should be provided for the waterproofing by the installation of concrete paving slabs on top. The joints in the paving slabs should not coincide with the movement joint, but a 25 mm joint should be provided alongside, filled with cellular neoprene to allow for the movement.

REFERENCES

BS 8110 *Structural use of concrete.* Part 1: 1985 *Code of practice for design materials and workmanship*

BS 6213: 1982 *Guide to the selection of construction sealants.*

Bickerdike, Allen. Rich & Partners in association with Turlogh O'Brien *Design failures in buildings. Second series* George Godwin Ltd.

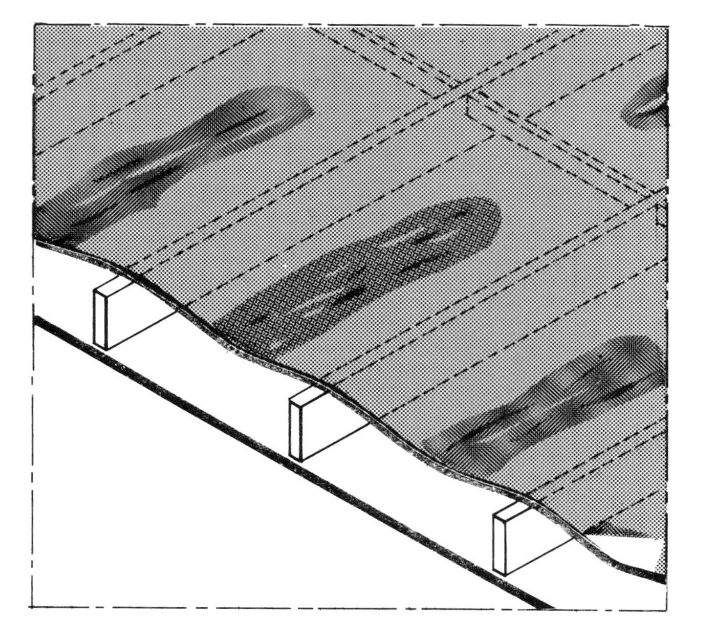

1 Characteristic degradation of moisture sensitive deck.

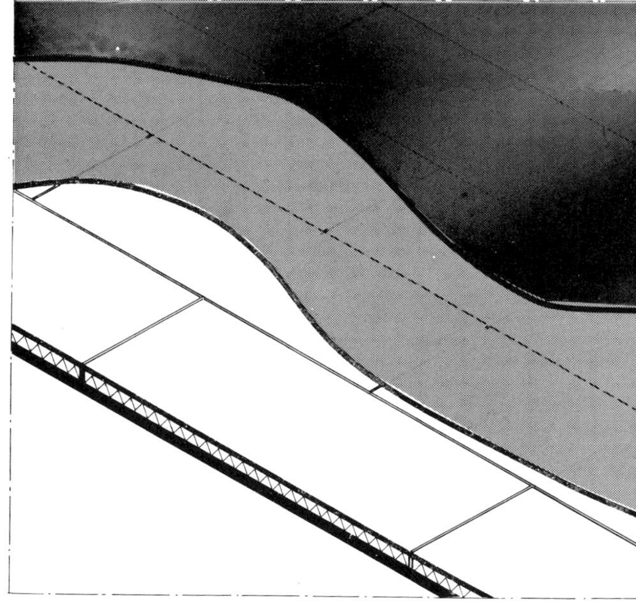

2 The weak (machine direction) in felt is different in each layer, consequently splits are more likely to coincide with board joints in one direction in one layer and with board joints in the other direction in another layer.

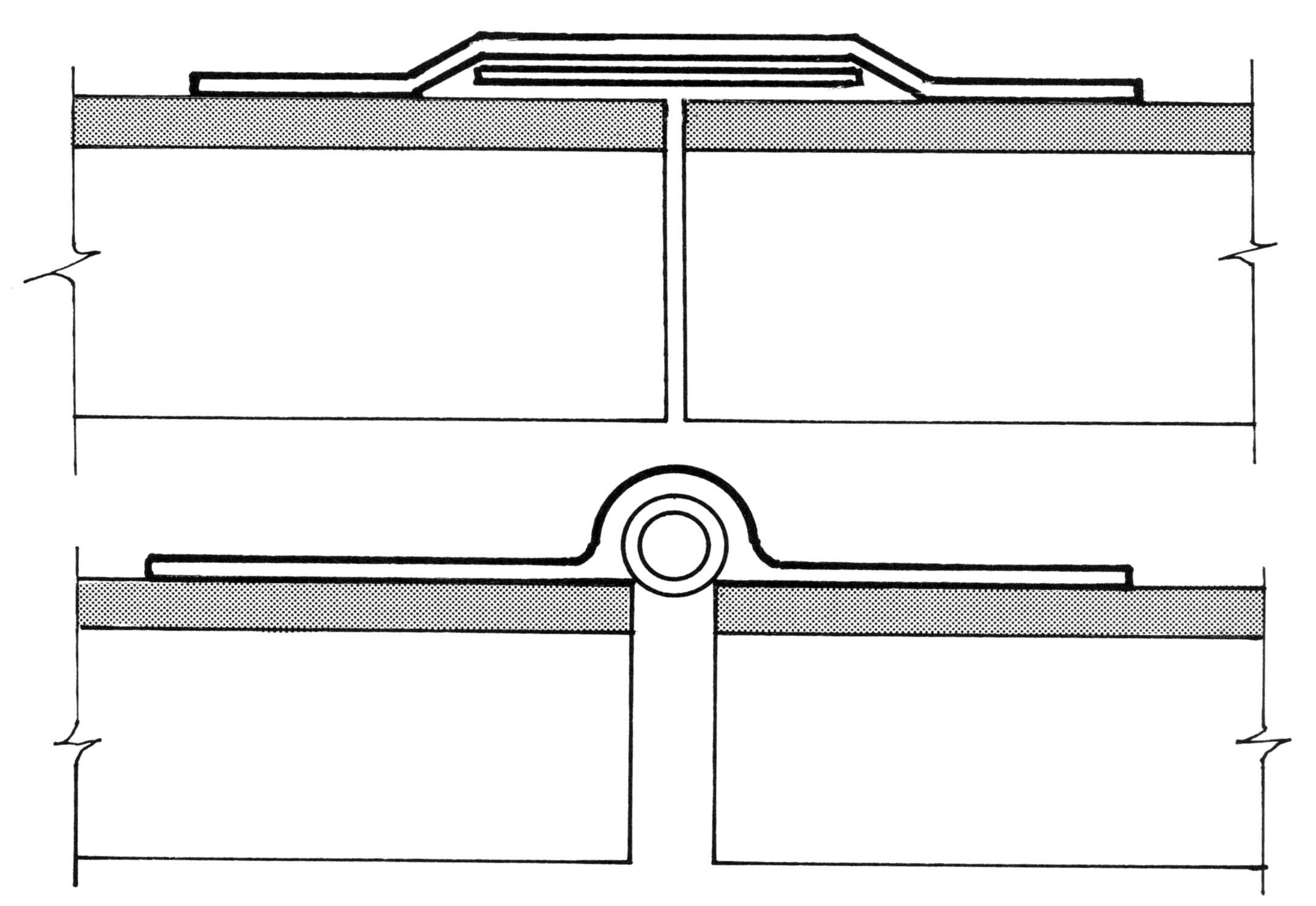

3, 4 Alternative repairs that incorporate small expansion joints. In 3 a strip of felt is loose laid over the joint, in 4 which will incorporate more movement than 3 a rubber or plastic pipe is used. Both these remedial joints should only be undertaken by specialist contractors.

8.3.2 FAILURE OF ROOF DECK BENEATH BUILT UP FELT MEMBRANE

In this defect, the failure of the roof membrane is the consequence of movement having occurred in the deck, and insufficient allowance having been made for movement of the membrane. This defect should not be confused with 8.3.17, which deals with defective felt membranes.

Symptoms

Leaks in wet weather, splits in the felt membrane. Narrow ridges in the felt covering, extending the length of the roof and at regular intervals.

Investigation

Establish the roof construction and compare the condition of the roof finish in the damaged areas with that elsewhere. Note any changes.

Note what insulation material has been installed (if any) its location, and what methods have been used to fix it.

Ascertain if leakage has occurred, and if it has, look for probable points of entry. The location of the leak inside is often removed from the failure in the membrane.

Try to discover the weather conditions that were prevailing at the time the roof was constructed. If moisture has been trapped in the deck during construction, the deck is more susceptible to cyclical thermal expansion and contraction.

If splits are present in the membrane, open up a typical area and ascertain if the split coincides with joints in the deck.

Diagnosis and cure

The most likely reason for this defect is water that has been trapped in either a woodwool, or a chipboard deck during its construction which causes the deck elements to swell. When this moisture dries out, the decking sheets shrink at the end joints. The felt membrane over these joints has stretched to accommodate this movement, so that when the subsequent inevitable thermal expansion happens, it causes the already stretched felt to buckle upwards to absorb the movement. Over time, this cyclical thermal expansion and contraction will cause fatigue and splits in the felt.

Remedial action will depend on the severity of the problem. On small roofs, if there are only a few splits, they can be located and repaired as they occur. On larger roofs, expansion joints can be incorporated into the felt membrane. This can be done by stripping back the membrane along the line of the joints for a width of at least 250 mm. A fresh length of felt of the same width can then be inserted into this gap, and two or three layers of new felt can be bonded down over this loose length to introduce a slip plane. This will allow the deck to move without splitting the felt.

An alternative is to lay a completely new covering. This can be done over the existing membrane if the following procedure is followed:

- Remove any stone chippings.
- Any large blisters and sharp ridges should be cut open and sealed flat with bitumen.
- A layer of 13 mm thick fibreboard should be fully bonded over the whole roof area. It is essential that this must be dry, laid in stages, and fully protected from rain and moisture. This is not intended as insulation but as a cushioning material to prevent future movements of the deck being transmitted to the new membrane.
- A new membrane of two or three layers of fully bonded BS 747 high performance material is laid over the fibreboard.

A similar thermal stress pattern can also occur in felts laid over polystyrene insulation. This material has twice the movement of any other common insulating board, which results in relatively rapid opening and closing of board joints. These cyclical movements can also be compounded by differential vertical movement of adjacent boards when they are mechanically fixed, and by maintenance foot traffic causing deflections in the deck. If the polystyrene boarding has been attached to the deck with adhesive and the membrane is not extensively split, a new BS 747 high performance material may be overlaid on a partially bonded first layer on top of the existing membrane.

Alternatively, the membrane and polystyrene can be removed, and replaced by a more stable insulant such as rigid urethene board or cork board. A new BS 747 high performance material can then be laid over it.

A third alternative is possible if the existing structure has sufficient upstand height, and is able to take the additional load. This is to repair the existing membrane which will become a vapour check, and to overlay it with a further layer of a more stable insulant covered with a BS 747 high performance material.

REFERENCES

BRE Digest 8: *Built up felt roofs*

BRE Digest 180: *Condensation in roofs*

BRE Digest 221: *Flat roof design technical options*

Construction Feedback Digest 36. *Wood chipboard flat roof decks*

PSA *Technical guide to flat roofing* Volumes 1 and 2. Department of the Environment/PSA 1987

BS CP 144 *Roof covering*: Part 3: 1970 *Built up bitumen felt*. Part 4: 1970 *Mastic Asphalt*

BS 747: 1977 (1986) *Specification for roofing felts*

Design Note 46: *Maintenance and renewal in educational buildings Flat Roofs Criteria and methods of assessment, repair and replacement.* Department of Education and Science Architects and Building Group

1 & 3 Cracks in asphalt typical of movement in the structural deck.

2 Crack in asphalt carried over lead drip into gutter.

8.3.3 CRACKS IN ASPHALT ROOF

This defect usually occurs with lightweight roof decks such as chipboard or woodwool slabs.

Symptoms

Cracking of the asphalt roof covering coincident with the pattern of the roof structure underneath.

Cracking of the asphalt at the edges coincident with joints in the lead flashing bedded in the asphalt.

Investigation

Establish the construction of the roof and of the edge details.

Investigate whether there is any potential movement in the roof decking. This may be occasioned by maintenance traffic.

Note whether measures have been taken to cut down thermal movements.

Where lead flashings are dressed into the edge of the asphalt, investigate how joints in the leadwork were carried out, and what measures were taken to fix the lead to the roof deck.

Diagnosis and cure

This failure is generally due to the roof deck flexing too much under load. A single person walking over the roof can be sufficient to cause overstressing of the asphalt. The structure should have been designed with a limiting deflection of 0.2% of the span.

The only long term remedial treatment is to strip off the asphalt and to relay the deck so that only minimum flexing occurs.

Cracks along the lines of the lead flashing are due to the structural flexing combined with inadequate fixing of the lead to the deck and the asphalt, so that when thermal expansion occurs the lead pulls and splits the asphalt. A timber fixing plate bolted to the structural frame should be provided for fixing the lead. In addition, the back edge of the flashing should have a welted turn-up to anchor the asphalt, and any joints should be welted, not lapped.

A solar reflective finish should be provided to help minimise thermal movement. This finish should be subject to regular inspections, as it will need renewing every few years.

REFERENCES

BRE Digest 144: *Asphalt and built up roofing felts: durability*

PSA Technical guide to flat roofing Volumes 1 and 2. Department of the Environment/PSA 1987

BS CP 144 *Roof covering*: Part 4: 1970 *Mastic Asphalt*

Roofing Handbook. Mastic Asphalt Council and Employers Confederation

Bickerdike, Allen, Rich & Partners in association with Turlogh O'Brien *Design failures in buildings. Second series* George Godwin Ltd.

1 Typical ponding on a flat roof.

2 Ponding has occurred on this roof because the falls to the outlets are insufficient.

3 In some cases ponding occurs because the outlets have been blocked with debris.

8.3.4 WATER PONDING ON FLAT ROOFS

On many flat roofs, rainwater is not shed completely, and areas of the roof may be covered with water until it is evaporated by an improvement in the weather conditions.

Symptoms

Ponds formed on flat roofs during rain, remain and fail to drain away. Ponds may be of variable size, and sometimes they may be restricted to areas around drainage outlets. Pondage on the roof may be a contributory factor to the occurrence of leaks.

Investigation

Establish the roof construction and compare the condition of the roof finish in the ponded areas with that elsewhere and note any changes.

Ascertain whether leakage has occurred, and if it has, look for probable points of entry.

If there is ponding around the drainage outlets, check whether they are blocked either at roof level, or within the drainage pipe.

Diagnosis and cure

Flat roofs are generally specified as requiring the top surface to have a fall of 1 in 80 or greater when forming the deck. However, in practice allowance is not always made for any deflection that will occur due to the self weight of the decking or the finish. In consequence, dishing occurs and ponding results. A wrongly sloped deck can also cause ponding. If the roof finish ripples or cockles, the extent of the ponding may increase.

Blocking of the drainage system is usually caused by debris which should be removed, if necessary by rodding. Guards should be provided at outlets to prevent further blockages. There should be regular inspections of the roof, especially in the autumn when the danger from leaves etc. is greatest. Ponding around a drainage outlet can be due to the outlet having been installed proud, in which case it should be properly installed, or to the surrounding roof deflecting away from it.

Ponding on asphalt roofs rarely leads to leakage. Surface crazing is usually superficial.

Persistent ponding on bitumen covered flat roofs at the same place will eventually lead to a deterioration of the felt, especially if it is old. Immediate remedial work is not essential for ponded areas that do not leak. However, the roof should be subject to regular inspection to check whether the felt is becoming damaged. If it is, and the area affected is small, localised relaying may be sufficient to effect a repair. With larger areas, where the ponding is shallow, the roof can be levelled with insulation boards by placing them on top of the existing cover, and protecting them with an additional finish. Alternatively, the existing cover could be stripped, adequate falls provided in the substrate, and a new covering laid. Tapered cork fillers can be purpose made to correct the falls on individual flat roof decks.

REFERENCES

BRE Digest 8: *Built up felt roofs*

BRE Digest 144: *Asphalt and built up felt roofings: durability*

BS CP 144: Roof covering: Part 3: 1970 *Built up bitumen felt*. Part 4: 1970 *Mastic Asphalt*

Design Note 46: *Maintenance and renewal in educational buildings Flat roofs Criteria and methods of assessment, repair and replacement*. Department of Education and Science Architects and Building Group

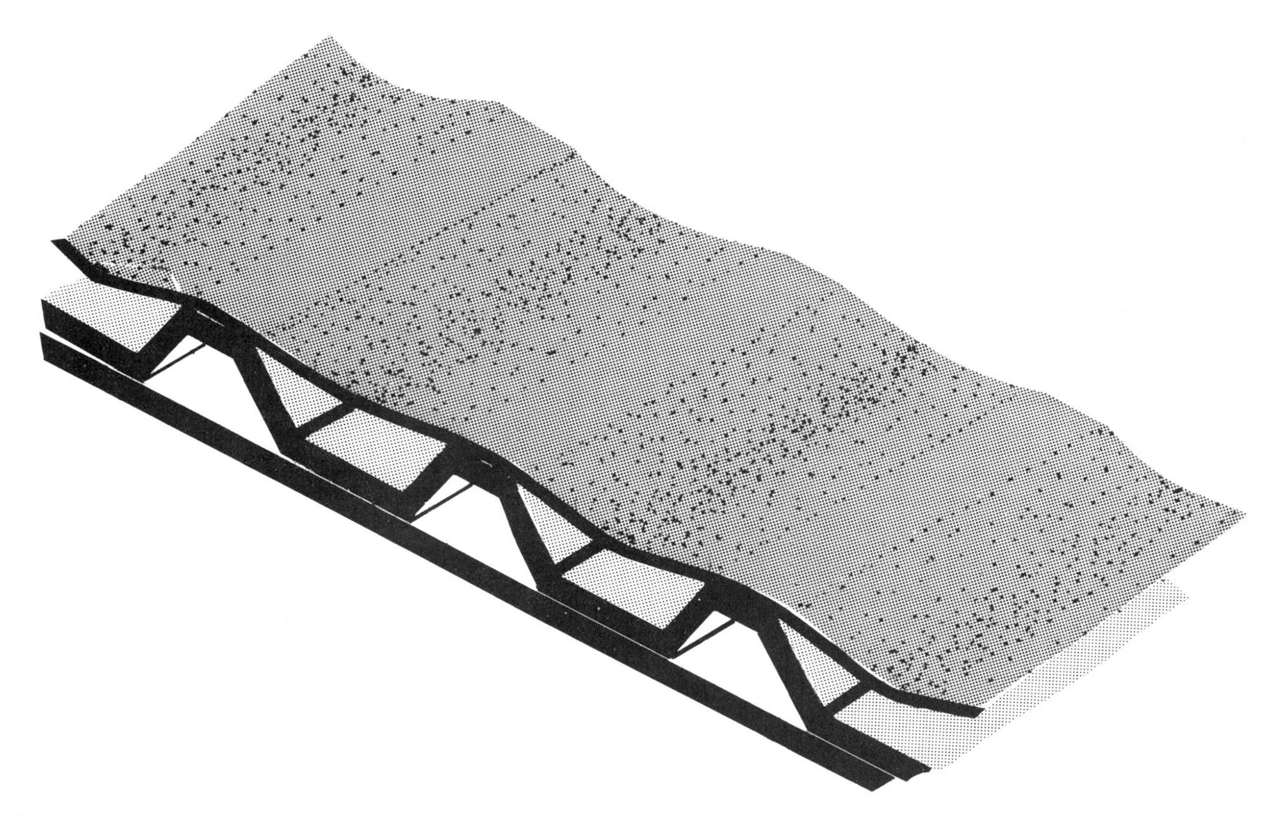

1 Characteristic degradation of fibreboard on metal decking. The degraded fibreboard and vapour check sink into the troughs and the roof has a striped appearance, with the areas over the deck troughs looking darker.

2 An example of degraded fibreboard.

8.3.5 DEGRADATION OF FIBREBOARD ON METAL DECKING

This defect often shows itself first at the perimeter area of the roof, or over areas where high humidity is present.

Symptoms

Water leakage through the roof which may occur at some distance from the point of entry. The metal decking can carry the water to a deck lap or fixing point before it will find its way out.

A striped effect on the roof surface coincident with the deck profile. The darker areas occur over the troughs due to contamination of the chippings (if they exist) and to membrane surface ponding effects. If the fibreboard is badly degraded, the membrane may actually sink into the trough lines. This is more common with asphalt membranes owing to their greater weight and flow characteristics.

The roof surface has a distinctly spongy feel when walked upon, and may have suffered damage from foot or other traffic.

Investigation

Establish the construction of the roof and ascertain if a vapour check has been provided beneath the fibreboard.

If the roof has been constructed for less than five years, investigate the weather conditions at the time the roof was constructed.

Ascertain whether additional insulation has been installed under the roof deck at ceiling level.

Expose the extent of degradation that has taken place by opening up typical affected areas.

Diagnosis and cure

Fibreboard is very moisture sensitive; it will move laterally, swell and degrade when wet. The two most likely causes for the moisture content of fibreboard reaching a critically high level are:

- Boards becoming wet during the construction process. The presence of a vapour barrier beneath the fibreboard has led to the damp board being sandwiched between the impermeable layers which has prevented drying out and allowed degradation to occur.
- Condensation from the areas under the roof deck has saturated the boards. If the roof cavity has been ventilated and there is no vapour check, the trough form of the metal decking will have encouraged air movement along the soffit of the board. This assists the drying out process and can prevent retained condensation from reaching damaging levels. However, where there is insulation at ceiling level (so that during winter the fibreboard is maintained at a colder temperature), and there is poor or obstructed ventilation of the roof cavity, interstitial condensation could reach damaging levels. The insertion of cavity barriers may have reduced air movement sufficiently to provoke this problem.

If the fibreboard is badly degraded, replacement of the board and membrane are the only effective solution. The fibreboard should be replaced with an insulant that is not vulnerable to condensation.

If degradation of the boards is limited, and if measures can be taken to reduce the amount of condensation occurring in the roofspace, consideration could be given to overlaying the repaired existing roof with new insulation and membrane. The additional insulation, by reducing the heat loss from the fibreboard, may either remove the critical condensation level existing in the fibreboard, or reduce it to an acceptable retained level. This latter course is unlikely to be appropriate for roofs with an asphalt membrane, as the problems are rarely discovered until they are widespread.

REFERENCES

Design Note 46: *Maintenance and renewal in educational buildings Flat roofs Criteria and methods of assessment, repair and replacement.* Department of Education and Science Architects and Building Group.

PSA *Technical guide to flat roofing Volumes 1 and 2.* Department of the Environment/PSA 1987.

BS CP 144 *Roof covering*: Part 3: 1970 *Built up bitumen felt.* Part 4: 1970 *Mastic Asphalt*

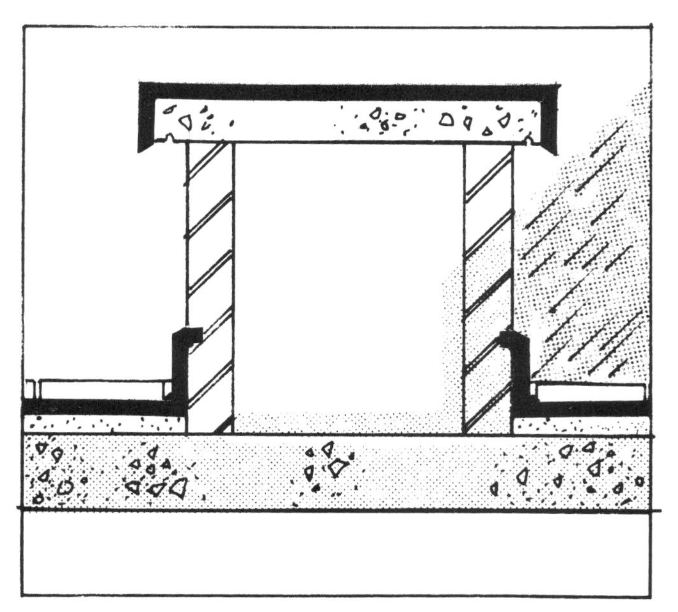

1 Rain penetrating the single skin brickwork duct enters the slab and leaks.

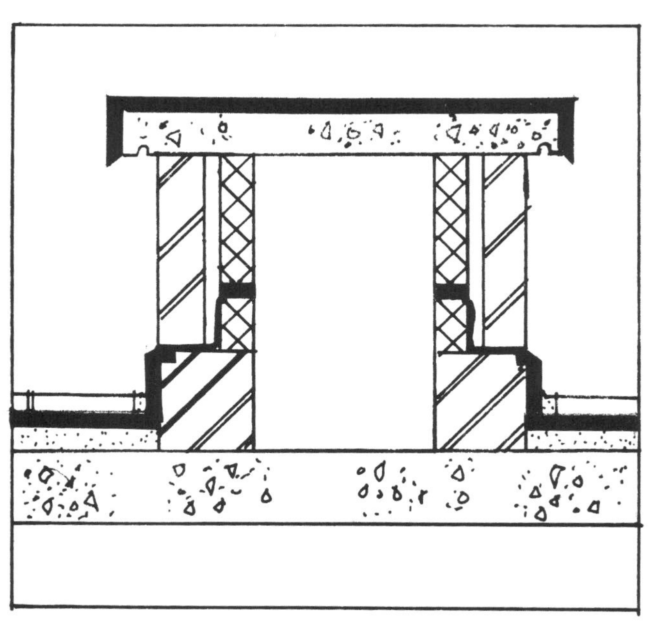

2 Waterproofing the duct using cavity construction for the walls.

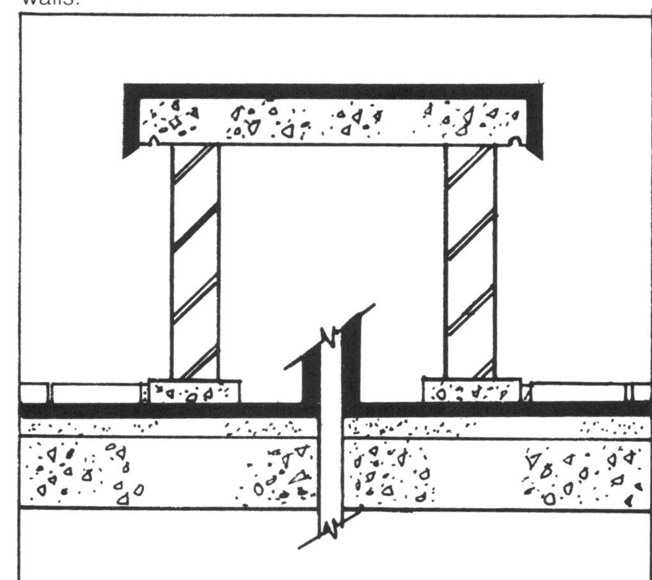

3 An alternative is for the waterproof roof membrane to continue under the duct. Any service penetrations of the slab will need to be in a waterproof sleeve.

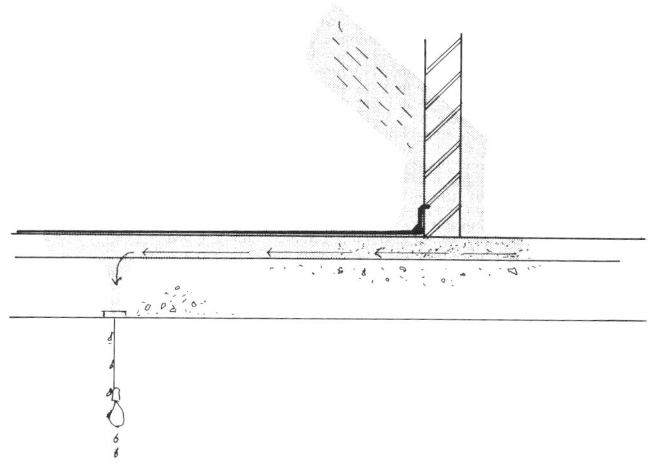

4 Water entry through a duct may travel some distance in the slab before manifesting itself as a leak at a weak point.

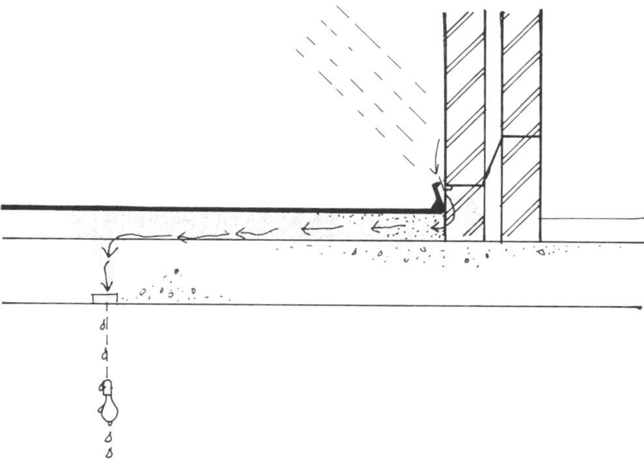

5 Workmanship around service ducts can sometimes be of a low standard so dampness may enter the screed or slab through badly formed skirtings.

8.3.6 RAIN PENETRATION OF FLAT ROOF IN VICINITY OF DUCTS

This defect relates to ducts running above the roof level.

Symptoms

Dampness showing on the ceiling, which is either weather related or occurring soon after the building has been completed.

Investigation

Establish the construction of the roof and of the service ducts.

If the roof has been recently laid, try to discover the weather conditions at the time that laying was undertaken.

Ascertain if the leaks are weather related. Investigate whether rain can penetrate the duct walls and whether or not the floor of the duct is waterproofed.

Diagnosis and cure

This defect can be attributed to one of two causes. If it happens soon after construction, and if it was known that rain had penetrated the screed prior to the waterproof membrane being installed, the dampness may be entrapped water coming out. If this is the case, the dampness is likely to be persistent until all the water has dried out.

If, however, the dampness appears to be weather related, water penetration of the structure through the ducts, or the junction of the duct with the roof, should be suspected. This is most likely due to the failure to provide a continuous waterproof membrane under the duct, and to the walls of the duct being porous. If the problem is found to be related to the junction, it can often be dealt with by a local repair. However, if the walls of the duct are found to be porous, they can only be satisfactorily remedied by rebuilding the ducts. This may be achieved by either making them of a totally impervious construction appropriately detailed with a waterproof joint at the roof, or the waterproof membrane on the roof can be made fully continuous under the duct. If this latter option is chosen it is essential that service entry points are sleeved, and a waterproof joint made between the sleeve and the membrane.

REFERENCES

Bickerdike, Allen, Rich and Partners in association with Turlogh O'Brien *Design failures in buildings 2nd Series.* Building Failure Sheet No 1. *Light fitting filled with water.* Building Failure Sheet No 2. *Roof duct.* George Godwin Ltd

PSA *Technical guide to flat roofing* Volumes 1 and 2. Department of the Environment/PSA 1987

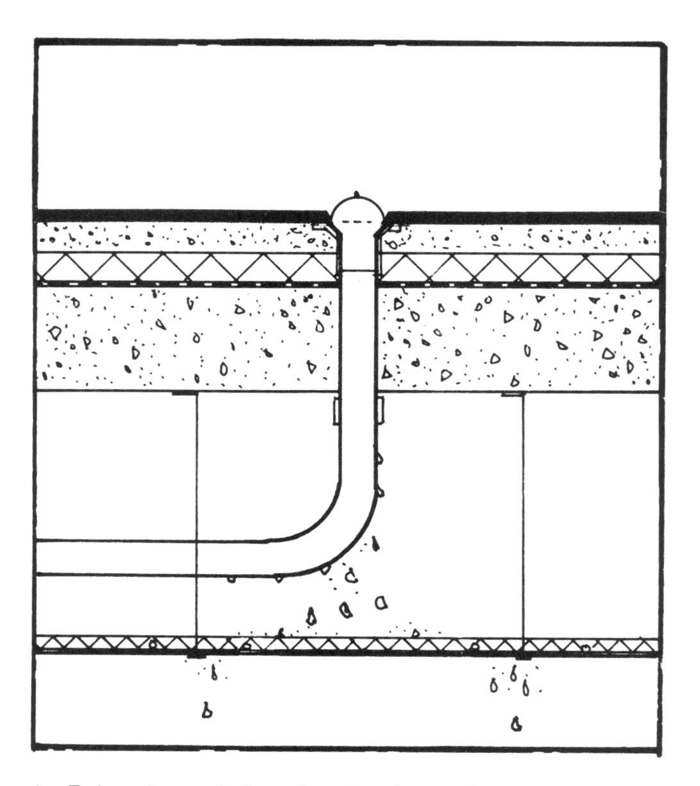

1 Rainwater cools the rainwater pipe so that warm moist air in the ceiling void condenses on the pipe and drips off.

8.3.7 CONDENSATION ON RAINWATER PIPES IN ROOF VOID

This failure may be wrongly presented as roof leakage.

Symptoms

Water dripping through the ceiling at the time of heavy rainfall, particularly in winter.

Investigation

Establish the construction of the roof and note the position of service pipes that may be present in the roof void. Check whether insulation has been installed around the pipes.

Ascertain if the leaks are weather related.

Examine the roof membrane for any signs of failure, and consider opening up a small section to discover whether interstitial moisture is present.

Examine the underside of the roof deck to ascertain whether there is evidence of water penetration, eg stalactites from a concrete deck or staining. Measure the level of humidity normally present in the roof void.

Investigate the nature of the occupancy of the building, and note the number of people normally present. In addition note the presence of water vapour producing activities such as showers etc.

Diagnosis and cure

If there is no evidence of water leakage through the roof deck, condensation forming within the roof void is a likely consideration. If the roof void is inadequately ventilated, moisture will collect within the void and may saturate the roof deck from underneath. If uninsulated rainwater downpipes pass through an inadequately ventilated roof void, condensation may form on the pipes when it rains. This is because the pipes will cool to the temperature of the rain and the moist air will condense on the cool surface of the pipes and eventually build up sufficiently to drip onto the ceiling below.

The number of persons regularly occupying the building in the spaces beneath the roof is an important factor. Sufficient water vapour to cause interstitial condensation can be generated within highly populated rooms (such as well used library reading rooms). It is worth remembering that every individual breathes out half a litre of water every eight hours.

The problem can be cured by providing thermal insulation to rainwater pipe, thereby excluding intermittently cold surfaces, and by providing sufficient ventilation to the roof void to prevent the build up of moist air.

REFERENCES

BS 5250: 1975 *Code of basic data for the design of buildings: The control of condensation in dwellings*

Bickerdike, Allen, Rich & Partners in association with Turlogh O'Brien *Design failures in buildings. 2nd series* Building failure sheet No 6 *Roof with heated ceiling.* George Godwin Ltd

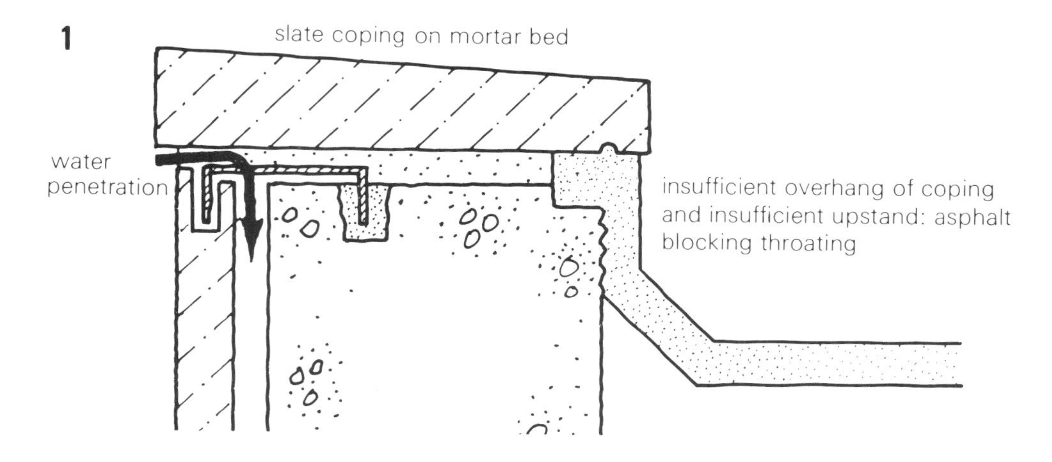

1 Leaking parapet detail.

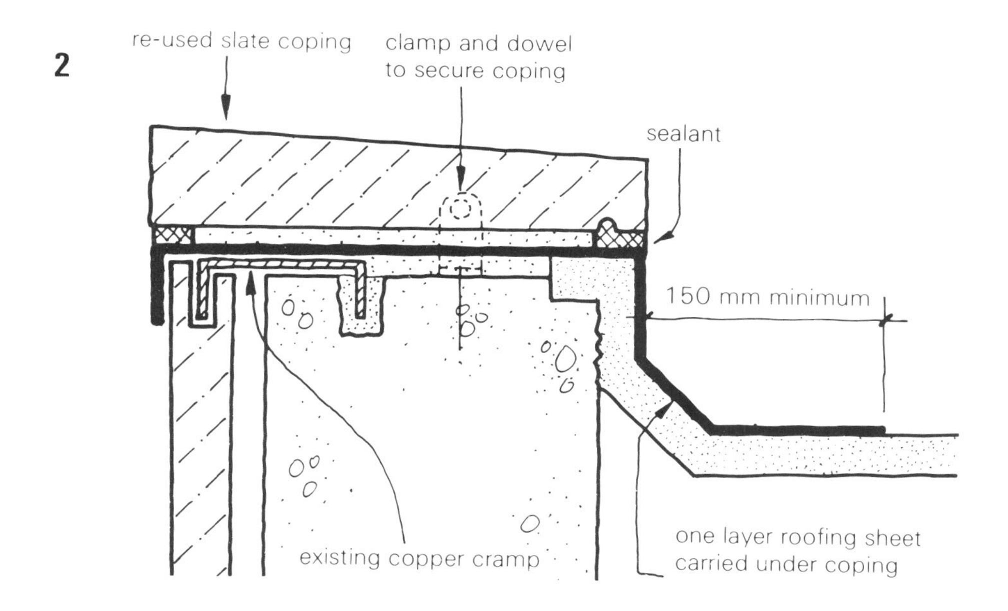

2 Coping laid to avoid leaks.

8.3.8 FAILURE OF COPING AT EDGE OF ROOF UPSTAND

This defect is unrelated to the material used for the roof covering.

Symptoms

Leaking at the roof perimeter.

Investigation

Establish the construction of the roof and the detail at the parapet.

Examine the coping, particularly how it has been bedded and the provisions that have been made for it to overhang the upstand.

Check the height of the upstand from the roof finish.

Diagnosis and cure

This failure is caused by an inappropriate edge detail, compounded by poor workmanship. It is most likely to be due to an undersized coping being constructed with insufficient overhang for the drip channel to be effective. Water penetrates beneath the coping, causing cracking in the bedding, with the result that the water percolates into the building through the top of the wall.

If there is too small an upstand between the roof finish and the coping (ie less than 150 mm), it will be easier for water to reach the coping upstand joint.

The most effective method of repair will be to remove the existing coping, to scrape off the existing bedding being careful not to cause damage to the membrane bedding, and then to installing a new coping of sufficient width to have effective drip channels. Where the upstand is small, a sealant should be incorporated at the edge of the upstand, in addition to the normal bedding of the coping.

If it is considered desirable for the existing coping to be retained for aesthetic reasons, it should be carefully removed and cleaned. All the existing bedding should be cleared away and the coping replaced on a layer of roofing sheet which must be bonded to the existing roofing and carried right over the parapet. The mortar bed should be protected by incorporating a sealant on either side, and additional mechanical fixings used to hold the coping down onto the layer of roofing sheet.

REFERENCES

Construction Feedback Digest 44. *Slate coping and cladding to parapet.* B & M Publications (London) Ltd

1 Typical spalling Eaves.

Eaves and verges on concrete deck (warm roof)

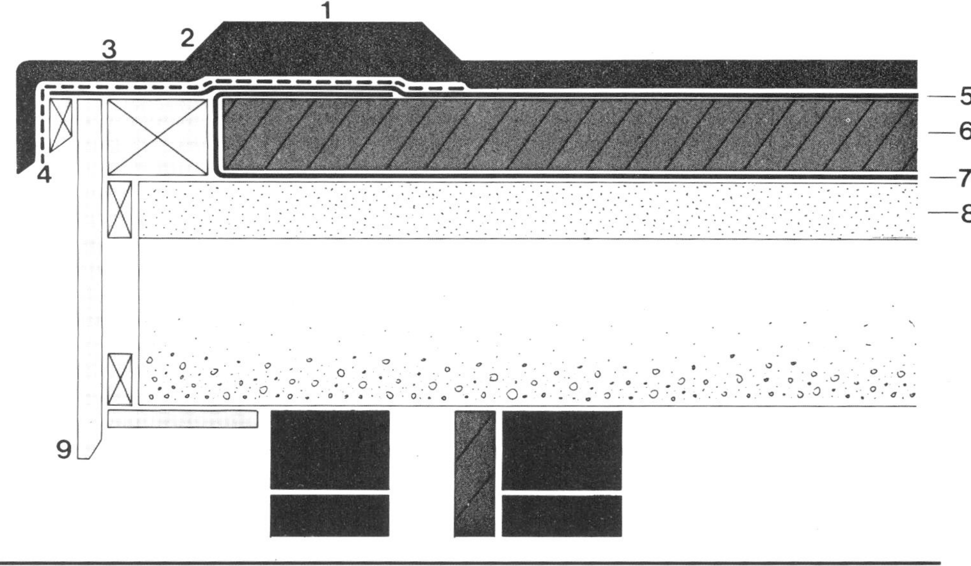

1 Optional solid mastic asphalt water check to size and position as required
2 Solar paint
3 20mm (minimum) three-coat mastic asphalt
4 Expanded metal lathing
5 Isolating layer
6 Insulation
7 Vapour barrier/check
8 Screed on concrete deck
9 Fascia

8.3.9 SPALLING CONCRETE EAVES SOFFITS

Symptoms

A reinforced concrete flat roof badly spalled at the edges of the eaves overhang.

Investigation

Establish the construction of the roof.

Examine the concrete cover to the reinforcement, and how the asphalt is detailed to finish at the edge.

Note whether a drip channel has been provided and also whether there is an angle on the eaves soffit. If the soffit slopes, note if it is towards or away from the building and whether there is any evidence of rainwater tracking back along the soffit towards the building.

Diagnosis and cure

Spalling concrete gradually occurs because there is insufficient cover protection to the reinforcement steel. Exposed soffit edges are particularly vulnerable to this problem through bad detailing of the waterproof finish, often exacerbated by the architect's desire to keep the visual impact of the edge to a minimum. Water reaches the steel, which consequently rusts, expanding as it does so. Initially the concrete will crack and rust stains mark the surface, but in very little time the force of the expansion will be sufficient to break the concrete away completley (see also 5.1.23 which deals with carbonation of concrete). The damaged concrete should be cut out and repaired by specialists.

A turned down flush joint to an asphalt waterproof finish over the concrete, could also be the cause of water penetration through capillary action. The edge detail should be reworked to incorporate an asphalt apron, and consideration should be given to including a solid asphalt watercheck on the edge of the roof.

Inward sloping eaves often allow rainwater to jump the throat. They should be avoided.

REFERENCES

Construction Feedback Digest 55. *Spalling concrete soffits.* B & M Publications (London) Ltd

BS 8110 *Structural use of concrete.* part 1: 1985 *Code of practice for design materials and workmanship*

1–2 Felt splitting at abutment because of localised movement between the deck and the wall.

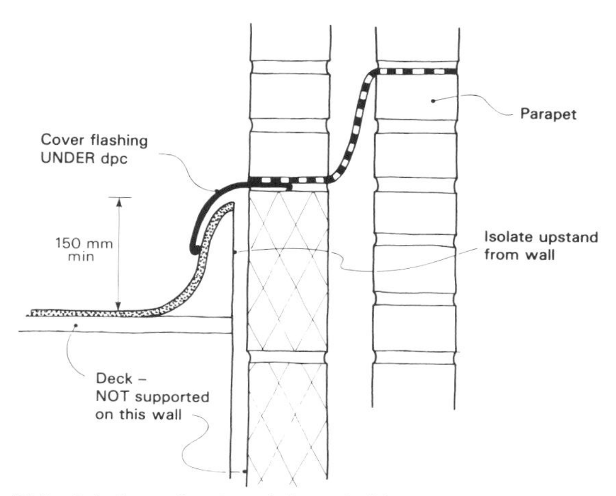

This detail applies to a felt roof although the shape of the upstand is somewhat outdated (the shape illustrated in Figure 5 below is now more usual).

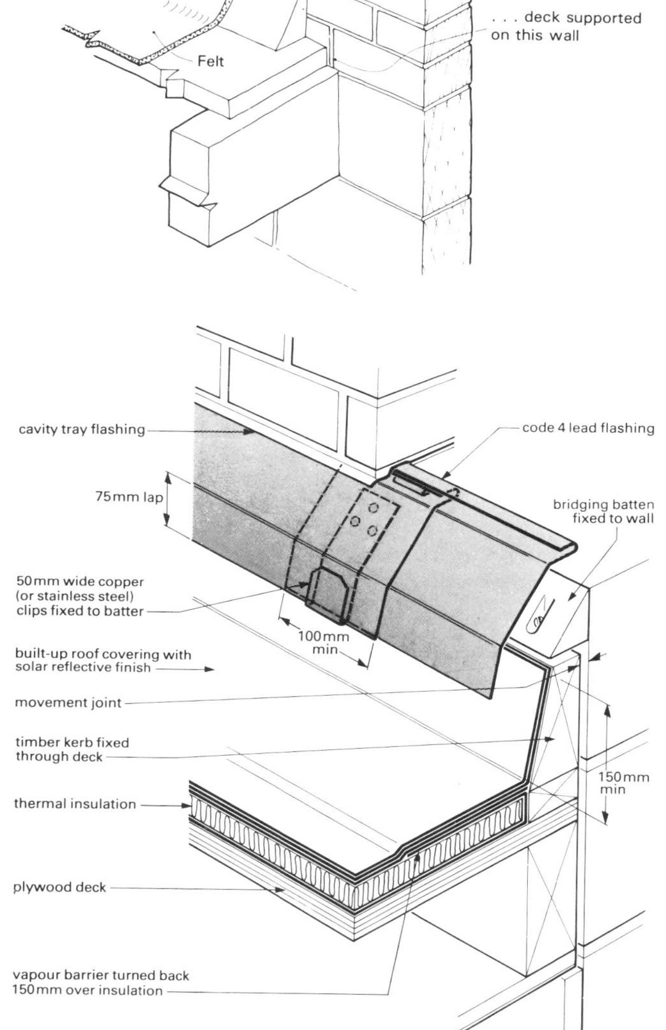

Failure
Felt, turned through 90°
splits even if no
roof movement

Failure
Splash-up overtops
inadequate upstand

Failure
Felt torn by
roof movement

Failure
Tray dpc drains
into inner leaf

3–5 Alternative details that separate the deck from the wall.

8.3.10 FAILURE OF ROOF PARAPET ABUTMENT FLASHING

This detail is commonly used with lightweight flat roofs where movement of the deck relative to the surrounding walls must be allowed for.

Symptoms

The lead flashing sags between fixing clips and water penetration occurs.

Investigation

Establish the construction of the detail and investigate whether any provision has been made to support the lead independently from the deck.

Examine the width of the lap joints and the number and location of the fixing clips.

Diagnosis and cure

This defect is usually attributable to lack of support to the lead and insufficient fixing clips.

The lead flashing should be reworked, incorporating a triangular bridging batten between the wall and the top of the kerb. The batten should be fixed only to the wall and a small gap should be left between the batten and the kerb so that the deck can move. Copper on stainless steel retaining clips should be fixed to the batten to retain the lead, at not more than 500 mm centres, with a fixing clip provided at every lap joint. All lap joints should be a minimum of 100 mm.

REFERENCES

Construction Feedback Digest 36 *Flat roof/Parapet abutment detail.* B & M Publications (London) Ltd

PSA *Technical guide to flat roofing* Volume 2. Section 3.1. Department of the Environment/PSA 1987

1–2 Typical example of cracked asphalt upstand.

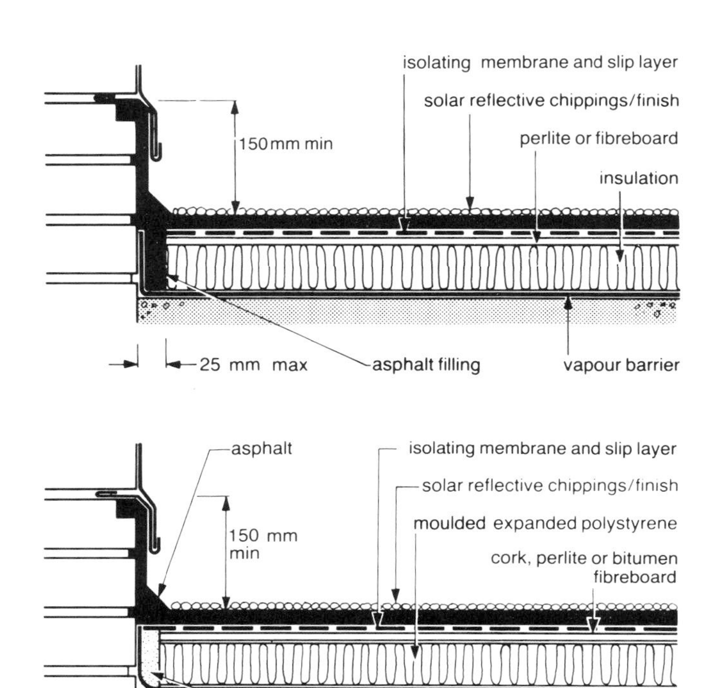

3–4 Alternative details for asphalt upstands.

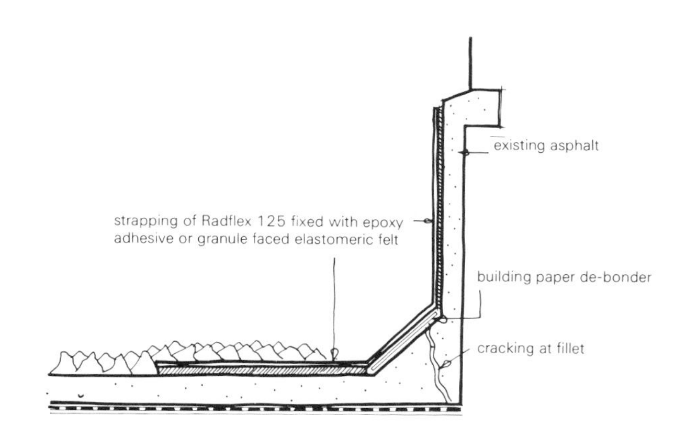

5 Remedial patching for cracks that are confined to fillets, where the upstands are in good condition.

8.3.11 CRACKING OF ASPHALT AT PARAPET

This defect occurs near junctions, where walls meet, and at parapets and verges.

Symptoms

Cracks in or around the asphalt fillet.

Investigation

Ascertain whether leakage has occurred, and if it has, look for probable points of entry.

Establish the roof construction, and cut out typical cracks to clarify the distance into the structure that the cracks penetrate.

Ascertain whether there has been deflection of the roof deck by looking for signs such as ponding (see also 8.3.3).

Look for signs of mechanical damage.

If insulation boards have been used, check that the gaps between the insulation and the edges of the upstands to which they abut have been filled.

Diagnosis and cure

Differential movement is frequently the root cause of cracking on asphalt roofs. Poor workmanship and design commonly aggravate the problem. Cracking to asphalt fillets at the base of upstands may have several causes that may be acting independently or in combination.

Ineffective cleaning of asphalt prior to the application of the fillet will result in poor adhesion.

Deflection of the roof deck can cause the asphalt to pull away from the upstand. Provided that the structure has completely settled, the asphalt work should be repaired and strap reinforcement applied to all the fillets.

Where the cracking has been found to be due to settlement in the screed (perhaps as a result of poor compaction), the cracks should be cut out and made good. Where water penetration has occurred, the affected areas should be taken down to the vapour barrier and all materials found to be damp should be removed and replaced. Strap reinforcement should be applied to all fillets.

If cutting open cracks reveals that there has been insufficient support beneath the asphalt fillet, eg if the gap between the insulation and the upstand has not been filled, the asphalt to the upstands should be removed with the aid of a hot asphalt poultice, (gas flame guns and cold chisels are not to be permitted), for a minimum distance of 225 mm from the edge. Wetted fibreboard should be cut out and replaced and all edge gaps completely filled and compacted using a damp cement/sand (1:4) mix. The repair should be completed by forming a new 50 mm two coat angle fillet to the upstands. The fillet and upstand may be protected from solar radiation by application of a solar reflective paint. All solar reflective finishes should be inspected at regular intervals to check they are still effective.

REFERENCES

Construction Feedback Digest 38 *Asphalt repairs* B & M Publications (London) Ltd (shows correct procedure for poulticing an area to be cut out)

Construction Feedback Digest 44 *Maintenance of cracked asphalt roofs.* B & M Publications (London) Ltd

Construction Feedback Digest 46 *Lack of support ends in tears.* B & M Publications (London) Ltd

Roofing Handbook. Mastic Asphalt Council and Employers Confederation.

1 The debris accumulated on this roof has blocked the outlet causing ponding. The asphalt is also bubbling due to entrapped moisture.

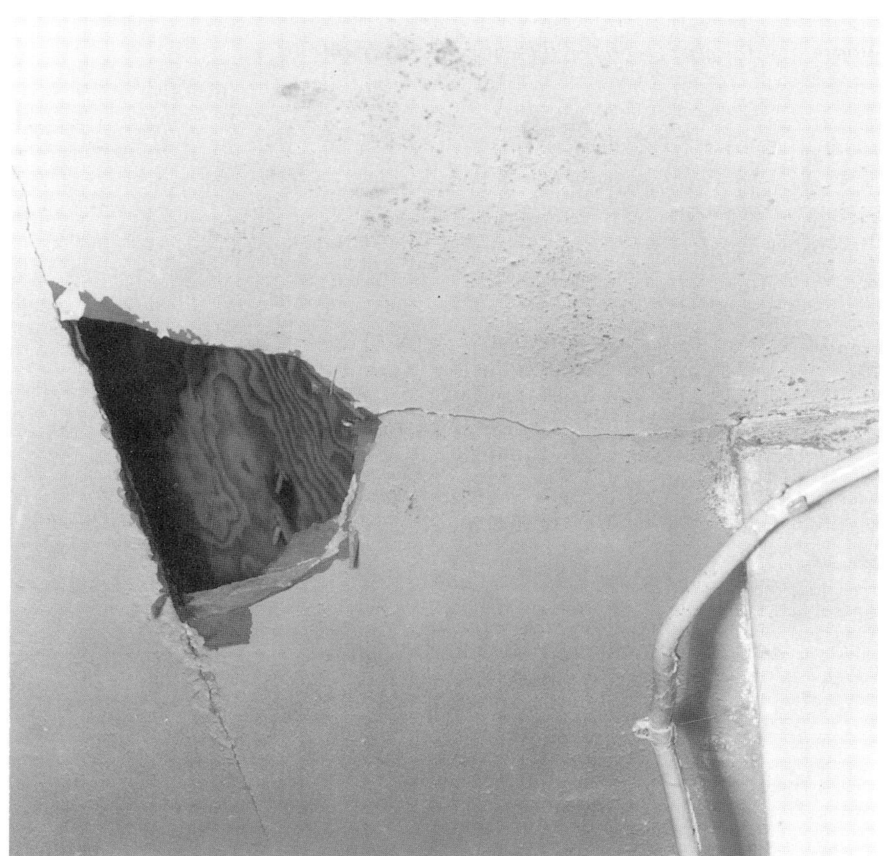

2 Inside the building of photo 1 showing leakages through the plasterboard ceiling and the saturated permanent plywood shuttering of the roof above.

8.3.12 DISRUPTION OF ASPHALT ROOF SURFACE

Thermal movement can cause blisters, cracks or slumping and cockling in asphalt where it is laid in large exposed areas.

Symptoms

Blistering of the asphalt.

Longitudinal cracks in the asphalt.

Slumping, cockling and rippling of the asphalt both at upstands and within the roof area leading to leaks.

Investigation

Establish the construction of the roof. If the roof is relatively new, try to discover the weather conditions when the roof was laid.

Note whether a solar reflective finish has been provided.

Look for signs that indicate that impact damage has occurred. For example, are there any indentations where scaffolding poles have rested on the asphalt?

Diagnosis and cure

Because asphalt is a dark material which easily softens with heat, it is particularly susceptible to the effects of thermal radiation. Many problems with asphalt roofs are exacerbated by (if not caused by) thermal change.

Blistering of asphalt is most common in summer. It can happen during, or soon after the laying of the roof, and is particularly liable if no solar protection is provided. It is caused by entrapped water being expanded by solar radiation that has also softened the asphalt. Where the blisters are cut out and repaired, those caused by water entrapped during construction are not likely to reform.

However, if the blisters are caused by leaks or interstitial condensation, their formation will very likely continue. These underlying problems should be dealt with first.

Long splits in the body of the asphalt, by contrast, are more likely to occur in extremely cold weather. During an exceptionally cold clear night the temperature of the asphalt may fall several degrees below the air temperature through heat loss by radiation. If the asphalt has no protective coating, and such a night is followed by a sunny day, the consequent rise in temperature can cause thermal shock to the asphalt, and splits result.

The splits should be repaired by cutting them out and applying a protective coating. This problem tends to be confined to particularly exposed locations, and is not very common in conurbations.

Lack of solar protection, either in the form of chippings or of a proprietary protective coating, can also exacerbate small localised cracks. These may have been caused by impact damage which did not show at the time of happening. The amount of thermal movement possible in an unprotected roof is enough to open the crack sufficiently for water penetration. Whenever such a crack is found it must be repaired to prevent further damage.

When asphalt is placed on non-horizontal surfaces, and is unrestrained, it will move under the force of gravity if sufficiently softened by heat. This movement will cause slumping, particularly on upstands. On occasions where the surface of the substrate of the roof is smooth enough to allow the sheathing felt to slip, creep of the asphalt down the falls of the roof may also occur. Both of these conditions can be helped, if not totally prevented, by the application of the solar protection. Vertical surfaces should be protected with a white solar reflective coating. Proprietary coatings have a relatively short life and will need to be renewed on vertical and pitched surfaces on a regular basis. Routine inspections should be made to determine that the coating is still effective.

REFERENCES

GLC Development and Materials Bulletin 143 *Solar reflective coatings for asphalt roofs*

Construction Feedback Digest 38 *Asphalt repairs* B & M Publications (London) Ltd (shows correct procedure for poulticing an area to be cut out)

Roofing Handbook. Mastic Asphalt Council and Employers Confederation

3 Badly slumped asphalt.

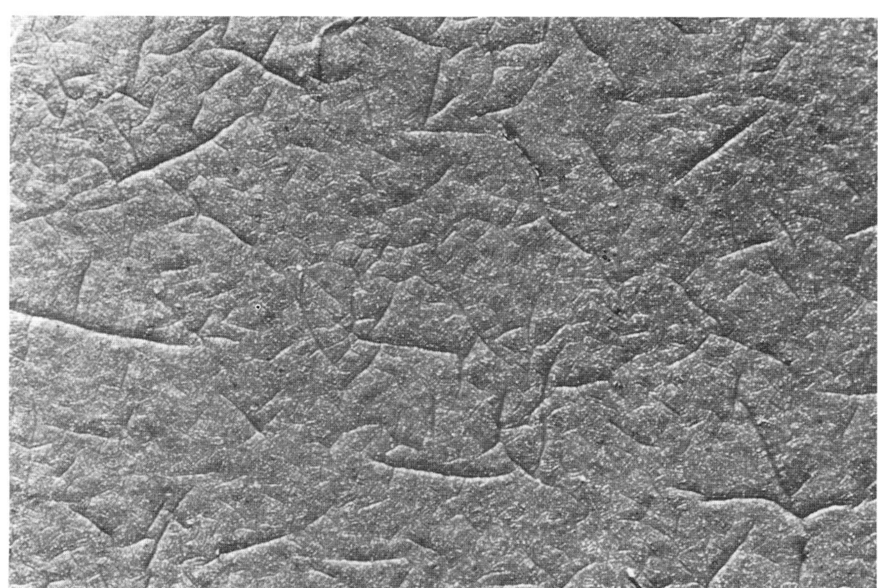
4 Crazed Asphalt.

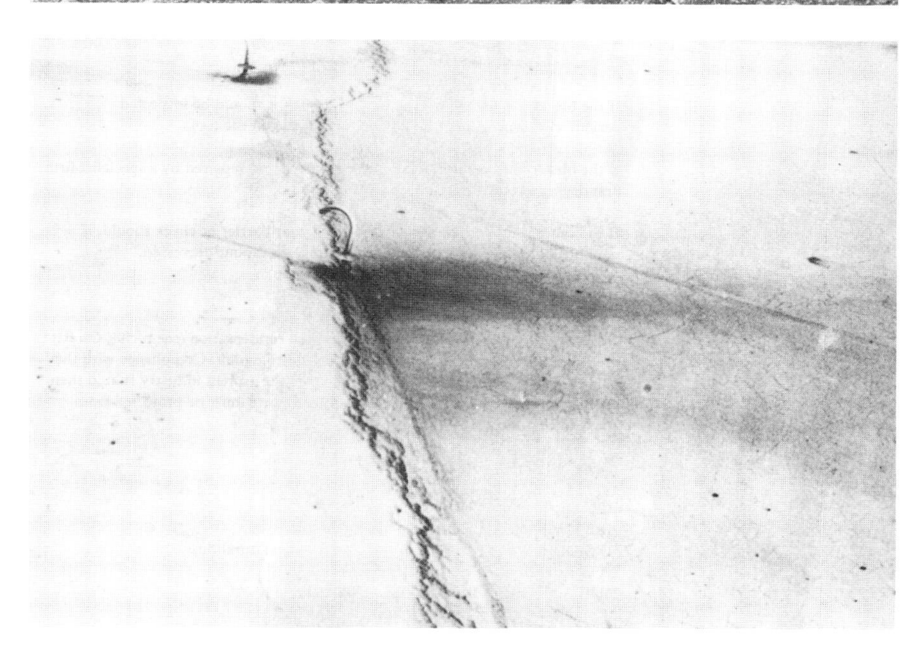
5 Longitudinal crack in asphalt roof.

1 General view of roof.

2 Close up of area affected by condensed flue gasses.

8.3.13 LOCALISED SURFACE DEGRADATION OF ASPHALT

This defect has been found where flue gasses continually or regularly discharge close to unprotected asphalt.

Symptoms

The degradation and breaking up of asphalt accompanied by leaks.

Investigation

Establish the construction of the roof and note whether the deterioration is generalised or local.

Note the position of any chimneys and their outlet height from the asphalt, and also whether the deterioration could be related to the discharge of flue gasses.

Send a section of damaged asphalt for laboratory analysis.

Diagnosis and cure

Asphalt is a very durable material, therefore degradation is likely to be the result of attack by a foreign agent. The presence of a chimney suggests that the flue gasses are the likely cause, particularly if the chimney is sited near the roof level. Raising the chimney a metre higher would ensure the exhaust is diluted before it reaches the asphalt and that the fumes are dispersed over a wider area.

The damaged asphalt should be cut out and renewed.

REFERENCES

Construction Feedback Digest 38 *Asphalt repairs* B & M Publications (London) Ltd (shows correct procedure for poulticing an area to be cut out)

Construction Feedback Digest 47 *Prevailing wind provides clue to asphalt failure.* B & M Publications (London) Ltd

1 Poor drainage on roof terraces increases the likelihood of leakage.

2 Mopped on bitumen is prone to cracking following drying shrinkage of the slab, so water will penetrate through the grouting and cause leakage.

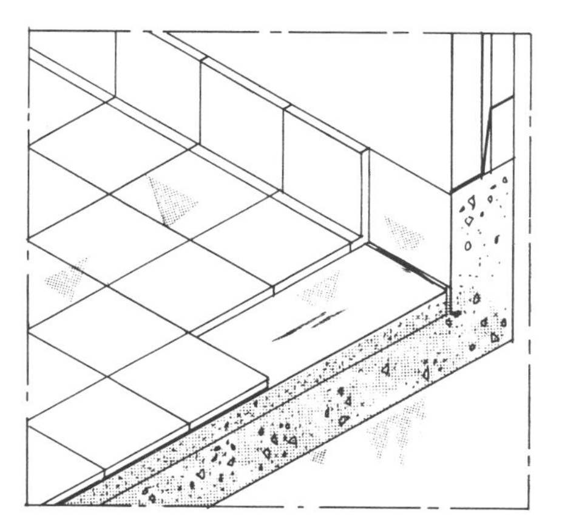

3 Caustic alkali dissolved out of cement lime bedding will attack tar in board fillers used for movement joints which will run out and stain the facade.

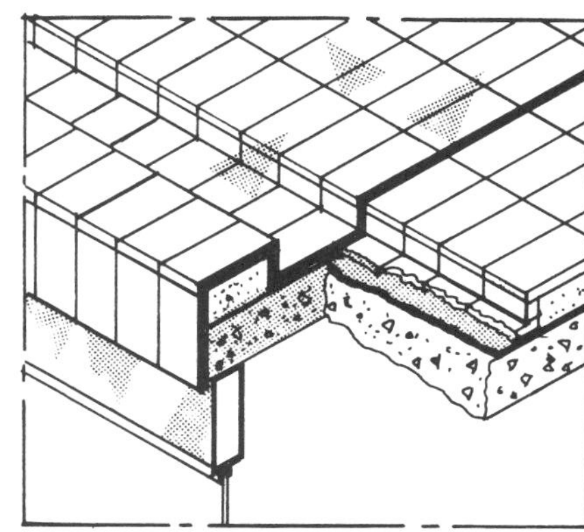

8.3.14 LEAKAGES ASSOCIATED WITH TILED ROOF TERRACES

Symptoms

Leaking from an overhead balcony or roof terrace.

The heads of windows (or the wall) beneath a roof terrace become streaked with lime deposits, and with tar as well where an expansion joint comes to the edge.

Investigation

Establish the construction of the roof terrace, the material used for the waterproofing, and the detail at the junction of the outside wall.

Check the provisions that have been made to drain water from the terrace. Check the bedding material of the tiles.

Examine the grouting between the tiles for cracks.

Look for evidence of frost damage causing loss of adhesion to the tiles.

Diagnosis and cure

Water dripping from the underside of an overhead terrace is most likely due to failure of the waterproofing. Mopped bitumen is not an effective waterproofer as it is prone to shrinkage cracking following drying-out.

Where the window heads (or wall) beneath a roof terrace have become stained, the bedding and grouting of the tiles may be the cause of the failure. No matter how carefully grouting is carried out, shrinkage cracks are almost bound to occur. Where tiles are bedded in cement/lime/sand mortar, and no provision has been made for drainage at slab level, water that seeps through into the tile bedding will cause saturation and dissolve the lime. This entrapped water is likely to seep out at the edge of the slab and cause staining. The problem is exacerbated if the terrace has been designed without drainage falls. The water entrapped in the tile bedding is also likely to contain caustic alkali. This will dissolve the tar out of any board filler in expansion joints so that it too will run out onto the wall. Such entrapped water will also probably be the cause of frost damage in the future.

The only satisfactory cure to both these problems is to take up the tiles on the terrace and relay them. In the case of the ineffective waterproofer, a new waterproof sheet membrane (of a minimum thickness of 0.5 mm) should be laid and sealed at all laps and, where appropriate, turned up at the wall and outer edges.

Where the terrace lacks falls, these should be provided in both the tiled surface and the waterproofed surface below. This can be done either by screeding to falls or by installing proprietary tapered cork insulation boards. Drainage should be provided to both the waterproofing and to the tiled surface.

An air entrained mortar should be used for bedding the tiles.

It is always preferable to use structural movement joint fillers that do not contain any constituents that can be leached out, even when in theory water will not get to them.

If lifting the tiles and carrying out structural repairs is considered inappropriate (or too expensive), the problem may be alleviated by providing a canopy to the terrace to protect it from the rain.

REFERENCES

Bickerdike, Allen, Rich and Partners in association with Turlogh O'Brien *Design failures in buildings 2nd Series.* Building Failure Sheet No 13 Building Failure Sheet No 14 *Balcony waterproofing. Terraced roof above windows.* George Godwin Ltd.

Duell John *Hard paving for terraces roof gardens and decks.* Architects' Journal 24.6.87 Architectural Press Ltd

1 Aluminium face lifting from felt base.

2 Aluminium felt cut to rainwater outlet, rather than being dressed into it.

8.3.15 FAILURE OF ALUMINIUM FACED FELT

Symptoms

Delamination of protective aluminium foil from bitumen backing sheet and associated entrapment of water between roofing felt plys.

Failure of capping sheet at rainwater outlets.

Investigation

Establish the construction of the roof, the edge detail(s), and rainwater outlet detail(s).

Investigate whether the manufacturers recommendations have been followed, particularly with regard to chamfering of cut edges.

Establish the extent of delamination of the capping foil from the felt membrane.

Diagnosis and cure

The most likely cause of failure with this material is that the manufacturer's instructions for laying have not been properly complied with.

Loose foil in general can usually be reapplied by using either a bitumen adhesive or by peeling back the foil, softening the bitumen coating with a small blow torch and pressing the facing back. Following the rebonding of loose facings, any bare patches should be coated with a bitumen/aluminium flake paste which will re-establish solar protection and improve the appearance.

When failure has occurred in a gutter bed, the old cap sheet should be removed and the underfelt prepared to receive a new cap sheet. It is essential that the sheets be correctly dressed to rainwater outlets. Those that have an external clamping ring introduced into the outlets after the felts have been formed are to be preferred.

REFERENCES

Construction Feedback Digest 59. *Defects of aluminium faced roofing felts, parapet gutters, and associated works.* B & M Publications (London) Ltd

1 Gravel blown into heap leaving areas of roof bare.

8.3.16 WIND SCOUR OF GRAVEL BALLAST

Loose laid gravel ballast may be specified to provide protection from ultra violet degradation to a bitumen felt membrane on a flat roof, or to anchor insulation slabs against wind uplift. Occasionally the wind can also scour chippings laid on top of felt roofings for solar protection.

Symptoms

The gravel (or chippings) has been moved by the wind leaving areas of roofing without their covering. During exceptionally high winds, some of the gravel may be blown right off the roof and damage to property below (such as parked cars) or to people may result.

Investigation

Ascertain the form of construction, the geography of the roof, and the purpose for the presence of the gravel or chippings. (White chippings are usually specified for solar reflectance and ordinary gravel for ballast.)

Note the extent and location(s) of the scour.

Measure the size of the gravel or chippings. In exposed locations smaller sizes are unlikely to be suitable.

Ascertain whether there have been unusually strong winds immediately prior to the problem appearing. The stronger the wind the greater the likelihood of scour occurring.

Diagnosis and cure

The windspeed required to blow gravel off a flat roof depends on three factors

- The size of the stones
- The aerodynamic force exerted by the wind on the stones. (This depends on the detailed structure of the airflow over the stones and the nature of the terrain upwind of the building. The orientation and geography of the building, in addition to the nominal wind speed also have an effect.)
- The height and distance the stones have to travel to leave the roof top.

BRE Digest 311 gives methods to calculate whether wind scour is likely and calculations to ascertain the appropriate sizes for stones and need to be replaced with paving stones, providing the roof structure can accommodate the additional loading.

REFERENCES

BRE Digest 119: *The assessment of wind loads*

BRE Digest 283: *The assessment of wind speeds over topography*

BRE Digest 284: *Wind loads on canopy roofs*

BRE Digest 295: *Stability under wind loads of loose-laid external roof insulation boards*

BRE Digest 311: *Wind scour of gravel ballast on roofs*

1 Working in bad weather may produce a permanently bad roof.

2 Felt cracked and lap joint opening.

3 Badly blistering felt.

8.3.17 CRACKED FELT MEMBRANE AND LOSS OF ADHESION OF FELT MEMBRANE

Symptoms

Cracked or cockled felt membrane accompanied by blisters and patches of membrane unattached to the deck.

Investigation

Establish the form of construction, and the type of membrane.

If a screed has been incorporated beneath the membrane, discover the provisions that were made to allow it dry out, and the measures taken to protect it from the weather prior to the membrane being laid.

Try to ascertain the manufacturer's recommendations for laying the membrane.

Investigate the weather conditions immediately prior to and during the laying of the membrane.

Diagnosis and cure

Most surface cocklings of felt membranes are due to one of two causes; either movement in the deck beneath (see 8.3.2) or the presence of water beneath the membrane or trapped within the different layers of the membrane. The latter cause can be due either to construction water that has not drained or dried out, or to rainwater that has fallen during the application of the membrane. In summer the heat of the sun will cause the entrapped moisture to vaporise and expand, and thus create blisters, ripples and cockles. Remedial treatment will depend on the extent of the problem. If the defects have not led to leakage, there is no need to repair the felt immediately. However, it should be kept under observation. Entrapped water in a screed or roof deck should be dried out. This can be assisted by the installation of proprietory screed vents or of box ventilators.

If it is suspected that the membrane was laid in cold weather conditions (ie less than 5°C), the membrane may be poorly adhered, or even damaged due to the cold. If a large area of the roof is affected, complete stripping and recovering may be necessary.

REFERENCES

BRE Digest 8: *Built up roof felts*

BRE Defect Action Sheet 63: *Flat or low pitched roofs: laying flexible membranes when weather may be bad*

BS CP 144 Roof covering: Part 3: 1970 *Built up bitumen felt*

Movement joints: detail at brick dividing wall between two timber decks

1 Large plastic headed screws with washers
2 Non-ferrous or grp cover flashing
3 Code 4 lead flashing
4 Pre-treated timber battens
5 Non-combustible filler cavity closer (fire stop)
6 Large-headed felt nails
7 Granule-faced HP elastomeric felt strapping
8 Angle fillet 50 × 50mm (minimum)
9 Insulation
10 Vapour-barrier/check
11 Plywood
12 Timber joists

Note: Roof movement joints must coincide with parapet wall joints

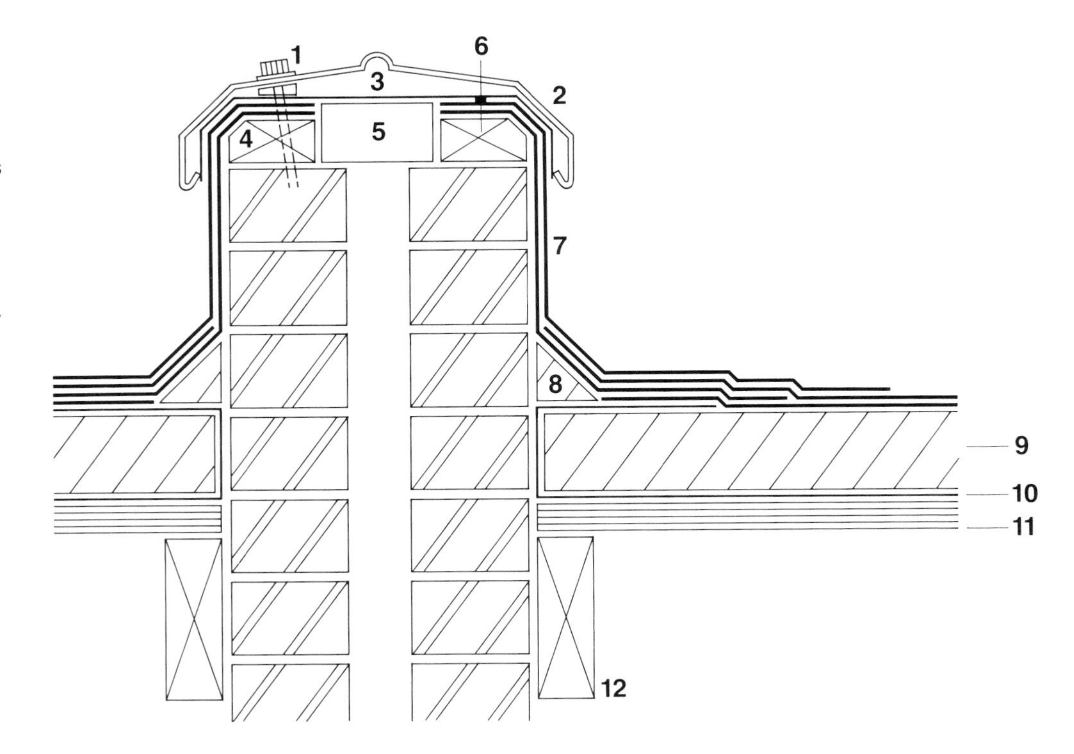

Flush movement joint on concrete deck (inverted roof)

1 Concrete paving slabs on pads
2 Insulation
3 Granule-faced HP elastomeric felt strapping
4 Dip formed in capping sheet
5 100mm wide debonding building paper
6 Built-up bitumen felt
7 Pre-treated timber blocks bolted to concrete
8 Non-combustible insulation
9 Concrete deck

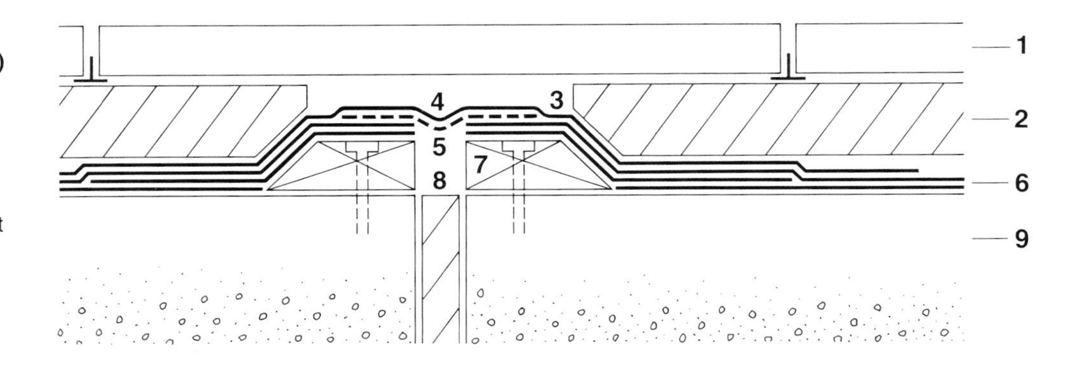

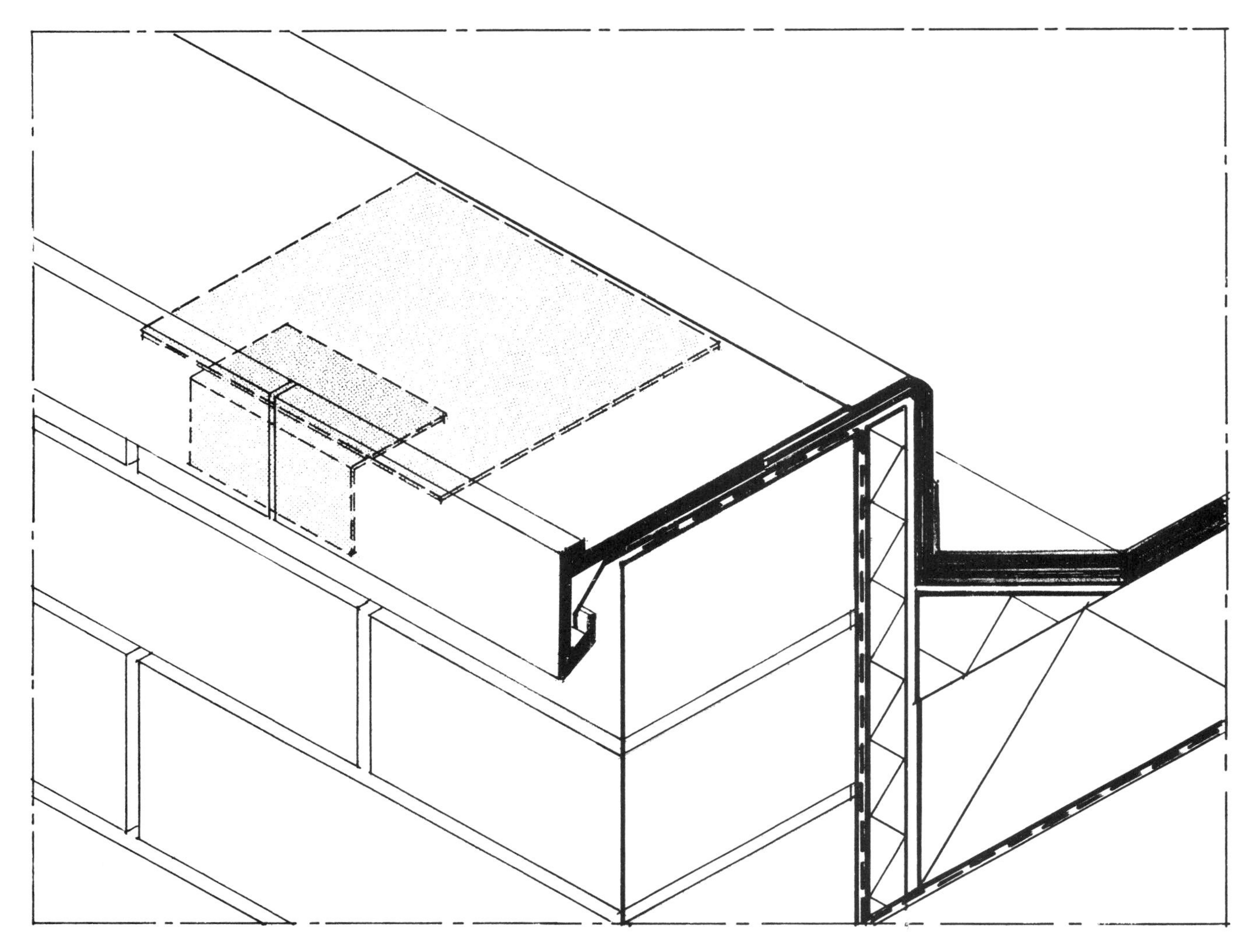

1 Detail of junction in aluminium trim with felt patches reinforcing the junction.

8.3.18 SPLITS IN ROOF MEMBRANES AT TRIM

Symptoms

Splitting of felt membrane at junctions in aluminium edge trim and at the back edge of the trim.

Investigation

Establish the form of construction and the type of membrane.

Examine the fixing of the aluminium trim and note at what centres it has been secured. Ascertain if any reinforcement has been provided at junctions in the trim.

Diagnosis and cure

This defect is caused by differential thermal expansion movement of the aluminium trim relative to that of the felt. It is exacerbated when there are insufficient fixings to the trim.

Remedial treatment will depend on the extent of the problem. If there are only a few splits at junctions of the trim, a simple patch of high performance felt partially bonded over the butt joints may be sufficient to effect a cure.

Alternatively, particularly if the trim was initially installed with fixings at more than 300 mm centres, the trim could be removed and refitted. Make sure the first layer of felt is dressed into the base of the trim, and strips of felt are inserted under each butt joint to reinforce it. It is essential that the trim be fixed at not less than 300 mm centres and a 3 mm gap for expansion be allowed at each butt joint.

If the condition of the edge is bad enough to warrant removal of the trim, consideration should also be given to replacing the trim detail with that of a welted drip.

REFERENCES

Design Note 46: *Maintenance and renewal in educational buildings Flat roofs Criteria and methods of assessment, repair and replacement.* Department of Education and Science Architects and Building Group.

BS CP 144 Roof Covering: Part 3: 1970 *Built up bitumen felt*

1 Characteristic degradation and splitting of felt under junctions of paving slab.

8.3.19 DEGRADATION OF FELT ROOF UNDER CRACKS IN PAVING SLABS

This defect occurs when a felt membrane has been covered with concrete paving slabs.

Symptoms

Water penetration internally. The presence of the paving slabs means that it is difficult to observe this defect externally before the slabs are removed.

Investigation

Establish the form of construction, and the type of membrane.

Carefully remove the paving slabs ensuring that no secondary damage to the membrane occurs.

Diagnosis and cure

This defect occurs where the narrow strips of felt under the junctions of the paving slabs are subjected to both different exposure conditions and to movement from the slabs. This leads to the felt first losing its strength and flexibility followed second by splitting. The only long term remedial treatment is to dry out and seal all the splits in the felt with bitumen, and then to overlay the existing membrane with a BS 747 high performance material. If the existing roof has poor thermal insulation, it would be appropriate to consider incorporating additional insulation when renewing the membrane.

REFERENCES

Design Note 46: *Maintenance and renewal in educational buildings Flat roofs Criteria and methods of assessment, repair and replacement.* Department of Education and Science Architects and Building Group

BS CP 144 Roof Covering: Part 3: 1970 *Built up bitumen felt*

BS 747: 1977 (1986) *Specification for roofing felts*

CHAPTER 9 BUILDING SERVICES

CHAPTER 9 BUILDING SERVICES Contents

SERVICES

Henry Eldridge's original book included a few defects related to services. In this revision several more have been added, but no attempt has been made to tackle the full range of defects that occur in this area. Rather the aim has been, to pick those that an architect or building surveyor is most likely to face without the help of specialist contractors and services engineers. In particular to select defects relating to the interaction of services and the building fabric.

The diagnosis of faults in service installations can be approached as a step by step process. Each part of an installation is connected and faults can often be followed through in a logical sequence. However, information from outside the installation may also help with the diagnosis – for example, periods of peak usage or external temperature extremes, can cause breakdowns.

Another aspect of services defects is the predictability of breakdown of the components, especially some mechanical and electrical components. Such items as immersion heaters, lift door gear, escalators, pumps and lamps have known life expectancy. They may exceed this without failing, but after a known period of time at a certain intensity of use, the risk of failure becomes so great that general replacement of all components of a particular type becomes a more economical proposition than piecemeal replacement in response to breakdowns. Where replacement involves shutdown of the installation, or time consuming and expensive access arrangements, general replacement before failure occurs is more likely to be acceptable.

It is fairly common to find that defects in service installations originate from careless workmanship. Building materials may be left in the drainage system, the insulation of electrical circuits may be faulty, concealed heating pipes may weep. Such faults can be picked up before the building is in use if tests are carried out when the installation is first commissioned. Details of such tests, and what was done to correct any faults that may have been discovered, may be valuable evidence for the investigation of defects that subsequently come to light.

One area in which architects and building surveyors tend to be drawn into when assessing the performance of services installations, is problems associated with condensation. The temperature and relative humidity of the air within the building becomes an essential part of such an investigation. Although mention is made of condensation in several chapters of this book, no attempt has been made to suggest ways of curing condensation by carefully balanced alterations to the heating and ventilating installations. Specialists advice is likely to be needed. Some of the references given here may help to focus on what improvement should be considered.

REFERENCES

Condensation in Dwellings Part 1 *A Design guide* Part 2 *Remedial measures* HMSO

BRE Digest 110: 1972 *Condensation*

BRE Digest 297:1985 *Surface condensation and mould growth in traditionally built dwellings*

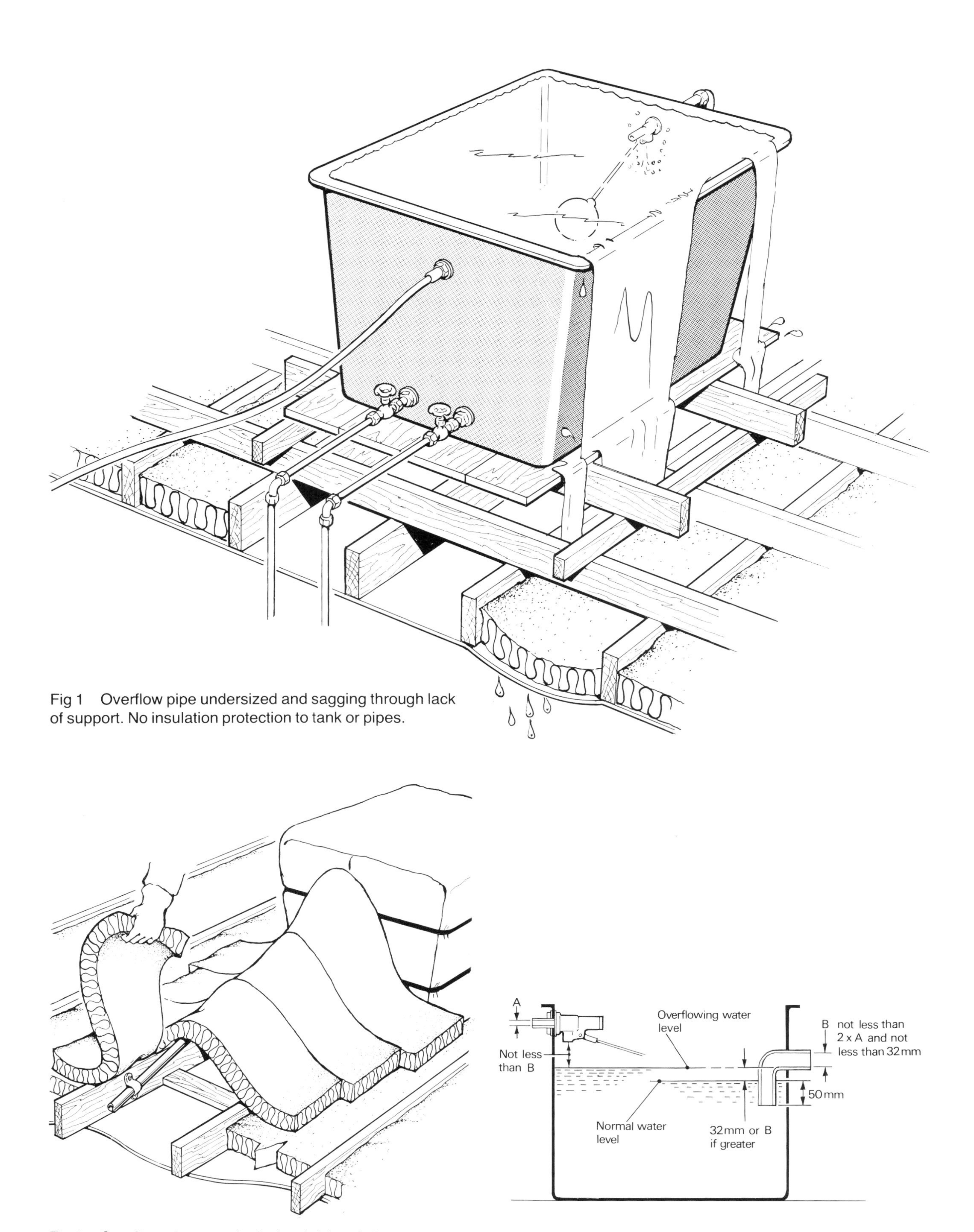

Fig 1 Overflow pipe undersized and sagging through lack of support. No insulation protection to tank or pipes.

Fig 3 Overflow pipe must be below loft insulation to prevent freezing.

Fig 2 Diagrammatic presentation of minimum dimensional requirements for cold water storage tank overflow.

9.1.1 COLD WATER STORAGE CISTERNS. OVERFLOWING OR NOT FILLING

Major damage to ceilings, decorations, furnishings and electrical wiring can occur through cisterns overflowing.

Symptoms

Water pours through overflow pipes or water supply becomes disrupted. Water drips through the ceiling and in severe cases the ceiling may collapse with extensive flooding into the room below. Fuses might blow and ceiling lights fail to illuminate.

Investigation

Turn off the mains water supply. Remove the float and shake it, listening for any movement of water inside. Check the inlet valve, examining it for any restriction of movement that may be caused by grit, sand or metal filings. Examine the washer for any splits or signs of incomplete contact with the nozzle from which the water flows. Examine the end of the nozzle for any wear of the seating.

Measure the inside diameter of the overflow pipe and compare it with that of the inlet pipe, check the level of the overflow to determine the level that water in the cistern will reach before excess water will discharge. Is the cistern adequately insulated and is the insulation continued over the overflow pipe? Is the overflow pipe installed with sufficient fall?

Diagnosis and cure

The inlet valve

There are various reasons why ball valves fail to function correctly. The float, if made of copper, may corrode and fill with water or in certain cases may even fall off. In either case the valve won't shut, allowing water to flow freely into the cistern causing it to overflow. Brass nozzles may corrode and make incomplete contact with their washer causing the valve to leak. The washer may wear or split. Grit or other material may cause the plunger to stick or it may become coated with scale.

Defective floats should be replaced by plastics floats; worn nozzle seatings can be reseated by means of a special tool or be replaced, preferably by one made of nylon; grit or other loose material can be flushed out or dislodged with the aid of a bottle brush; lime scale deposits can be removed with the aid of dilute acid, make sure the valve is thoroughly washed with clean water before refitting it; washers can be readily replaced.

The overflow pipe

The pipe may not be of sufficient diameter to handle a full flow of water into the cistern; it may be placed at the wrong level; it may have insufficient gradient or be inadequately supported to prevent it sagging; it may not be properly lagged.

It is now accepted that the overflow pipe in a new installation should have a bore at least twice that of the inlet pipe. In existing installations it should not be less than 32 mm; it is good practice for it to be turned down in the cistern to terminate 50 mm below normal water level to prevent the entry of cold outside air and floating debris. The pipe should have the maximum possible slope, preferably not less than 1 in 10 and should be adequately supported to prevent deflection. Water discharging from an overflow has only the force of gravity to motivate it, whereas the inlet flow will be under mains water supply pressure. The invert level of the overflow should be 32 mm above normal water level or equal to the outlet bore diameter if this is greater. The discharge height of the ball valve must not be less than the outlet bore diameter at overflowing water level. The overflow pipe should be protected from freezing and this is best achieved by laying quilt over the pipe making sure that it overlaps the loft insulation. Make sure that the overflow outlet projects from the building by enough to prevent any water from hitting the side of the building. It is unwise to have a hinged flap at the outlet point as they have a tendency to bind with age and are liable to freeze shut. If it is necessary to reposition the overflow pipe on the cistern, make sure that the original outlet point is capped and sealed. Check it to make sure that it does not leak.

REFERENCES

BS 6700: 1987 *Specification for design, installation, testing and maintenance of services supplying water for domestic use within buildings and their curtilages*

BRE Defect Action Sheet 61 *Cold water cisterns; overflow pipes*

Department of the Environment, Housing development note V *Water services in housing.* Part 1 *Performance requirements for cold water services.* Part 2 *Cold water installations*

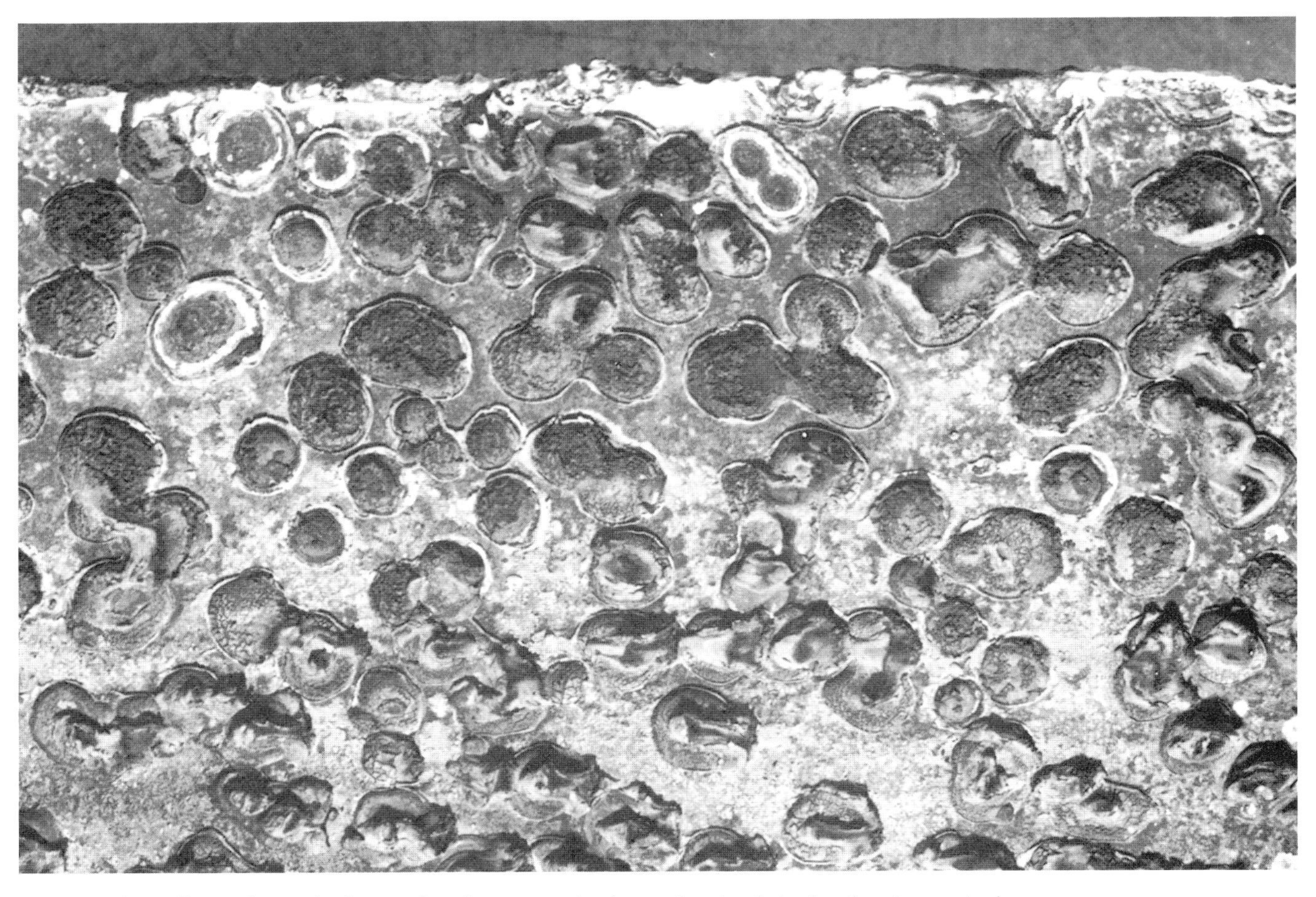

Photograph 1 Extensive and advanced surface corrosion in a galvanised steel water storage tank.

9.1.2 WATER SUPPLY. GALVANISED COLD WATER STORAGE TANKS. LEAKS

This can occur in tanks of all sizes from those supplying a single domestic unit to those supplying whole blocks of flats.

Symptoms

Water leaks from the bottom or sides of the tank, with possible rust stains on the inside of the tank and on the outside where the leaks are occurring. There may also be signs of rust on the inside, isolated from the points of leakage. Water drawn from the tank could be discoloured and may taste unpalatable.

Investigation

Turn off the mains water supply and drain all the water from the tank, first noting any obvious external leakage points. These may be in the form of very fine sprays of water which can be detected by passing the hand close to the surface of the tank in areas where dampness is occurring.

Examine both outside and inside surfaces of the tank noting all areas of rust. Are there any foreign objects in the tank? Look for the obvious, such as metal tools as well as the not so obvious, such as iron, brass and copper filings, though at a late stage they may not be very recognisable. Are there any scratch marks on the inside?

Diagnosis and cure

Acid or soft waters may attack zinc so that it no longer provides a protective coating to steel. This attack will be made much worse by the presence of any extraneous metal items that are allowed to remain in the tank. If the rusting is only localised, it may be possible to effect a patch repair. If the rusting is more extensive, but not weakening the tank structure, it should be possible to remove it and paint the inside of the tank with a non-tainting bitumen solution. Alternatively the tank can be lined with a plastics liner that forms a continuous bag inside the tank.

If the tank is badly corroded, there is no alternative but to replace it. Consider using a different material such as plastics or fibreglass.

REFERENCES

BS 417: Part 2:1973 *Specification for galvanised mild steel cisterns and covers, tanks and cylinders*

BS 6700:1987 *Specification for design, installation, testing and maintenance of services supplying water for domestic use within buildings and their curtilages*

BRE Digest 98: *Durability of metals in natural waters*

Photo 1 Galvanic corrosion occurring at the joint of a steel pipe fitted with a brass nipple.

Photo 2 Steel nipple and bend corroded to the point of failure when fitted to a copper cylinder.

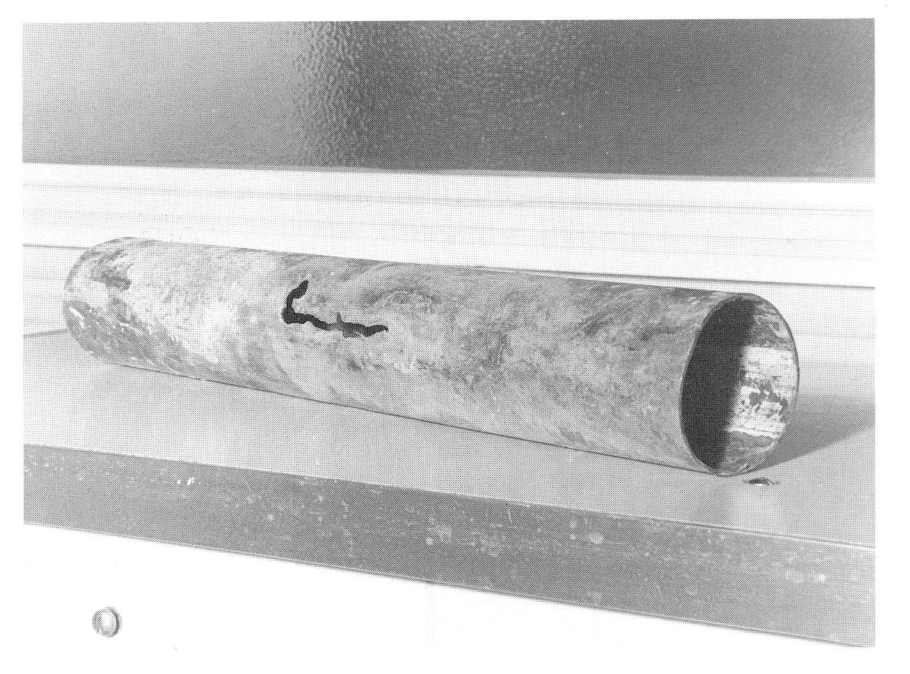

Photo 3 A failed section of copper pipe caused by use of bleach cleaning powders and acid-based descaling fluids.

Photo 4 Samples of isolating fittings to prevent surface to surface contact between different metals.

9.1.3 WATER SUPPLY. MIXED METAL SYSTEMS. LEAKS

This occurs in water supply systems where two different metals are in close contact with each other. For other types of leaking joints refer to 9.1.4.

Symptoms

Leakage of water from the system, more commonly through perforations in the tank (see item 9.1.2), in the vicinity of joints in the system or by valves, taps etc. The fact that there may be more than one type of metal in the system may not be obvious until investigation.

Investigation

Mark any points of the system where leaks are occurring before turning off the water supply at an isolating valve or at the mains stopcock. It will probably be necessary to remove suspect defective parts so that they can be sent to a specialist for chemical or metallurgical examination since the main evidence is likely to be on the inside of the fitting. Check out the rest of the system to see if similar fittings have been used elsewhere. There may not be any leaks at present but there probably will be in the future. Check any metal washers that may be of a different metal to the main system.

Contact the local water authority to find out the cupro-solvency of the water in the area.

Diagnosis and cure

If two or more different metals are present in a system there will always be a risk of bi-metallic corrosion occurring. The risk is especially high when copper or copper-bearing alloys are associated with steel or zinc-coated metals or alloys. It is fairly common to see hot water systems using a steel tank with copper pipes and brass flanges, nuts etc. Corrosion may develop at any point that these dissimilar metals are in contact, but in areas where the water is cupro-solvent the dissolved copper may affect another metal in the system at a considerable distance from the sources of the copper.

Metal filings may have got into the system when it was assembled, but these should be easy to identify and dispose of.

Ideally the system should be converted to using the same base metal throughout, but where this is not practicable it may be possible to minimize metal to metal contact by using plastics washers and pipe-connectors. If this is done, check that the system was not being used as the earth carrier for the electrical circuit in the building. Plastics washers and pipe connectors will not be effective if the water proved to be cupro-solvent or if the corrosion has reached an advanced stage. In closed circuit hot water systems it will be a useful precaution to include a corrosion inhibitor in the water.

REFERENCES

BS 6700:1987 *Specification for design, installation, testing and maintenance of services supplying water for domestic use within buildings and their curtilages*

BRE Digest 98: *Durability of metals in natural waters*

Copper Development Association Information Sheet 36 *Dezincification resistant brass*

Photo 1 Very obvious signs of water leakage at the joints of this steel pipe.

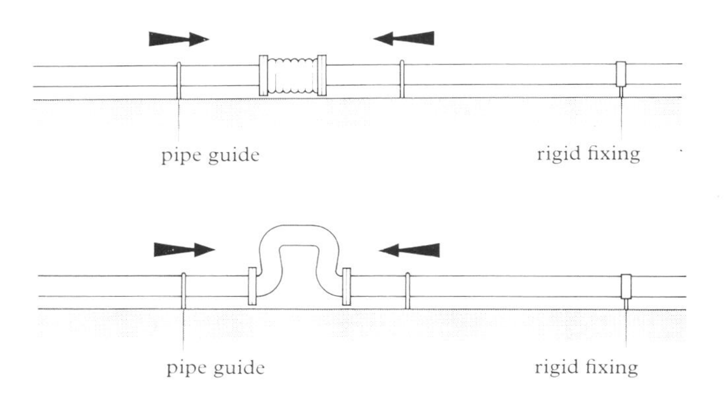

Figs 1 & 2 Two different methods of providing for expansion in long pipe runs.

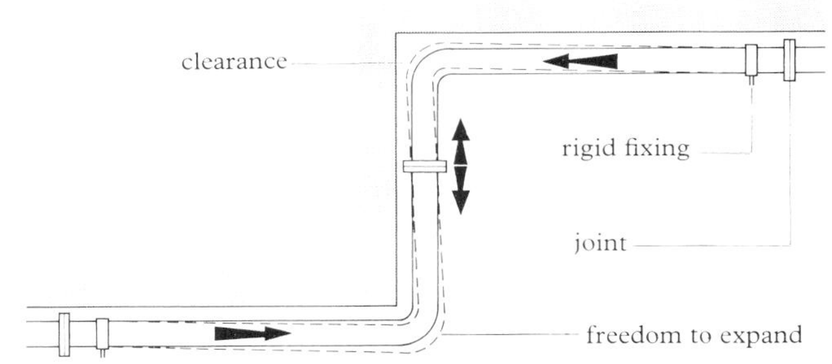

Fig 3 Allowance must also be made for pipe expansion at internal and external corners.

9.1.4 WATER SUPPLY. PIPES LEAKING AT JOINTS

Although the majority of leaks at the joints in a plumbing system will be due to deficiencies in workmanship, some may be due to the design.

Symptoms

Leaks may be very small and only visible as rust or corrosion marks on the outside of the pipes. Larger leaks tend to be very obvious from clues such as puddles on the floor.

Investigation

Any investigation required should generally be confined to noting the material of the pipes and fittings, the type and frequency of joint, the spacing of fixing points and the straight line length of pipe runs. It may help to pinpoint a remedy if the extent and history of the leak can be established.

Diagnosis and cure

Pipe joints leak for several different reasons and sometimes they occur in combination.

Older pipe installations were almost exclusively made out of cast or spun iron. These pipes are impossible to bend therefore any change in direction can only be achieved by using screwed elbows, angles, tees etc. Straight pipe runs are jointed with screw-on sleeves. When the joints leak it can usually be attributed to insufficient allowance for expansion and contraction, particularly where the pipes are conveying hot fluids. The strain on joints caused by long straight pipe runs should be relieved by either forming expansion loops, introducing changes of direction or by fitting proprietory expansion bellows. Any minor leaks at joints may be cured by tightening or remaking the joint.

If two disimilar metals have been used to make the joint, a leak may occur due to bi-metallic corrosion (see item 9.1.3).

In copper tube based systems the joints may be formed using capillary, compression and very occasionally plastics 'push-fit' connectors.

Capillary fittings may leak because the joint wasn't made properly, but leaks of this nature should be discovered when the system is tested. A far more serious problem is the chemical action of fluxes used when making the joints. They may cause the pipe to become perforated and any affected part must be cut out and replaced, taking care to remove all traces of any harmful material.

Compression joints may leak if not thoroughly tightened, especially if they are located in areas with limited access. More commonly they leak through thermal contraction of the pipework particularly if the pipes are too tightly restrained causing them to pull out of the joint. If the original olives have become deformed, the joint should be remade using new ones; any restraint of the pipework should be relieved by allowing more clearance at the obstruction.

Plastics pipes have not until very recently been recommended for use with hot water systems because of their high coefficient of expansion and their tendency to soften. Where they are used for cold water supply they are normally jointed with 'push-fit' connectors which will remain watertight except under severe mechanical shock conditions. Any faulty connector or fitting can be replaced with comparative ease.

REFERENCES

BSCP 342: Part 2: 1974 *Code of practice for centralised hot water supply*

BS 6700: 1987 *Specification for design, installation, testing and maintenance of services supplying water for domestic use within buildings and their curtilages*

Copper Development Association Technical Notes TN22 *Design and installation guide for copper water services in building*

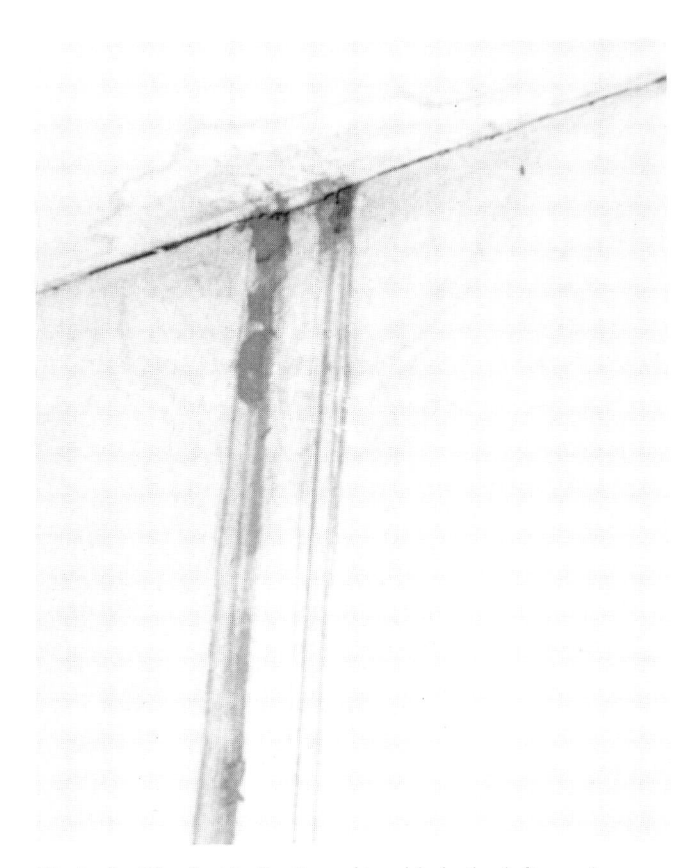

Photo 1 The first indication of trouble in the loft may be water running down the walls below.

Photo 2 A typical burst pipe which has blown at its weakest point i.e. the outside of a bend.

Outside diameter of pipe mm (nominal)	Thermal conductivity of insulating materials not exceeding: 0.035 W/(m.K)		0.04 W/(m.K)		0.055 W/(m.K)		0.07 W/(m.K)	
	Indoor mm	Outdoor mm	Indoor mm	Outdoor mm	Indoor mm	Outdoor mm	Indoor mm	Outdoor mm
Up to 15	22	27	32	38	50	63	89	100
Over 15 up to 22	22	27	32	38	50	63	75	100
Over 22 up to 42	22	27	32	38	50	63	75	89
Over 42 up to 54	16	19	25	32	44	50	63	75
Over 54 up to 76.1	13	16	25	25	32	44	50	63
Over 76.1 and flat surfaces	13	16	19	25	25	32	38	50
typical insulating materials	Polyurethane foam or foamed or expanded plastics, including rigid and flexible preformed pipe insulation of these materials		Corkboard			Exfoliated vermiculite (loose fill)		

Table 1 Minimum thicknesses of pipe insulation materials as specified in BS 6700 : 1987.

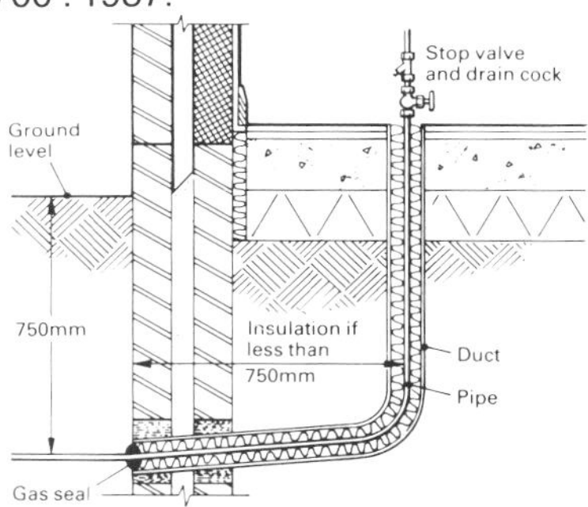

Figure 1 Precautions required for cold water mains supply pipes for dwellings with a solid ground floor.

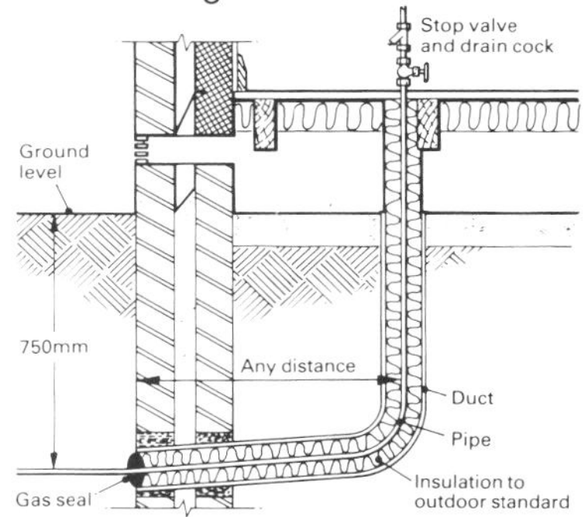

Figure 2 Precautions required for cold water mains supply pipes for dwellings with a suspended ground floor.

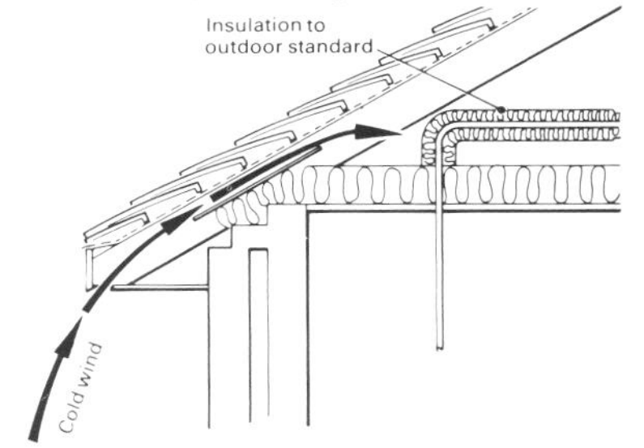

Figure 3 Exposed pipes in well insulated lofts must be insulated to outdoor standards.

9.1.5 WATER SUPPLY. FROZEN PIPES

The cost of providing protection against freezing in pipes is small when compared with the cost of repairs to damaged property caused by a burst pipe.

Symptoms

The water supply to taps is interrupted; cisterns fail to fill (see also 9.1.1). The disruption can affect either or both of the mains supply and internal services. There could be flooding or leaking water from pipes or joints and fittings. This is most likely to occur after a period of sub-zero temperature when the weather gets warmer or when a building is reheated after being left cold.

Investigation

First establish whether the disruption is localised – ask a neighbour if they have water, or phone the water authority in case they have needed to shut off the supply (in the telephone directory under 'WATER'). If the external mains pipe has frozen, shut off the supply at the stopcock which should be found just outside the boundary line. Check the depth of the supply pipe below ground level at all points along its length.

Internally, check the point of entry of the water main. It will often be in an unheated cellar or underfloor space therefore check on any insulation provision. Check any standpipe in the garden, garage or outhouse. Examine all pipework and fittings in the building, especially in areas that are cold for long periods such as well insulated lofts.

Diagnosis and cure

During the winter, water in mains pipes is likely to be fairly close to freezing point when it arrives at a consumers premises. Pipes that are outside the insulated fabric of the building will be vulnerable to freezing unless they are insulated to outdoor standard. However, insulation will only slow down the rate of heat loss, therefore the water in pipes that is not regularly drawn off may eventually freeze. External mains supply pipes must be buried deeply enough to avoid freezing. This should be 750 mm minimum at all points but may have to be greater up to a maximum of 1350 mm if local experience has shown it to be necessary. If future landscaping works are anticipated, they must be planned to avoid decreasing the depth of cover over the main pipe. The ground over the pipe must always be easy to excavate therefore avoid constructing driveways, paths etc. in the vicinity of the pipe.

Where the mains pipe enters the building, it should be maintained in an insulated duct up to inside floor level. With solid ground floors the insulation only needs to be to indoor standard but suspended ground floors will require the insulation to be to outdoor standard. All outdoor standard insulation must be wrapped with waterproof tape to avoid it getting wet.

Some pipes inside the building may be in a position that makes them more vulnerable to freezing, ie close to permanent vents, windows and external doors and in chases on external walls. All pipework in an insulated loft should be placed beneath the loft insulation and where it comes close to any eaves ventilation it will need to be insulated to outdoor standard.

When water does freeze in a pipe, it can be difficult to locate the actual point of blockage. Turn on an outlet tap that is connected to the pipe with disrupted supply. Water flowing from the tap will be an indication of successful defrosting action. Starting closest to the outlet, work on the more obvious areas of pipework applying gentle heat with the aid of cloths that have been dipped in hot water or better still by playing an electric hot air paint stripper over the pipe. The use of blowtorches is not advisable because of the fire hazards involved, (especially if there is no water to put the fire out). The more slowly the temperature is raised the less likely that a burst will occur.

Thawing out the system may reveal damage to the pipework and fittings; the supply must then be turned off until repairs have been made. As a precaution against future freezing the following actions should be considered.

- Fit heat trace cables to pipes that are at risk.
- Increase the thickness of insulation in unheated areas.
- Fit a frost stat to the heating system in buildings that are likely to be unoccupied for long periods.
- Fit easily accessible drain cocks to both hot and cold water services.
- Relocate vulnerable pipework away from high risk positions.

REFERENCES

BRE Defect Action Sheet 103: *Hot and cold water systems – protection against frost*

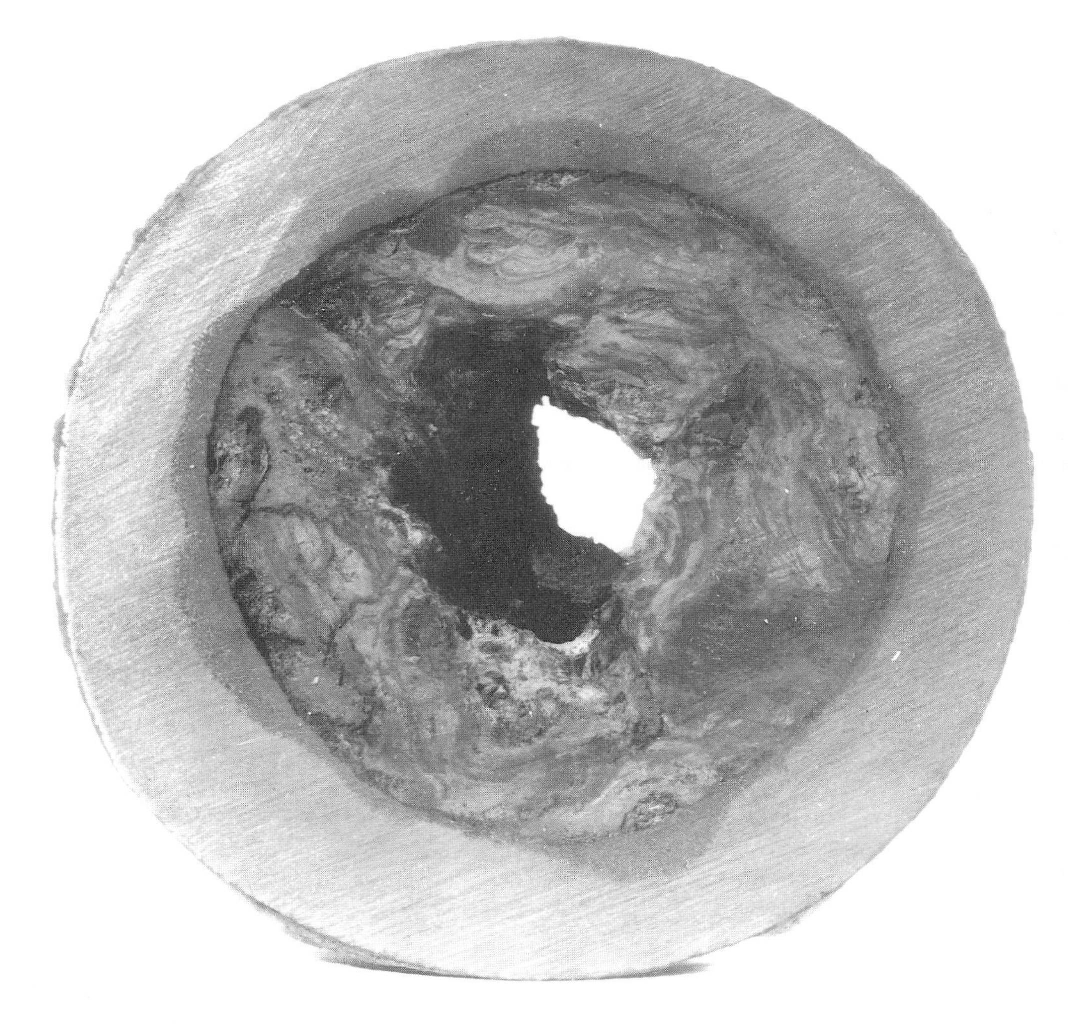

Photo 1 Typical cross section through an extremely furred-up pipe. The water flow had become very restricted at this stage.

Photo 2 Longitudinal section through a furred-up pipe where the water flow has ceased completely.

9.1.6 HOT WATER SUPPLY. RESTRICTED FLOW, FURRING OF PIPES

The risk of furring-up in hard water areas can be minimised by keeping the temperature of the hot water below 60°C and by the fitting of water softening plant or equipment.

Symptoms

Over a period of years the quantity of hot water available gradually decreases and may even cease entirely.

Investigation

If hot water is still available but in a reduced quantity, note the length of time that the boiler operates when heating the water. Note the cut-out temperature of the boiler thermostat.

If circulating pumps are fitted, check that they are functioning correctly and check that all valves that should be open, are fully open.

Remove the first length of pipe on the flow side next to the boiler. If it is not at least partially filled with scale the whole system may have to be dismantled to find the point of blockage.

Contact the local water authority to establish the hardness rating for the water.

Diagnosis and cure

In hard water areas (chalk or limestone) there may be deposition of calcium carbonate in the pipes leading from the boiler to the hot water storage tank. The bore will be gradually reduced almost to the point of complete blockage and consequently only very small amounts of hot water will be transferred to the tank.

If it is suspected that furring-up is occurring but has not reached an advanced stage, it is possible to use a chemical scale removing treatment. Because of the nature of the chemicals used, the work is best entrusted to a specialist contractor. Where treatment is not possible, it will be necessary to replace the affected pipework.

The problem of scale build-up only occurs in direct feed systems, ie where the heated water is replaced with fresh cold water. In hard water areas, consideration should be given to changing to an indirect system with a calorifier, ie where the feed pipe from the boiler passes through the hot water storage cylinder in a coiled pipe and then returns to the boiler. Alternatively it may be preferable to install a water softening treatment plant (although they are expensive to purchase and to maintain.)

If a direct system is still employed, the build-up of scale can be retarded by fitting a thermostat that restricts the boiler outlet flow to 60°C maximum.

REFERENCES

BS 2486: 1978 *Recommendations for treatment of water for land boilers*

BS 6700: 1987 *Design, installation, testing and maintenance of services supply water for domestic use within buildings and their curtilages*

GLC Development and Materials Bulletin 131: *Item 3 Cleaning and descaling of boilers in low pressure primary heating systems*

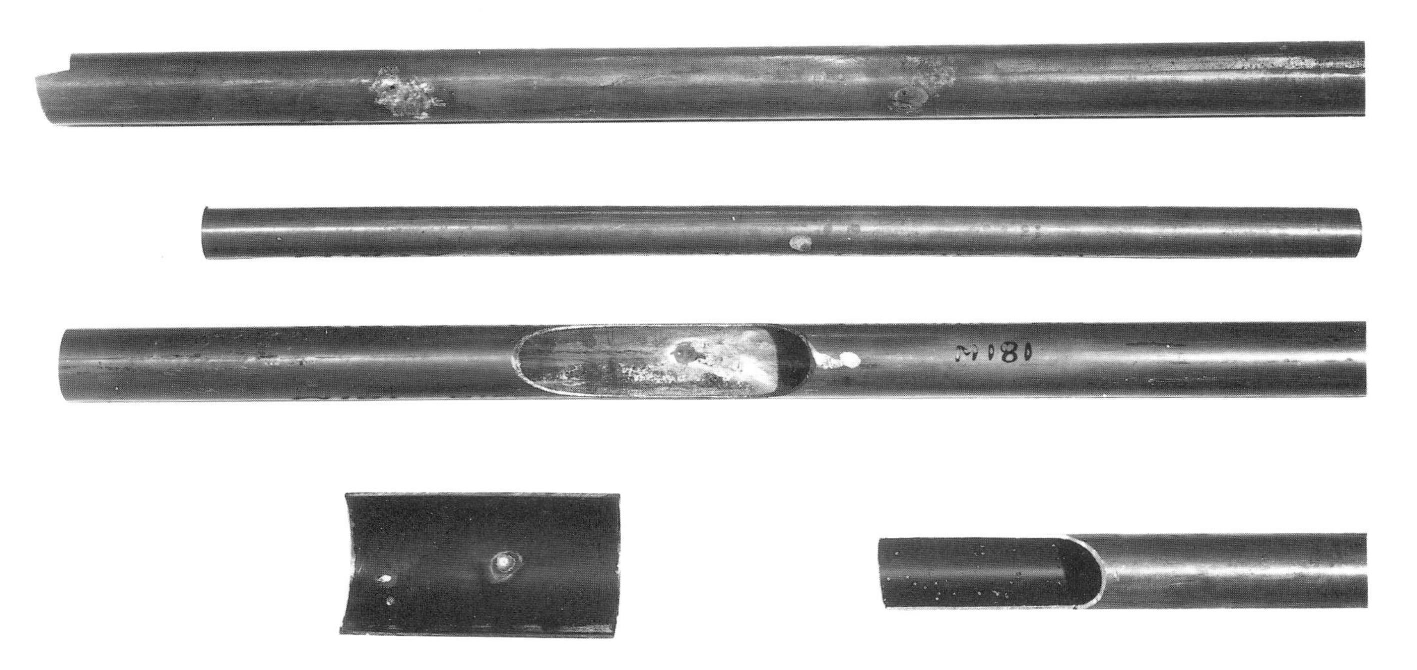

Photo 1 Pitting of copper pipes due to an internal film of carbon or oxide can cause sufficient obstruction to produce whistling and hissing noises.

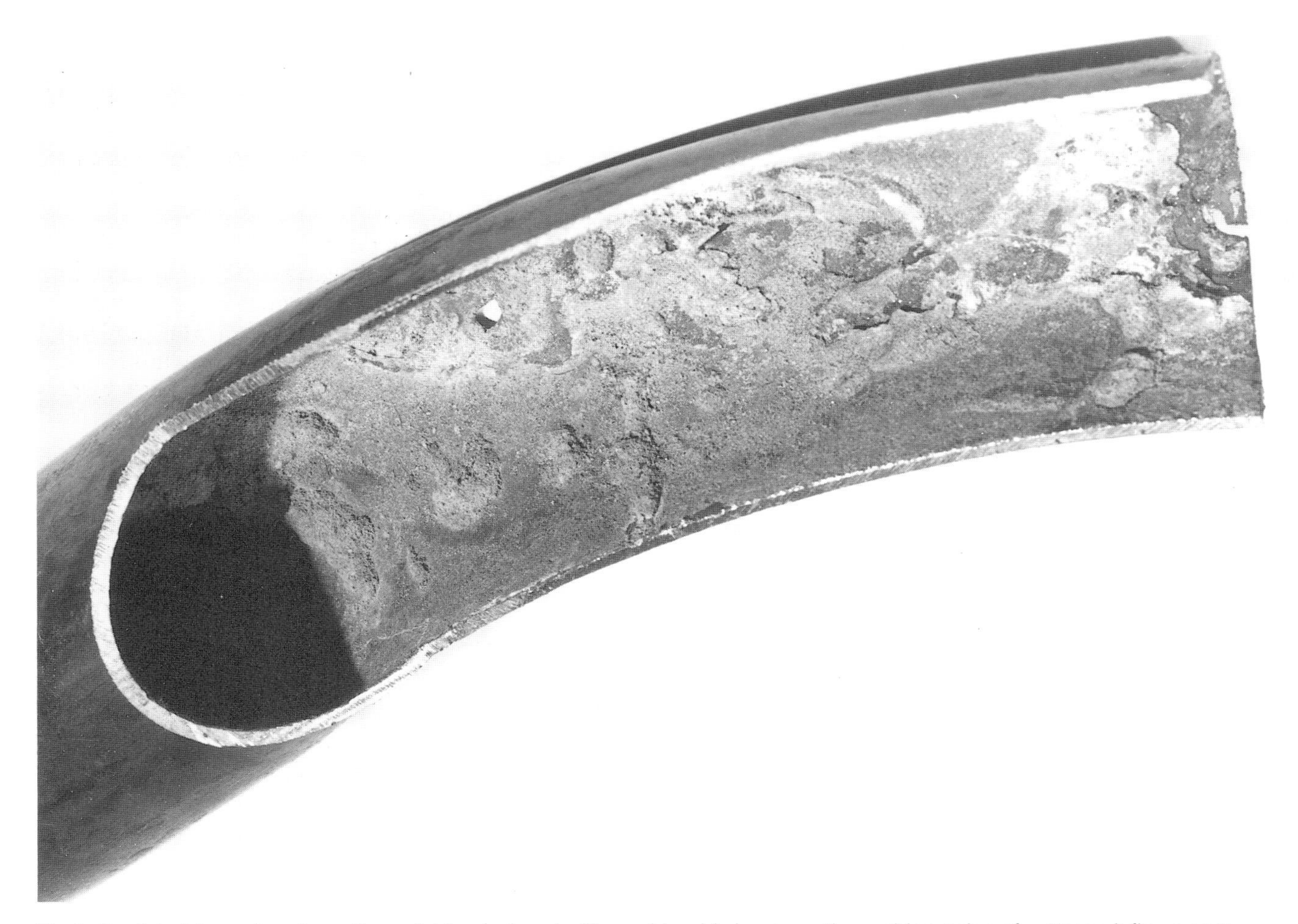

Photo 2 A build-up of scale on the outside of a bend will considerably increase the rushing noise of water as it flows past.

9.1.7 WATER SUPPLY. NOISE AND VIBRATION

There are three basic types of noise that affect water supply systems:

1 Water hammer 2 Water flow noise 3 Vibration and hum

Symptoms

1 When a tap or a valve is closed suddenly, a loud bang similar to a hammer blow on the pipework is heard. Occasionally there may be a rapid succession of bangs.

2 When a tap or a valve is opened, water can be heard rushing through the pipes. It is sometimes accompanied by a whistling or singing noise.

3 Vibration and hum can be felt and heard through the building structure. The problem occurs most commonly in large systems such as blocks of flats serviced by a central boiler house.

Investigation

1 Turn each tap on and off and operate each valve in a scheduled sequence to determine which are causing the water hammer. Examine the washer of each control that causes it. Check the operating mechanism of float operated valves for oscillation.

2 Measure the internal diameter of the pipe and calculate the flow rate of water in the pipe by drawing off a measured volume of water and noting the length of time that it takes. Check the pipework for any long vertical drops that may be causing cavitation.*

3 Examine the pipework system for any vibration isolators such as metal bellows or sections of rubber hose. Where pumps are used to circulate the water, check them for vibration and note whether they are mounted on the building structure.

Diagnosis and cure

1 The primary cause of water hammer is a sudden increase in pressure in the system caused by the rapid closure of a tap or valve. It is a common problem with self-closing taps of the kind fitted in public conveniences to prevent water wastage. With normal domestic-style taps a cure will probably be effected by simply renewing the tap washer. Old washers tend to lose their pliancy thereby shutting off the water flow too rapidly.

When there is a rapid series of bangs the culprit may be a worn jumper washer in a tap which vibrates violently, or it may be a float valve oscillating due to wear in the valve mechanism or the fitting of too small a float. Either will be cured by proper maintenance procedures.

2 Flow noise is attributable to the speed at which the water travels through the pipes. It is not significant at velocities below 3 metres per second and any whistling or hissing noises at normal water pressures will probably be caused by obstructions within the pipes. At high velocities cavitation may occur which will greatly increase flow noise. Water speeds in long vertical drops can be reduced by introducing changes of direction in the pipework.

3 Faulty pumps and pump mountings are nearly always to blame for vibration and hum in pipework. Metal pipes are extremely good transmitters of sound, therefore perceived noise may be quite some distance from the source. Small domestic installations will normally have a pump which is located close to the boiler on the flow side and mounted directly on the pipe without any contact with the structure. Vibration from these pumps can be effectively eliminated by breaking the pipe run and inserting a short reinforced rubber hose section. On larger installations metal bellows type vibration isolators will usually effect a cure. Consider also using resilient mountings for the pump and vibration-isolating pipe clips and brackets to help reduce structure-borne sound.

* Cavitation
BS 4118 describes cavitation as 'the formation and collapse of cavities in flowing water following a sharp drop in pressure'.

REFERENCES

BS 4118:1981 *Glossary of sanitation terms*

BS 6700: 987 *Design, installation, testing and maintenance of services supplying water for domestic use within buildings and their curtilages*

BRE Digest 143: *Sound insulation: basic principles*

CIBSE Guide B12 *Sound control*

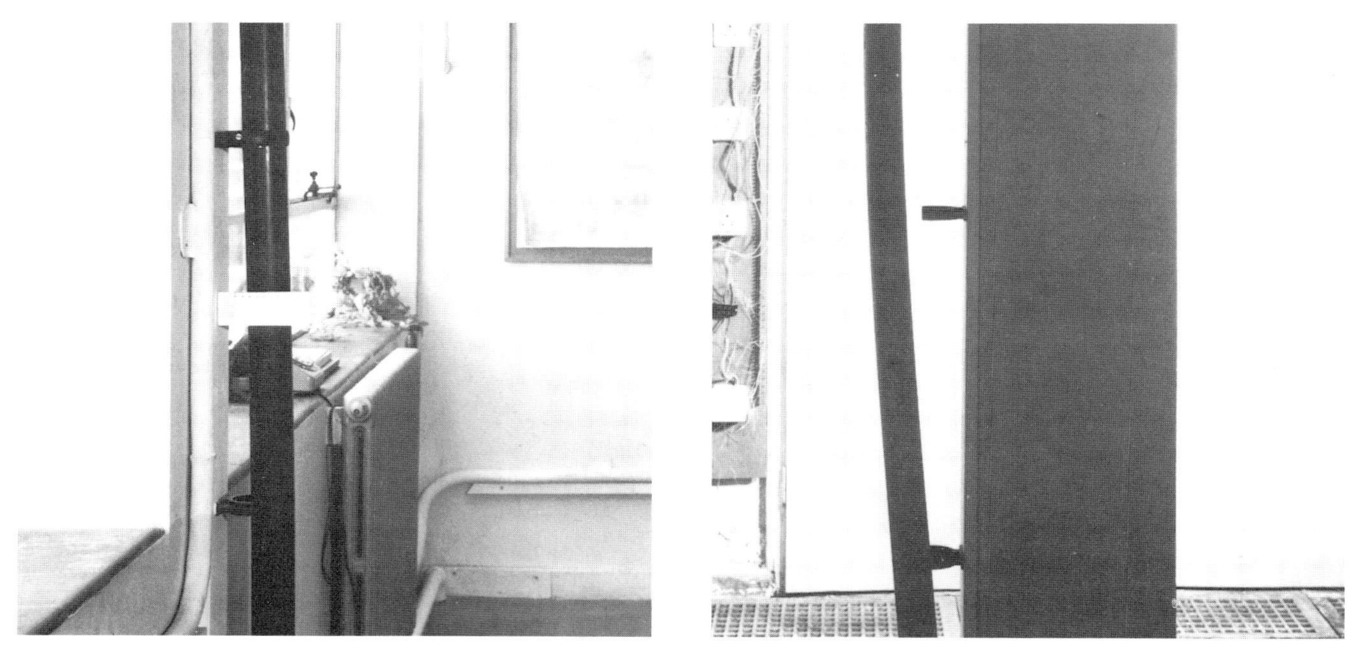

Photo 1 & 2 Plastics pipes may expand to the degree that they bow and pull away or out of their fixing clips.

Material	Density kg/m^3	Linear expansion per °C Coefficient	mm/m	Max temperature recommended for continuous operation °C	Short-term tensile strength MN/m^2	Behaviour in fire
Polythene* low density	910	20×10^{-5}	0.2	80	7–16	Melts and burns
high density	945	14×10^{-5}	0.14	104	20–38	like paraffin wax
Polypropylene	900	11×10^{-5}	0.11	120	34	Melts and burns like paraffin wax
Polymethyl methacrylate (acrylic).	1185	7×10^{-5}	0.07	80	70	Melts and burns readily
Rigid PVC (UPVC)	1395	5×10^{-5}	0.05	65	55	Melts but burns only with great difficulty
Post-chlorinated PVC (CPVC)	1300–1500	7×10^{-5}	0.07	100	55	Melts but burns only with great difficulty
Plasticised PVC	1200–1450	7×10^{-5}	0.07	40–65	10–24	Melts, may burn, depending on plasticiser used
Acetal resins	1410	8×10^{-5}	0.08	80	62	Softens and burns fairly readily
ABS	1060	7×10^{-5}	0.07	90	40	Melts and burns readily
Nylon	1120	8×10^{-5}	0.08	70–110	50–80	Melts, burns with difficulty
Polycarbonate	1200	7×10^{-5}	0.07	110	55–70	Melts, burns with difficulty
Phenolic laminates	1410	3×10^{-5}	0.03	110	80	Highly resistant to ignition
GRP laminates	1600	$2\text{–}4\times10^{-5}$	0.02	90–150	100	Usually inflammable. Relatively flame-retardant grades are available

Key: UPVC = unplasticised polyvinyl chloride GRP = glass-reinforced polyester PVC = polyvinyl chloride ABS = acrylonitrile/butadiene/styrene copolymer

*High density and low density polythene differ in their basic physical properties, the former being harder and more rigid than the latter. No distinction is drawn between them in terms of chemical properties or durability. The values shown are for typical materials but may vary considerably, depending on composition and method of manufacture.

Table 1 Typical properties of plastics used in building taken from Building Research Establishment digest 69.

9.2.1 DRAINAGE PIPEWORK. SOFTENING OF PLASTICS PIPES AND TRAPS, ABOVE GROUND

Plastics wastes and traps are now commonplace and most prove adequate under normal use, but care should be taken to ensure that the right plastics is chosen for the conditions.

Symptoms

Waste pipes bow and pull away from clip fixing brackets. Waste traps become distorted, they may start to leak and the free flow of water becomes restricted.

Investigation

A quick visual inspection will establish the extent of bowing and distortion. It will be necessary to inspect the full run of the waste pipe to check whether joints are leaking and to assess the total extent of damage.

Remove the waste trap and examine the inside for signs of pitting or scour marks.

Find out the type of plastics used – this is often printed or moulded on the components.

Check how frequently boiling water is poured down the waste and whether chemical cleaners, paint strippers, etc. could have been disposed of down the system.

Diagnosis and cure

Misunderstandings about the limitations of plastics materials in use has led to the blame for failure being aimed incorrectly at the product manufacturers. The kitchen sink has always been a favourite place to dispose of unwanted fluid waste. Discarded molten fats tend to solidify in the system causing a blockage which may be cleared by pouring boiling water down the waste. The more traditional metal waste systems could tolerate such abuse, but the new plastics systems soften and distort. ABS products have a higher softening point than uPVC products and could be considered a suitable replacement for damaged parts. The greatest hazard to plastics wastes are paint strippers, cleaning fluids and similar solvents. Although these products should never be poured down a domestic drain, they often are when decorating works are in progress. The solvents attack the inside of the trap. If left there long enough they will completely dissolve it.

All distorted and failed parts will need to be replaced with either a different plastics or another material better able to cope with its intended use.

REFERENCES

BRE Digest 60: *Durability and application of plastics*

BS 3943: 1979 *Specification for plastics waste traps*

BS 5255: 1976 *Specification for plastics waste pipe and fittings*

BS 5572: 1978 *Code of practice for sanitary pipework*

BS CP 312: Part 1: 1973 *Code of practice for plastics pipework (thermoplastics material). Part 1. General principles and choice of material*

Photo 1 A kinetic water ram drain clearer in action. The kit comes with a variety of fittings to suit different outlet sizes. Must only be used by experienced personel.

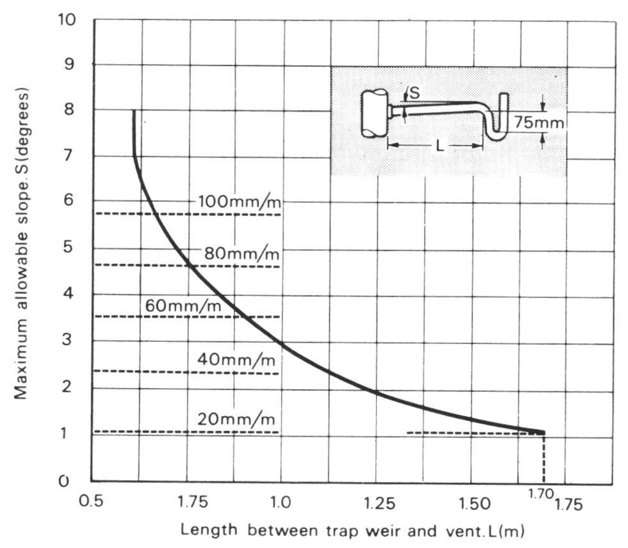

Figure 1 Design chart for branch pipes for a single wash basin showing minimum desirable slope.

9.2.2 DRAINAGE PIPEWORK. BLOCKAGE OF WASTE TRAPS, ABOVE GROUND

The gradual increase in use of mechanical sink waste disposers is helping eliminate this problem.

Symptoms

The symptoms are obvious: the flow of water down a waste outlet slows and eventually stops altogether. The defect is most commonly encountered with kitchen sinks.

Investigation

Examine the outside of the waste trap. If it is distorted refer to item 9.2.1.

Open or remove the trap to reveal the contents that may be causing the blockage (put a plug in the sink outlet first). Examine the contents to determine the nature of the obstructing material.

In the case of copper waste traps, a thorough check should be made externally and internally for corrosion products.

Diagnosis and cure

The build-up of corrosion products inside a trap will reduce its efficiency and act as a catch point for solids passing through. The reasons for the corrosion should be established and faulty components replaced.

In bath and hand wash basins, blockages can occur due to a build-up of soap and hair in the waste. With kitchen sinks the blockage is more likely to be due to an accumulation of grease and waste vegetable matter etc. Most blockages will be partially cleared with the aid of a plunger but BS 5572 recommends that the system is flushed with a strong soda crystal solution. This is done by dissolving the soda crystals in very hot water in the basin/sink at the rate of 1 Kg crystals to every 9 litres of water. When the crystals have completely dissolved, remove the plug to allow the solution to flush the trap and the discharge pipe.

In hard water areas, lime scale deposits can build up in discharge pipes and stacks, eventually causing a blockage. This can only effectively be removed with acid based descaling fluid and is therefore best entrusted to an experienced plumber. Make sure the pipes are well flushed with water to remove all traces of acid when the work has been completed. Take care to protect people with special clothing, gloves and eye shields whilst the work is in progress, and thoroughly clean down all exposed surfaces of sanitary ware with detergent cleanser afterwards.

It may be necessary to consider resiting appliances or rerouting pipework to have a greater fall giving a better self-cleansing action. Periodic flushing with soda crystal solution will help prevent further residue build-up.

REFERENCES

BS 5572: 1978 *Code of practice for sanitary pipework* (Formerly CP 304)

BRE Digest 248: *Sanitary pipework:* Part 1 *Design basis*

BRE Digest 249: *Sanitary pipework:* Part 2 *Design of pipework*

CIBSE Guide B8 *Sanitation and waste disposal*

DOE Housing Design Bulletin 30 *Services for housing: Sanitary plumbing and drainage*

Photo 1 Unprotected inspection chamber that has become completely blocked by mortar and debris during the construction period.

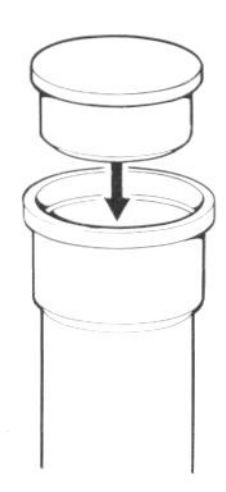

Figure 1 Standing drainage pipes should always be fitted with a temporary stopper to prevent the entry of builders' debris.

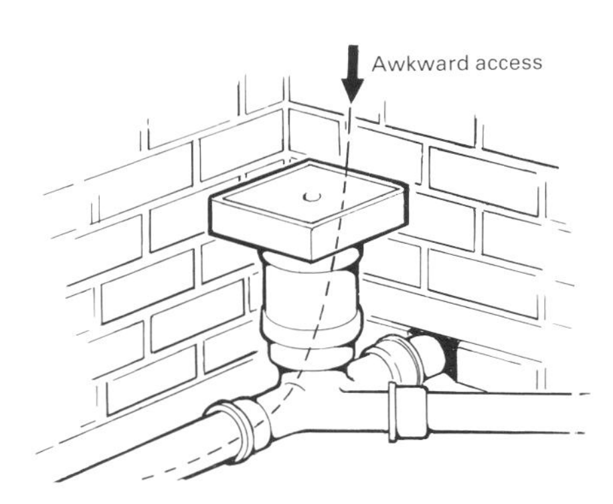

Figure 2 Access points should be positioned to provide working space for rodding.

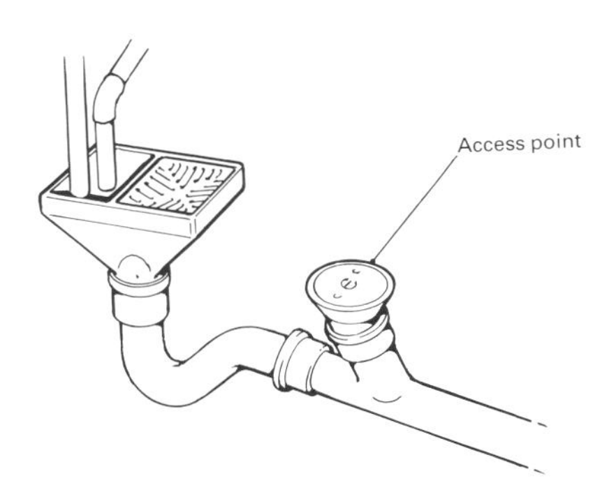

Figure 3 Access point must be correctly sited in relation to the drain run.

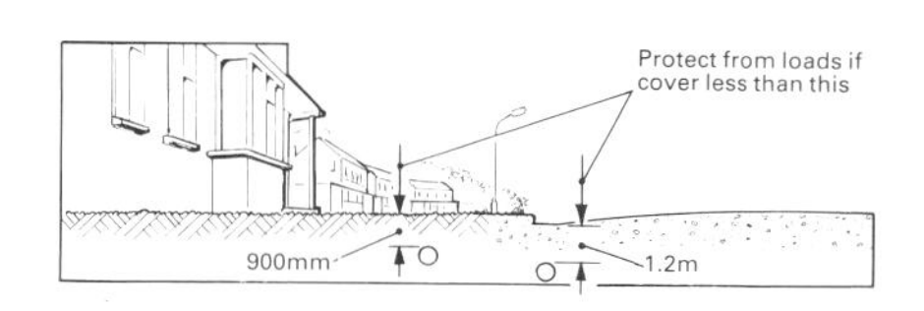

Figure 4 Drainage pipes require protection from loads particularly where vehicle parking is not strictly controlled.

Photo 2 An example of a mechanical rotating drain clearing machine in use.

9.2.3 DRAINAGE PIPEWORK. BLOCKAGE, BELOW GROUND

Blocked drains often occur quite soon after buildings are first occupied.

Symptoms

The symptoms vary depending on the severity of the blockage; liquids may be leaking from an inspection pit or the cover may even be lifting; gullies may overflow; in extreme cases sanitary appliances may not empty or, worse still, sewage may back-up in appliances where the drains are branched before the blockage point.

Investigation

Lift all inspection chamber covers and check for deposits of builders rubble etc. especially if the system is new. Try to obtain plans of the drainage layout although it is often possible to sight between inspection chamber covers to determine pipe runs.

Note the position of trees in relation to the drain pipes.

Diagnosis and cure

There are many ways that a drain can get blocked and equally there are several methods of clearance.

The single most common cause of blockage is the accidental or intentional dumping of builders rubbish down sanitary appliances and into manholes, inspection pits and inspection chambers. Because the drains are laid at a very early stage of the construction programme they are vulnerable to damage and abuse. Temporary stoppers and covers to blank off open pipes and chambers as recommended by the manufacturers are very rarely used on site. The dumping of bulky items into W.C.s may happen at any time and can very easily form a blockage.

Other causes are pipe fractures or collapse, invasion by tree roots through pipe joints and fractures, build-up of silt in pipes with insufficient fall, growth of bulky fungi and deformation of flexible pipes due to excessive compression. The actual cause of the blockage will generally become apparent during the unblocking procedure.

There are several different pieces of equipment available for unblocking drains.

The most simple is traditional rodding canes which interconnect with locking joints and are pushed down the drain via an access point such as a rodding eye. The canes are now more commonly made of steel or polypropylene and can be fitted with different types of head to suit the needs of the particular blockage. Most specialist drain clearing contractors favour mechanically rotated rods which can be fitted with a wide variety of tools including saws for root cutting and cutting blades for reducing more solid objects.

Water jetting is also carried out by specialists. The water is pumped at high pressure through a nozzle attached to a hose that is fed down into the drain. The nozzle has backward pointing jets to drive it down the drain as well as forward facing jets to remove deposits of compacted or loose debris. The success of this equipment is largely dependant on the skill of the operator. It has sufficient power to move objects as large as housebricks and the added bonus of cleaning the drain as well as clearing it.

Large quantities of debris can be removed by the technique of winching. This involves dragging various pieces of equipment through the drain on a wire which is either hand or mechanically winched.

Where pipes are believed to be fractured or to have collapsed, it may be wise to have a closed circuit TV camera survey conducted before proceeding with costly excavation and repairs. Surveys of this nature are often the only way of revealing deformation of plastics pipes which could otherwise return to their intended profile as the pressure on them is relieved during excavation. Broken sections of pipe must be replaced and invading tree roots should be cut back. It may even be necessary to remove trees that are likely to cause further problems. Deformed plastics pipes should be replaced with new sections provided they are capable of withstanding the loads imposed on them. The most likely cause for deformation is the use of incorrect backfilling material therefore this will need to be replaced by one which is more suitable.

If a bulky fungal growth is found, it would be wise to have it identified so that its source of food can be ascertained and subsequently removed. Some such growths have been found to be associated with the roots of elm trees, even though the trees themselves have been removed.

Where the drains have subsided, causing

accumulations of silt, they will need to be excavated and relaid to the correct falls. Lengths of drain can sometimes be replaced without excavation by breaking and expanding the bore of the old pipes and drawing in a new pipe along the same line.

REFERENCES

BS 8301: 1985 Code of practice for *Building drainage* (formerly CP 301)

BS CP 312: Part 1: 1973 Code of practice for *Platics pipework (thermoplastics material)* Part 1. *General principles and choice of material*

BRE Defect Action Sheet 89: *Domestic foul drainage systems: avoiding blockages – specification*

BRE Defect Action Sheet 90: *Domestic foul drainage systems: avoiding blockages– installation*

BRE Digest 292: *Access to domestic underground drainage systems*

Payne, Rolf *Drain maintenance: estate management* Construction Press

Photo 1 Care needs to be taken on site to ensure that unsuitable backfill material does not cover drainage pipes.

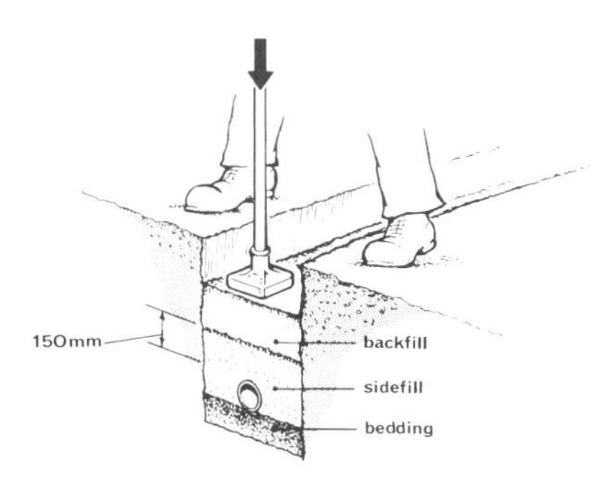

Figure 1 With clayware pipes, correct sidefill material should be laid around the pipe and hand tamped in 100 mm layers until 150 mm cover to the top of the pipe is attained.

Photo 2 Pipes must be laid on suitable bedding material throughout their length and never be supported on bricks or similar objects.

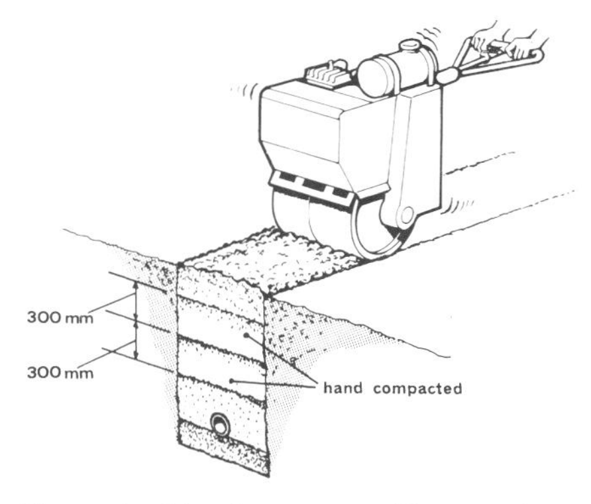

Figure 2 Hand compacted layers of backfill material should be a maximum of 300 mm each and mechanical vibratory compaction excluded until a minimum of 600 mm backfill has been laid.

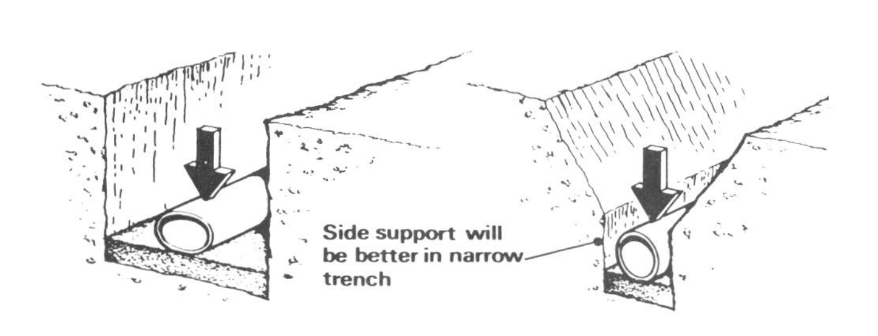

Figure 3 Narrow trenches are more likely to protect plastics drain pipes against distortion from loads above.

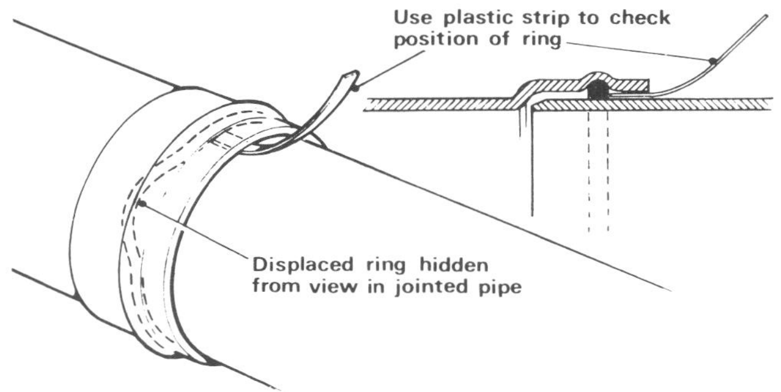

Figure 4 When placing plastics drain pipes check that the ring seal has not been displaced.

9.2.4 DRAINAGE PIPEWORK. LEAKAGE, BELOW GROUND

Although the Building Regulations require that the joints of a drainage system remain watertight, it is possible that many do leak without showing any symptoms, prior to being tested.

Symptoms

Patches of ground close to the drain run become waterlogged. Puddles on the surface show discolouration. There is an unpleasant smell. Where the leakage is severe, the bedding material and the back-filling may be washed away causing subsidence of the ground.

During routine maintenance of the system, clear water may be observed running through the pipes more or less constantly. The system could be blocked (see 9.2.3).

Investigation

Establish what material the pipes are made of and the type and jointing method employed. This can be done by digging a small trial pit to expose a length of pipe.

Carry out leak tests in accordance with the requirements of the Building Regulations – Approved Document H1, section 1.32 or BS 5572, section 12.

If the pipes are leaking below ground level (the smoke test will establish this), it may be worthwhile to have an underground TV survey done. This will identify visually obvious defects and a video tape recording can be made for records purposes.

Where the ground has subsided, carefully excavate the remaining material and take samples of the soil, fill and bedding material for analysis. Examine the pipe for cracks and splits, check the collars for signs of fracture and note whether the joints are visibly leaking.

Diagnosis and cure

Leakage of water from a drainage system may be via cracks in damaged pipes, via the joints or a combination of both. In older systems, inspection chambers may be leaking due to cracking of the benching. This can be relatively simple to repair by replacement but it will be necessary to check for signs of deterioration of the brickwork and mortar joints.

Before flexible jointing methods gained universal acceptance, pipes were jointed with a strong cement mortar mix. Minor ground movements, particularly in areas with a high clay content, put a great deal of stress on the pipes which caused fractures to occur at the collar of the pipes. The fractures then leaked, allowing ground water to enter the system and facilitated tree root invasion. These joints were also susceptible to sulphate attack.

Flexible jointed pipes can tolerate minor ground movements without leaks occurring, provided that the sealing rings were properly seated in the first place. Where minor leaks do occur and assuming falls are maintained at their proper slopes, it may be possible to seal the pipes by pressure grouting or with one of a number of different types of proprietary internal sleeving systems. Ultimately the best cure will be to replace the faulty sections making sure that correct bedding and backfilling procedures are carried out.

REFERENCES

BS 8301:1985 *Code of practice for building drainage* (formerly CP 301)

BRE Defect Action Sheet 30: *Plastics drainage pipes: laying, jointing and backfilling*

The Building Regulations 1985 *Approved document H Drainage and waste disposal*

Payne, Rolf *Drain maintenance: estate management* Construction Press

Photo 1 Where horizontal pipes lack sufficient support they will sag and joints will start to leak.

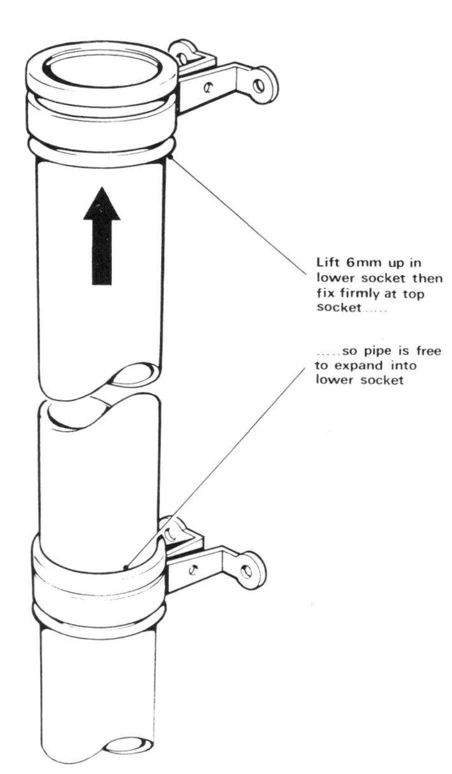

Figure 2 Every length of soil vent pipe must be supported at the collar provided, but allow space for expansion below.

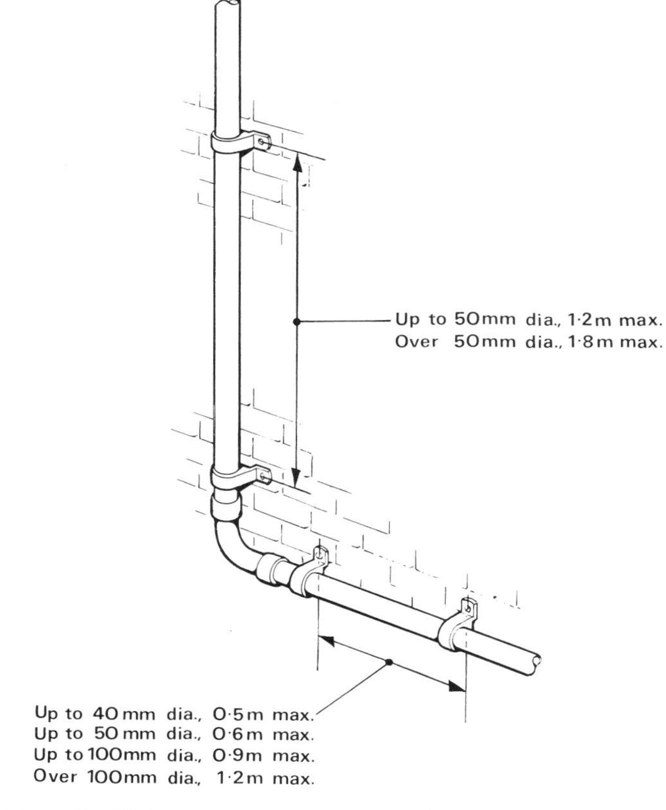

Figure 1 Minimum support spacing horizontally and vertically for plastics pipes.

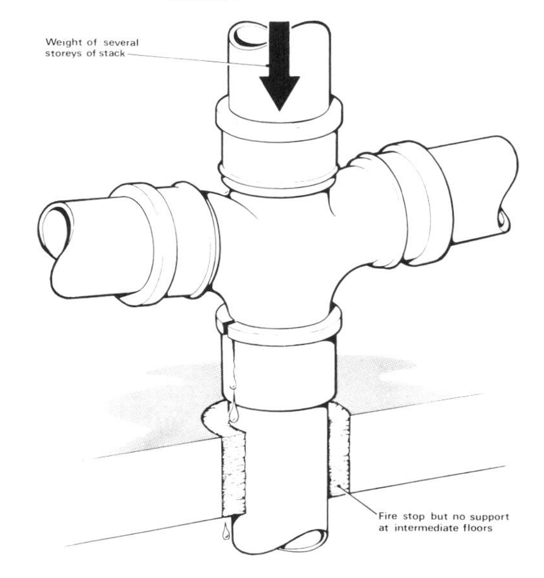

Figure 3 An unsupported stack will settle with its own selfweight and break the lower collars.

Pipe diameter	Support spacing distance
Less than 50mm	1.2m maximum
More than 50mm	1.8m maximum

Table 1 Support spacing for vertical pipes.

Pipe diameter	Support spacing distance
Less than 40mm	0.5m maximum
Less than 50mm	0.6m maximum
Less than 100mm	0.9m maximum
More than 100mm	1.2m maximum

Table 2 Support spacing for horizontal pipes.

9.2.5 DRAINAGE PIPEWORK. POOR DRAINAGE, LEAKAGE. ABOVE GROUND, PLASTICS PIPES

Different principles need to be adopted for the installation of plastics pipework to prevent failure.

Symptoms

Drainage of water from sanitary appliances becomes sluggish. Joints start to leak and pipes distort between supports on horizontal runs. Unsupported vertical stacks settle downwards, their joints open up and socket collars break.

Investigation

Ascertain the type of plastics that has been used in the system.

Measure the distance between and note the position of all support brackets on both vertical and horizontal runs of pipework.

Note whether horizontal pipes are sagging between supports.

Check joints and sleeves for movement and measure the distance between the moving joints.

Diagnosis and cure

When plastics pipes are heated they expand a lot more than their metal or pitch-impregnated fibre counterparts. Different plastics expand at different rates, that for polypropylene being almost twice the rate for unplasticized polyvinyl chloride (uPVC). Plastics also soften when heated and may distort permanently if not supported at frequent enough intervals on horizontal or near horizontal runs.

Vertical soil vent pipes also expand and contract when subjected to varying temperatures the effect of which is to make them settle downwards thereby imposing considerable loads on the socket collars. In multi-storey blocks the downward creep can allow joints to open at inlet points to the stack. All vertical pipes over 75 mm diameter must be supported by fixings below the socket collars at centres not exceeding 1.8 m (see accompanying table). Plastics pipe manufacturers usually incorporate a bracket sleeve moulded onto the pipe to help locate the fixing bracket. When tightening these brackets it is essential to first fully insert the spigot into the lower socket and then lift the pipe by 6 mm to allow expansion of the pipe into the lower socket.

All cracked or distorted pipework must be replaced with new. When assembling the system ensure that pipes are adequately supported. Where solvent welded joints are used, provide sufficient 'push-fit' joints to allow for the necessary thermal movements. Before inserting them into movement joints ensure that pipe ends are lubricated with the manufacturers recommended lubricant (never use 'boss white' with plastics). This lubricant is not only designed to aid assembly but also to ease movement in service.

REFERENCES

BS 4514: 1983 *Specification for unplasticized PVC soil and ventilating pipes, fittings and accessories*

BS 5572: 1978 Code of practice for *Sanitary pipework* (formerly CP 304)

BS CP 312: Part 1: 1973 Code of practice for *Plastics pipework (thermoplastics material)* Part 1 *General principles and choice of material*

BRE Defect Action Sheet 41: *Plastics sanitary pipework: jointing and support – specification*

BRE Defect Action Sheet 42: *Plastics sanitary pipework: jointing and support – installation*

Photo 1 Ultra violet light has degraded these sections of unprotected pipe to the point of disintegration.

Photo 2 Where the spigot on this pipe has been protected at the joint it has remained unaffected.

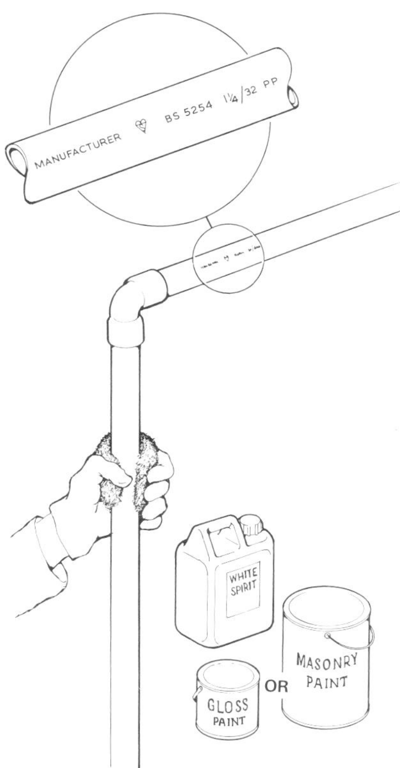

Figure 1 Exposed polypropylene pipes should be protected with paint. New pipework should be degreased and roughened prior to painting. Sound weathered pipework should just be wiped clean.

9.2.6 DRAINAGE PIPEWORK. PIPE DISINTEGRATION, ABOVE GROUND EXTERNALLY

Plastics pipework is comparatively stable and unaffected by the elements except in one particular circumstance.

Symptoms

Plastics drainage pipes exposed to sunlight become brittle and eventually collapse.

Investigation

Attempt to establish which type of plastics is involved in the failure – this may be very difficult as any British Standard identity markings could have disappeared with the collapsed portion. Identify the nature of any adjacent pipework and check on its condition.

Note whether any attempt had been made to protect the pipes with a paint coating.

Find out whether any chemicals are disposed of through the pipes.

Diagnosis and cure

Most plastics products manufactured for the construction industry have proved to be very durable and unaffected by most chemical and climatic conditions. However, time has proved that care must be exercised to install these products in the correct manner and to use the correct plastics for the circumstances.

Provided aggressive chemical attack is not involved, the most likely cause for the disintegration is ultra-violet photo oxidation of the pipe which was undoubtedly made from polypropylene. These pipes have proved to be unsuitable for external use unless they are protected with some form of coating. This can be achieved by applying a coat of exterior masonry paint or gloss paint without an undercoat or primer having first degreased the surface with white spirit and roughened it with fine wire wool. Alternatively it may be much easier to substitute pipes that remain unaffected by sunlight such as uPVC (unplasticized polyvinyl chloride).

Polypropylene has proved reliable for internal fittings such as waste traps etc.

REFERENCES

BS 4514:1983 *Specification for unplasticized PVC soil and ventilating pipes, fittings and accessories*

BS 5254:1976 *Specification for polypropylene waste pipe and fittings (external diameter 34.6 mm, 41.0 mm and 54.1 mm)*

BS 5255:1976 Specification for plastics waste pipe and fittings

BRE Defect Action Sheet 101: *Plastics sanitary pipework: specifying for outdoor use*

BRE Digest 69: *Durability and application of plastics*

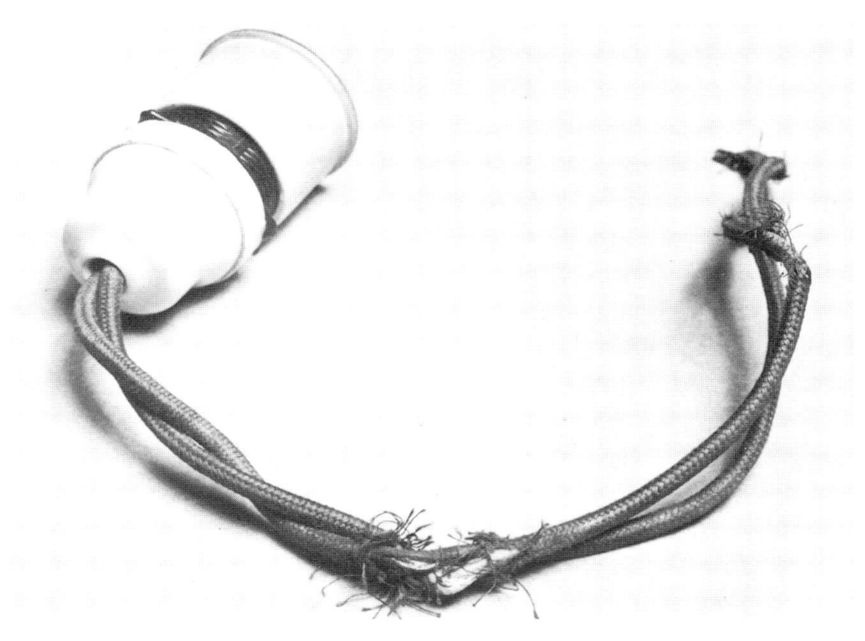

Photo 1 Old style lighting flex is affected by the heat of the lamp and becomes brittle with age.

Photo 2 Face mounted socket outlets at low level are vulnerable to impact damage.

9.3.1 ELECTRICAL WIRING. FUSES BLOWING, INSULATION FAILURE

Older types of electrical cables and flexes have a limited life expectancy usually due to hardening of their insulation material.

Symptoms

Fuses blow, or other protective devices such as circuit breakers operate, cutting off the electricity supply to all or part of the system.

The fault may be localised such as at a plug top, or it may occur at the consumer unit. It may be accompanied by a distinctive burning smell which can cause some plastics items to melt.

Investigation

If the problem appears to be major or the incoming mains fuse has blown, no attempt should be made to tackle any investigation work without first contacting a specialist electrician or the local electricity board.

If the fault is localised and confined to a plug-in appliance, first check the plug top for loose connections or an incorrectly rated fuse. Inspect the appliance cable for signs of wear or breaks in the insulation layer. Test the socket by plugging in an appliance that is known to be without fault. If the socket is dead check the circuit fuse/contact breaker at the consumer unit. Call an electrician to test the circuit if doubts persist about its integrity.

Diagnosis and cure

In older properties that have not had their electrical circuits modernised there is always a danger that cables will decay through a breakdown of the insulation. This happens through age embrittlement which is accelerated through exposure to heat and light. Physical disturbance of the cable when attempting to change lighting fittings or sockets etc. will often crack the insulation which is at most risk near its terminals. Newer cables have more flexible plastics insulation which makes them less vulnerable. If the insulation is found to be defective at any one place or there is evidence of overloading it will be wise to have the whole system checked and if necessary, renewed.

REFERENCES

Regulations for electrical installation Fifteenth edition 1981 Institution of Electrical Engineers

Property Services Agency *Standard specification (M & E) No. 1 Electrical installations* HMSO

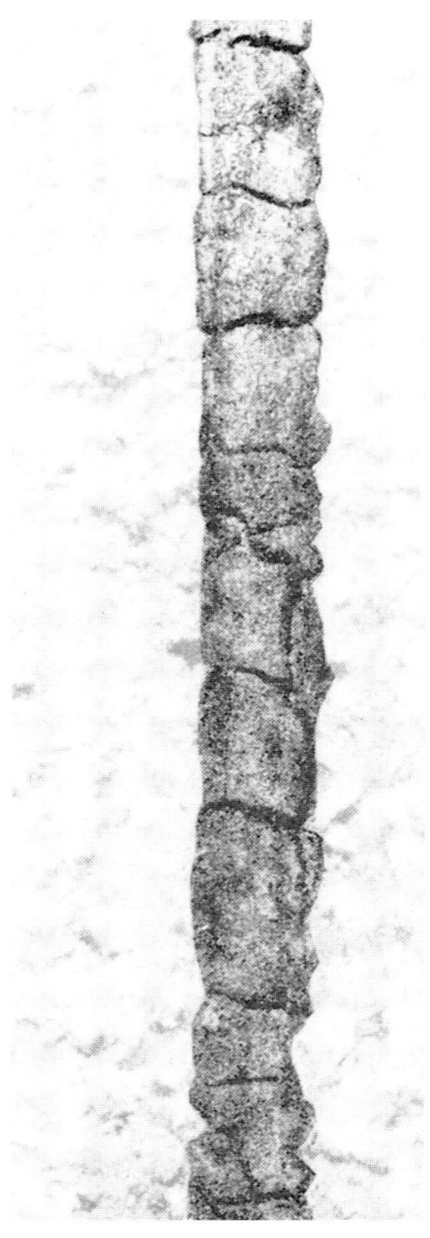

Photo 1 Cable that has overheated to a dangerous state with the risk of short circuit and fire.

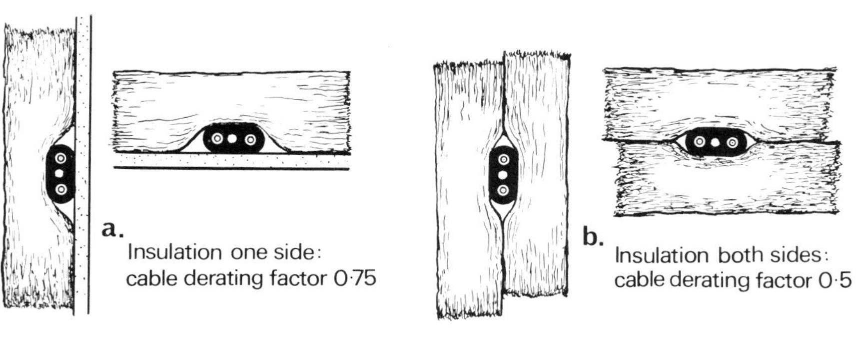

Figure 1 Cables need to be derated in these circumstances.

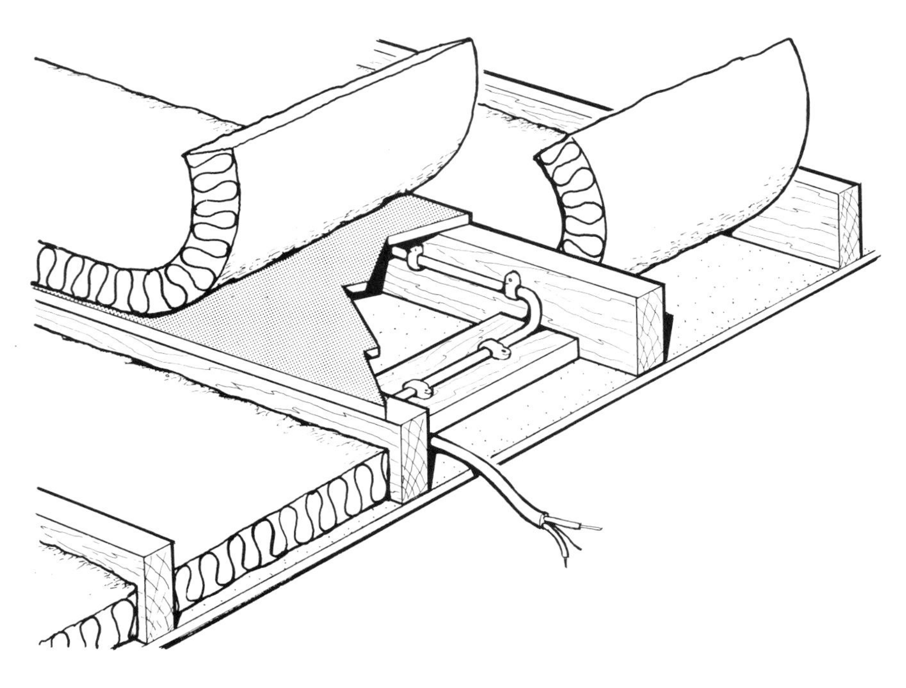

Figure 2 If the cable can not be moved it is best to keep it isolated from the insulation.

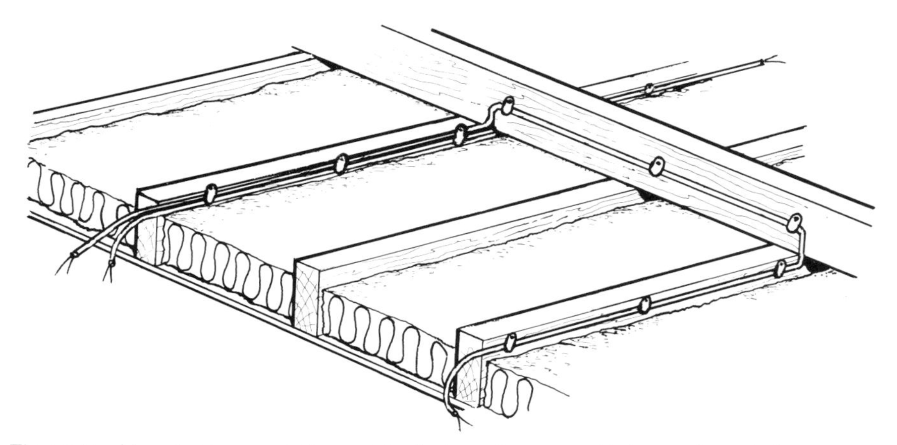

Figure 3 New cable or existing cable that can be moved is best clipped above the insulation but not on top of the joist where it could be trodden on or be damaged by goods stored in the loft.

9.3.2 ELECTRICAL WIRING. CABLES OVERHEATING, INSULATED LOFTS ETC.

Electricity supply cables overheat, chiefly in lofts that have been insulated after the wiring has been installed.

Symptoms

Electrical circuits fail and fuses blow. Cables get very hot and can cause fires.

Investigation

Do not attempt an investigation that could endanger personal safety (see also 9.3.1).

Check all cable runs that may have been buried by thermal insulation or subjected to localised heat sources. Establish the power rating of the cables and note the loads that they are required to carry.

Diagnosis and cure

When power is run through an electrical cable, heat is generated in the cable, the quantity of heat being dependent upon the load. Under normal circumstances this heat is safely dissipated to the surrounding air provided that the cable is not being overloaded. With lighting cables this is not likely to present a problem but power cables are quite often loaded to their full working capacity. If cables are covered with thermal insulation, they lose some of the ability to shed excess heat and may overheat, eventually failing as the sheathing breaks down. A similar problem occurs where cables are run close to, or are connected to, equipment that is a localised source of heat such as hot water cylinders, immersion heaters and ducted pipes.

The IEE wiring Regulations require that a derating factor be applied to cables in these circumstances. Where cables are covered by thermal insulation on one side only, such as in a loft when they are clipped to a joist, the factor is 0.75. Where they are surrounded on both sides the factor is 0.5. It will be necessary to apply a further derating factor for high temperature surroundings – it is not unreasonable to expect a roof loft to reach 35°C in summer – and another factor where cables are bunched together. Similar principles apply to cables that are run inside the cavity of an outside wall that has been cavity filled with insulation.

Hot water cylinder cupboards will be serviced satisfactorily with PVC cable, except for the cable from the isolation switch to the immersion heater which will have to be mineral insulated heat-resisting quality.

Cables that have failed will need to be replaced, preferably by an electrician who can calculate the required derating factors. Consideration should be given to relocating the cable runs to avoid overheating problems.

PVC covered cable must be kept clear of expanded polystyrene insulation which can damage it.

REFERENCES

Regulations for Electrical installation Fifteenth edition 1981 Institution of Electrical Engineers

BRE Defect Action Sheet 62: *Electrical services: avoiding cable overheating*

Photo 1 Modern day smokeless fuels are unlikely to produce this degree of smoke. A small quantity of water thrown carefully on the fuel should produce sufficient steam to confirm a problem.

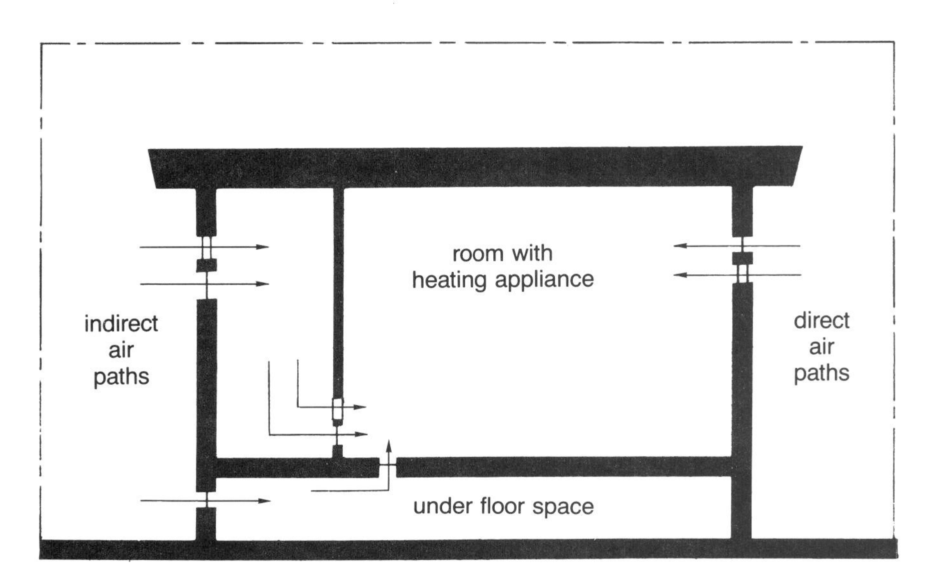

Figure 1 The source of combustion may be either by a direct or indirect path or a combination of both.

9.4.1 OPEN FIREPLACES. SMOKING INTO ROOM

A common enough problem, experienced by most people at some time in their lives. With the advent of modern smokeless fuels the effects may not be so visible, but the unpleasant smell of combustion fumes will give it away.

Symptoms

Solid fuel fires release smoke or fumes into the room instead of up the chimney. The frequency of occurrence can vary significantly; it may only be on occasional brief spells; or at times when the wind is high or gusting; or at certain times of the year; it may even be persistent.

Investigation

The various stages of investigation will dictate the diagnosis and cure.

1 Have the flue swept to determine whether it is partially restricted or blocked.

2 Check the flue for 'pull' either with the aid of a monometer or more simply by holding a flaming rolled newspaper torch up in the lower end of the flue.

3 Measure the size of the fireplace opening and the chimney throat. Verify the effectiveness of the throat by placing a smouldering object in the grate and noting whether the smoke is drawn naturally to the throat or tends to drift out into the room.

4 Note whether there is a sufficient supply of air in the room to feed the combustion requirements of the fire. Look for air vents in the walls and see whether there are any draughts round the windows and doors, even the internal room door, and note the extent of weatherstripping.

5 Record the incidence of smoking and note the wind speed and direction at the time. Measure the height and position of the chimney relative to the roof and establish the orientation of the building.

6 Ascertain the airflow direction in the room with the aid of tracer smoke particularly near windows, doors and ventilators. Look for the existence of power extractor fans.

7 Note the position of the building relative to the surrounding landscape. Particular attention should be paid to tall buildings, trees, etc.

Diagnosis and cure

1 The action of sweeping the chimney will remove most restrictions. In the case of a stubborn blockage it may be necessary to lower a suitably sized cast metal or concrete coring ball through the chimney terminal down the flue to the fireplace. Do not drop objects down the system or deliberately 'fire' the chimney as this may cause considerable structural damage. Proper maintenance of the chimney should include annual sweeping or even more frequent cleaning depending on the types of fuel being burnt.

2 Lack of pull in the flue may be due to insufficient length or there may be too many bends in the flue or they could be too acute. The flue may be too small. BS 6461 recommends a flue size of 185 mm nominal square or 225 mm nominal diameter with an absolute minimum size for circular flues of 200 mm diameter.

It will be a comparatively simple exercise to increase the length of the flue to produce a natural chimney effect by adding to its overall height. The straightening of the flue may well prove uneconomic due to the extensive structural work involved in rebuilding it.

3 The gathering to the throat, the throat itself or the fireplace opening may be incorrectly shaped or sized. The Solid Fuel Advisory Service recommends that the gathering should be smooth, and shaped to lead the smoke to the throat which should be sized 100 mm front to back by about 300 mm width. With the use of smooth flue linings, this throat size may induce an oversupply of air over the top of the fire possibly causing the fire to struggle for combustion air. This can be cured by fitting an adjustable throat restrictor which should be removable to facilitate sweeping. The fireplace opening itself for an open fire need not be more than 550 mm high by 500 mm wide. If the height is greater than this, try holding a suitable board over the upper part of the opening and observe whether a better draught is induced. A more permanent closure can then be fitted.

4 The fire will be starved of oxygen if it can not get a plentiful supply of combustion air. This can happen when windows and doors are fitted with weatherstripping. Even the fitting of carpet in the room can be enough to cause a problem by cutting off the air supply through gaps in the floorboards and under the door.

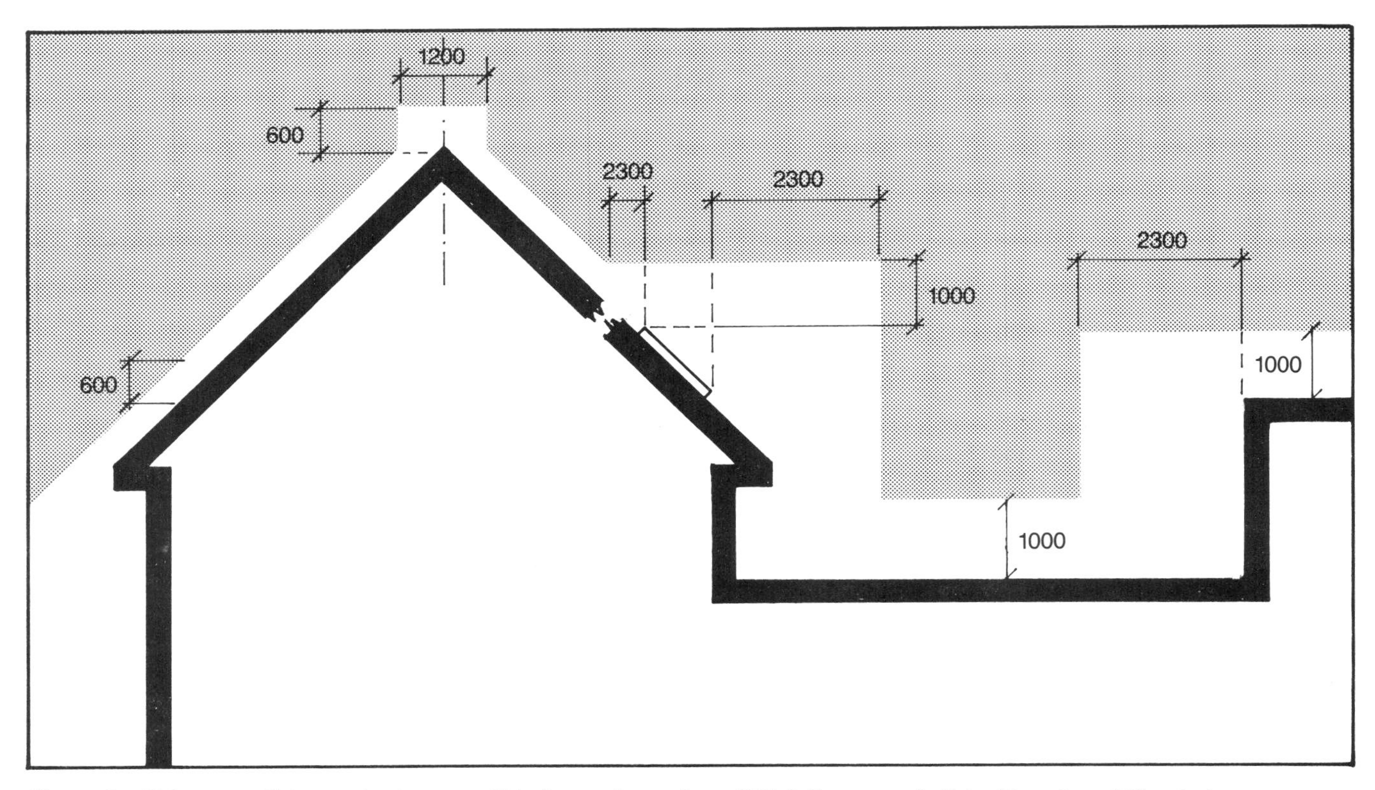

Figure 2 Chimney outlets must not occur within these dimensions. BSI defines a roof pitch of less than 10° as being considered flat.

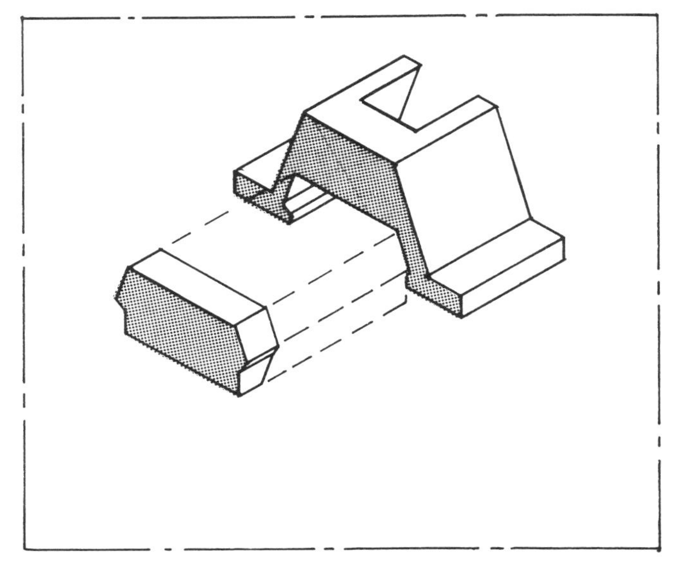

Figure 3 The Coal Board promotes a standard precast concrete throat gather unit.

Air should be supplied from a source which will not create an unwanted draught. A ventilator could be fitted in an external wall or window (or an existing one unblocked). If the room is at ground floor level and has a suspended timber floor, it would be possible to fit ventilation grilles in the floor provided that the air bricks are not blocked. A more popular solution which does not introduce cold air from the outside, is to fit a ventilation box in an internal wall, to borrow air from another room or hallway that has an adequate supply.

5 The wind speed and direction will set up pressure and suction zones on the outside of the building which will vary according to the shape and angle of the roof. This will have a twofold effect on the fire. Firstly, it is essential that the chimney terminates at a high enough level to be unaffected by the zones. Secondly, if the normal air supply point is in a suction zone, it will create negative pressure in the flue. An alternative air supply route will have to be provided if the problem is persistent.

6 The room may be subjected to negative pressure by air extraction. The most obvious and easiest to pinpoint is a powered extractor fan, which if switched on will draw air from the easiest source, often the fireplace. Fuel appliances in adjacent rooms may also 'steal' the greater proportion of the supply air.

7 Surrounding structures can have a dramatic effect on the airflow pattern round the building. Even on a comparatively calm day a tall structure may cause a substantial downdraught of air. Trees and hills can have a considerable influence on airflow patterns, channelling winds into unexpected gusts. Protection against downdraughts may be given by a proprietary chimney cowl, one type of which is available with an electrically powered extractor fan.

REFERENCES

BS 1181:1971 *Clay flue linings and flue terminals*

BS 1251:1987 *Open-fireplace components*

BS 3376:1982 *Solid mineral fuel open fires with convection, with or without boilers*

BS 4834:1972 *Inset open fires without convection*

BS 6461:1984 *Installation of chimneys and flues for domestic appliances burning solid fuel (including wood and peat)*

Part 1 *Code of practice for masonry chimneys and flues pipes*

Part 2 *Code of practice for factory-made insulated chimneys for internal applications*

BS 8303:1986 *Code of practice for installation of domestic heating and cooking appliances burning solid mineral fuels*

BDA Design Guide 9 *Brickwork domestic fireplaces and chimneys.* Brick Development Association

The Building Regulations 1985 *Approved Document F. Ventilation Approved Document J. Heat producing appliances*

Photo 1 Flue terminal much too close to eaves soffit and it should also be 600 mm from external corner.

Photo 2 The terminal should be at least 300 mm beneath this openable window.

TERMINAL POSITION	MINIMUM DISTANCE	
	Natural Draught	**Fanned Draught**
A Directly below an openable window or other opening e.g. air brick	300 mm	300 mm
B Below gutters, soil pipes or drain pipes	300 mm*	75 mm*
C Below eaves	300 mm*	200 mm*
D Below balconies	600 mm	200 mm
E Beside vertical drain pipe and soil pipes	75 mm	75 mm
F Beside internal or external corners	600 mm	300 mm
G Above ground or balcony level	300 mm**	300 mm**
H From a surface facing a terminal	600 mm	600 mm
I Where a terminal faces a terminal	600 mm	1.2 m

**If the terminal is fitted within 850 mm of a plastic or painted gutter or 450 mm of painted eaves, an aluminium shield of at least 750 mm long should be fitted to the underside of the gutter or painted surface.*

***If a terminal is fitted less than 2 m above either a balcony or ground level, or a flat roof to which people have access, then a suitable terminal guard should be provided.*

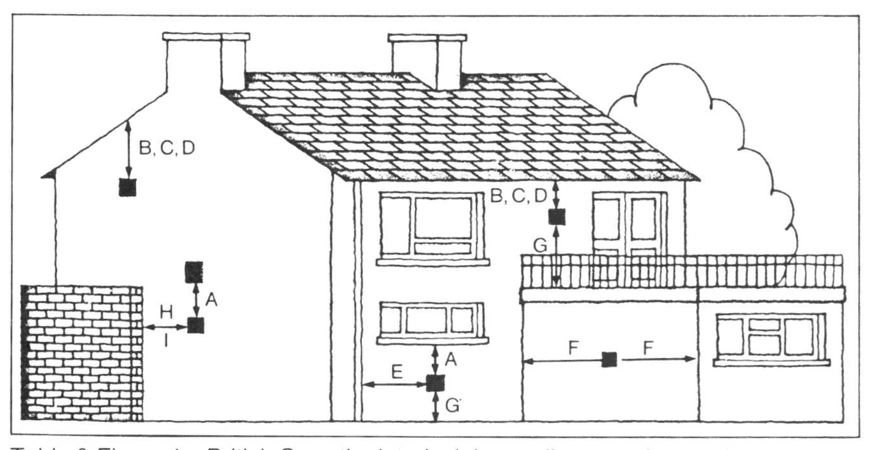

Table & Figure 1 British Gas stipulated minimum distances for terminal location.

9.4.2 BALANCED FLUE TERMINAL. PHYSICAL DAMAGE, OUTSIDE WALLS

Balanced flue terminals need to be correctly sited.

Symptoms

Flue terminals incorrectly sited include those that oversail public walkways causing a hazard to pedestrians; those that are positioned where they can be struck by an opening door or window; or those that are sited too close to features which could be scorched.

Investigation

Note whether any form of guard has been installed over the terminal.

Measure the distance of the flue from projecting features such as rainwater downpipes and soffits etc. also opening features ie doors and windows, and external and internal corners. Check the height of the terminal above ground level and also its distance from any adjacent terminal.

Diagnosis and cure

Lack of thought at the design stage or incorrect installation procedures, could result in a badly sited balanced flue terminal.

With a balanced-flue appliance the position of the flue must comply with certain criteria. It should be positioned on an outside wall such that:

- there is a free flow of fresh air over the terminal to allow unrestricted intake of air and dispersion of products of combustion.
- It is sited at least 300 mm from any opening into the building (including air vents)
- It is protected with a terminal guard if it is sited in a position where people could come into contact with it.
- It is designed so as to prevent the entry of matter which may restrict the flue.

These requirements will ensure that the terminal satisfies the provisions of the Building Regulations. British Gas goes further than this and also stipulates that the terminal is sited a minimum distance of:

- 300 mm below gutters, soil pipes or drain pipes
- 300 mm below eaves
- 600 mm below balconies
- 75 mm from vertical drain pipes or soil pipes
- 600 mm from an internal or external corner
- 300 mm above ground or balcony level
- 600mm from a surface facing a terminal
- 600 mm from a terminal facing a terminal

Furthermore they suggest that if the terminal is fitted within 850 mm of a plastics or painted gutter or within 450 mm of painted eaves, an aluminium shield at least 750 mm long should be fitted to the underside of the gutter or painted surface.

The repositioning of the flue outlet to comply with all these stipulations may prove to be very difficult as the appliance will have to be moved to suit. It may prove to be most economical to replace the appliance with one that uses a fanned draught flue. These are much more compact for a given output, can be positioned closer to openings and projections (see table) and need not necessarily be immediately adjacent to the appliance.

REFERENCES

BS 715:1986 *Metal flue pipes, fittings, terminals and accessories for gas-fired appliances with a rated input not exceeding 60 kW*

BS 5440: *Code of Practice for Flues and air supply for gas appliances of rated input not exceeding 60 kW (1st and 2nd family gases)* Part 1:1978. *Flues* Part 2:1976. *Air supply*

BRE Defect Action Sheet 92. *Balanced flue terminals: location and guarding*

The Building Regulations 1985. Approved document J. *Heat producing appliances*

British Gas. *Gas in Housing. A technical guide*

INDEX

D

E

F

G

H

I

J

L

M

O

P

R

S

T

Printed in the United Kingdom for Her Majesty's Stationery Office
Dd 290541 C50 11/89 3840